Recent Developments in Photosynthesis Research

Recent Developments in Photosynthesis Research

Edited by **Agatha Wilson**

R CALLISTO
REFERENCE

New York

Published by Callisto Reference,
106 Park Avenue, Suite 200,
New York, NY 10016, USA
www.callistoreference.com

Recent Developments in Photosynthesis Research
Edited by Agatha Wilson

International Standard Book Number: 978-1-63239-535-1 (Hardback)

Printed in the United States of America.

Contents

Preface

This book has been a concerted effort by a group of academicians, researchers and scientists, who have contributed their research works for the realization of the book. This book has materialized in the wake of emerging advancements and innovations in this field. Therefore, the need of the hour was to compile all the required researches and disseminate the knowledge to a broad spectrum of people comprising of students, researchers and specialists of the field.

Photosynthesis is one of the most essential reactions on Earth and is a scientific area that is the topic of several research projects. The purpose of this book is to present the basic aspects of photosynthesis, and the outcomes collected from various research groups. It comprises of major topics like the path of carbon in photosynthesis, and special topics in photosynthesis. Important sub-topics of these subjects such as the path of carbon in photosynthesis, high-CO2 response mechanisms in microalgae and role of C to N balance in the regulation of photosynthetic process have been broadly discussed and analyzed by experts in the book.

At the end of the preface, I would like to thank the authors for their brilliant chapters and the publisher for guiding us all-through the making of the book till its final stage. Also, I would like to thank my family for providing the support and encouragement throughout my academic career and research projects.

Editor

Part 1

The Path of Carbon in Photosynthesis

The Path of Carbon in Photosynthesis – XXVIII – Response of Plants to Polyalkylglucopyranose and Polyacylglucopyranose

Arthur M. Nonomura, Barry A. Cullen and Andrew A. Benson
Scripps Institution of Oceanography, University of California San Diego
USA

1. Introduction

A series of discoveries began 63 years ago through the collaboration of Melvin Calvin and Andrew Benson 1948. Paving the path (Benson 2002a) has since included the publications that have described the initial products of carbon fixation, from phosphoglycerate onward. From its inception, the program has followed a design that has been based on far-reaching interdisciplinary discourse, quoting the philosophy of Alexander Graham Bell, that, "Great discoveries and improvements invariably involve the cooperation of many minds." For example, investigations three decades ago by Wolf, Nonomura & Bassham 1985, established characteristics of the alga, "Showa," that accumulated the highest *in vitro* concentrations of hydrocarbons, 40% botryococcenes; from which gasoline and aviation fuels may be derived by catalytic hydrocracking in a conventional petroleum refinery. Studies of algal metabolism were as critical to the development of concepts of the Prize from 1948 through 1961, as they are today--special reference given to the lollipop (Nobelprize, 2011), a flat panel glass chamber that was designed by Benson and built by Harry Powell for controlled culture of *Chlorella* to track radiolabeled carbon metabolism in the laboratory --and thereby, paths converged, as Nonomura & Benson 1992 developed experimental methods for the feeding of single-carbon (C_1) fragments to "Showa." That led to foliar delivery into angiosperms of up to 15 M C_1 formulations supplemented with normally phytotoxic levels of ammoniacal nitrogen. When these applications were undertaken in sunlit fields, adjacent controls wilted by mid-afternoon, but rows treated with C_1 formulations remained fully turgid, showing no signs of wilt. Moreover, Benson & Nonomura 1992 discovered that treatments with these C_1 formulations inhibited glycolate formation, confirming their observations of long durations of reduced photorespiration. In addition to replicating increased yields, Ligocka et al. 2003 discovered corresponding increases of nitrate reductase and alkaline phosphatase by C_1 formulations. In the meantime, Gout et al. 2000 followed metabolism of the C_1 formulations by NMR to the identification of methyl-β-D-glucopyranoside; be that the case, little is known about glycosylation although it is a natural process in the metabolism of C_1 fragments. Thus, in our programs of experimental biology, we investigated responses of plants to substituted glycopyranosides (Benson, 2002a); and we recently showed that not only do substituted glycopyranosides improve productivity, they are transported in plants and metabolized (Benson et al., 2009; Biel et al., 2010;

Nonomura et al., 2011). For our current studies, we selected a polyalkylglycopyranose and we manufactured polyacylglycopyranoses as candidate compounds. Plant responses to formulations of 0.5 to 10 mM polysubstituted glucopyranoses were similar to those of growth enhancements from treatment with foliar formulations of 0.3 M methylglucopyranosides. Therefore, we conclude that when highly substituted sugars are appropriately formulated for application at optimal dosages, visibly discernible enhancement of vegetative productivity may be achieved that is statistically significant and highly consistent. Treatments with polyacylglucopyranoses resulted in significant enhancements of root and shoot growth without toxicity and we tracked putative metabolites of polyacylglucopyranoses and ammoniacal nitrogen. Finally, we gather evidence of mechanisms in which substituted glycopyranoses compete with sugars to release them from lectins.

2. Materials and methods

Plants were cultured in research facilities according to previously described methods (Benson et al., 2009). General supplementation of foliar formulations included the following: 10 - 50 mM ammonium salt; 1 - 6 ppm manganese, Mn-EDTA; and 1 - 6 ppm iron, Fe-HEDTA. For example, foliar solutions of 1,2,3,4-tetramethyl-β-D-glucopyranose, hereafter referred to as tetramethylglucopyranose, were formulated as follow: 23 mM ammonium sulfate, $(NH_4)_2SO_4$, 6 ppm Mn and 6 ppm Fe; and Nutrient Control contained 23 mM $(NH_4)_2SO_4$, 6 ppm Mn and 6 ppm Fe. We found the above supplements to be effective with polyalkylglucopyranoses and polyacylglucopyranoses, particularly, with Mn adding consistency of response to treatments. Moreover, our preliminary tests showed that without ammoniacal and Mn supplementation, the compounds were inactive. Solutions for foliar applications included phytobland surfactant blends applied at a concentration of 1 g/L made from a random block copolymer (Pluronic® L-62, BASF) with a polysiloxane wetter (Q2-5211, Dow Corning®) compounded at a 3:1 ratio. Controls were placed in the same location and all plants were given identical irrigation, fertigation, and handling. Plants were cultured in trays containing soil-less MetroMix® 560 media. Grasses were cultured in MetroMix® 560 blended with 25% calcined clay (Turface®). Plants were matched to control populations, treated after emergence of cotyledon and true leaves, and later harvested within one or two weeks for analysis. For biomass, plants were dried completely in ovens heated to 80°C and weighed. The performance of compounds was evaluated by comparing statistical means of individual weights of shoots and roots. Over the course of taking replicated corresponding measurements, we found that dry biomass gains were proportional to fresh wet gains, allowing us to exclude cell enlargement responses such as to gibberellin and to undertake exploratory surveys and dose response curves with wet weights. When the dosage was sufficiently narrowed, we obtained dry weights, as well. Individual plantings were cultured in plastic flats of identical volumes as needed for optimal growth for size and age within the same planting cycle. All potted plants were regularly given water-culture nutrients (Hoagland & Arnon, 1950). Foliar spray applications of identical volumes, either 100 or 186 liters/hectare (L/ha), were mechanically applied. Manual sprays were spray-to-drip volumes of approximately 800 L/ha. Isolation of metabolites of mixed polyacetylglucopyranoses was undertaken by methods of Biel et al. 2010 and modified by collecting numerous two dimensional chromatography strips, followed by C_{18} reverse phase column chromatography, and yielding 10 milligrams (mg) of the concentrate from 25% methanol:5% formic aqueous eluants. Isolates showed a chromatographic metabolite, R_f 0.34, identified by staining with ninhydrin. The isolated metabolite was dissolved in 100 ml of aqueous solution for treatment of roots. For all populations, means of different treatment groups were compared using Student's t-test with significance at the 95% probability level. Confidence intervals of the population means are "p" values, counts of population numbers are "n" values, and ± standard error is denoted, "SE." Specialty chemicals were from Sigma (St. Louis, MO), including the following: N^6-benzyladenine glucoside; tetramethylglucopyranose (TMG); tetraacetylglucopyranose (TAG); and methylglucopyranosides (MeG). A mixture of polyacetylglucopyranoses (MPG) was made by the authors with a modified chemical synthesis method of Hyatt & Tindall 1993, in which the extent of reaction was controlled by heating cycles. Vascular plants included Canola Nexera 500, *Brassica napus* L., a shoot crop; radish 'Cherry Bell' *Raphanus sativus* L., a root crop; rice, *Oryza sativa* L., a cereal crop; and corn TMF 114, *Zea mays* L. ssp. *Mays;* and these species were maintained as previously described (Benson et al., 2009).

3. Results

We initiated this investigation by surveying polysubstituted glycopyranoses formulated with nutrients to establish discernable trends of growth responses without deficiency. We started with tetramethylglucopyranose because of a consistency of response that we had experienced with methylglucopyranoside. Manual spray-to-drip foliar treatments were applied to even stands of 5 cm tall radish, formulated as follows: Nutrient Control 15 mM $(NH_4)_2SO_4$, 1 ppm Fe, 1 ppm Mn, 1 g/L surfactants; and 0.3 mM tetramethylglucopyranose and 1 mM tetramethylglucopyranose supplemented with 15 mM $(NH_4)_2SO_4$, 1 ppm Fe, 1 ppm Mn, and 1 g/L surfactants. The growth of radish shoots was not affected within the two week trial period and radish roots treated with foliar applications of formulations of Nutrient Control or 0.3 mM tetramethylglucopyranose showed no significant difference   (n=36; SE 0.05; p=0.8) from controls. However, foliar applications of 1 mM tetramethylglucopyranose to radish shoots resulted in a significant (n=36; SE 0.07; p=0.05) 27% enhancement of mean weights of roots over those of Nutrient Control. Results of root analyses are displayed in Figure 1.

Following our establishment of an effective foliar dose that improved root yields, we developed a formulation for row crops that delivered 186 L/ha, less than a quarter the volume of manual applications to shoots. Foliar treatments to even stands of corn were of the following: Nutrient Control, 15 mM $(NH_4)_2SO_4$, 1 ppm Fe, 1 ppm Mn, 1 g/L surfactants; and 3 mM tetramethylglucopyranose, identically supplemented with the above nutrients. The results of this experiment can be seen in Figure 2, where the application of foliar 3 mM tetramethylglucopyranose resulted in a significant (n=18; p=0.03) increase over the Nutrient Control.

Tetraacetylglucopyranose (TAG) is similar to tetramethylglucopyranose except that the sugar is substituted around the pyranose-ring with four acyl-groups instead of alkyl-groups. The range of activity for polyacylglycopyranoses was unknown, therefore, 10 mM TAG was explored in these trials. Foliar 10 mM TAG, formulated in the same nutrient solution as Nutrient Control, was applied to shoots of radish and harvested a week later. Results of foliar application (Fig. 3) showed a significant (n=11, p=0.004) 27% increase of root mean dry weight as compared to Nutrient Control. The growth response of the roots of radish to foliar tetraacetylglucopyranose, therefore, was similar to that of tetramethylglucopyranose.

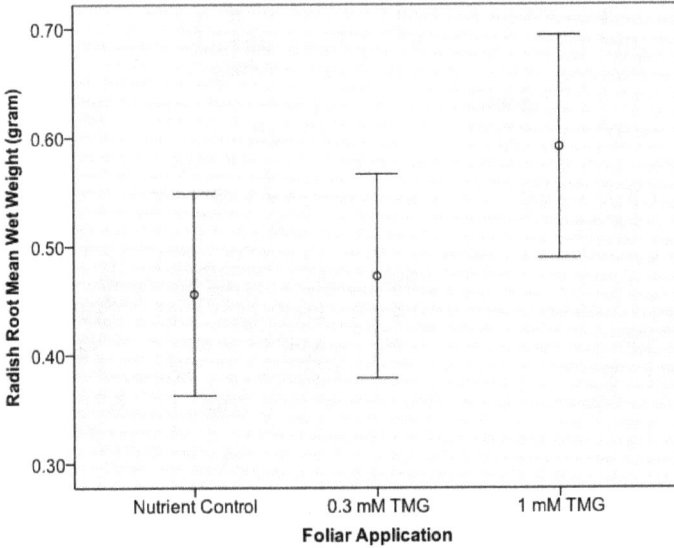

Fig. 1. Initial surveys of radish showed that applications of high volumes of 1 mM tetramethylglucopyranose formulations, 1 mM TMG, to shoots, significantly (n=36; p=0.05) enhanced root mean dry weights as compared to Nutrient Control. Means are marked as small open circles at the midpoint of error bars that indicate ±SE.

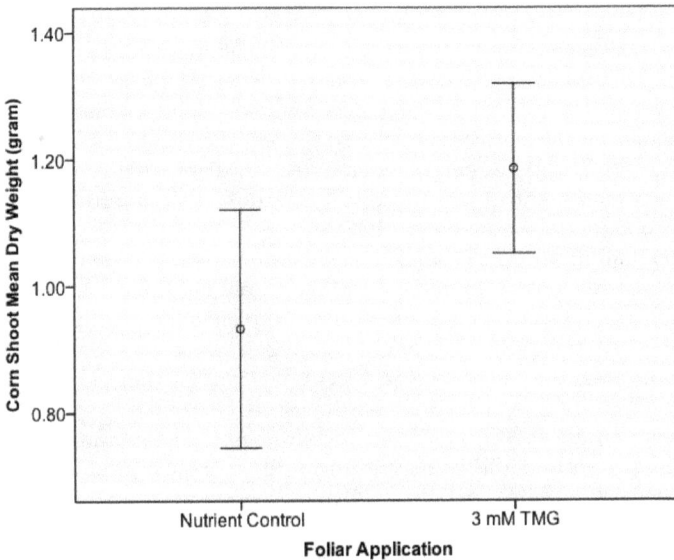

Fig. 2. As a result of foliar applications of 3 mM tetramethylglucopyranose, 3 mM TMG applied at 186 L/ha, a volume typical for row crops, shoot mean dry weights of corn improved significantly (n=11; p=0.004) as compared to Nutrient Control. Error bars indicate ±SE.

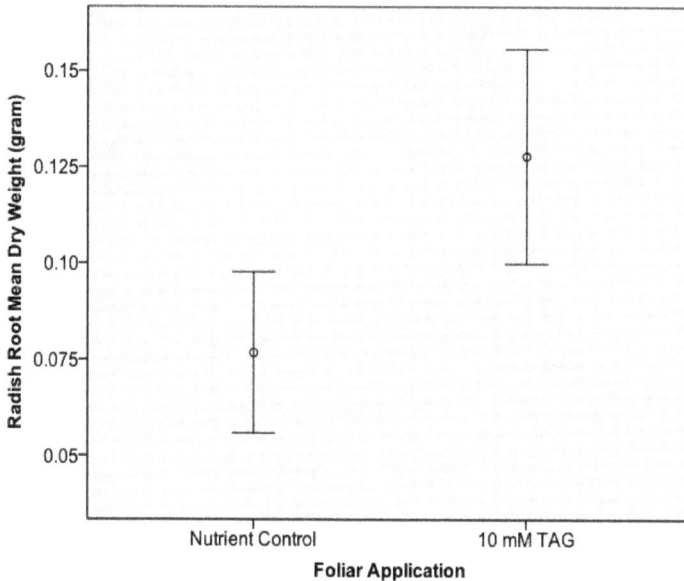

Fig. 3. Foliar application of tetraacetylglucopyranose, 10 mM TAG, to shoots of radish resulted in a significant (n=11; p=0.004) increase of root mean dry weight as compared to Nutrient Control. Error bars indicate ±SE.

We investigated a process for the chemical synthesis of our mixed polyacylglucopyranoses (MPG) at different temperatures to control the reaction; and results are summarized in Figure 4. Between 60° and 72° C, we succeeded in stocking supplies that we manufactured in 500 gram batches. We incorporated 4 mM and 8 mM tetraacetylglucopyranose (TAG) into our tests as positive controls to compare against the different 60° and 72° C batches of 4 mM, 8 mM and 12 mM MPG. Treatments with untreated Control, Nutrient Control, and 12 mM MPG resulted in no difference of growth. In contrast, 4 mM and 8 mM concentrations, showed significantly (n=36; p<0.05) higher vegetative yields than controls. With similar results from these two different batches comparable to tetraacetylglucopyranose, we extended tests to concentrations on various species of plants.

On Canola, responses to foliar applications of 3 mM mixed polyacetylglucopyranoses, 4 mM tetraacetylglucopyranose and 309 mM methylglucopyranoside were compared. All treatments contained 40 mM ammonium nitrate and surfactant blend, including Nutrient Control, and results are graphically depicted in Figure 5. Three treated populations each showed significant (p=0.000) shoot wet weight increases over Nutrient Control, as follow: 3 mM mixed polyacetylglucopyranoses, n=37, 18% increase; 4 mM tetraacetylglucopyranose, n=35, 20% increase; and 309 mM methylglucopyranoside, n=36, 14% increase. We have clearly demonstrated that the far lower concentrations of 3 - 4 mM polyacetylglucopyranoses than the ~100-fold higher dose of 309 mM methylglucopyranoside resulted in comparable growth increase responses from Canola.

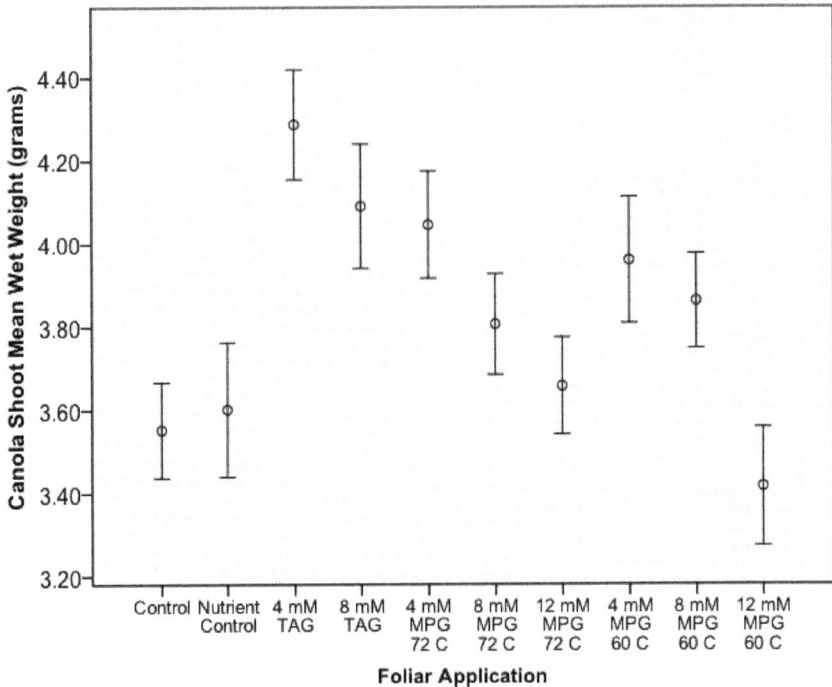

Fig. 4. Similar to nutrient-supplemented formulations of 4 mM and 8 mM tetraacetylglucopyranose (4 mM and 8 mM TAG), responses of populations of Canola to foliar applications of two different batches of mixed polyacetylglucopyranoses (4 mM MPG 72 C & 4 mM MPG 60 C) showed significantly higher (n=36; p<0.05) mean shoot weights than the untreated Control and the Nutrient Control. Error bars indicate ±SE.

A dose response curve of 1 – 3 mM mixed polyacetylglucopyranoses (MPG) was applied to roots of corn with treatments that were supplemented with identical nutrients to the control. Roots were saturated *in situ* with 5 ml of respective formulations per plant. Two weeks later, shoots were harvested and weighed. Statistical analyses showed trends in Figure 6, as follow: 1 mM MPG showed a signficant (n=21; p=0.006) increase in yield over Nutrient Control; 2 mM MPG showed a positive trend that was not significant (n=21; p=0.07); and 3 mM MPG showed a negative trend that may have resulted from exposure of roots to a critical concentration of the acidic mixed polyacetylglucopyranoses.

Treatments of roots of rice with mixed polyacetylglucopyranoses was compared to a treatment with a high concentration of methylglucopyranosides. All solutions were supplemented with identical nutrients including the following: 17 mM $(NH_4)_2SO_4$, 5 mM $(NH_4)_2HPO_4$ and 3 ppm Mn. Roots were saturated *in situ* with 10 ml of respective formulations per 100 cc culture vessel of each individual plant for two weeks. Roots exposed to formulations of 0.5 mM mixed polyacetylglucopyranoses and 50 mM methylglucopyranosides showed significant (n=27; p=0.000) increases in shoot yields of approximately 15% over controls. Results are graphically summarized in Figure 7.

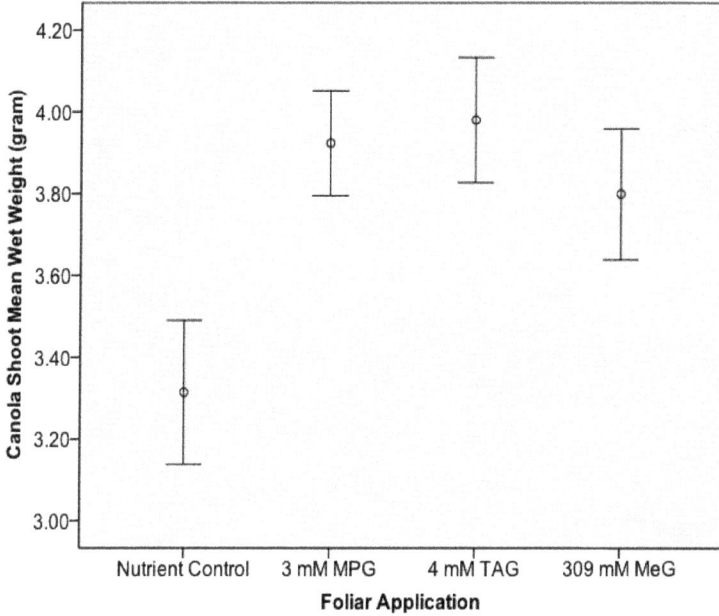

Fig. 5. Foliar applications with low concentrations of polyacetylglucopyranoses, 3 mM MPG and 4 mM TAG, were comparable to treatments with high methylglucopyranosides, 309 mM MeG, and resulted in significant shoot enhancements over Nutrient Control.

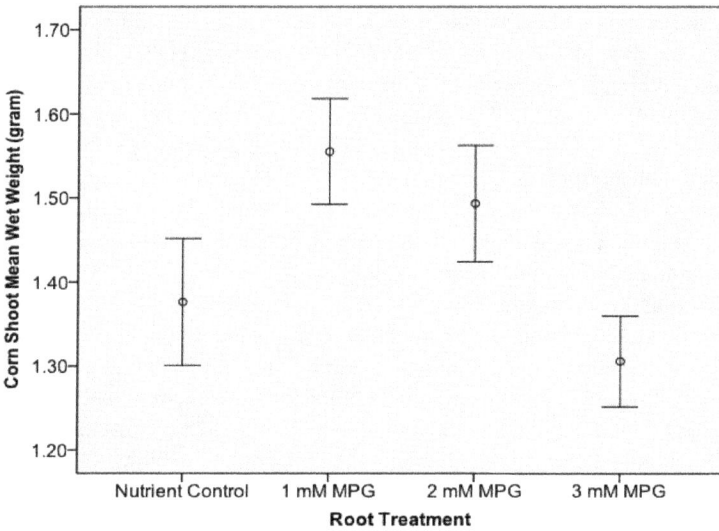

Fig. 6. Dose response of corn roots exposed to various concentrations of mixed polyacetylglucopyranoses (MPG) showed best results at 1 mM MPG with significant (n=2; p=0.006) shoot enhancement over Nutrient Control. Error bars indicate ±SE.

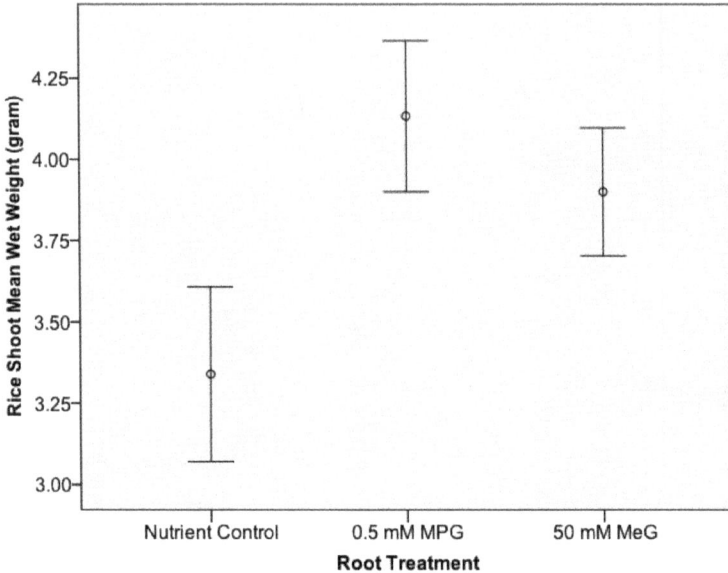

Fig. 7. Roots of whole rice plants exposed to 0.5 mM mixed polyacetylglucopyranoses (0.5 mM MPG) or 50 mM methylglucopyranosides (50 mM MeG), showed significant (n=27; p=0.000) shoot increases over Nutrient Control. Error bars indicate ±SE.

Roots of corn immersed in 1 mM mixed polyacetylglucopyranoses formulated with phosphate buffer to avoid artifacts associated with acidity, were additionally supplemented with the following: 23 mM $(NH_4)_2SO_4$, 5 mM K_2HPO_4, 3 mM KH_2PO_4 and 3 ppm Mn. Roots were saturated with 5 ml of respective formulations per 100 cc soil-less media per plant. The individual shoots of each of the rice plants were harvested after two weeks. Results are displayed in Figure 8 and show significant (n=21; Mean Wet Weight p=0.000; Mean Dry Weight p=0.006) increases of 12% in vegetative yields of shoots over the population of the Nutrient Control. Buffering the solution with phosphates may have nutritionally contributed to the rapid growths of the Nutrient Control and the active treatments while safening the solutions at the same time.

Treatment of roots with mixed polyacetylglucopyranoses resulted in enhancement of shoots, therefore, our hypothesis was that 100 L/ha foliar applications would similarly accelerate development of shoots. Foliar 1 mM and 7 mM mixed polyacetylglucopyranoses were compared to Nutrient Control, both solutions containing, 1 g/L surfactant blend, 23 mM $(NH_4)_2SO_4$ and 3 ppm Mn. Populations included the following: untreated control, n=20; Nutrient Control, n=20; 1 mM MPG, n=20; and 7 mM MPG, n=19; and with no significant difference between the untreated and Nutrient Controls. Foliar treatments of 1 mM and 7 mM mixed polyacetylglucopyranoses resulted in a significant (p=0.003) increase of shoot mean wet weight and a significant (p=0.002) 20% increase of mean dry weight as compared to the Nutrient Control. Shoot dry weights are summarized in Figure 9.

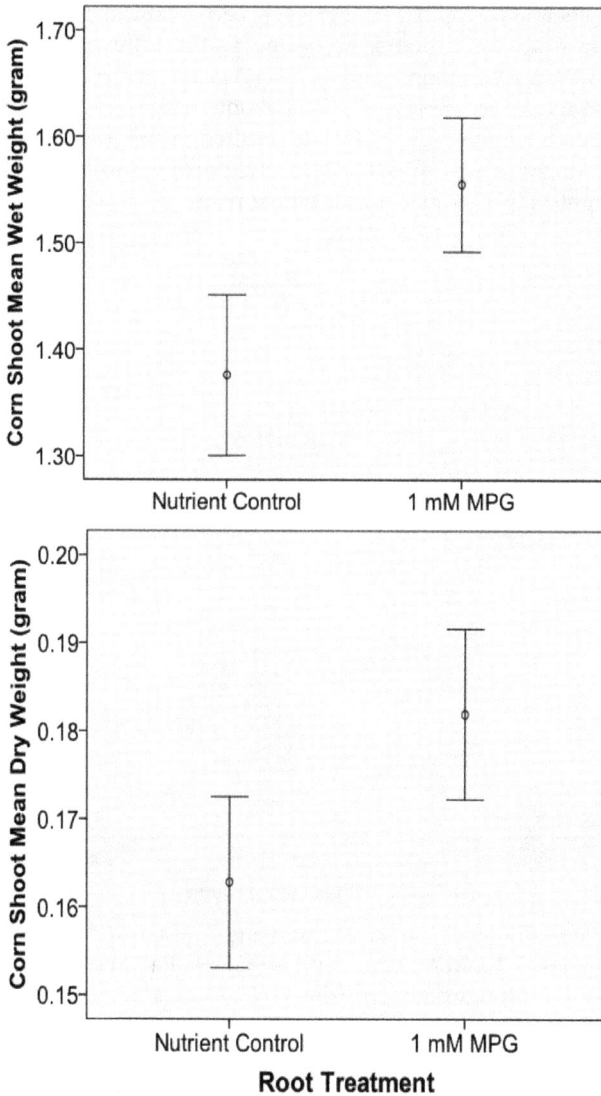

Fig. 8. Top: Roots of corn exposed to 1 mM mixed polyacetylglucopyranoses (1 mm MPG) showed significantly (n=21; p=0.000) increased mean shoot wet weights. Bottom: Dry weights corresponded to wet weights, showing a significant (n=21; p=0.006) 12% increase of mean dry weight over Nutrient Control. Error bars indicate ±SE.

We collected sufficient samples of chromatographic isolates of putative metabolites (Isolate) of mixed polyacetylglucopyranoses (MPG) from aqueous extracts of host plants to undertake a comparison of activities. Mixed polyacetylglucopyranoses, 5 mM MPG, were positive controls. All aqueous solutions for the treatment of roots were adjusted to pH 6.5 and were made up with the following nutrients: 23 mM $(NH_4)_2SO_4$; 5 mM

(NH$_4$)$_2$HPO$_4$; and chelated 3 ppm Mn. For each corn plant, 5 ml of test solution was applied to saturate roots for 6 hours, according to the following doses: 5 mg mixed polyacetylglucopyranoses per plant; and 0.5 mg Isolate per plant. Entire plants were harvested after one week for whole weights of shoots with roots. A summary of data in Figure 10 shows treatment with 0.5 mg Isolate resulted in significant (n=21; p=0.005) 11% increase and treatment with mixed polyacetylglucopyranoses (MPG) resulted in significant (n=21; p=0.000) 15% increase of whole plant mean wet weight over Nutrient Control.

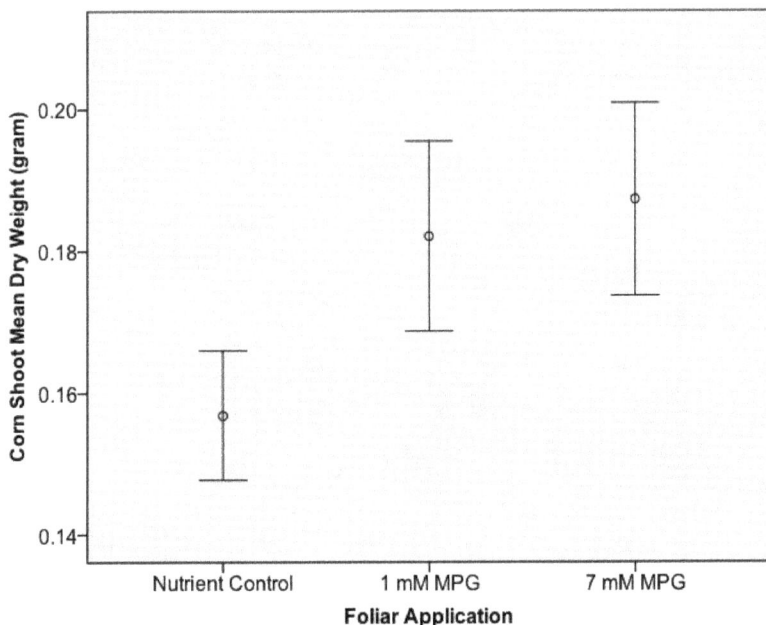

Fig. 9. Foliar applications of formulations of 1 mM and 7 mM mixed polyacetylglucopyranoses (1 mM MPG & 7 mM MPG) resulted in significant (p=0.002) increased shoot mean dry weights of approximately 20% over the population of the Nutrient Control. Error bars indicate ±SE.

The photograph of corn plants exhibited in Figure 11, was taken immediately prior to harvest, one week after treatment with the metabolite of mixed polyacetylglucopyranoses, 0.5 mg Isolate. The Isolate, right, in which the leaf tip reaches the top of the photograph, is visibly distinguishable from the Nutrient Control, left, of which the shoots reach approximately 1 cm below the top border of the photograph. This visual comparison of the generally larger plants of the Isolate, corroborates the 15% increase in the total mean weight of the plant as compared to the Nutrient Control.

Finally, N[6]-benzyladenine glucoside chromatographed similarly to our metabolic Isolate of mixed polyacetylglucopyranoses. Therefore, these chromatographic results may suggest that supplementation of formulations with ammoniacal nitrogen may be suited to the incorporation of a nitrogen moiety into polyalkylglycopyranoses and polyacylglycopyranoses.

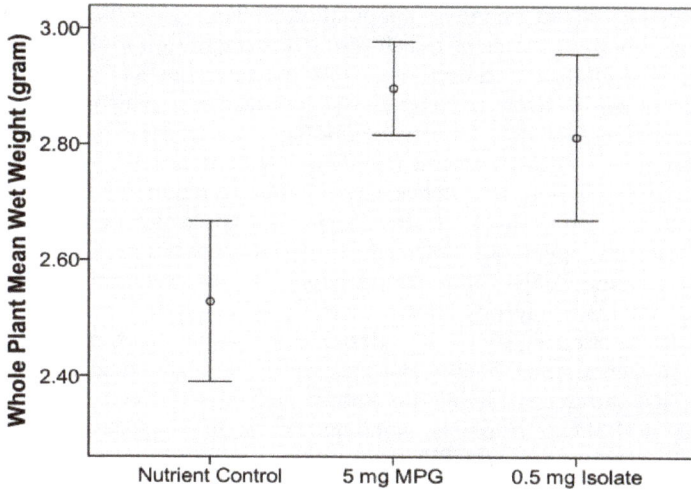

Fig. 10. Treatment of roots of corn with a metabolite of mixed polyacetylglucopyranoses, 0.5 mg Isolate, resulted in significant increases comparable to effects of treatments with ten times the quantity of mixed polyacetylglucopyranoses, 5 mg MPG. Error bars indicate ±SE.

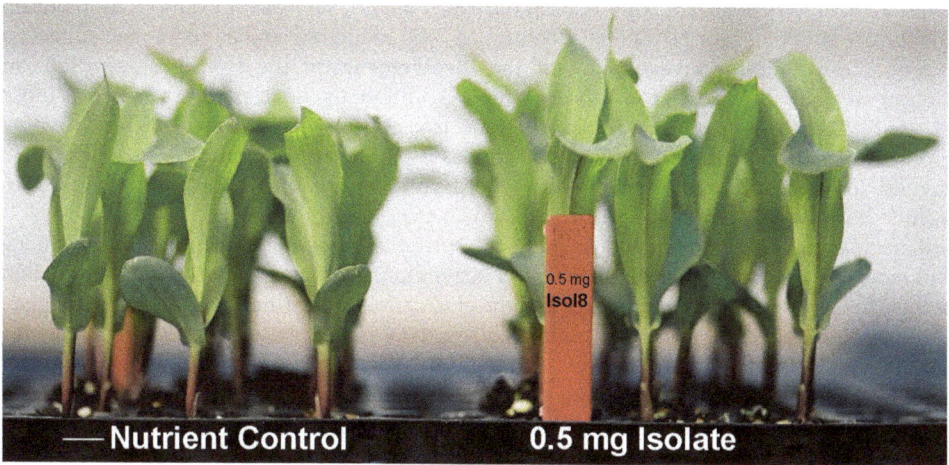

Fig. 11. A chromatographically isolated metabolite of mixed polyacylglycopyranoses (Isolate) was applied to roots of young corn plants and, after a week, the experiment was photographed to show that the enhanced growth of plants treated with the Isolate was visibly discernable from Nutrient Controls. The population of 0.5 mg Isolate, right, is also tagged "Isol8," and Nutrient Control is separated, to the left. Scale = 1 cm.

4. Discussion

Whether to shoots or roots, nutrient applications of relatively low concentrations of polyalkylglycopyranose or polyacylglycopyranoses enhanced vegetative productivity of plants as compared to control populations. For example, applications of polyalkylglucopyranose to shoots of radish resulted in significant root enhancements over controls; and, furthermore, applications of polyacylglucopyranoses to roots of corn resulted in significant increases of shoots as compared to controls, all of which were supplemented with identical carrier solutions. Similar to findings of our previous experiments with alkylglycopyranosides and C_1 fragments (Nonomura & Benson, 1992; Benson & Nonomura, 2009), polyalkylglycopyranose and polyacylglycopyranose required supplementation with nitrogen for active improvements of growth and our discovery of a ninhydrin-stained product of polyacylglycopyranose indicated incorporation of nitrogen into a highly active metabolite. To our knowledge, little is understood of the structure or functions of the metabolites of polyalkylglycopyranoses and polyacylglycopyranoses, however, a metabolized fraction of alkylglycopyranosides similarly stained with ninhydrin (Biel et al., 2010). These nitrogenous metabolites may be related to sugar-conjugated plant growth regulators (SPGRs) and in consideration of known pathways to conjugation and content of as much as 100 mM SPGRs in plants, Nonomura et al. 2011 reported significant growth enhancements from treatments with SPGRs, including 0.3 mM N^6-benzyladenine glucoside. Therefore, it may be useful to undertake similar investigations of polyacylglycopyranoses in the presence of various soluble nutrients to further elucidate roles of their metabolites (Nonomura and Benson, 1992), understand involvements of molecular networks regulated by carbon and nitrogen (Nonomura et al., 2011), and determine the nature of manganese requirement (Benson et al., 2009) for their potential and for their capacity as vehicles of crop improvement to be realized. Ammoniacal nitrogen sources, such as for example, ammonium sulfate, are substrates for NADPH:Cytochromes P450 reductase, and as such, may suggest a potential direction for investigation of the involvements of the Cytochromes P450, an offering of a tantalizing area for future research. Additionally, other pathways, such as those of site-directed alkylation and transmembrane transporters, have been shown to involve glycopyranosides (Nonomura et al., 2011) and, as it pertains throughout the eukaryotes, we suggest looking into similar models for photosynthetic organisms that are expanded to include polyalkylglucopyranoses (Kaback et al., 2007).

Inasmuch as the development and tracking of appropriate probes in our future investigations will most certainly shed light on the mode of activity of substituted sugars, the determination of their functions in diverse complexes and the elucidation of their mechanisms remain a goal. Thus, upon consideration of the accumulated results of our recent series of experiments, we have sought a system that involves competitive binding of sugars that would be both abundant and ubiquitous. That is, the system should occur in C_3 and C_4 plants because we have found that Canola and corn respond to treatments of MeG and TAG. Moreover, we have been looking for a system that could bind α-D-glycopyranosides and β-D-glycopyranosides. Upon careful examination, we found that plant lectins possess just these suitable features. In plants, up to a quarter of the protein content of seeds may be attributable to lectins as well as up to ten percent of the protein content of leaves, but even with such prominence, it had been held that vacuolar lectins served no endogenous role in plants (Lannoo et al., 2007). Notably, a number of lectins have been structurally defined to the extent that it has been generally established that manganese is required for competitive binding of methyl-α-D-glucopyranoside and glucose to occur (see, for example, Brewer et al., 1973), fulfilling yet another requirement of the system. Furthermore, based on Biel et al. 2010 showing that radiolabeled methylglucopyranoside is transported into leaves intact with a large portion of label in the protein fraction, we expect that our application of 309 mM methylglucopyranoside to a shoot (Fig. 5) should be sufficient to saturate the system. Therefore, we propose that under conditions in which the cellular sugar concentration of a plant is diminished, chemical competition against substituted sugars acts to release sugar from lectin—and this is an essential process to sustain viability. Also, this act of competitive binding by lectins may naturally displace sugars on a regular basis allowing energy to be reapportioned for growth resulting from metabolism of the freed sugars. For example, we may assume that in the field, the concentration of methyl-β-D-glucopyranoside remains nearly constant in the plant (Aubert et al., 2004) and as a result of midday photorespiratory depletion of the concentration of glucose, competition for binding to lectin by the methyl-β-D-glucopyranoside arises and glucose is released. To an extent, the timely release of this free glucose may mitigate the effects of any impoverishment of glucose. Afterwards, under conditions more conducive to photosynthesis, perhaps later in the afternoon and morning, critical concentrations of glucose are rebuilt to sufficiently high levels that a surfeit of glucose outcompetes methyl-β-D-glucopyranoside, causing the substituted sugar to be displaced, while glucose wins by binding to the lectin. This cycle repeats itself on a daily basis, releasing sugar at each lengthy deprivation event, followed by the capture of fresh sugar upon resuming photosynthesis. Indeed, Nature's response to major environmental stimuli by means of chemical competition is well known. For example, photosynthesis turns to photorespiration strictly as a result of oxygen outcompeting carbon dioxide for binding to Rubisco. In our case, the higher the quantity of lectins residing in the plant, the more capable it may be of capturing and releasing sugars to endure prolonged periods of photorespiration. In contrast, when exogenous chemical competitors for binding sites on lectins are applied to plants, especially by the input of substrates, such as methyl-α-D-glucopyranoside, that do not naturally occur in plants, the duration of the effect may be substantially extended precisely because such foreign compounds may be selected for competitive advantages. Therefore, responses to treatments with substituted sugars must be carefully measured against the conformation of binding sites, biochemical structure, and their orders of preferences for prospective sugars. From another perspective, empirically formulated dosages of crops may possibly reflect the content and binding determinations of major lectins in a cultivar. For example, in the present investigation, our hypothesis is that the relatively low concentrations of tetraacetylglucopyranose required for the improved growth response of treated plants over controls evidently suggests a proportionally higher order of binding to lectins than methyl-β-D-glucopyranoside. Therefore, our search in the future will be focused on the details of descriptions of the functions of substituted sugars in relation to defining suitability to our proposed actions of lectins in the path of carbon, from which we are looking forward to elucidation of the most appropriate competitors, their quantities of application, and when to bind them. Across the broad field of photosynthetic ecosystem management, there exist numerous lectins, each variant with an array of binding characteristics that, hereafter, will further elucidate a competitive path of carbon in photosynthesis (Benson, 2002a & b).

5. References

Aubert, S., Choler, P., Pratt, J., Douzet, R., Gout, E. & Bligny, R. 2004. Methyl-β-D-glucopyranoside in higher plants: accumulation and intracellular localization in *Geum montanum* L. leaves and in model systems studied by ^{13}C nuclear magnetic resonance. *Journal of Experimental Botany* 55(406): 2179–2189

Benson, A. A. 2002a. Paving the Path. *Annual Review of Plant Biology* 53: 1-25.

Benson, A. A. 2002b. Following the Path of Carbon in Photosynthesis: A personal story. *Photosynthesis Research* 73: 29-49.

Benson, A. A. & Nonomura, A. M. (1992) The Path of Carbon in Photosynthesis: Methanol inhibition of glycolic acid accumulation, *in* Murata, N. (ed.), *Research in Photosynthesis 1, Proceedings of the IX International Congress on Photosynthesis*, Kluver, Nagoya, Japan, P-522.

Benson, A. A., Nonomura, A. M. & Gerard, V. A. (2009) The Path of Carbon in Photosynthesis. XXV. Plant and Algal Growth Responses to Glycopyranosides, *Journal of Plant Nutrition* 32(7): 1185-1200.

Biel, K. Y., Nonomura, A. M., Benson, A. A. & Nishio, J. N. (2010) The Path of Carbon in Photosynthesis. XXVI. Uptake and transport of methylglucopyranoside throughout plants. *Journal of Plant Nutrition* 33(6): 902–913.

Brewer, C. F., Sternlicht, H. , Marcus, D. M. & Grollman, A. P. (1973) Binding of ^{13}C-Enriched α–Methyl-D-Glucopyranoside to Concanavalin A as Studied by Carbon Magnetic Resonance. *Proceedings of the National Academy of Sciences, USA* 70 (4): 1007-1011.

Calvin, M. and Benson, A. A. (1948) The Path of Carbon in Photosynthesis, Science, 107: 476.

Gout, E., Aubert, S., Bligny, R., Rebeille, F., Nonomura, A., Benson, A. A. & Douce, R. (2000) Metabolism of methanol in plant cells. Carbon-13 nuclear magnetic resonance studies. *Plant Physiology* 123(1): 287-296.

Hoagland, D. R. and D. I. Arnon. 1950. The Water-Culture Method for Growing Plants without Soil. California Agricultural Experiment Station Circular 347, The College of Agriculture, University of California, Berkeley: 32 pp. URL: http://plantbio.berkeley.edu/newpmb/faculty/arnon/Hoagland_Arnon_Solution.pdf

Hyatt, J. A. & Tindall, G. W. (1993) The Intermediacy of Sulfate Esters in Sulfuric Acid Catalyzed Acetylation of Carbohydrates. *Heterocycles* 35(1): 227-234.

Kaback, H. R., Dunten, R., Frillingos, S., Venkatesan, P., Kwaw, I. , Zhang, W. & Ermolova, N. (2007) Site-directed alkylation and the alternating access model for LacY. *Proceedings of the National Academy of Sciences, USA* 104: 491-494.

Ligocka, A., Zbiec, I, Karczmarczyk, S. & Podsialdo, C. (2003) Response of some cultivated plants to methanol as compared to supplemental irrigation. *Electronic Journal of Polish Agricultural Universities*, Agronomy, Volume 6, Issue 1. URL: http://www.ejpau.media.pl

Lannoo, N., Vandenborre, G., Miersch, O., Smagghe, G., Wasternack, C., Peumans, W. J. & Van Damme, Els J. M. (2007) The Jasmonate-Induced Expression of the *Nicotiana tabacum* Leaf Lectin. *Plant Cell Physiology* 1–12 URL: http://www.pcp.oxfordjournals.org

Nobelprize.org. (2011) The Nobel Prize in Chemistry 1961. URL: http://nobelprize.org/nobel_prizes/chemistry/laureates/1961/

Nonomura, A. M. & Benson, A. A. (1992) The Path of Carbon in Photosynthesis: Improved crop yields with methanol. *Proceedings of the National Academy of Sciences, USA* 89(20): 9794-9798.

Nonomura, A. M., Benson, A. A. & Biel, K. Y. (2011) The Path of Carbon in Photosynthesis. XXVII. Sugar-conjugated plant growth regulators enhance general productivity. *Journal of Plant Nutrition* 34:(5): 653 – 664.

Wolf, F. R., Nonomura, A. M. & Bassham, J. A. (1985) Growth and branched hydrocarbon production in a strain of *Botryococcus braunii* (Chlorophyta). *Journal of Phycology* 21(3): 388-396.

High-CO$_2$ Response Mechanisms in Microalgae

Masato Baba[1,2] and Yoshihiro Shiraiwa[1,2]
[1]Graduate School of Life and Environmental Sciences,
University of Tsukuba, Tsukuba, Ibaraki,
[2]CREST, JST,
Japan

1. Introduction

The concentrations of atmospheric CO$_2$ and aquatic inorganic carbon have decreased over geologic time with minor fluctuations. In contrast, O$_2$ concentration has increased through the actions of photosynthetic organisms. Therefore, photosynthetic organisms must adapt to such dramatic environmental change. Aquatic photosynthetic microorganisms, namely eukaryotic microalgae, cyanobacteria, and non-oxygen-evolving photosynthetic bacteria, have developed the ability to utilize CO$_2$ efficiently for photosynthesis because CO$_2$ is a substrate for the primary CO$_2$-fixing enzyme ribulose-1,5-bisphosphate carboxylase/oxygenase (Rubisco) and its related metabolic pathways such as the Calvin–Benson cycle (C$_3$ cycle). As the Rubisco carboxylase reaction is suppressed by elevated O$_2$ concentrations via competition with CO$_2$, photosynthetic organisms have developed special mechanisms for acclimating and adapting to changes in both CO$_2$ and O$_2$ concentrations. Examples of such mechanisms are the microalgal CO$_2$-concentrating mechanisms (CCM), the facilitation of "indirect CO$_2$ supply" with the aid of carbonic anhydrase and dissolved inorganic carbon (DIC)-transporters (see Section 3), and C$_4$-photoysnthesis (for review, see Giordano et al., 2005; Raven, 2010). Many reports on low-CO$_2$-acclimation/adaptation mechanisms have been published, particularly in relation to certain cyanobacteria and unicellular eukaryotes. However, knowledge of high-CO$_2$-acclimation/adaptation mechanisms is very limited. We recently identified an acceptable high-CO$_2$-inducible extracellular marker protein, H43/Fea1 (Hanawa et al., 2007) and a *cis*-element involved in high-CO$_2$-inducible gene expression in the unicellular green alga *Chlamydomonas reinhardtii* (Baba et al., 2011a). We also identified other high-CO$_2$-inducible proteins in the same alga using proteomic analysis (Baba et al. 2011b). In this chapter, we briefly introduce low-CO$_2$-inducible phenomena and mechanisms as background and then review recent progress in elucidating the molecular mechanisms of the high-CO$_2$ response in microalgae.

2. Aquatic inorganic carbon system

The CO$_2$ concentration dissolved in aqueous solution (dCO$_2$) is equilibrated with the partial pressure of atmospheric CO$_2$ (pCO$_2$) by Henry's law and depends on various environmental factors such as temperature, Ca^{2+} and Mg^{2+} levels, and salinity (e.g., Falkowski & Raven,

2007). The dCO_2 dissociates into bicarbonate (HCO_3^-), and carbonate (CO_3^{2-}) and these three species of DIC attain equilibrium at a certain ratio depending on pH, ion concentrations, and salinity (Fig. 1). HCO_3^- is the dominant species at physiological pH (around 8), which is similar to that in the chloroplast stroma where photosynthetic CO_2 fixation is actively driven (for review, see Bartlett et al., 2007). However, Rubisco [E.C. 4.1.1.39] reacts only with dCO_2, not bicarbonate or carbonate ions. At a pH of 8, the dCO_2/HCO_3^- ratio becomes extremely small (approximately 1/100) resulting in a high bicarbonate concentration and an increase in the total DIC pool size. The dCO_2 concentration equilibrates with atmospheric CO_2 at approximately 10 μM, whereas the bicarbonate concentration is approximately 2 mM at the surface of the ocean (Falkowski & Raven, 2007).

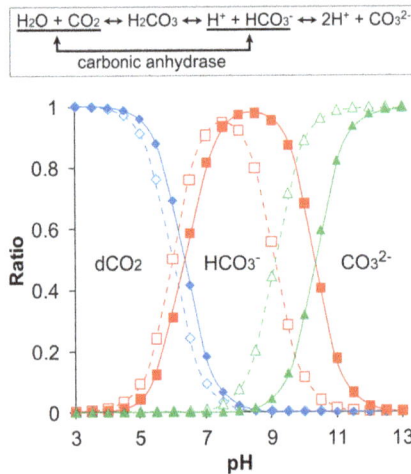

Fig. 1. Equilibration of dissolved inorganic carbon species in freshwater and seawater. Parameters used were as follows (at 25°C): For freshwater, pKa_1 = 6.35, pKa_2 = 10.33; for seawater, pKa_1 = 6.00, pKa_2 = 9.10 (Table 5.2, Falkowski & Raven, 2007). Filled symbols and solid line, freshwater; clear symbols and dotted line, seawater; diamonds, dCO_2; squares, bicarbonate; triangles, carbonate.

CO_2 must be supplied rapidly when it is actively fixed by Rubisco in the chloroplast stroma during photosynthesis. CO_2 is supplied by both diffusion from outside of cells and the conversion of bicarbonate. However, these processes are very slow and become limiting for photosynthetic CO_2 fixation. In the former case, CO_2 must be continuously transported from outside of the cells via the cytoplasm through the plasmalemma and the chloroplast envelope. The diffusion rate of CO_2 in water is approximately 10,000-fold lower than that in the atmosphere (Jones, 1992). In the latter case, bicarbonate accumulated in the stroma can be a substrate when the dehydration rate to convert bicarbonate to CO_2 is comparable to Rubisco activity. However, the rate of chemical equilibration between CO_2 and the bicarbonate ion is very slow relative to photosynthetic consumption of CO_2 (Badger & Price, 1994; Raven, 2001); the first-order rate constants of hydration (CO_2 to bicarbonate) and dehydration (bicarbonate to CO_2) are 0.025–0.04 s^{-1} and 10–20 s^{-1}, respectively, at 25°C (Ishii et al., 2000). Such CO_2-limiting stress becomes a motive for photosynthetic organisms to develop unique CO_2-response mechanisms.

3. The CO$_2$-concentrating mechanism and phenomena induced by CO$_2$ limitation

The atmospheric CO$_2$ level has gradually decreased over recent geological time with some fluctuations (Condie & Sloan, 1998; Falkowski & Raven, 2007; Giordano et al., 2005; Inoue, 2007), although it has been increasing rapidly due to CO$_2$ emissions from fossil fuels since the industrial revolution. Thus, photosynthetic organisms have adapted to utilize CO$_2$ efficiently for photosynthesis. Generally, eukaryotic microalgae and cyanobacteria have developed efficient CO$_2$-utilization mechanisms and exhibit high photosynthetic affinity for CO$_2$ when grown under CO$_2$-limiting conditions. Under elevated CO$_2$ conditions, they exhibit low affinity for CO$_2$, as enough CO$_2$ is available for photosynthesis. These properties can change over hours when photosynthetic microorganisms are grown under various CO$_2$ conditions (for review, see Miyachi et al., 2003) (Fig. 2).

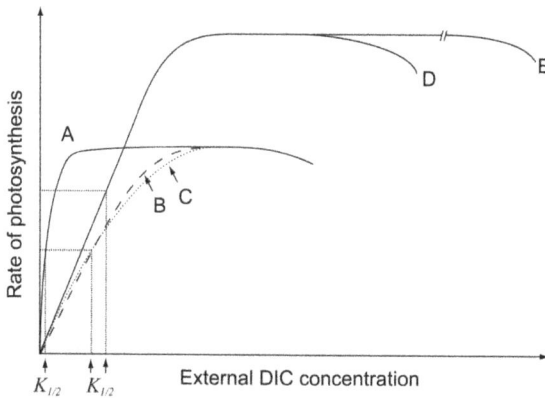

Fig. 2. Relationship between photosynthetic rate and external dissolved inorganic carbon (DIC) concentration in microalgae grown under low-, high-, and extremely high-CO$_2$ conditions. A, low-CO$_2$-acclimated cells (grown in air with 0.04% CO$_2$); B, low-CO$_2$-acclimated cells treated with a carbonic anhydrase inhibitor (e.g., *Chlorella*); C, high-CO$_2$-acclimated cells (grown in air containing 1–5% CO$_2$) (e.g., *Chlamydomonas*); D, high-CO$_2$-acclimated cells (e.g., *Chlorella*); D, extremely high-CO$_2$-acclimated cells grown under >40% CO$_2$ conditions. This figure is modified from Miyachi et al. (2003).

High photosynthetic activity of low-CO$_2$-grown cells under a CO$_2$-limiting concentration is due to the CCM, which is induced by cellular acclimation to limiting CO$_2$ (e.g., Aizawa and Miyachi, 1986). Two main factors are essential for CCM: inorganic carbon transporters that facilitate DIC membrane transport of CO$_2$ and/or bicarbonate through the plasmalemma and the chloroplast envelope, and carbonic anhydrases (CAs), which facilitate diffusion by stimulating the indirect supply of CO$_2$ from outside of cells to Rubisco. CA catalyzes the equilibration reaction of the hydration and dehydration of CO$_2$ and bicarbonate, respectively. The rapid equilibration catalyzed by CA stimulates the increase in bicarbonate concentration at physiological pH and augments the contribution of bicarbonate for diffusion. Finally, the processes driven by CA induce increases in the amount of bicarbonate carried near Rubisco and then CO_2 produced from bicarbonate is immediately supplied to Rubisco when CA is located near Rubisco (see also Fig. 4). The relative specificity to CO_2/O_2 and affinity to CO_2 of Rubisco became more efficient over evolutionary time, indicating that Rubisco in eukaryotic microalgae is more efficient for CO_2 fixation than that in cyanobacteria (Falkowski & Raven, 2007). Such species-specific properties remain unchanged in present living organisms. However, even in eukaryotic algae, the affinity of Rubisco for CO_2 is insufficient to saturate activity at present atmospheric CO_2 concentrations. Therefore, the cells continuously activate mechanisms such as CCM to increase their affinity for CO_2. For further information on CCM, we recommend reading several previously published reviews (e.g., Aizawa & Miyachi, 1986; Badger et al., 2006; Giordano et al., 2005; Kaplan & Reinhold, 1999; Miyachi et al., 2003; Moroney & Ynalvez, 2007; Raven et al., 2008; Raven, 2010; Spalding, 2008; Yamano & Fukuzawa, 2009).

CCM is reversibly induced/suppressed by the decrease/increase in CO_2 concentration, respectively, in cyanobacteria and eukaryotic microalgae when the duration of acclimation is on an hour- or day-order length. However, in the unicellular green alga *Chlamydomonas reinhardtii* in which CCM is the most characterized among eukaryotic microalgae, cells grown for 1000 generations under high-CO_2 conditions are unable to re-acclimate to low-CO_2 conditions, exhibiting low photosynthetic affinity for CO_2 even when the cells are re-exposed to low CO_2 conditions (Collins & Bell, 2004; Collins et al., 2006). This suggests that CCM can be irreversibly lost when cells undergo prolonged acclimation/adaptation to high-CO_2 conditions. Such adaptation has been suggested to occur in natural populations (Collins & Bell, 2006). Although CCM-deficient mutants of cyanobacteria and the green alga *C. reinhardtii* are lethal, such lethality is prevented by elevated CO_2 concentration (e.g., Price & Badger, 1989; Spalding et al., 1983; Suzuki & Spalding, 1989). In *C. reinhardtii*, CCM is induced under either 1.2% CO_2 in air at 1000 µmol photons m^{-2} s^{-1} or 0.04% CO_2 in air at 120 µmol photons m^{-2} s^{-1}, suggesting that CCM induction can be regulated by not only external CO_2 concentration but also other signals derived from photorespiratory and/or excess photoenergy stresses, although the detailed mechanisms are not yet known (Yamano et al., 2008). CCM can be induced by an artificially produced strong limitation in CO_2 supply in large-scale photobioreactors where dCO_2 is consumed via photosynthesis (Yun & Park, 1997).

In some microalgae, the supply of CO_2, not bicarbonate, is strongly limited at alkaline pHs in closed culture systems and such a limitation may be a factor or signal for inducing CCM (Colman & Balkos, 2005; Diaz & Maberly, 2009; Verma et al., 2009). The euglenophyte *Euglena mutabilis* and an acid-tolerant strain of *Chlamydomonas* do not induce CCM under any conditions (Balkos & Colman, 2007; Colman & Balkos, 2005), suggesting that photosynthetic carbon fixation is not limited by CO_2 supply even under ambient atmospheric conditions. These results indicate that there is species-specific variation in the induction mechanism of CCM depending on physiological and ecological conditions (for review, see Giordano et al., 2005; Raven, 2010).

4. High-CO_2 response phenomena

The atmospheric CO_2 level is presumed to have been very high during the ancient geological era (Condie & Sloan, 1998; Falkowski & Raven, 2007; Giordano et al., 2005; Inoue, 2007), so microalgae are believed to have been high-CO_2-adapted/acclimated cells. Microalgae preserve their ancient physiological properties at present, and the relative specificity of Rubisco is a typical example. Even in the present environment, high-CO_2 conditions occur in soil where CO_2 concentration changes drastically between the atmospheric level and $\geq 10\%$ (v/v) (for review, see Buyanovsky & Wagner, 1983; Stolzy, 1974). Accordingly, phenomena that are induced under high-CO_2 conditions, such as high-CO_2 acclimation, remain important for microalgae to survive in various environments.

Among the various phenomena induced by high CO_2 concentrations, keenly interesting topics are how to maximize inorganic CO_2 fixation and organic production by microalgae for CO_2 mitigation and mass cultivation. The most frequently used species for studies on fast growth and tolerance to high CO_2 levels is *Chlorella sp.*, followed by *Scenedesmus sp.*, *Nannochloropsis sp.*, and *Chlorococcum sp.* The CO_2 concentration used for such studies varies from atmospheric levels to 100% (Kurano et al., 1995; Maeda et al., 1995; Olaizola, 2003; Seckbach et al., 1970). Appropriate CO_2 supply for saturation of microalgal growth is approximately 5% in the unicellular green alga *Chlorella* (Nielsen 1955). The growth of microalgae and cyanobacteria is generally inhibited under very high concentrations of CO_2. Some species isolated from extreme environments can grow rapidly with tolerance to very high and extremely high CO_2 conditions such as >40% (for review, see Miyachi et al., 2003). Even in a high-CO_2-tolerant microalga, growth is suppressed at > 60% CO_2 in air (Satoh et al., 2004). The rate of maximum photosynthesis per packed cell volume increases in some species, such as *Chlorella*, but not in other species, such as *Chlamydomonas*, even when cells are acclimated to high-CO_2 conditions (Miyachi et al., 2003) (Fig. 2). However, the detailed mechanism on such high CO_2 tolerance needs to be clarified.

Many reports have focused on lipid biosynthesis for biofuel production, and response surface methodology (Box & Wilson, 1951) has been used very effectively to evaluate multiple factors associated with total biomass production. Excellent review articles on large-scale cultivation for biofuel production by microalgae and cyanobacteria have focused on how to obtain the best productivity under high-CO_2 conditions (Ho et al., 2011; Kumar et al., 2010; Lee J.S. & Lee J.P., 2003), but not on the underlying mechanisms of how cells provide high productivity under fine regulation.

One of the best examples of sequential analysis was performed systematically in the high-CO_2-tolerant unicellular green alga *Chlorococcum littorale* (for review, see Miyachi et al., 2003). *C. littorale* is a unicellular marine chlorophyte that was isolated from a saline pond in Kamaishi City, Japan; it grows rapidly under extremely high CO_2 conditions (e.g., 40%, and even at 60% CO_2; Chihara et al., 1994; Kodama et al., 1993; Satoh et al., 2004). Several experiments have revealed that cellular responses, namely the regulation of photosystem (PS) I and PS II, the production of ATP, and pH homeostasis are well maintained particularly in *C. littorale*, but not in high-CO_2-sensitive species such as the green soil alga *Stichococcus bacillaris*, during a lag period when cells are transferred from low to extremely high levels of CO_2 (Demidov et al., 2000; Iwasaki et al., 1996, 1998; Pescheva et al., 1994; Pronina et al., 1993; Sasaki et al., 1999; Satoh et al., 2001, 2002). However, many of the processes that make it possible for cells to grow under such extremely high-CO_2 conditions remain to be understood.

Photosynthesis in acidic environment, the influence by ocean acidification, and the effect of O_2 on photorespiration are also deeply associated with high-CO_2-induced phenomena. Some microalgal species have been isolated mainly from acidic environments where only CO_2 is predominant and supplied to algal cells as a substrate for photosynthesis (Balkos & Colman, 2007; Colman & Balkos, 2005; Diaz & Maberly, 2009; Verma et al., 2009; for review see Raven, 2010). Three synurophyte algae, *Synura petersenii*, *Synura uvella*, and *Tessellaria volvocina*, have been studied in detail for the DIC uptake mechanism and show unique photosynthetic properties (Bhatti & Coleman, 2008). These species have no external carbonic anhydrase on the cell surface, no bicarbonate uptake ability, and exhibit a low affinity for DIC during photosynthesis, indicating a lack of CCM as in high-CO_2-grown/acclimated cells. However, their Rubisco shows a relatively high affinity for CO_2, and cells such as *S. petersenii* accumulate large amounts of internal DIC via diffusive uptake of CO_2 facilitated by a pH gradient across the cell membranes, as reported previously in spinach chloroplasts (Heldt et al., 1973). These data suggest that the affinity of Rubisco for CO_2 and the homeostasis of the pH gradient play key roles in the whole-cell affinity for CO_2 and the pH-tolerance of microalgae. Under high-CO_2 conditions, Rubisco can get enough CO_2 supply although CCM is usually lost in high-CO_2 cells. The physiological status of synurophyte algae living at acidic pH may be similar to cells that are exposed to high-CO_2 conditions even under low-CO_2 conditions.

Increasing pCO_2 induces a decrease in oceanic pH and causes gradual equilibrium shifts from bicarbonate ions to CO_2 in seawater. Therefore, ocean acidification is said to be another high-CO_2 problem (for review, see Doney et al., 2008). Coccolithophorids, marine phytoplankton that form cells covered with $CaCO_3$, are very sensitive to calcium carbonate saturation and pH shifts in seawater. The effects of ocean acidification on algal physiology have been studied in several coccolithophorid species such as *Emiliania huxleyi* and *Pleurochrysis carterae*, although some conflicting results have been reported (Fukuda et al., 2011; Igresiaz-Rhodorigez et al., 2008; Riebesell et al., 2000). Hurd et al. (2009) indicated the importance of maitaining pH in experiments and demonstrated that doing so via high-CO_2 bubbling creates conditions that are much closer to actual ocean acidification than acidification by adding HCl. The effects of high-CO_2 conditions on calcification and photosynthesis would be closed up in later analyses. Fukuda et al. (2011) reported that the coccolithophorid *E. huxleyi* possesses alkalization activity, which helps compensate for acidification when photosynthesis is actively driven. Furthermore, when oceanic acidification is caused by the bubbling of air with elevated CO_2, coccolithophorid cells increase both photosynthetic activity and growth and are not damaged because of the stimulation of photosynthesis (unpublished data by S. Fukuda, Y. Suzuki & Y. Shiraiwa). These results suggest that ocean acidification will not immediately harm coccolithophorids. However, long-term experimental evidence is strongly required on this topic.

Badger et al. (2000) described how low-CO_2-grown microalgae tend to have low photorespiratory activity, as determined by photosynthetic O_2 uptake in C_4 plants because of the function of CCM. O_2 uptake under illumination is relatively insensitive to changes in CO_2 concentration, because the activity depends predominantly on the activity of non-photorespiratory reactions probably such as the Mehler reaction and oxidizing reaction in the mitochondria (Badger et al., 2000). CO_2 insensitivity is also observed in *C. reinhardtii* (Sültemeyer et al., 1987) although photosynthetic O_2 uptake increases considerably with increasing light intensity (Sültemeyer et al., 1986). Accordingly, the photosynthetic productivity of microalgae may not be significantly enhanced by suppressing photorespiration. The rate of maximum photosynthesis, calculated on a cell volume, increases clearly in *Chlorella* but not so in *Chlamydomonas* when cells are acclimated to high-CO_2 conditions (Miyachi et al., 2003) (Fig. 2). In *C. reinhardtii*, growth rate is only slightly higher (1.3–1.8-fold) in cells grown under high-CO_2 than in those grown under ordinary air (Baba et al., 2011b; Hanawa, 2007). These results suggest that low-CO_2-acclimated/grown cells have a very highly efficient carbon-fixation mechanism for maintaining high growth rates even under atmospheric CO_2 levels, so we need to carefully optimize growth conditions when we want to obtain high algal growth and production using CO_2 enrichment (see also section 5).

5. Molecular mechanisms for high-CO_2 responses

Microalgae can acclimate to high-CO_2 conditions by changing their photosynthetic properties such as CCM. The half-saturation concentration of CO_2 for changing cellular photosynthetic characteristics, i.e., CO_2 affinity, is 0.5% in the unicellular green alga *Chlorella kessleri* 211-11h (formerly *C. vulgaris* 11h; Shiraiwa & Miyachi, 1985). CCM-related proteins are also degraded simultaneously when cells are transferred from low- to high-CO_2 conditions (see references in section 3). Yang et al. (1985) found that, during acclimation to high-CO_2 conditions, CA, an essential component of CCM, was passively degraded and thus the process took almost 1 week.

C. reinhardtii cells in freshwater and in soil are exposed to drastically fluctuating concentrations of CO_2 between atmospheric level and ≥10% (v/v) (for review, see Stolzy, 1974; Buyanovsky & Wagner, 1983). To grow in such habitats and maintain optimum growth, the alga needs to rapidly change its physiology. Such rapid acclimation was in fact observed in *C. reinhardtii* cells that were successfully acclimated to 20% CO_2 within a few days (Hanawa, 2007). The specific growth rate (μ) of *C. reinhardtii* was 0.176 in ordinary air containing 0.04% CO_2 where dCO_2 and total DIC were 1.62 and 6.19 μM, respectively, at pH 6.8 (Hanawa, 2007) (Fig. 3). Although dCO_2 and total DIC concentrations in the culture media, which were equilibrated with 0.3, 1.0, and 3.0% CO_2 (v/v) in air, were 28-, 121-, and 489-fold higher than that in ordinary air, respectively, alga-specific growth rates under the respective conditions were only 1.3-, 1.8-, and 1.7-fold higher than that in air (Hanawa, 2007) (Fig. 3). In a wall-less mutant of *C. reinhardtii* CC-400 (same as *CW-15*), the growth rate and the amount of total proteins increased only 1.5-fold even when the CO_2 concentration was increased from atmospheric level to 3% (Baba et al., 2011b). These results clearly indicate that, in *C. reinhardtii*, CO_2 enrichment is not advantageous to increase in growth rate, as the fully low-CO_2-acclimated cells acquire CCM and grow quickly with a near-maximum growth rate even under atmospheric levels of CO_2. These results are true when cells are growing logarithmically at low cell density to prevent self-shading. However, when cell density is quite high, the ratio of growth at high to low CO_2 is usually quite high. This is probably due to the decrease in growth under air conditions. Under such conditions, CO_2 supply is strongly limited resulting in very low growth rates under air-level CO_2. Nevertheless, the growth rate does not exceed the specific growth rate obtained at the logarithmic growth stage.

High-CO_2-grown *C. reinhardtii* declines CCM physiologically by losing CA and active DIC transport systems in order to avoid secondary inhibitory effects caused by excess DIC accumulation (for review, see Miyachi et al., 2003; Spalding, 2008; Yamano & Fukuzawa, 2009) but no other significant responses have been reported until recently. Recently, we found drastic changes in extracellular protein composition (Baba et al., 2011b) including induction of the H43/Fea1 protein (Hanawa et al., 2004, 2007; Kobayashi et al., 1997). The wall-less mutant of *C. reinhardtii*, *CW-15*, releases a large amount of extracellular matrix, including periplasm-locating proteins, named as extracellular proteins, into the medium (Hanawa et al., 2007; Baba et al., 2011b). Our previous studies clearly showed that the extracellular protein composition changes drastically when *C. reinhardtii* cells are transferred

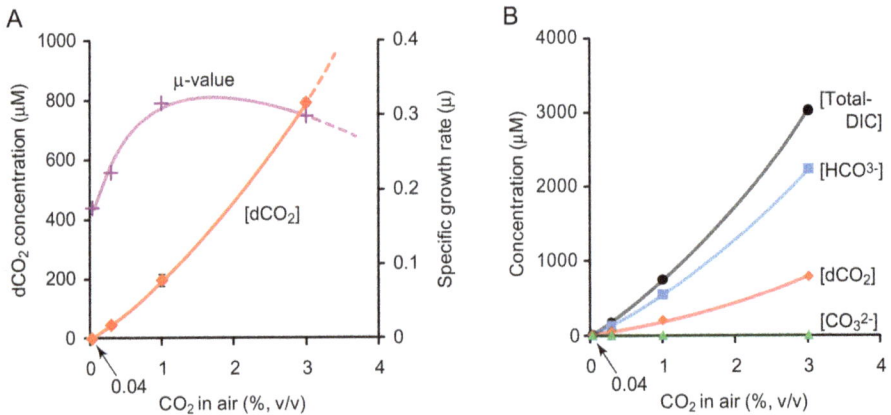

Fig. 3. Relationship between the specific growth rate and dCO_2 concentration in an air-bubbled culture of *Chlamydomonas reinhardtii* (A), and the concentrations of three dissolve inorganic carbon (DIC) species in the culture (B). The concentration of dCO_2 was experimentally determined. Each DIC species was calculated by Henley's law and the Henderson–Hasselbalch equation, respectively. The parameters were as follows (for freshwater at 25°C): pKa_1 = 6.35, pka_2 = 10.33. The culture medium used was a high salt medium supplemented with 30 mM MOPS (pH 6.8). Crosses, specific growth rate; diamonds, dCO_2; circles, total DIC; squares, bicarbonate; triangles, carbonate. Fig. 3A is modified from Hanawa, 2007.

from atmospheric air to 3% CO_2 in air (Hanawa et al., 2004, 2007; Kobayashi et al., 1997), whereas an SDS-PAGE profile of intracellular-soluble and -insoluble proteins showed no clear difference (Baba et al., 2011b). Recently, we analyzed 129 proteins by proteomic analysis and identified 22 high-CO_2-inducible proteins from *C. reinhardtii* cells transferred from low- to high-CO_2 conditions (Baba et al., 2011b). These high-CO_2-inducible proteins are multiple extracellular hydroxyproline-rich glycoproteins (HRGPs), such as nitrogen-starved gametogenesis (NSG) protein (Abe et al., 2004), inversion-specific glycoprotein (ISG) (Ertl et al., 1992), and cell wall glycoprotein (GP) (Goodenough et al., 1986), together with sexual pherophorin (PHC) (Hallmann, 2006), gamete-specific (GAS) protein (Hoffmann & Beck, 2005), and gamete-lytic enzymes (Buchanan & Snell, 1988; Kinoshita et al., 1992; Kubo et al., 2001). Both GP and ISG are classified as HRGPs together with PHC, GAS, and sexual agglutinin with a shared origin (Adair, 1985). HRGPs are generally involved in sexual recognition of mating-type, plus or minus gametes, in the *Chlamydomonas* lineage (Lee et al., 2007). Among these proteins, NSG, GAS, and gamete-lytic enzymes are generally known to be induced during the gametogenetic process. The sexual program, including gametogenesis in *Chlamydomonas*, is strictly regulated by nitrogen availability (for review, see Goodenough et al., 2007). Drastic changes in the expression of gametogenesis-related extracellular proteins were clearly observed in *C. reinhardtii* cells in response to high-CO_2 but not to environmental nitrogen concentrations, because the experiment was performed under nitrogen-sufficient conditions (Baba et al., 2011b). No visible effect of high-CO_2 signal alone was observed on mating (Baba et al., 2011b). From these results, we concluded that the high-CO_2 signal induced gametogenesis-related proteins but that the signal was not strong enough or was still missing some necessary factors to trigger mating. Otherwise, these gametogenesis-related protein families and/or HRGPs may have another function under high-CO_2 conditions.

The biological meaning of the expression of gametogenesis-related proteins at the stage of vegetative growth is quite mysterious. CCM may be differentially regulated by changes in nitrogen availability, depending on the species (for review, see Giordano et al., 2005). In *C. reinhardtii*, mildly limited nitrogen availability suppresses CCM and mitochondrial β-CA expression (Giordano et al., 2003) and the increase in NH_4^+ concentration promotes the efficiency of photosynthetic CO_2 utilization (Beardall & Giordano, 2002). From these results, Giordano et al. (2005) suggested that the induction of CCM and related phenomena induced by CO_2 limitation is regulated to satisfy an adequate C/N ratio. Basically, cells growing under high-CO_2 conditions may require more nitrogen, at least no less than low-CO_2-acclimated cells, and tend to attain nitrogen-limitation status easily. In contrast, Giordano et al. (2005) suggested that activating CCM may reduce the loss of nitrogen through the photorespiratory nitrogen cycle. Namely, NH_4^+ produced by converting Gly to Ser through the C_2 cycle in mitochondria is transported to and re-fixed in the chloroplasts by the GS2/GOGAT cycle where chloroplastic GS2 is induced in response to CO_2 concentration in *C. reinhardtii* (Ramazanov & Cárdenas, 1994). In previous works, the NH_4^+ excretion rate from algal cells was lower in high-CO_2 cells than in low-CO_2 cells when monitored in the presence of 1 mM 1-methionine sulfoximine, a specific inhibitor of GS activity, to prevent re-fixation of NH_4^+ in *C. reinhardtii* CW-15 (Ramazanov & Cárdenas, 1994) and similarly in *C. vulgaris* 211-11h (Shiraiwa & Schmid, 1986). A decrease in the intracellular NH_4^+ level was first reported to induce gametogenesis-related genes in *C. reinhardtii* (Matsuda et al., 1992). Thus, it is reasonable to hypothesize that gametogenesis is triggered by a decrease in intracellular NH_4^+ levels under high-CO_2 conditions when photorespiration is suppressed. However, further study is required, as photorespiratory activity in *C. reinhardtii* is very low (Badger et al., 2000).

Another report suggested the close participation of CO_2 in inorganic nitrogen assimilation (for review, see Fernández et al., 2009). LCIA, or NAR1.2, is involved in the bicarbonate transport system in chloroplasts (Duanmu et al., 2009) but is not regulated by nitrogen availability, and has been identified as a low-CO_2-inducible gene by expressed sequence tag (EST) analysis (Miura et al., 2004). However, NAR1 genes generally involve members of the formate/nitrite transporter family (Rexach et al., 2000). In fact, LCIA-expressing *Xenopus* oocytes display both low-affinity bicarbonate transport and high-affinity nitrite transport activities (Mariscal et al., 2006), suggesting that LCIA is involved in both bicarbonate uptake and nitrite uptake induced under low-CO_2 conditions. In other words, the suppression of LCIA by high-CO_2 conditions may reduce nitrogen availability in the chloroplast. Additionally, the molecular structure of the high-affinity-bicarbonate transporter *cmpABCD* is very similar to that of the nitrate/nitrite transporter *nrtABCD* in *Synechococcus* sp. PCC7942 (for review, see Badger & Price, 2003). The expression of high-affinity nitrate and nitrite transporter (HANT/HANiT) system IV is triggered by a sensing signal of low CO_2 but not NH_4^+ (Galván et al., 1996; Rexach et al., 1999). These data suggest that changes in CO_2 concentration may also affect intracellular nitrogen availability. Further study should be conducted to identify the cooperative effect of CO_2 and nitrogen availability on the expression of CO_2, nitrogen, and gametogenesis-responsive proteins.

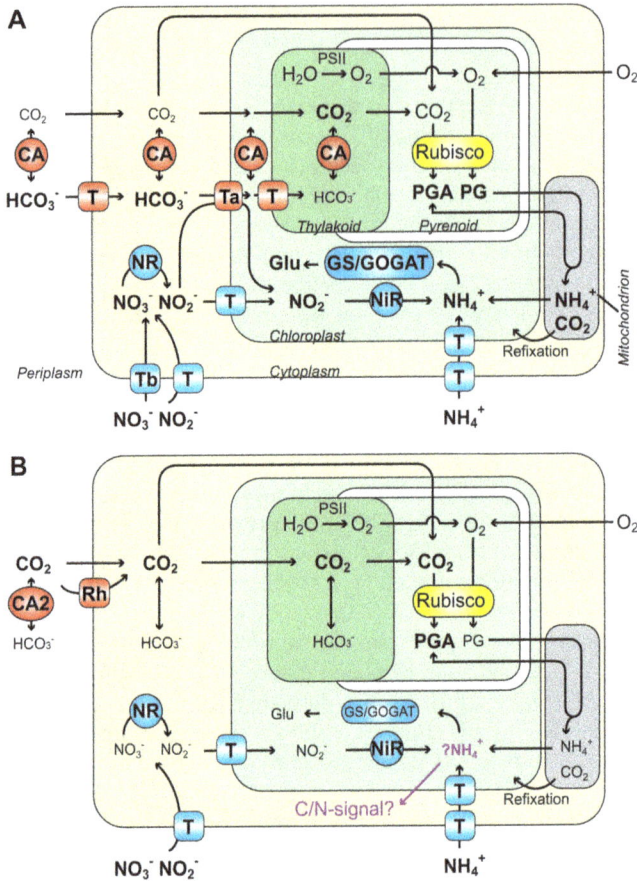

Fig. 4. Schematic illustration of a C/N-status model in low- (A) and high-CO_2-acclimated cells (B) under respective CO_2 conditions produced during acclimation in *C. reinhardtii*. Dissolved inorganic carbon and nitrogen species drawn in bold dominate. CA, carbonic anhydrases; CA2, CAH2 (Fujiwara et al., 1990; Rawat & Moroney, 1991; Tachiki et al., 1992); NiR, nitrite reductase; NR, nitrate reductase; PG, 2-phosphoglycolate; PGA, 3-phosphoglycerate; PSII, photosystem II; Rh, Rh1 (Soupene et al., 2002; Yoshihara et al., 2008); T, (putative) transporters; Ta, LCIA (Duanmu et al., 2009; Mariscal et al., 2006); Tb, HANT/HANiT system IV (Galván et al., 1996; Rexach et al., 1999). CCM models of WT/LC cells, inorganic nitrogen assimilation, and photorespiratory carbon oxidation in *C. reinhardtii* are modified from Yamano et al. (2010), Fernández et al. (2009), and Spalding (2009), respectively.

CAH2 was first reported as an active α-type carbonic anhydrase induced under high-CO_2 conditions and light (Fujiwara et al., 1990; Rawat & Moroney, 1991; Tachiki et al., 1992), but it is poorly expressed and located in the periplasmic space (Rawat & Moroney, 1991). However, the physiological roles and expressional regulation of high-CO_2-inducible CAH2 are not well understood. Another high-CO_2-inducible protein, Rh1, has been identified as a human Rhesus protein in a homology search and is a paralog of the ammonium and/or CO$_2$ channels (Soupene et al., 2002). The lack of Rh1 impairs cell growth in *C. reinhardtii* under high-CO$_2$ conditions (Soupene et al., 2004). Fong et al. (2007) proposed that Rh proteins served as H$_2$CO$_3$ transporters in *Escherichia coli* under high-CO$_2$ conditions. Rh1 was originally expected to be located on the chloroplast envelope *in silico* but the Rh1-GFP fusion protein is located in the plasma membrane in transgenic *C. reinhardtii* cells (Yoshihara et al., 2008).

Some mechanisms of CCM, the photorespiratory nitrogen cycle, and the nitrate/nitrite transport system, and the interactions among them, are summarized in relation to high- and low-CO$_2$-acclimated cells in Figure 4.

6. High-CO$_2$ signaling

How can microalgal cells sense the CO$_2$ signal and respond to changes in CO$_2$ concentration? The most abundant extracellular carbonic anhydrase, CAH1, in low-CO$_2$ cells is replaced by high-CO$_2$-inducible extracellular 43 kDa protein/Fe-assimilation 1 (H43/FEA1) when low-CO$_2$-cells are transferred to high-CO$_2$ conditions (Allen et al., 2007; Baba et al., 2011a; Hanawa et al., 2004, 2007; Kobayashi et al., 1997). We found that H43/FEA1 was the most abundant extracellular soluble protein, which occupied about 26% of the total extracellular proteins of high (3%)-CO$_2$-grown cells for 3 days (Baba et al., 2011b). *H43/FEA1* homologous genes are found in the genomic sequences of the chlorophytes *Scenedesmus obliquus*, *Chlorococcum littorale*, and *Volvox carteri*, and the dinoflagellate *Heterocapsa triquerta* (Allen et al., 2007). This suggests that the *H43/FEA1* orthologs may be widely distributed among at least chlorophyte algae.

The function of H43/FEA1 is not completely understood but one possible role may be in iron assimilation (Allen et al., 2007; Rubinelli et al., 2002). Allen et al. (2007) identified *FEA1*, *FEA2*, and a candidate ferrireductase (*FRE1*) are expressed coordinately with iron assimilation components, and it was hypothesized that the proteins may facilitate iron uptake with high affinity by concentrating iron in the vicinity of the cells (Allen et al., 2007). *FEA1* and *FRE1* homologs were previously identified as the high-CO$_2$-responsive genes *HCR1* and *HCR2* in the marine chlorophyte *C. littorale*, suggesting that the components of the iron-assimilation pathway are responsive to changes in CO$_2$ concentration (Sasaki et al., 1998). A homology search of DNA sequences showed that *H43, FEA1,* and *HCR1* are identical (Allen et al., 2007; Hanawa et al., 2007), indicating that *H43/FEA1* expression was also induced by iron deficiency with transcriptional regulation. Therefore, we proposed that the gene is expressed as *H43/FEA1* (Baba et al., 2011a, 2011b).

In *C. reinhardtii*, 0.3% (v/v) CO$_2$ in air is sufficient to trigger the expression of the high-CO$_2$-inducible *H43/FEA1* and expression is correlated linearly between 0.04% and 0.3% (Hanawa et al., 2007). *H43/FEA1* can also be induced under heterotrophic conditions in the presence of acetate as an organic carbon source even under low-CO$_2$ conditions (Hanawa et al., 2007). In a previous study, the dCO$_2$ concentration in a cell suspension increased about 28 times from 1 to approximately 28 μM, which was identical to that equilibrated under the bubbling of 0.22% CO$_2$ in light, when cells were incubated in the presence of acetate and 3-(3,4-dichlorophenyl)-1,1-dimethylurea (DCMU) (Hanawa et al., 2007). From these data, the authors concluded that the induction of *H43/FEA1* is triggered by the CO$_2$ signal, even CO$_2$ generated from respiration, but not acetate itself or the change in carbon metabolite abundance. Thus, *H43/FEA1* expression can be regulated by a high-CO_2 signal at the transcriptional level, irrespective of high-CO_2 conditions. H43/FEA1 is highly reliable as a high-CO_2 response marker. The signal for *H43/FEA1* expression might be sensed by putative proteins localized on the cell membrane, which are influenced by protein modifiers and send the signal for *H43/FEA1* expression (Hanawa et al., 2007).

H43/FEA1 expression is induced under excessive levels of Cd (>25 µM) or iron-deficient conditions (<1 µM) (Allen et al., 2007; Rubinelli et al., 2002). Fei et al. (2009) reported two transcriptional *cis*-elements that are responsive to the Fe-deficient signal (FeREs) for *H43/FEA1* expression, namely FeRE1 and FeRE2, which are located at -273/-259 and -106/-85 upstream from the *H43/FEA1* transcriptional initiation site. The conserved sequence motif was identified from some iron-deficiency-inducible genes (Fei et al., 2009). However, according to our recent study, the two *cis*-elements are not necessary for the high-CO_2-induced expression of the *H43/FEA1* gene (Baba et al., 2011a). The high-CO_2-responsive *cis*-element (HCRE) was located at a -537/-370 upstream region from the *H43/FEA1* transcriptional initiation site, although the precise location has not yet been determined (Baba et al., 2011a). These results show that *H43/FEA1* expression is regulated by the high-CO_2 signal alone via the HCRE, which is located distantly from the iron-deficient-responsive element. This observation indicates that *H43/FEA1* is a multi-signal-regulated gene (Fig. 5). We have not yet determined whether all of these signals may affect the expression of other high-CO_2-inducible proteins (Baba et al., 2011b). Allen et al. (2007) reported some proteins that are iron-deficient-responsive but not CO_2-responsive, so those proteins are considered components of the iron-assimilation system. In addition, an iron-assimilation component was not found among high-CO_2-inducible extracellular proteins analyzed experimentally (Baba et al., 2011b). The expression by either high-CO_2 or iron-deficient signals is a unique feature of *H43/FEA1*.

The regulation of CCM-related gene expression, which is positively induced by a low-CO_2 signal and negatively induced by a high-CO_2 signal, has been well characterized in *C. reinhardtii*. A zinc-finger protein named CCM1/CIA5 has been identified as a candidate of the CCM master regulator (Fukuzawa et al., 2001; Miura et al., 2004; Xiang et al., 2001). CCM1/CIA5 is a protein complex with a molecular mass of approximately 290–580 kDa that is induced independently by DIC availability; Zn is necessary for its enzymatic function (Kohinata et al., 2008). One of the CCM1/CIA5-mediated signaling systems functions in the expression of *CAH1*, which encodes a low-CO_2-inducible periplasmic carbonic anhydrase (Fukuzawa et al., 1990) and the signaling is mediated by a Myb-type transcriptional regulator named LCR1 (Yoshioka et al., 2004). CCM1/CIA5 may possibly function as an amplifier for the CO_2 signaling cascade (Yamano et al., 2008). A direct signaling factor for CCM induction has not been identified, although some candidates have been reported (Giordano et al., 2005; Kaplan & Reinhold, 1999; Yamano et al., 2008). The *CCM1/CIA5* mutant lacks suppression of *H43/FEA1* expression under both low-CO_2 and iron-sufficient conditions (Allen et al., 2007), suggesting that *H43/FEA1* expression is regulated by the CCM1/CIA5-dependent signaling cascade. However, the regulatory mechanism seems to be complex. The responses of CAH1 and H43/FEA1 expression are not an all-or-none type to the signals for a change in environmental CO_2 concentration, acetate concentration, and light intensity (Hanawa et al., 2007). Signaling for *H43/FEA1* expression may be partially associated with CCM1/CIA5 signaling, although additional signals may also exist (Fig. 5). CAH2 is continuously expressed in a *CCM1/CIA5* mutant independent of CO_2 concentration

(Rawat & Moroney, 1991), suggesting that CAH2 expression is regulated by CCM1/CIA5 in the wild type. However, another high-CO$_2$-inducible protein, Rh1, is not likely regulated by CCM1/CIA5 (Wang et al., 2005).

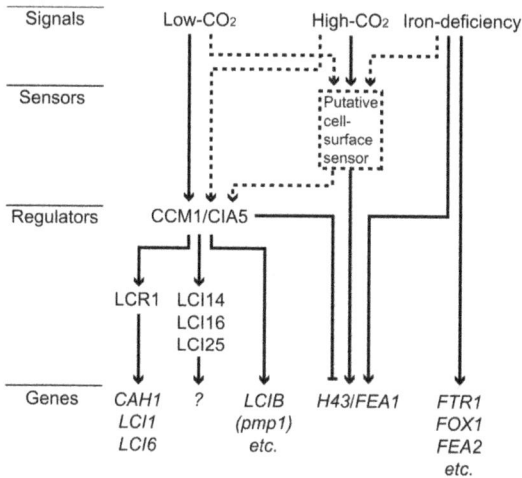

Fig. 5. Schematic model of high-CO$_2$ signaling for *H43/FEA1* induction. Solid and broken lines are expective and putative signaling flows, respectively. Low-CO$_2$ signaling is modified from Miura et al. (2004) and Yamano and Fukuzawa (2009). Iron-deficient-inducible genes are according to Allen et al. (2007). Cd signaling on *H43/FEA1* induction, proposed by Rubinelli et al. (2002), is not drawn because little about it is known.

7. Conclusion

Compared to low-CO$_2$-inducible mechanisms that are well understood, analyses of high-CO$_2$-responsive mechanisms in microalgae at the molecular level have just started using the unicellular green alga *C. reinhardtii*. An accurate characterization of the acclimation mechanisms to high-CO$_2$ conditions will be important for both a detailed understanding of sensing and responding to environmental CO$_2$ changes and maximizing algal biomass productivity in mass cultivation. H43/FEA1, the most abundant extracellular protein in high-CO$_2$-acclimated cells, is expressed in response to multiple signals, including high-CO$_2$, iron-deficiency, or Cd-stress conditions. This suggests that, in addition to the high-CO$_2$ signal itself, abnormally stressful conditions such as strong nutrient depletion caused by rapid growth under high-CO$_2$ conditions may trigger expression of the gene. Targeted proteomics of whole *C. reinhardtii* established by Wienkoop et al. (2010) and a cDNA array (Yamano et al., 2008) or transcriptomics (Yamano & Fukuzawa, 2009), which has been applied to an expression analysis of CCM-associated genes,, would be useful for further detailed analysis of high-CO$_2$ response phenomena. Our recent data indicate that the expression of gametogenesis-related proteins, which are strictly regulated by nitrogen availability, is triggered by high-CO$_2$ signals with a drastic change in extracellular proteins. These gametogenesis-related proteins in the periplasmic space of *C. reinhardtii* cells may play novel and crucial roles when *C. reinhardtii* is grown under high-CO$_2$ conditions.

8. Acknowledgments

This work was financially supported, in part, by a Grant-in-Aid for Scientific Research (Basic Research Area (S), No. 22221003 to YS) from the Japan Society for the Promotion of Science, the Core Research of Evolutional Science & Technology program (CREST) from the Japan Science and Technology Agency (JST) (to MB & YS), another fund from the Japan Science and Technology Agency (CREST/JST, to YS), and by the Global Environment Research Fund from the Japanese Ministry of Environment (FY2008-2010) to YS.

9. References

Abe, J.; Kubo, T.; Takagi, Y.; Saito, T.; Miura, K.; Fukuzawa, H. & Matsuda, Y. (2004) The Transcriptional Program of Synchronous Gametogenesis in *Chlamydomonas reinhardtii*. *Current Genetics*, Vol.46, No.5, (November 2004), pp. 304–315, ISSN 0172-8083.

Adair, W.S. (1985) Characterization of *Chlamydomonas* Sexual Agglutinins. *Journal of Cell Science*, Vol.2, Supplement, (1985), pp. 233–260, ISSN 0269-3518

Aizawa, K. & Miyachi, S. (1986) Carbonic Anhydrase and CO_2 Concentrating Mechanisms in Microalgae and Cyanobacteria. *FEMS Microbiology Letters*, Vol.39, No.3, (August 1986), pp. 215–233, ISSN 0378–1097.

Allen, M.D.; del Campo, J.A.; Kropat, J. & Merchant, S.S. (2007) *FEA1*, *FEA2*, and *FRE1*, Encoding Two Homologous Secreted Proteins and a Candidate Ferrireductase, Are Expressed Coordinately with *FOX1* and *FTR1* in Iron-Deficient *Chlamydomonas reinhardtii*. *Eukaryotic Cell*, Vol.6, No.10, (October 2007), pp. 1841–1852, ISSN 1535-9778.

Baba, M.; Hanawa, Y.; Suzuki, I. & Shiraiwa, Y. (2011a) Regulation of the Expression of *H43/Fea1* by Multi-Signals. *Photosynthesis research*, Epub ahead of print, (January 2011), ISSN 1573–5079.

Baba, M.; Suzuki, I. & Shiraiwa, Y. (2011b) Proteomic Analysis of High-CO_2-Inducible Extracellular Proteins in the Unicellular Green Alga, *Chlamydomonas reinhardtii*. *Plant & cell physiology*, Epub ahead of print, (June 2011), ISSN 0032-0781.

Badger, M.R. & Price, G.D. (1994) The Role of Carbonic Anhydrase in Photosynthesis. *Annual Review of Plant Physiology and Plant Molecular Biology*, Vol.45, (June 1994), pp. 369–392, ISSN 1040-2519.

Badger, M.R. & Price, G.D. (2003) CO_2 Concentrating Mechanisms in Cyanobacteria: Molecular Components, Their Diversity and Evolution. *Journal of experimental botany*, Vol.54, No.383, (February 2003), pp. 609–622, ISSN 0022-0957.

Badger, M.R.; von Caemmerer, S.; Ruuska, S. & Nakano, H. (2000) Electron Flow to Oxygen in Higher Plants and Algae: Rates and Control of Direct Photoreduction (Mehler reaction) and Rubisco Oxygenase. *Philosophical transactions of the Royal Society of London. Series B, Biological sciences*, Vol.355, No.1402, (October 2000), pp. 1433–1446, ISSN 0962-8436.

Badger, M.R.; Price, G.D.; Long, B.M. & Woodger, F.J. (2006) The Environmental Plasticity and Ecological Genomics of the Cyanobacterial CO_2 Concentrating Mechanism. *Journal of experimental botany*, Vol.57, No.2, (2006), pp. 249–265, ISSN 0022-0957.

Balkos, K.D. & Colman, B. (2007) Mechanism of CO_2 Acquisition in an Acid-Tolerant *Chlamydomonas*. *Plant, cell & environment*, Vol.30, No.6, (June 2007), pp. 745–752, ISSN 0140-7791.

Bartlett, S.G.; Mitra, M. & Moroney, J.V. (2007) CO_2 Concentrating Mechanisms, In: *The Structure and Function of Plastids.*, Edited by Wise, R.R & Hoober J.K., pp. 253–271. Springer, ISBN 978-1-4020-4061-0, Dordrecht, The Netherlands.

Beardall, J. & Giordano, M. (2002) Ecological Implications of Microalgal and Cyanobacterial CCMs and Their Regulation. *Functional Plant Biology*, Vol.29, No.2-3, (2002), pp. 335–347, ISSN 1445-4408.

Bhatti, S. & Colman, B. (2008) Inorganic Carbon Acquisition by Some Synurophyte Algae. *Physiologia plantarum*, Vol.133, No.1, (February 2008), pp. 33–40, ISSN 0031-9317

Box, G.E.P. & Wilson, K.B. (1951) On the Experimental Attainment of Optimum Conditions. *Journal of the Royal Statistical Society Series B*, Vol.13, No.1, (1951), pp. 1–45, ISSN 0035-9246.

Buchanan, M.J. & Snell, W.J. (1988) Biochemical Studies on Lysin, a Cell Wall Degrading Enzyme Released During Fertilization in *Chlamydomonas*. *Experimental cell research*, Vol.179, No.1, (November 1988), pp. 181–193, ISSN 0014-4827.

Buyanovsky, G.A. & Wagner, G.H. (1983) Annual Cycles of Carbon Dioxide Level in Soil Air. *Soil Science Society of America Journal*, Vol.47, No.6, (1983), pp. 1139–1145, ISSN 0361-5995.

Chihara, M.; Nakayama, T.; Inouye, I. & Kodama, M. (1994) *Chlorococcum littorale*, a New Marine Green Coccoid Alga (Chlorococcales, Chlorophyceae). *Archiv für Protistenkunde*, Vol.144, No.3, (1994), pp. 227–235, ISSN 0003-9365.

Collins, S. & Bell, G. (2004) Phenotypic Consequences of 1,000 Generations of Selection at Elevated CO_2 in a Green Alga. *Nature*, Vol. 431, No.7008, (September 2004), pp. 566–569, ISSN 0028-0836.

Collins, S. & Bell, G. (2006) Evolution of Natural Algal Populations at Elevated CO_2. *Ecology letters*, Vol.9, No.2, (February 2006), pp. 129–135, ISSN 1461-023X.

Collins, S.; Sültemeyer, D. & Bell, G. (2006) Changes in C Uptake in Populations of *Chlamydomonas reinhardtii* Selected at High CO_2. *Plant, cell & environment*, Vol.29, No.9, (September 2006), pp. 1812–1819, ISSN 0140-7791.

Colman, B. & Balkos, K.D. (2005) Mechanisms of Inorganic Carbon Acquisition by *Euglena* species. *Canadian journal of botany. Journal canadien de botanique*, Vol.83, No.7, (July 2005), pp. 865–871, ISSN 0008-4026.

Condie, K.C. & Sloan, R.E. (January 1998) *Origin and Evolution of Earth: Principles of Historical Geology.* Prentice-Hall, ISBN 978-0134918204, NJ.

Demidov, E.; Iwasaki, I.; Satoh, A.; Kurano, N. & Miyachi, S. (2000) Short-term Responses of Photosynthetic Reactions to Extremely High-CO_2 Stress in a "High-CO_2" Tolerant Green Alga, *Chlorococcum littorale* and an Intolerant Green Alga *Stichococcus bacillaris*. *Russian journal of plant physiology : a comprehensive Russian journal on modern phytophysiology*, Vol.47, No.5, (September 2000), pp. 622–631, ISSN 1021-4437.

Diaz, M.M. & Maberly, S.C. (2009) Carbon-Concentrating Mechanisms in Acidophilic Algae. *Phycologia*, Vol.48, No.2, (March 2009), pp. 77–85, ISSN 0031-8884.

Doney, S.C.; Fabry, V.J.; Feely, R.A. & Kleypas, J.A. (2009) Ocean Acidification: the Other CO_2 Problem, *Annual review of marine science*, Vol.1, (January 2009), pp. 169–192, ISSN 1941-1405.

Duanmu, D.; Miller, A.; Horken, K.; Weeks, D. & Spalding, M.H. (2009) Knockdown of Limiting-CO_2-Induced Gene *HLA3* Decreases HCO_3^- Transport and Photosynthetic Ci Affinity in *Chlamydomonas reinhardtii*. *Proceedings of the National Academy of*

        Sciences of the United States of America, Vol.106, No.14, (April 2009), pp. 5990–5995, ISSN 0027-8424.

Ertl, H.; Hallmann, A.; Wenzl, S. & Sumper, M. (1992) A Novel Extensin That May Organize Extracellular Matrix Biogenesis in *Volvox carteri*. *The EMBO journal*, Vol.11, No.6, (June 1992), pp. 2055–2062, ISSN 0261-4189.

Falkowski, P.G. & Raven, J.A. (2007) *Aquatic Photosynthesis, 2nd Ed.* Princeton University Press, ISBN 978-0691115511, Princeton, NJ.

Fei, X.; Eriksson, M.; Yang, J. & Deng, X. (2009) An Fe Deficiency Responsive Element with a Core Sequence of TGGCA Regulates the Expression of *Fea1* in *Chlamydomonas reinharditii*. *Journal of biochemistry*, Vol.146, No.2, (August 2009), pp. 157–166, ISSN 0021-924X.

Fernández, E.; Llamas, Á. & Galván, A. (2009) Nitrogen Assimilation and its Regulation, In: *The Chlamydomonas Sourcebook, 2nd Ed, Vol. 2.*, Edited by Stern D.B., pp. 69–113. Academic Press, ISBN 978-0-12-370875-5, San Diego, CA.

Fong, R.N.; Kim, K.-S.; Yoshihara, C.; Inwood, W. & Kustu, S. (2007) The W148L Substitution in the *Escherichia coli* Ammonium Channel AmtB Increases Flux and Indicates That the Substrate Is an Ion. *Proceedings of the National Academy of Sciences of the United States of America*, Vol.104, No.47, (November 2007), pp. 18706–18711, ISSN 0027-8424.

Fujiwara, S.; Fukuzawa, H.; Tachiki, A. & Miyachi, S. (1990) Structure and Differential Expression of Two Genes Encoding Carbonic Anhydrase in *Chlamydomonas reinhardtii*. *Proceedings of the National Academy of Sciences of the United States of America*, Vol.87, No.24, (December 1990), pp. 9779–9783, ISSN 0027-8424.

Fukuda, S.; Suzuki, I.; Hama, T. & Shiraiwa, Y. (2011) Compensatory Response of the Unicellular-Calcifying Alga *Emiliania huxleyi* (Coccolithophoridales, Haptophyta) to Ocean Acidification. *Journal of Oceanography*, Vol.67, No.1, (February 2011), pp. 17–25, ISSN 0916-8370.

Fukuzawa, H.; Fujiwara, S.; Yamamoto, Y.; Dionisio-Sese, M.L. & Miyachi, S. (1990) cDNA Cloning, Sequence, and Expression of Carbonic Anhydrase in *Chlamydomonas reinhardtii*: Regulation by Environmental CO_2 Concentration. *Proceedings of the National Academy of Sciences of the United States of America*, Vol.87, No.11, (June 1990), pp. 4383–4387, ISSN 0027-8424.

Fukuzawa, H.; Miura, K.; Ishizaki, K.; Kucho, K.I.; Saito, T.; Kohinata, T. & Ohyama, K. (2001) *Ccm1*, a Regulatory Gene Controlling the Induction of a Carbon-Concentrating Mechanism in *Chlamydomonas reinhardtii* by Sensing CO_2. *Proceedings of the National Academy of Sciences of the United States of America*, Vol.98, No.9, (April 2001), pp. 5347–5352, ISSN 0027-8424.

Galvan, A.; Quesada, A. & Fernández, E. (1996) Nitrate and Nitrite Are Transported by Different Specific Transport Systems and by a Bispecific Transporter in *Chlamydomonas reinhardtii*. *The Journal of biological chemistry*, Vol.271, No. 4, (January 1996), pp. 2088–2092, ISSN 0021-9258.

Giordano, M.; Beardall, J. & Raven, J.A. (2005) CO_2 Concentrating Mechanisms in Algae: Mechanisms, Environmental Modulation, and Evolution. *Annual review of plant biology*, Vol.56, (June 2005), pp. 99–131, ISSN 1543-5008.

Giordano, M.; Norici, A.; Forssen, M.; Eriksson, M. & Raven, J.A. (2003) An Anaplerotic Role for Mitochondrial Carbonic Anhydrase in *Chlamydomonas reinhardtii*. *Plant Physiology*, Vol.132, No.4, (August 2003), pp. 2126–2134, ISSN 0032-0889.

Goodenough, U.W.; Gebhart, B.; Mecham, R.E. & Heuser, J.E. (1986) Crystals of the *Chlamydomonas reinhardtii* Cell Wall: Polymerization, Depolymerization, and Purification of Glycoprotein Monomers. The Journal of cell biology, Vol.103, No.2, (August 1986), pp. 405–417, ISSN 0021-9525.

Goodenough, U.; Lin, H. & Lee, J. (2007) Sex Determination in *Chlamydomonas. Seminars in cell & developmental biology*, Vol.18, No.3, (February 2007), pp. 350–361, ISSN 1084-9521.

Hallmann, A. (2006) The Pherophorins: Common, Versatile Building Blocks in the Evolution of Extracellular Matrix Architecture in *Volvocales. The Plant journal : for cell and molecular biology*, Vol.45, No.2, (January 2006), pp. 292–307, ISSN 0960-7412.

Hanawa, Y. (2007) Study on a CO₂ Sensing Mechanism by the Expression Analysis of a High CO₂ Inducible *H43* Gene in *Chlamydomonas reinhardtii*. A Dissertation Submitted to the Graduate School of Life and Environmental Sciences, the University of Tsukuba.

Hanawa, Y.; Iwamoto, K. & Shiraiwa, Y. (2004) Purification of a Recombinant H43, a High-CO₂-Inducible Protein of *Chlamydomonas reinhardtii*, Expressed in *Escherichia coli. Japanese Journal of Phycology*, Vol.52, Supplement, (2004), pp. 95–100, ISSN 0038-1578.

Hanawa, Y.; Watanabe, M.; Karatsu, Y.; Fukuzawa, H. & Shiraiwa, Y. (2007) Induction of a High-CO₂-Inducible, Periplasmic Protein, H43, and Its Application as a High-CO₂-Responsive Marker for Study of the High-CO₂-Sensing Mechanism in *Chlamydomonas reinhardtii. Plant & cell physiology*, Vol.48, No.2, (January 2007), pp. 299–309, ISSN 0032-0781.

Heldt, H.; Werdan, K.; Milovancev, M. & Geller, G. (1973) Alkalization of the Chloroplast Stroma Caused by Light-Dependent Proton Flux into the Thylakoid Space. *Biochimica et Biophysica Acta*, Vol.314, No.2, (August 1973), pp. 224–241, ISSN 0006-3002.

Ho, S.H.; Chen, C.Y.; Lee, D.J. & Chang, J.S. (2011) Perspectives on Microalgal CO₂-Emission Mitigation Systems - A Review. *Biotechnology Advances*, Vol.29, No.2, (March-April 2011), pp. 189–198, ISSN 0734-9750.

Hoffmann, X.K. & Beck, C.F. (2005) Mating-Induced Shedding of Cell Walls, Removal of Walls from Vegetative Cells, and Osmotic Stress Induce Presumed Cell Wall Genes in *Chlamydomonas. Plant Physiology*, Vol.139, No.2, (October 2005), pp. 999–1014, ISSN 0032-0889.

Hurd, C.L.; Hepburn, C.D.; Currie, K.I.; Raven, J.A. & Hunter, K.A. (2009) Testing the Effects of Ocean Acidification on Algal Metabolism: Considerations for Experimental Designs. *Journal of Phycology*, Vol.45, No.6, (November 2009), pp. 1236–1251, ISSN 0022-3646.

Iglesias-Rodriguez, M.D.; Halloran, P.R.; Rickaby, R.E.M.; Hall, I.R.; Colmenero-Hidalgo, E.; Gittins, J.R.; Green, D.R.H.; Tyrrell, T.; Gibbs, S.J.; Dassow, P.; Rehm, E.; Armbrust, E.V. & Boessenkool, K.P. (2008) Phytoplankton Calcification in a High-CO₂ World. *Science*, Vol.320, No.5874, (April 2008), pp. 336–340, ISSN 0036-8075.

Inoue, I. (2007) *The Natural History of Algae: Second Edition: Perspective of Three Billion years Evolution of Algae, Earth and Environment.* Tokai University Press, ISBN 978-4-486-01777-6, Japan.

Ishii, M.; Yoshikawa, H. & Matsueda, H. (March, 2000) *Coulometric Precise Analysis of Total Inorganic Carbon in Seawater and Measurements of Radiocarbon for the Carbon Dioxide in the Atmosphere and for the Total Inorganic Carbon in Seawater. Technical Reports of the Meteorogical Research Institute No. 41.*, Meteorological Research Institute, Japan, Retrieved from <http://www.mri-jma.go.jp/Publish/Technical/index_en.html>

Iwasaki, I.; Hu, Q.; Kurano, N. & Miyachi, S. (1998) Effect of Extremely High-CO_2 Stress on Energy Distribution Between Photosystem I and Photosystem II in a High-CO_2 Tolerant Green Alga, *Chlorococcum littorale* and the Intolerant Green Alga *Stichococcus bacillaris. Journal of photochemistry and photobiology. B, Biology,* Vol.44, No.3, (July 1998), pp. 184–190, ISSN 1011-1344.

Iwasaki, I.; Kurano, N. & Miyachi, S. (1996) Effects of High-CO_2 Stress on Photosystem II in a Green Alga, *Chlorococcum littorale*, Which Has a Tolerance to High CO_2. *Journal of photochemistry and photobiology. B, Biology,* Vol.36, No.3, (December 1996), pp. 327–332, ISSN 1011-1344.

Jones, H.G. (1992) *Plants and Microclimate: A Quantitative Approach to Environmental Plant Physiology, 2nd edition.*, Cambridge University Press, ISBN 9780521425247, Cambridge, UK.

Kaplan, A. & Reinhold, L. (1999) CO_2 Concentrating Mechanisms in Photosynthetic Microorganisms. *Annual Review of Plant Physiology and Plant Molecular Biology,* Vol.50, (June 1999), pp. 539–570, ISSN 1040-2519.

Kinoshita, T.; Fukuzawa, H.; Shimada, T.; Saito, T. & Matsuda, Y. (1992) Primary Structure and Expression of a Gamete Lytic Enzyme in *Chlamydomonas reinhardtii*: Similarity of Functional Domains to Matrix Metalloproteases. *Proceedings of the National Academy of Sciences of the United States of America,* Vol.89, No.10, (May 1992), pp. 4693–4697, ISSN 0027-8424.

Kobayashi, H.; Odani, S. & Shiraiwa, Y. (1997) A High-CO_2-Inducible, Periplasmic Polypeptide in an Unicellular Green Alga *Chlamydomonas reinhardtii* (abstract no. 493). *Plant Physiology*, Vol.114, Supplement, p. 112, ISSN 0032-0889.

Kodama, M.; Ikemoto, H. & Miyachi, S. (1993) A New Species of Highly CO_2-Tolerant Fast-Growing Marine Microalga Suitable for High-Density Culture. *Journal of Marine Biotechnology,* Vol. 1, (1993), pp. 21–25, ISSN 0941-2905.

Kohinata, T.; Nishino, H. & Fukuzawa, H. (2008) Significance of Zinc in a Regulatory Protein, CCM1, Which Regulates the Carbon-Concentrating Mechanism in *Chlamydomonas reinhardtii. Plant & cell physiology*, Vol.49, No.2, (January 2008), pp. 273–283, ISSN 0032-0781.

Kubo, T.; Saito, T.; Fukuzawa, H. & Matsuda, Y. (2001) Two Tandemly-Located Matrix Metalloprotease Genes with Different Expression Patterns in the *Chlamydomonas* Sexual Cell Cycle. *Current Genetics*, Vol.40, No.2, (September 2001), pp. 136–143, ISSN 0172-8083.

Kumar, A.; Ergas, S.J.; Yuan, X.; Sahu, A.K.; Zhang, Q.; Dewulf, J.; Malcata, F.X. & an Langenhove, H.V. (2010) Enhanced CO_2 Fixation and Biofuel Production via Microalgae: Recent Developments and Future Directions, *Trends in biotechnology*, Vol.28, No.7, (June 2010), pp. 371–380, ISSN 0167-7799.

Kurano, N.; Ikemoto, H.; Miyashita, H.; Hasegawa, T.; Hata, H. & Miyachi, S. (1995) Fixation and Utilization of Carbon Dioxide by Microalgal Photosynthesis. *Energy Conversion and Management*, Vol.36, No.6–9, (June-September 1995), pp. 689–692, ISSN 0196-8904.

Lee, J.S. & Lee, J.P. (2003) Review of Advances in Biological CO_2 Mitigation Technology. *Biotechnology and Bioprocess Engineering*, Vol.8, No.6, (2003), pp. 354–359, ISSN 1226-8372.

Lee, J.H.; Waffenschmidt, S.; Small, L. & Goodenough, U. (2007) Between-Species Analysis of Short-Repeat Modules in Cell Wall and Sex-Related Hydroxyproline-Rich Glycoproteins of *Chlamydomonas*. *Plant Physiology*, Vol.144, No.4, (August 2007), pp. 1813–1826, ISSN 0032-0889.

Maeda, K.; Owada, M.; Kimura, N.; Omata, K. & Karube, I. (1995) CO_2 Fixation from the Flue Gas on Coal-Fired Thermal Power plant by Microalgae. *Energy Conversion and Management*, Vol.36, No.6–9, (June-September 1995), pp. 717–720, ISSN 0196-8904.

Mariscal, V.; Moulin, P.; Orsel, M.; Miller, A.J.; Emilio Fernández, E. & Galvána, A. (2006) Differential Regulation of the *Chlamydomonas Nar1* Gene Family by Carbon and Nitrogen. *Protist*, Vol.157, No.4, (October 2006), pp. 421–433, ISSN 1434-4610.

Matsuda, Y.; Shimada, T. & Sakamoto, Y. (1992) Ammonium Ions Control Gametic Differentiation and Dedifferentiation in *Chlamydomonas reinhardtii*. *Plant & cell physiology*, Vol.33, No.7, (1992), pp. 909–914, ISSN 0032-0781.

Miura, K.; Yamano, T.; Yoshioka, S.; Kohinata, T.; Inoue, Y.; Taniguchi, F.; Asamizu, E.; Nakamura, Y.; Tabata, S.; Yamato. K.T.; Ohyama, K. & Fukuzawa, H. (2004) Expression Profiling-Based Identification of CO_2-Responsive Genes Regulated by CCM1 Controlling a Carbon Concentrating Mechanism in *Chlamydomonas reinhardtii*. *Plant Physiology*, Vol.135, No.3, (July 2004), pp. 1595–1607, ISSN 0032-0889.

Moroney, J.V. & Ynalvez, R.A. (2007) A Proposed Carbon Dioxide Concentration Mechanism in *Chlamydomonas reinhardtii*. *Eukaryotic Cell*, Vol.6, No. 8, (August 2007), pp. 1251–1259, ISSN 1535-9778.

Nielsen, E.S. (1955) Carbon Dioxide As Carbon Source and Narcotic in Photosynthesis and Growth of *Chlorella pyrenoidosa*. *Physiologia Plantrum*, Vol.8, No.2, (April 1955), pp. 317–335, ISSN 0031-9317.

Olaizola, M. (2003) Microalgal Removal of CO_2 from Flue Gases: Changes in Medium pH and Flue Gas Composition Do not Appear to Affect the Photochemical Yield of Microalgal Cultures. *Biotechnology and Bioprocess Engineering*, Vol.8, No.6, (November-December 2003), pp. 360–367, ISSN 1226-8372.

Pesheva, I.; Kodama, M.; Dionisio-Sese, M.L. & Miyachi, S. (1994) Changes in Photosynthetic Characteristics Induced by Transferring Air-Grown Cells of *Chlorococcum littorale* to High-CO_2 Conditions. *Plant & cell physiology*, Vol.35, No.3, (1994), pp. 379–387, ISSN 0032-0781.

Price, G.D. & Badger, M.R. (1989) Isolation and Characterization of High CO_2-Requiring-Mutants of the Cyanobacterium *Synechococcus* PCC7942, Two Phenotypes that Accumulate Inorganic Carbon but Are Apparently Unable to Generate CO_2 within the Carboxysome. *Plant Physiology*, Vol.91, No.2, (October 1989), pp. 514–525, ISSN 0032-0889.

Pronina, N.A.; Kodama, M. & Miyachi, S. (1993) Changes in Intracellular pH Values in Various Microalgae Induced by Raising CO_2 Concentrations. *XV International Botanical Congress*, ISBN 9783921800386, Yokohama, Japan, August, 1993.

Ramazanov, Z. & Cárdenas, J. (1994) Photorespiratory Ammonium Assimilation in Chloroplasts of *Chlamydomonas reinhardtii*, *Physiologia Plantarum*, Vol.91, No.3, (July 1994), pp. 495–502, ISSN 0031-9317.

Raven, J.A. (2001) A Role for Mitochondrial Carbonic Anhydrase in Limiting CO_2 Leakage from Low CO_2-Grown Cells of *Chlamydomonas reinhardtii*. *Plant, Cell & Environment*, Vol.24, No.2, (February 2001), pp. 261–265, ISSN 0140-7791.

Raven, J.A. (2010) Inorganic Carbon Acquisition by Eukaryotic Algae: Four Current Questions. *Photosynthesis research*, Vol.106, No.1-2, (June 2010), pp. 123–134, ISSN 1573-5079.

Raven, J.A.; Cockell, C.S. & De La Rocha, C.L. (2008) The Evolution of Inorganic Carbon Concentrating Mechanisms in Photosynthesis. *Philosophical transactions of the Royal Society of London. Series B, Biological sciences*, Vol.363, No.1504, (August 2008), pp. 2641–2650, ISSN 0962-8436.

Rawat, M. & Moroney, J.V. (1991) Partial Characterization of a New Isoenzyme of Carbonic-anhydrase Isolated from *Chlamydomonas reinhardtii*. *The Journal of biological chemistry*, Vol.266, No.15, (May 1991), pp. 9719–9723, ISSN 0021-9258.

Rexach, J.; Fernández, E. & Galván, A. (2000) The *Chlamydomonas reinhardtii* Nar1 Gene Encodes a Chloroplast Membrane Protein Involved in Nitrite Transport. *The Plant Cell*, Vol.12, No.8, (August 2000), pp. 1441–1453, ISSN 1040-4651.

Rexach, J.; Montero, B.; Fernández, E. & Galván, A. (1999) Differential Regulation of the High Affinity Nitrite Transport Systems III and IV in *Chlamydomonas reinhardtii*. *The Journal of biological chemistry*, Vol.274, No.39, (September 1999), pp. 27801–27806, ISSN 0021-9258.

Riebesell, U.; Zondervan, I.; Rost, B.; Tortell, P.D.; Zeebe, R.E. & Morel, F.M.M. (2000) Reduced Calcification of Marine Plankton in Response to Increased Atmospheric CO_2. *Nature*, Vol.407, No.6802, (September 2000), pp. 364–367, ISSN 0028-0836.

Rubinelli, P.; Siripornadulsil, S.; Gao-Rubinelli, F. & Sayre, R.T. (2002) Cadmium- and Iron-Stress-Inducible Gene Expression in the Green Alga *Chlamydomonas reinhardtii*: Evidence for H43 Protein Function in Iron Assimilation. *Planta*, Vol.215, No.1, (May 2002), pp. 1–13, ISSN 0032-0935.

Sasaki, T.; Kurano, N. & Miyachi, S. (1998) Induction of Ferric Reductase Activity and of Iron Uptake Capacity in *Chlorococcum littorale* Cells under Extremely High-CO_2 and Iron-Deficient Conditions. *Plant & cell physiology*, Vol.39, No.4, (1998), pp. 405–410, ISSN 0032-0781.

Sasaki, T.; Pronina, N.A.; Maeshima, M.; Iwasaki, I.; Kurano, N. & Miyachi, S. (1999) Development of Vacuoles and Vacuolar ATPase Activity under Extremely High-CO_2 Conditions in *Chlorococcum littorale* Cells. *Plant Biology*, Vol.1, No.1, (January 1999), pp. 68–75, ISSN 1435-8603.

Satoh, A.; Kurano, N.; Harayama, S. & Miyachi, S. (2004) Effects of Chloramphenicol on Photosynthesis, Protein Profiles and Transketolase Activity under Extremely High CO_2 Concentration in an Extremely-high-CO_2-tolerant Green Microalga, *Chlorococcum littorale*. *Plant & Cell Physiology*, Vol.45, No.12, (December 2004), pp. 1857–1862, ISSN 0032-0781.

Satoh, A.; Kurano, N. & Miyachi, S. (2001) Inhibition of Photosynthesis by Intracellular Carbonic Anhydrase in Microalgae under Excess Concentrations of CO_2. *Photosynthesis research*, Vol.68, No.3, (2001), pp. 215–224, ISSN 1573-5079.

Satoh, A.; Kurano, N.; Senger, H. & Miyachi, S. (2002) Regulation of Energy Balance in Photosynthesis in Response to Changes in CO_2 Concentrations and Light Intensities During Growth in Extremely High-CO_2-Tolerant Green Microalgae. *Plant & cell physiology*, Vol.43, No.4, (April 2002), pp. 440–451, ISSN 0032-0781.

Seckbach, J. Baker, F.A. & Shugarman, P.M. (1970) Algae Thrive under Pure CO$_2$. *Nature*, Vol.227, No.5259, (1970), pp. 744–745, ISSN 0028-0836.

Shiraiwa, Y. & Miyachi, S. (1985) Effect of Temperature and CO$_2$ Concentration on Induction of Carbonic Anhydrase and Changes in Efficiency of Photosynthesis in *Chlorella vulgaris* 11h. *Plant & cell physiology*, Vol.26, No.3, (1985), pp. 543–549, ISSN 0032-0781.

Shiraiwa, Y. & Schmid, G.H. (1986) Stimulation of photorespiration by the carbonic anhydrase inhibitor ethoxyzolamide in *Chlorella vulgaris*. *Zeischrift für Naturforschung* Vol.41, No.5-6, (1986), pp.564–570, ISSN 0341-0382.

Soupene, E.; Inwood, W. & Kustu, S. (2004) Lack of the Rhesus Protein Rh1 Impairs Growth of the Green Alga *Chlamydomonas reinhardtii* at High CO$_2$. *Proceedings of the National Academy of Sciences of the United States of America*, Vol.101, No.20, (May 2004) :7787–7792, ISSN 0027-8424.

Soupene, E.; King, N.; Feild, E.; Liu, P.; Niyogi, K.K.; Huang, C.-H. & Kustu, S. (2002) Rhesus Expression in Green Alga Is Regulated by CO$_2$. *Proceedings of the National Academy of Sciences of the United States of America*, Vol.99, No.11, (May 2002), pp. 7769–7773, ISSN 0027-8424.

Spalding, M.H. (2008) Microalgal Carbon-Dioxide-Concentrating Mechanisms: *Chlamydomonas* Inorganic Carbon Transporters. *Journal of experimental botany*, Vol.59, No.7, (2008), pp. 1463–1473, ISSN 0022-0957.

Spalding, M.H. (2009) The CO$_2$-Concentrating Mechanism and Carbon Assimilation, In: *The Chlamydomonas Sourcebook, 2nd Ed, Vol. 2.*, Edited by Stern D.B., pp. 257–301. Academic Press, ISBN 978-0-12-370875-5, San Diego, CA.

Spalding M.H.; Spreitzer, R.J. & Ogren, W.L. (1983) Reduced Inorganic Carbon Transport in a CO$_2$-Requiring Mutant of *Chlamydomonas reinhardtii*. *Plant Physiology*, Vol.73, No.2, (October 1983), pp. 273–276, ISSN 0032-0889.

Stolzy, L.H. (1974) Soil Atmosphere. In: *The Plant Root and Its Environment.*, Edited by Carson E.W., pp 335–362. University Press of Virginia, ISBN 978-0813904115, Charlottesville, VA.

Suzuki, K. & Spalding, M.H. (1989) Adaptation of *Chlamydomonas reinhardtii* High-CO$_2$-Requiring Mutants to Limiting CO$_2$. *Plant Physiology*, Vol.90, No.3, (July 1989), pp. 1195–1200, ISSN 0032-0889.

Sültemeyer, D.F.; Klug, K. & Fock, H.P. (1986) Effect of Photon Fluence Rate on Oxygen Evolution and Uptake by *Chlamydomonas reinhardtii* Suspensions Grown in Ambient and CO$_2$-Enriched Air. *Plant Physiology*, Vol.81, No.2, (June 1986), pp. 372–375, ISSN 0032-0889.

Sültemeyer, D.F.; Klug, K. & Fock, H.P. (1987) Effect of Dissolved Inorganic Carbon on Oxygen Evolution and Uptake by *Chlamydomonas reinhardtii* Suspensions Adapted to Ambient and CO$_2$-Enriched Air. *Photosynthesis research*, Vol.12, No.1, (1987), pp. 25–33, ISSN 1573-5079.

Tachiki, A.; Fukuzawa, H. & Miyachi, S. (1992) Characterization of Carbonic-anhydrase Isozyme CA2, Which Is the CAH2 Gene-product, in *Chlamydomonas reinhardtii*. *Bioscience, Biotechnology, & Biochemistry*, Vol.56, No.5, (May 1992), pp. 794–798, ISSN 0916-8451.

Verma, V.; Bhati, S.; Huss, V.A.R. & Colman, B. (2009) Photosynthetic Inorganic Carbon Assimilation in a Free-Living Species of *Coccomyxa* (Chlorophyta). *Journal of Phycology*, Vol.45, (2009), pp. 847–854, ISSN 0022-3646.

Wang, Y.; Sun, Z.; Horken, K.M.; Im, C.S.; Xiang, Y.; Grossman, A.R. & Weeks, D.P. (2005) Analysis of CIA5, the Master Regulator of the Carbon-concentrating Mechanisms in *Chlamydomonas reinhardtii*, and Its Control of Gene Expression. *Canadian Journal of Botany*, Vol.83, No.7, (July 2005), pp. 765–779, ISSN 0008-4026.

Wienkoop, S.; Weiss, J.; May, P.; Kempa, S.; Irgang, S.; Recuenco-Munoz, L.; Pietzke, M.; Schwemmer, T.; Rupprecht, J.; Egelhofer, V. & Weckwerth, W. (2010) Targeted Proteomics for *Chlamydomonas reinhardtii* Combined with Rapid Subcellular Protein Fractionation, Metabolomics and Metabolic Flux Analyses. *Molecular BioSystems*, Vol.6, No. 6, (June 2010), pp. 1018–1031, ISSN 1742-2051.

Xiang, Y.; Zhang, J. & Weeks, D.P. (2001) The *Cia5* Gene Controls Formation of the Carbon Concentrating Mechanism in *Chlamydomonas reinhardtii*. *Proceedings of the National Academy of Sciences of the United States of America*, Vol.98, No. 9, (April 2001),5341–5346, ISSN 0027-8424.

Yamano, T. & Fukuzawa, H. (2009) Carbon-Concentrating Mechanism in a Green Alga, *Chlamydomonas reinhardtii*, Revealed by Transcriptome Analyses. *Journal of basic microbiology*, Vol. 49, No.1, (February 2009), pp. 42–51, ISSN0233-111X.

Yamano, T.; Miura, K. & Fukuzawa, H. (2008) Expression Analysis of Genes Associated with the Induction of the Carbon-Concentrating Mechanism in *Chlamydomonas reinhardtii*. *Plant Physiology*, Vol.147, No.1, (May 2008), pp. 340–354, ISSN 0032-0889.

Yamano, T.; Tsujikawa, T.; Hatano, K.; Ozawa, S.; Takahashi, Y. & Fukuzawa, H. (2010) Light and Low-CO_2-Dependent LCIB-LCIC Complex Localization in the Chloroplast Supports the Carbon Concentrating Mechanism in *Chlamydomonas reinhardtii*. *Plant & cell physiology*, Vol.51, No.9, (July 2010), pp. 1453–1468, ISSN 0032-0781.

Yang, S.Y.; Tsuzuki, M. & Miyachi, S. (1985) Carbonic Anhydrase of *Chlamydomonas*: Purification and Studies on Its Induction Using Antiserum Against *Chlamydomonas* Carbonic Anhydrase. *Plant & cell Physiology*, Vol.26, No.1, (1985), pp. 25–34, ISSN 0032-0781.

Yoshihara, C.; Inoue, K.; Schichnes, D.; Ruzin S.; Inwood, W. & Kustu, S. (2008) An Rh1–GFP Fusion Protein Is in the Cytoplasmic Membrane of a White Mutant Strain of *Chlamydomonas reinhardtii*. *Molecular Plant*, Vol.1, No.6, (November 2008), pp. 1007–1020, ISSN 1674-2052.

Yoshioka, S.; Taniguchi, F.; Miura, K.; Inoue, T.; Yamano, T. & Fukuzawa, H. (2004) The Novel Myb Transcription Factor LCR1 Regulates the CO_2-Responsive Gene *Cah1*, Encoding a Periplasmic Carbonic Anhydrase in *Chlamydomonas reinhardtii*. *The Plant Cell*, Vol.16, No.6, (June 2004), pp. 1466-1477, ISSN 1040–4651.

Yun, Y.S. & Park, J.M. (1997) Development of Gas Recycling Photobioreactor System for Microalgal Carbon Dioxide Fixation. *Korean Journal of Chemical Engineering*, Vol.14, No.4, (1997), pp. 297–300, ISSN 0256–1115.

3

Oscillatory Nature of Metabolism and Carbon Isotope Distribution in Photosynthesizing Cells

Alexander A. Ivlev

Russian Agrarian State University – "MSKHA of K.A.Timirjazev"
Russian Federation

1. Introduction

A study of carbon isotopic characteristics of plants and animals, such as, shifts in carbon isotope ratio of plant biomass relative to environmental CO_2, $\delta^{13}C$ values of biochemical fractions and individual metabolites, different isotopic patterns of biomolecules and diurnal isotopic changes of respired CO_2, evidences that in a living cell carbon isotope fractionation takes place.

The above characteristics might be the source of valuable information on cell metabolism and regulation of metabolic processes, on assimilate transport, and different aspects of "organism – habitat" interactions. The efficiency of the involving this information in living organism studies greatly depends on the validity of carbon isotope fractionation model used for the interpretation. The validity of the model first of all is determined by the adopted view on the nature of isotope effect origin.

Two alternative points of view have been suggested in the literature. One of them asserts (Galimov, 1985; Schmidt, 2003) that carbon isotope effect and isotope distribution in biomolecules are of thermodynamic order. It means that isotope distribution of metabolites doesn't depend on biosynthesis pathway but is determined by the properties of the molecules themselves, i.e. by their structure and energy characteristics. According to the second point, supported by most of the researchers, the metabolic isotope effects are of the kinetic nature and carbon isotope distributions in metabolites are determined by mechanisms and pathways of their formation.

A lot of facts accumulated till now allow saying with confidence that thermodynamic concept is erroneous and the rare casual coincidences only simulate thermodynamic equilibrium (O'Leary & Yapp, 1978; Monson & Hayes, 1982; Ivlev, 2004). In some publications it was shown that the "thermodynamic" idea is "incompatible with the concept of life as a fundamental phenomenon" (Varshavski, 1988; Buchachenko, 2003). So we'll concentrate on the kinetic concept.

Within the frame of the "kinetic" concept two different approaches have been developed. The first is the steady-state model assumes that all the processes in a living cell during photosynthesis proceed simultaneously in stationary conditions. It also means that carbon isotope fractionation proceeds in stationary conditions too. The approach was put forward by Park and Epstein (Park & Epstein, 1960, 1961) and was developed by Farquhar et al.

(1982), Vogel (1993) and others (Gillon & Griffiths, 1997). Hayes (2001) has extended this approach to the common case, including secondary metabolism (metabolism in glycolytic chain).

According to the steady-state model, carbon isotope fractionation in photosynthesis can be presented as follows:

$$\Delta = a + (b - a)\, p_i/p_a \qquad (1)$$

where Δ is a carbon isotope discrimination, equal to the difference between $\delta^{13}C$ of environmental CO_2 and that of biomass carbon; a is a carbon isotope effect of CO_2 diffusion from the space into a photosynthesizing cell; b is a carbon isotope effect of ribuloso-1,5-bisphosphate (RuBP) carboxylation appearing in CO_2 fixation; p_a and p_i are the CO_2 partial pressures in the atmosphere and in the leaf space.

This simple steady-state balance model was rather convenient to explain the coherence between physiological response of plants to changing environmental conditions that impact stomatal conductance and net photosynthesis. Especially positive results were obtained in the field of carbon – water relations (Farquhar et al., 1989). Nevertheless even such a simple expression (1) turned to be contradicting. According to (1), isotope discrimination Δ approaches a or b values dependent on what is rate controlling stage – diffusion either biochemical. Direct measurements of activation energy of mesophyll cell conductance (Laisk, 1977) showed that diffusion is a rate-limiting stage in CO_2 assimilation. Hence, according to model, most of C_3-plants should be "heavier" than they are and Δ values should approach 4 - 5‰, i.e. a, whereas in fact they are close to 29‰, i.e. b. Other discrepancies were described in (Ivlev, 2003). The more the equation (1) was used, the more inconsistencies were found. Numerous corrections were introduced into expression (1) to take into account other processes, where carbon isotope fractionation might be, and to remove inconsistencies. Entirely the expression (1) was transformed into expression like the following (Farquhar & Lloid, 1993):

$$\Delta = a\,\frac{p_a - p_i}{p_a} + a_i\,\frac{p_i - p_c}{p_a} + b\,\frac{p_c}{p_a} - \frac{1}{p_a}\left(\frac{eR_d}{k} + f \cdot \Gamma^*\right) \qquad (2)$$

where p_a, p_i and p_c refer to the partial pressure of CO_2 in the atmosphere, substomatal cavity and chloroplast, respectively, a is the fractionation during the diffusion in air, a_i is the combined fractionation during dissolution and diffusion through the liquid phase , b is the assumed net fractionation during carboxylation by ribuloso-1,5-bisphosphate carboxylase/oxygenase (Rubisco) and by phosphoenolpyruvate carboxylase (PEPC), k is the carboxylation efficiency, R_d is the day respiration rate, Γ^* is the CO_2 compensation point in the absence of day respiration, e and f are the fractionation during dark respiration and photorespiration, respectively.

The expression (2), unlike to (1), is inconvenient for isotope fractionation analysis in photosynthesis since contains too many parameters to be determined. Even more complex expressions are obtained when it is required to describe intramolecular isotope distribution (Tcherkez et al., 2004). Using theoretical analysis of carbon isotope fractionation in metabolic chain under stationary conditions Hayes (2001) have shown that it was impossible to predict isotope composition of metabolites and their isotopic patterns since they depend not only on the isotope characteristics of the prior metabolites in the chain, but on the partitioning of carbon fluxes at the down-stream cross-points. Thus the integrative steady-state approach is insufficient for the explanation of short-term or intramolecular carbon isotope fractionation processes.

Another approach of the kinetic concept is presented in the works of Ivlev and colleagues (Ivlev, 1989, 1993, 2008; Igamberdiev et al., 2001; Ivlev et al. 2004; Roussel et al, 2007). Opposite to steady-state idea, the authors put forward and developed the idea that metabolic processes are discrete and periodic ones. Periodicity of metabolic processes allows concluding that substrate pools in cells are periodically filled and depleted. It is well known fact that isotope fractionation accompanying the metabolic processes in case of depletion is followed by Raleigh effect (Melander&Sauders,1983). This effect establishes the dependence between isotope ratio of initial substrate ($\delta^{13}C_{init.substrate}$), reaction product ($\delta^{13}C_{product}$), isotope fractionation coefficient (α) and the extent of pool depletion F in accordance with the equation:

$$[\delta^{13}C_{product} \cdot 10^{-3} + 1] = [\delta^{13}C_{init.substrate} \cdot 10^{-3} + 1]\frac{1}{F}[1 - (1 - F)^{1/\alpha}], \qquad (3)$$

where $\alpha = {}^{12}k/{}^{13}k$, ${}^{12}k$ and ${}^{13}k$ are the rate constants of isotopic species of the molecules.

The Raleigh effect is the most essential feature of carbon metabolism in a living cell. It closely relates to filling/depletion regime of cell functioning and to oscillatory character of metabolic reactions. Another fundamental feature related to Raleigh effect is the strict temporal sequence of metabolic reactions, i.e. temporal organization in a cell. Using equation (3) and carbon isotopic composition of metabolites, it is possible to distinguish the temporal sequence of many metabolic events.

Kinetic nature of carbon isotope effect and participation of polyatomic carbon molecules in metabolic reactions give the evidences that most of the biomolecules in metabolic chains inherit their isotope distributions from the precursors thus proving there is no isotope exchange between carbon atoms within the carbon skeletons. Most frequent cases where isotopic shifts emerge are linked with C – C bond cleavage, especially at the cross-points of metabolic pathways. The kinetic nature of isotope effect is manifested by the fact that only those carbon atoms of skeleton disposed at the ends of broken bonds undergo isotopic shifts. These and specificity of enzymatic interactions determine individual isotopic pattern of the biomolecules. Taking into account the above factors in combination with Raleigh effect and the putative pathways of the metabolite synthesis allows reconstructing of isotopic patterns of the molecules and gives a fine tool for metabolism study.

Finally, the known regularities of inter- and intramolecular carbon isotope distribution in a cell indicate that metabolic oscillations are undamped and in-phase. Otherwise these isotopic regularities couldn't exist. The existence of the regularities on account of the Raleigh effect means, that at a given functional state of a cell, the metabolite syntheses within the repeated cell cycles occur at a certain level of substrate pool depletion. Moreover the functioning of different cells is synchronized.

2. Carbon isotope fractionation in photosynthesis and photosynthetic oscillation concept

The first step to the oscillation model was the emergence of the discrete model based on the experimentally observed data on ${}^{12}C$ enrichment of plant and photosynthesizing microorganism biomass relative to ambient CO_2 at different conditions. The model assumed that CO_2 assimilation is a discrete process and CO_2 enters the cells by separate batches (Ivlev, 1989), but not by continuous flow like in a steady-state model (Farquhar et al., 1982). On account of isotope effect in RuBP carboxylation discrete model explained different levels of ^{12}C enrichment of photosynthetic biomass by the Raliegh effect accepting that only part of the CO_2 batches is fixed. The observed isotopic difference between C_3 and C_4 plants (Smith & Epstein, 1971) was explained by the same manner. Indeed, due to anatomical peculiarities of C_4- plants (Edwards & Walker, 1983) they are capable to re-assimilate almost all respired CO_2 thus increasing the extent of CO_2 batches depletion (F in expression 3)

The question was - what's the reason making CO_2 flux to be discrete? It was assumed that CO_2 assimilation flux periodically is interrupted by the reverse flux of the respired CO_2 directed from the cell to the environments. It was also assumed that such a "ping -pong" mechanism is due to double function of the key photosynthetic enzyme – Rubisco, which is capable to function as carboxylase or oxygenase depending on CO_2/O_2 concentration ratio in a cell (Ivlev, 1992). Switching mechanism splits CO_2 flux entering and leaving the cell into separate batches. This hypothesis got strong support when new carbon isotope effect of photorespiration has been discovered (Ivlev, 1993).

Some facts known from the literature the traditional steady-state model failed to explain. In gas exchange experiments with the use of CO_2 enriched in ^{13}C the advantageous fixation of "heavy" molecules instead of "light" ones by leaves of different plants was observed (Sanadze et al., 1978). The similar results with the use of [14]CO_2 were obtained in the experiments with alga (Voznesenskii et al., 1982). Moreover the primary assimilates turned to be isotopically "heavier" relative to the ambient CO_2. In the experiments with photosynthesizing bacteria *Ectothiorhodospira shaposhnikovii* there was a change in the sign of isotope discrimination linked with the growth of ^{13}C content in the ambient CO_2 (Ivanov et al.,1978). To explain these facts it was assumed an existence of the new isotope effect related to photorespiration, and having the opposite sign to that in CO_2 assimilation. The analysis of the tentative points in photorespiration loop, where such an effect might emerge, showed that the most plausible point for its origin was glycine dehydrogenase reaction (Ivlev, 1993) (Fig.1), where decarboxylation of glycine occurs.

The following study of Calvin cycle and photorespiration biochemistry in virtue of carbon isotope composition of the primary assimilates allowed concluding on two phases of Calvin cycle functioning. In the first phase Calvin cycle produces glucose-6-phosphate (G6P) and other products from the fixed CO_2. During this phase the derived products are used to accumulate the reserve pool of starch to feed glycolytic chain in the dark and provide substrates for lipid, protein and lignin components syntheses. Carbon isotope fractionation in RuBP carboxylation results in ^{12}C enrichment of the cycle metabolites and finally the biomass as a whole relative to the ambient CO_2.

It takes some time to substantiate the experimental validity of the hypothesis and to prove that glycine decarboxylation is the very point *in vivo* where carbon isotope fractionation results in ^{13}C enrichment of biomass (Ivlev et al., 1996; 1999; Igamberdiev et al., 2001; 2004).

In the second phase Calvin cycle forms in combination with glycolate cycle the photorespiratory loop. The residual part of G6P produced in the previous phase converts into pentoses and then in form of phosphoglycolate leaves Calvin cycle and enters glycolate cycle where oxidative glycine decarboxylation occurs (Fig.1). After some transformations carbon flux in form of trioses returns back to Calvin cycle. Carbon isotope fractionation in glycine decarboxylation produces CO_2 enriched in ^{12}C evolving from the cell, whereas carbon substrates spinning in the loop are enriched in ^{13}C and result in corresponding enrichment of photorespiratory products and biomass. The level of ^{13}C enrichment depends on what how many turns carbon substrate flux makes in the loop, or what the extent of photorespiratory pool depletion is achieved (Raleigh effect).

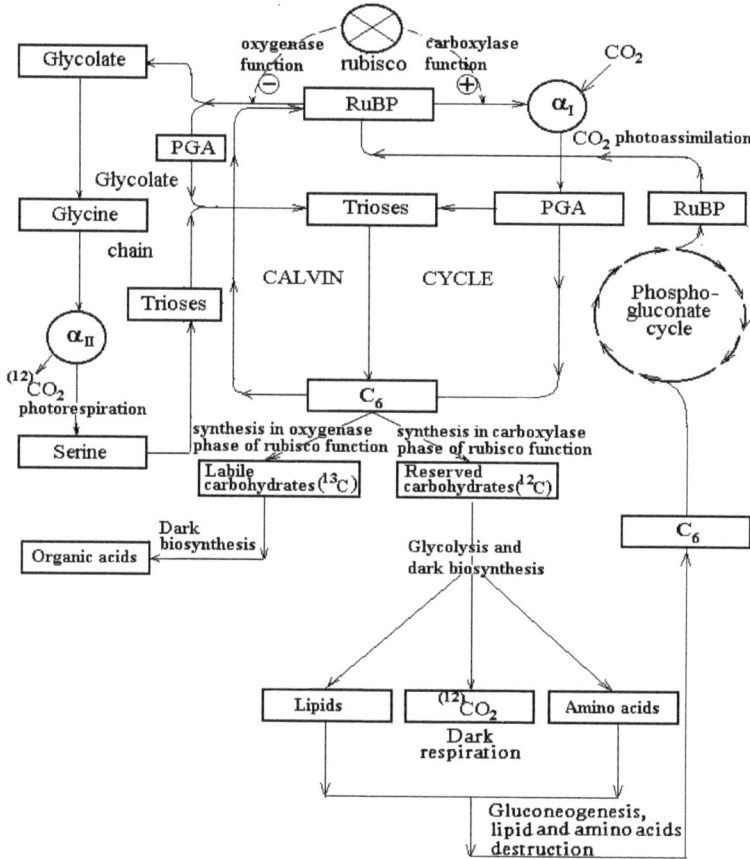

α_1 – a point of carbon isotope fractionation in CO_2 assimilation (carbon isotope effect in RuBP carboxylation), α_{11} – a point of carbon isotope fractionation in photorespiration (carbon isotope effect in glycine decarboxylation)

Fig. 1. Oscillating model of carbon isotope fractionation in photosynthesis.

2.1 Experimental facts support the presence of photosynthetic oscillations

The experimental data presented in Tables 1-3 show distinct differences in isotope composition of metabolites derived in the both phases of Calvin cycle oscillations. Table 1 illustrates ^{13}C enrichment of leaf oxalates of different C_3 and CAM-plants (Rivera & Smith, 1979; Raven et al., 1982). Their synthesis is mainly bound to glycolate cycle of photorespiratory loop.

Plant species	Plant type	Whole leaf	Oxalates	Reference
Spinaceae oleracea	C₃	- 27.5	- 11.9	Rivera&Smith, 1979
Pelagronium	C₃	- 31.0	- 12.4	Rivera&Smith, 1979
Mereurialis perennis	C₃	- 27.9	- 13.7	Rivera&Smith, 1979
Spinaceae oleracea	C₃	- 25.7	- 19.9	Raven et al., 1982
Echinomastus intertextus	CAM	- 13.4	- 7.3	Raven et al., 1982
Echinomastus horizonthalomus	CAM	- 13.0	- 7.8	Raven et al., 1982
Escobaria ruberoulosa	CAM	- 12.3	- 8.3	Raven et al., 1982
Opuntia euglemannii	CAM	- 13.3	- 8,5	Raven et al., 1982
Opuntia imbricata	CAM	- 14.1	- 8.7	Raven et al., 1982

Table 1. Carbon isotope ratio of leaf biomass and oxalates of some oxalate accumulating plants.

¹³C Distribution in protein fraction of some photosynthesizing microorganisms (Abelson & Hoering, 1961) gives more evidences in favor of the oscillation model (Table 2). Amino acids like serine, glycine, alanine and aspartic acid, whose pools, at least, in part are supplied from photorespiratory loop, have appeared more enriched in ¹³C as compared with those whose synthesis predominantly bound to glycolytic chain and Krebs cycle, like glutamic acid, leucine and lysine (Igambediev, 1988; 1991).

Microorganism Amino acid	*Chlorella* total carbon	*Anacystis* total carbon	*Gracilaria* Total carbon	*Euglena* total carbon
Serine	- 5,7	-	- 14,1	- 8,3
Glycine	- 14.3	- 10,0	- 10,2	- 10,0
Alanin	- 10,3	- 9,8	- 15,2	-14,3
Aspartic acid	- 6,6	- 9,7	- 14,4	-9,7
Glytamic acid	- 18,7	- 11,1	- 17,2	-17,3
Leucine	- 22,7	- 17,3	- 22,5	-23,5
Lysine	- 17,0		-	-22,8

Table 2. Carbon isotope distribution in amino acids from protein fraction from biomass of some photosynthesizing microorganisms. Isotopic shifts are given relative to nutrient CO_2 having $\delta^{13}C = 0‰$. Extract from Table 3 in (Abelson & Hoering, 1961).

Index	Concentration of NaCl in medium, mM		
	0	425	595
$\delta^{13}C$ of dry matter, ‰	- 61.6	- 59.0	- 64.5
$\delta^{13}C$ of lipids, ‰	- 66.0	- 65.0	- 63.8
$\delta^{13}C$ of proteins, ‰	- 42.1	- 40.9	- 47.3
$\delta^{3}C$ of labile sugars‰	- 30.0	-	- 30.5
$\delta^{13}C$ of proline, ‰	- 29.0	-	- 31.5

Table 3. Distribution of ¹³C in biomass and biochemical fractions of marine alga *Chorella stigmatophora*, grown under effect of different environmental factors. $\delta^{13}C$ of ambient CO_2 is - 21‰ (Ivlev & Kalinkina, 2001; Kalinkina & Udel'nova, 1990)

The same picture illustrated by the data on ^{13}C distribution in biomass of marine alga *Chlorella stigmatophora* grown under different environmental conditions on the CO_2 of the known carbon isotope ratio is presented in Table 3.

Lipids and proteins distinctly differ in carbon isotope ratio as compared with labile sugars and organic acids. In special experimental studies (Kalinkina &Udel'nova, 1991; Kalinkina & Naumova, 1992) the authors have proven that these components were the products of photorespiration pathway whereas most of lipids and proteins are synthesized via glycolytic chain and Krebs cycle (Metun, 1963; Strikland, 1963).

Quite another object, C_3-CAM tropical plant *Clusia minor*, grown under different environmental conditions is given in Table 4. Soluble sugars and organic acids, whose origin is linked with photorespiratory carbon flux, are enriched in ^{13}C as compared with amino acid and lipid fractions. Notably the latter fraction, besides lipids, contains pigments some of which, like chlorophyll (Ivlev, 1993), at least partially are formed at the expense of photorespiratory flux. It makes lipid fraction isotopically "heavier" than amino acid fraction is. Thus all the presented data confirmed the idea on two phases of Calvin cycle functioning and drew to the conclusion that the phases are alternating, i.e. are separated in time. In fact, if the processes proceeded simultaneously the isotopically different carbon fluxes couldn't arise. Passing the same pieces of Calvin cycle they would inevitably mix. Bearing in mind that the first phase of cycle functioning corresponds to carboxylase function of Rubisco, while the second – to oxygenase one, we called the first as carboxylase phase and the second as oxygenase. To confirm the oscillating idea we tried to get more independent arguments and examined isotopic patterns of metabolites derived in different phases of Calvin cycle.

N	FRACTION	WET SEASON		DRY SEASON	
		Exposed leaf	Shaded leaf	Exposed leaf	Shaded leaf
		Dawn	Dusk	Dawn	Dusk
1	Total carbon	-25.7	-30.3	-24.6	-29.1
2	Lipids and pigments	-28.7	-32.2	-27.7	-30.8
3	Amino acids	-31.7	-32.6	-31.3	-32.7
4	Soluble sugars	-21.2	-29.2	-17.9	-21.9
5	Organic Organic acids	-22.3	-27.7	-21.1	-24.5

Table 4. Carbon isotope composition of biochemical fractions isolated from leaves of *Clusia minor* under different environmental conditions (Borland et al., 1994). Samples were taken at dawn and dusk. $\delta^{13}C$ Values are given in per mille relative to PDB standard.

2.2 Intramolecular isotopic patterns of glucose, anomalous isotope composition of CO_2 evolved in light enhanced dark respiration, and some non-isotopic arguments support the oscillation hypothesis

As noted above, kinetic nature of isotope effects and specificity of enzymatic interactions provide specific carbon isotope distribution of many metabolites. Having compared isotopic patterns of G6P, formed in carboxylase and oxygenase phases of Calvin cycle we found they should be quite different. According to the theoretical estimates the synthesis of G6P in carboxylase phase results in practically uniform ^{13}C distribution along the molecule skeleton due to transaldolase and transketolase cycle reactions which randomize atoms with cycle turns growth. Carbon isotope distribution of G6P synthesized in photorespiration loop is characterized by ^{13}C enrichment of carbon atoms in C-3 and C-4 positions of glucose skeleton, slight ^{13}C enrichment in C-2 and C-5 positions, while atoms C-1 and C-6 are enriched in ^{12}C (Fig. 2). To understand this specific distribution let's follow what isotope fractionation in glycine dehydrogenase complex (GDC) occurs.

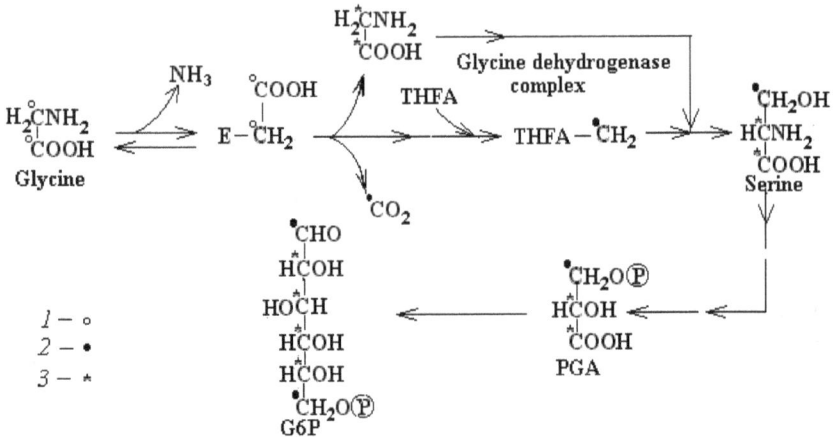

Empty circles (1) denote isotope composition of carbon atoms in the initial substrate, filled circles (2) denote atoms get enriched in ^{12}C, asterisks (3)

Fig. 2. The emergence of the isotope inhomogeneity in G6P as a result of kinetic carbon isotope effect in GDC.

As shown on Fig. 2, isotope distribution in G6P is determined by isotope distributions in glycine and C_2-fragment derived in decarboxylation. In glycine decarboxylation both atoms of residual glycine get enriched in ^{13}C while methylene carbon atom as well as CO_2 located at the ends of the cleaved C – C bond relative to the atoms in the initial substrate get enriched in ^{12}C (Melander & Saunders, 1983). In GDC the methylene fragment linked with the cofactor, tetrahydrofolic acid (THFA), is transferred to the residual glycine molecule thereby forming the serine (Oliver et al., 1990). Following transformations result in the specific isotopic pattern of G6P shown on Fig.2. Moreover at each turn of the carbon flux, spinning in photorespiratory loop, isotope distribution not only retains, but is reproduced again and again. So ^{13}C enrichment of G6P as well as intramolecular isotopic discrepancies increase with the number of turns (with the growth of photorespiration intensity) (Ivlev et al., 2010). Isotope pattern of G6P synthesized in carboxylase phase is not studied yet. But glucose from the starch of storage organs of some plants has been investigated (Table 5).

Bearing in mind that G6P is the main structural unit used for glucose synthesis and comparing data in Table 5 with the results of G6P modeling (Ivlev, 2005; Ivlev et al., 2010), it is easy to conclude they are strongly resembled. Hence the starch glucose is of photorespiratory origin. The assertion is supported by the fact that storage organs are formed in the period of ontogenesis when oxidative processes related to intensification of photorespiration sharply increase (Abdurachmanova et al., 1990; Igamberdiev, 1991). This fact correlates with the observed ^{13}C enrichment of seeds, fruits, and edible roots of plants as compared with the carbon isotopic composition of other plant organs (leaf, stem) (Lerman et al., 1974; White, 1993; Saranga et al., 1999; Ivlev et al., 1999).

Object	$\delta^{13}C$ of glucose	$\Delta^{13}C = \delta^{13}C - \delta^{13}C_{glucose}$, i - atom number					
		$OCH_{(1)}\text{-}HC_{(2)}OH\text{-}OHC_{(3)}H\text{-}HC_{(4)}OH\text{-}HC_{(5)}OH\text{-}C_{(6)}H_2OH$					
		C_1	C_2	C_3	C_4	C_5	C_6
Beta vulgaris, tuber (Rossman et al., 1991)	-25.2	-1.6	-0.4	+2.1	+6.3	-1.7	-5.1
Zea mays, seeds (Rossman et al., 1991)	-10.8	-1.7	-0.1	+ 1.1	+3.6	-0.2	-3.6
Zea mays, seeds (Ivlev et al., 1987)	-12.5	-3.1	+ 1.9				-1.9
Triticum aestivum, seeds (Galimov et al., 1977)	-23.1	-7.1	+3.5*				-7.1
Solanum tuberosum, tuber (Galimov et al., 1977)	-24.9	-9.1	+4.5*				-9.1
Oryza sativum, seeds (Galimov et al., 1977)	-26.1	-6.9	+3.5*				-6.9
Pisum sativum, seeds (Galimov et al., 1977)	-24.9	-4.1	+2.1*				-4.1

Note: The isotopic shifts of the carbon atoms $\Delta^{13}C$ are given relative to total glucose carbon. $\delta^{13}C$ values of glucose are given in PDB units. *The $\delta^{13}C$ values of C-3 and C-4 atoms were calculated according to Galimov et al. (1977) assuming that the isotopic composition of the other carbon atoms equals to that of the C – 1 and C – 6.

Table 5. Intramolecular carbon isotope distribution in the starch glucose of storage organs of various plants.

The uneven carbon isotope distribution in oxygenase G6P explains the recently established fact of anomalous ^{13}C enrichment of light enhanced dark respiration CO_2 relative to labile carbohydrates from phloem sup, the supposed respiratory substrate. Indeed, the consideration of labile carbohydrates (oxygenase G6P), accumulated in the light, as substrate for dark synthesis of organic acids (Borland et al, 1994), allows concluding the following way of the conversion (Ivlev & Dubinsky, 2011). At first glucose splits into two triose molecules. Then the latter are subjected to decarboxylation and derived C_2-fragments are used to form organic acid skeletons while the evolved CO_2 forms LEDR CO_2 which inherits atoms from the C-3 and C-4 positions. The atoms as shown above are enriched in ^{13}C. The level of ^{13}C enrichment depends on light intensity and confirms the existence of Raliegh effect in photorespiration. The increase in illumination intensifies photorespiration, thus implying the increase in number of turns of carbon flux in photorespiratory loop. This in turn leads to photorespiratory pool depletion and to ^{13}C enrichment of the respired CO_2. Barbour et al. (2007) have noticed the relationship of light intensity and LEDR CO_2 ^{13}C enrichment in the experiment.

The oscillatory model suggests a coherent explanation of the relative ^{13}C enrichment of heterotrophic tissue of plants (seeds, stem, roots) comparing with autothrophic ones (leaves)

(Cernusak et al, 2009). In fact, labile carbohydrates are the main carbon source for heterotrophic growth (Kursanov, 1976). On the other hand, labile carbohydrates, being photorespiratory products, are enriched with [13]C. Gessler et. al. (2008) has confirmed this assertion experimentally. The authors found that water soluble fraction of leaf organic matter mainly consisting of the labile carbohydrates is enriched with ^{13}C unlike to the insoluble fraction mainly consisting of proteins and lipids whose origin relates to starch formed in carboxylase phase. Similarly, the model explains the resemblance in $\delta^{13}C$ values of the leaf water soluble organic matter and that of the phloem sap. Hence the above isotopic data firmly support the oscillation hypothesis.

There was an endeavor to find a direct evidence of the photosynthetic oscillations (Roussel et al., 2007). By using a fast response CO_2 gas exchange system the authors measured CO_2 concentration fluctuations in the subcellular space in tobacco leaves at low CO_2 concentrations nearby the compensation point. The chosen condition provided an easier way to discover the assumed oscillations. Because of a background noise, a special mathematical procedure was required to isolate the periodic component in the temporal sequence and to build an attractor proving the existence of the real oscillatory regime. The CO_2 concentration pulses with a period of the order of a few seconds were explained by the feedback interactions between CO_2 assimilation and photorespiration.

Fig.3 shows the principal interactions between the main participants of photosynthesis process and key enzyme Rubisco having dual function. Since the process occurs in different compartments: CO_2 assimilation in chloroplasts, photorespiratory CO_2 release in mitochondria, a certain time interval is needed for the CO_2 depletion near Rubisco. The delay in CO_2 release, following RuBP oxygenation, and competition between CO_2 and O_2 provide the conditions for oscillations (Roussel & Igamberdiev, 2011).

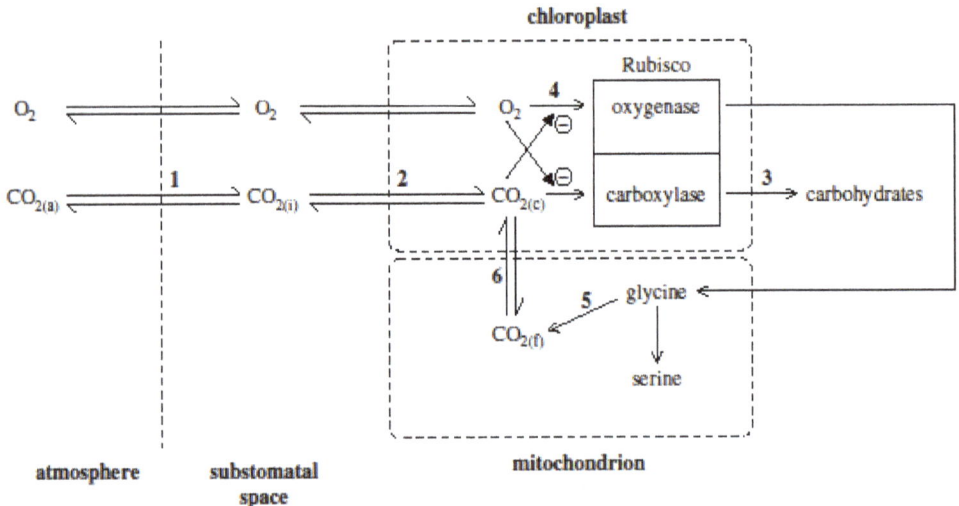

Fig. 3. Simplified scheme for carbon assimilation and photorespiration.

Dashed lines indicate compartmental boundaries. Carbon dioxide in the atmosphere passes through the stomata and enters the substomatal space (step 1). Eventually, it reaches the chloroplasts (transport step 2) where carbon is fixed (3). Under normal conditions, the leaf interior is well ventilated, leading to a reasonably uniform distribution of oxygen. Oxygen may participate in photorespiration (4), eventually leading to the appearance of glycine in the mitochondria and thus to photorespiration (5). The carbon dioxide produced by photorespiration is free to diffuse through the cytoplasm to the chloroplasts (6). Also shown is the inhibition of photorespiration by carbon dioxide and of carbon fixation by oxygen.

To check up this possibility we carried out the computational analysis of the scheme (Dubinsky & Ivlev, 2011). It can be presented as follows (Fig. 4). According to the scheme (Fig. 4), RuBP binds to the enzyme which is activated by Mg^{2+} and CO_2 (this is not considered here for simplification) and a quasi-equilibrium of the RuBP with the enzyme E is attained first (Tapia et al., 1995; Mauser et al., 2001). Then RuBP–enzyme complex reacts either with CO_2 or O_2 and the formation of the assimilation products occurs. The products are used either for further transformations in the cycle (the carboxylase phase) or for utilization in the photorespiration loop, comprising the Calvin cycle coupled with the glycolate cycle (initiated by the oxygenase phase).

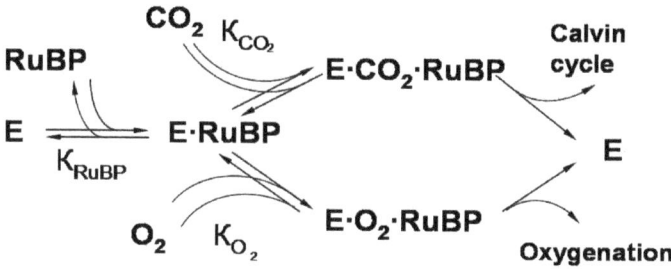

Fig. 4. The principal scheme of photosynthesis considering carboxylase and oxygenase functions of Rubisco.

The scheme on Fig. 4 is convenient for mathematical description and computational analysis. It was described by three differential equations:

$$\frac{dx}{dt} = \frac{1}{5} V_c \cdot \frac{(x/K_{RuBP}) \cdot (y/K_{CO_2})}{1 + (x/K_{RuBP}) + (x/K_{RuBP}) \cdot (y/K_{CO_2}) + (x/K_{RuBP}) \cdot (z/K_{O_2})}$$
$$- \frac{1}{10} V_{ox} \cdot \frac{(x/K_{RuBP}) \cdot (z/K_{O_2})}{1 + (x/K_{RuBP}) + (x/K_{RuBP}) \cdot (y/K_{CO_2}) + (x/K_{RuBP}) \cdot (z/K_{O_2})} - V_{out} \frac{(x/K_{out})}{1 + (x/K_{out})}$$

$$\frac{dy}{dt} = -V_c \cdot \frac{(x/K_{RuBP}) \cdot (y/K_{CO_2})}{1 + (x/K_{RuBP}) + (x/K_{RuBP}) \cdot (y/K_{CO_2}) + (x/K_{RuBP}) \cdot (z/K_{O_2})}$$
$$+ \frac{V_{ox}}{2} \cdot \frac{(x/K_{RuBP}) \cdot (z/K_{O_2})}{1 + (x/K_{RuBP}) + (x/K_{RuBP}) \cdot (y/K_{CO_2}) + (x/K_{RuBP}) \cdot (z/K_{O_2})} + k_{CO_2}(CO_{2out} - y)$$

$$\frac{dz}{dt} = V_c \cdot \frac{(x/K_{RuBP}) \cdot (y/K_{CO_2})}{1 + (x/K_{RuBP}) + (x/K_{RuBP}) \cdot (y/K_{CO_2}) + (x/K_{RuBP}) \cdot (z/K_{O_2})}$$
$$- \frac{3}{4} V_{ox} \cdot \frac{(x/K_{RuBP}) \cdot (z/K_{O_2})}{1 + (x/K_{RuBP}) + (x/K_{RuBP}) \cdot (y/K_{CO_2}) + (x/K_{RuBP}) \cdot (z/K_{O_2})} + k_{O_2}(O_{2out} - z) + J_{O_2}$$

where following notations are used: x, y and z are the RuBP, CO_2 and O_2 concentrations respectively, K_{CO2}, K_{O2} and K_{RuBP} are the equilibrium constants of the reactions E · RuBP ·$CO_2 \rightarrow$ E · RuBP + CO_2, E · RuBP ·$O_2 \rightarrow$ E · RuBP + O_2 and E · RuBP$\rightarrow$E + RuBP, respectively. Two first equations in their simplified form (Dubinsky et al., 2010) describe sugar (x) and CO_2 (y) concentration variations. The third describes variations of O_2 (z) concentration. In the set of equations V_c is the maximum rate of RuBP carboxylation, V_{ox} is the maximum rate of RuBP oxygenation. V_{out} is the maximum rate of sugar efflux, K_{out} is the Michaelis constant of the pseudoenzyme by means of which sugars are removed from the system (the real mechanism of sugars removal is certainly more complicated but it is the simplest way to describe the effect of sugar efflux saturation). k_{CO2} is the CO_2 diffusion coefficient from the surrounding medium into a cell, CO_{2out} is the CO_2 concentration in the medium, k_{O2} is the O_2 diffusion coefficient from the medium into the cell, O_{2out} is the O_2 concentration in the medium.

The solution of the system with cell parameters, taken from the literature (Dubinsky & Ivlev, 2011), results in establishing of counter-phase undamped oscillations with the period of 1 – 3 sec for CO_2 and O_2 and in respective oscillations of CO_2/O_2 ratio (Fig. 5). The oscillations could switch over Rubisco from carboxylase function to oxygenase and back.

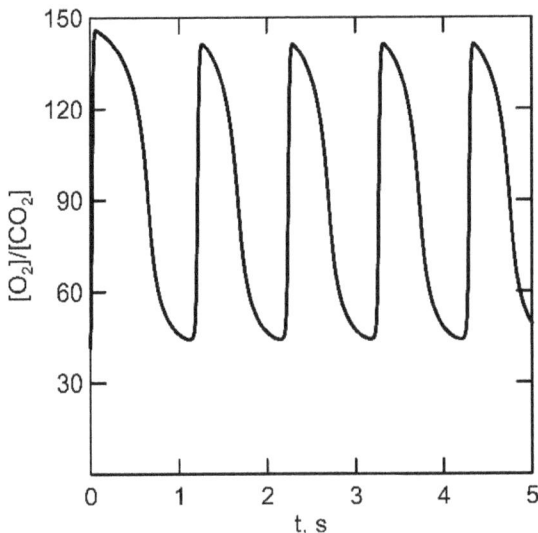

Fig. 5. The calculated photosynthetic oscillations of CO_2/O_2 concentration ratio according to the model described in the text.

Thus the theoretical calculations proved the principal possibility of the existence of sustained oscillations in carbon metabolism of a photosynthesizing cell.

Let's consider now how Calvin cycle works in different phases of photosynthetic oscillations from the point of ^{13}C isotope distribution in metabolites. In carboxylase phase of oscillations Calvin cycle works, as shown on Fig. 6. Due to carbon isotope effect in RuBP carboxylase complex all carbon atoms fixed happened to be enriched in ^{12}C relative to ambient CO_2. Transketolase and transaldolase reactions of the cycle randomize carbon atoms along carbon skeletons, i.e. Calvin cycle works as a mixer. It results in practically uniform ^{13}C distributions within metabolites. The pools of metabolites accumulated in this phase and utilized further in secondary metabolism to provide glycolytic chain, lignin synthesis and other metabolic needs with carbon source form so-called "light" (enriched in ^{12}C) carbon flux (see below).

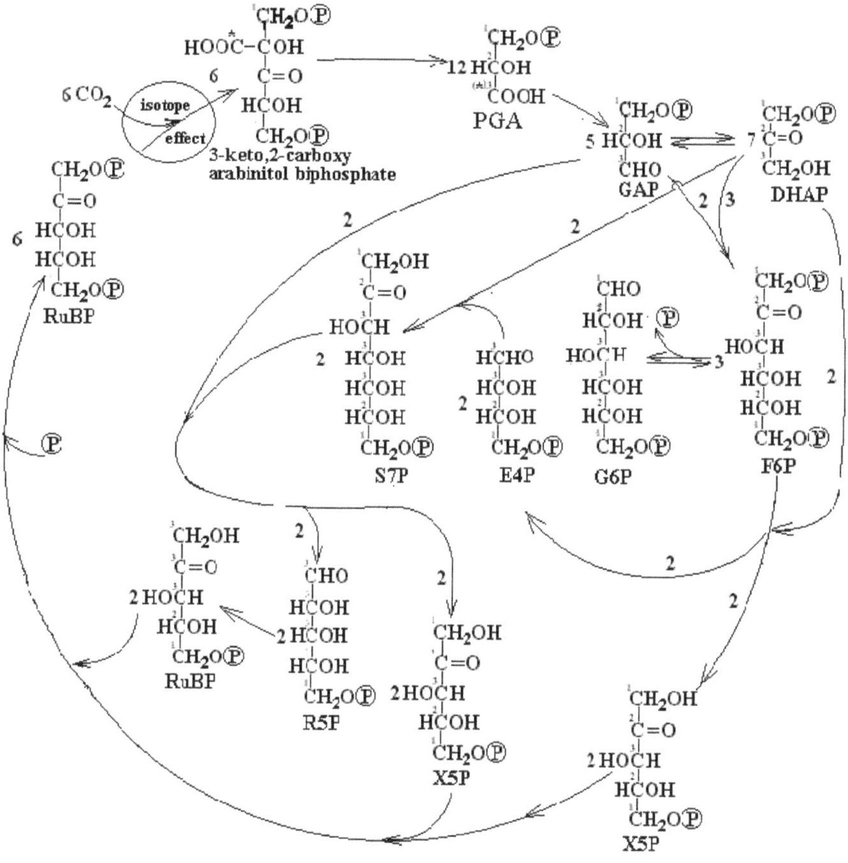

$$CH_2O\circledP$$
$$HOOC-\overset{*}{C}OH$$
6 $C=O$
$6 CO_2$ isotope $H\overset{}{C}OH$
effect $CH_2O\circledP$
3-keto,2-carboxy arabinitol biphosphate

$CH_2O\circledP$
$C=O$
6 $H\overset{}{C}OH$
$H\overset{}{C}OH$
$CH_2O\circledP$
RuBP

$$CH_2O\circledP$$
$$12\,H\overset{}{C}OH$$
$$COOH$$
PGA

CH_2OH
$C=O$
$HO\overset{}{C}H$
2 $H\overset{}{C}OH$
$H\overset{}{C}OH$
$H\overset{}{C}OH$
$CH_2O\circledP$
S7P

$H\overset{}{C}HO$
$H\overset{}{C}OH$
2 $H\overset{}{C}OH$
$CH_2O\circledP$
E4P

$CH_2O\circledP$
5 $H\overset{}{C}OH$
CHO
GAP

$CH_2O\circledP$
7 $C=O$
CH_2OH
DHAP

CHO
$H\overset{}{C}OH$ $\circledP$
$HO\overset{}{C}H$
$H\overset{}{C}OH$
$H\overset{}{C}OH$
$CH_2O\circledP$
G6P

$CH_2O\circledP$
$C=O$
$HO\overset{}{C}H$
$H\overset{}{C}OH$ 2
$H\overset{}{C}OH$
$CH_2O\circledP$
F6P

CH_2OH
$C=O$
2 $HO\overset{}{C}H$
$H\overset{}{C}OH$
$CH_2O\circledP$
RuBP

CHO
$H\overset{}{C}OH$
2 $H\overset{}{C}OH$
$H\overset{}{C}OH$
$CH_2O\circledP$
R5P

CH_2OH
$C=O$
2 $HO\overset{}{C}H$
$H\overset{}{C}OH$
$CH_2O\circledP$
X5P

CH_2OH
$C=O$
2 $HO\overset{}{C}H$
$H\overset{}{C}OH$
$CH_2O\circledP$
X5P

Fig. 6. Calvin cycle in carboxylase phase of Rubisco functioning.

The figures on the arrows and before the molecular formulas denote the number of the molecules involved in transformations of the cycle and formed in them; the figures before the atoms denote the number of carbon atoms in the PGA molecule; asterisks * are the exogenous carbon atoms attached to the carboxyl group of 2-carboxy-3-ketopentite and then to the C-3 position of PGA; P in a circle is the phosphate group in the molecules.

In oxygenase phase Calvin cycle works as shown on Fig. 7. Due to the isotope effect in GDC all metabolites formed in cycle transformations get enriched in ^{13}C relative to G6P, left after carboxylase phase. At that the specific intramolecular ^{13}C distributions determined by kinetic nature of the effect, by the specificity of enzymatic interactions and by the Raleigh effect appear. ^{13}C-Enrichment and heterogeneity of isotope distribution of metabolites becomes greater with the photorespiration intensity. The pools of metabolites mainly labile carbohydrates, some amino acids (glycine, serine, and related compounds) accumulated in oxygenase phase, like those formed in carboxylase phase, are utilized in secondary metabolism syntheses (organic acid, some parts of complex molecules, etc.)

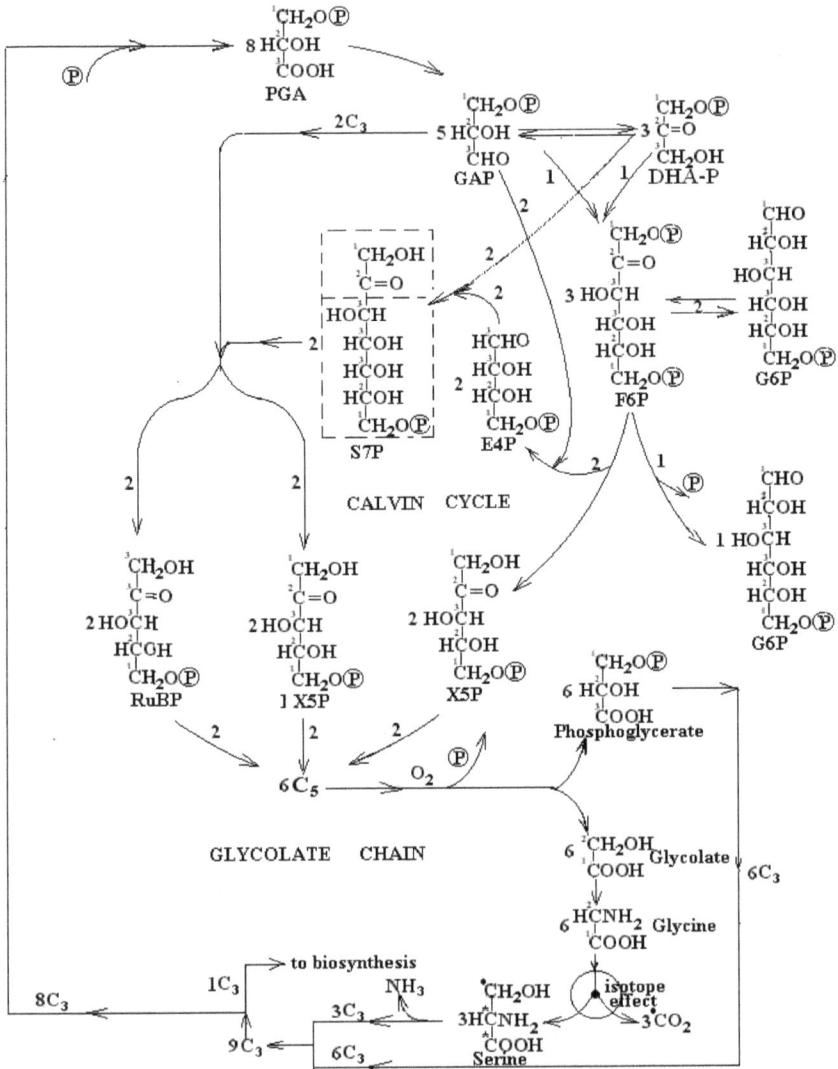

$CH_2O\textcircled{P}$
8 HCOH
COOH
PGA

$2C_3$

$CH_2O\textcircled{P}$
5 HCOH ⇌ 3 $CH_2O\textcircled{P}$
CHO   C=O
GAP CH_2OH
              DHA-P

CHO
HCOH
HOCH
HCOH
HCOH
$CH_2O\textcircled{P}$
G6P

CH_2OH
C=O
HOCH
2| HCOH
HCOH
HCOH
$CH_2O\textcircled{P}$
S7P

HCHO
HCOH
2 HCOH
$CH_2O\textcircled{P}$
E4P

$CH_2O\textcircled{P}$
C=O
3 HOCH
HCOH
HCOH
$CH_2O\textcircled{P}$
F6P

CALVIN CYCLE

CHO
HCOH
1 HOCH
HCOH
HCOH
$CH_2O\textcircled{P}$
G6P

CH_2OH
C=O
2 HOCH
HCOH
$CH_2O\textcircled{P}$
RuBP

CH_2OH
C=O
2 HOCH
HCOH
$CH_2O\textcircled{P}$
1 X5P

CH_2OH
C=O
2 HOCH
HCOH
$CH_2O\textcircled{P}$
X5P

$CH_2O\textcircled{P}$
6 HCOH
COOH
Phosphoglycerate

$6C_5$ O_2

GLYCOLATE CHAIN

6 CH_2OH Glycolate $6C_3$
     COOH

6 HCNH_2 Glycine
    COOH

to biosynthesis
$1C_3$
$3C_3$ NH_3
$9C_3$ $6C_3$

3 CH_2OH
3 HCNH_2
    COOH
Serine

isotope effect
$3CO_2$

$8C_3$

Fig. 7. Calvin cycle in oxygenase phase of Rubisco functioning.

All symbols denote the same as on Fig. 6.
Strict temporal organization of metabolism in a cell prevent from complete mixing of the above carbon fluxes (see below) and allows to use isotopic characteristics to investigate metabolic relations, pathways, assimilate transport, etc. More arguments evidencing in favor of photosynthetic oscillations were given in the work of Ivlev (2010).

3. Carbon isotope fractionation in secondary metabolism of photosynthesizing cell

The idea on the existence of energy and carbon oscillations in glycolytic chain was firstly proved in respect to heterotrophic organisms (Sel'kov, 1975, 1978). We have accepted this

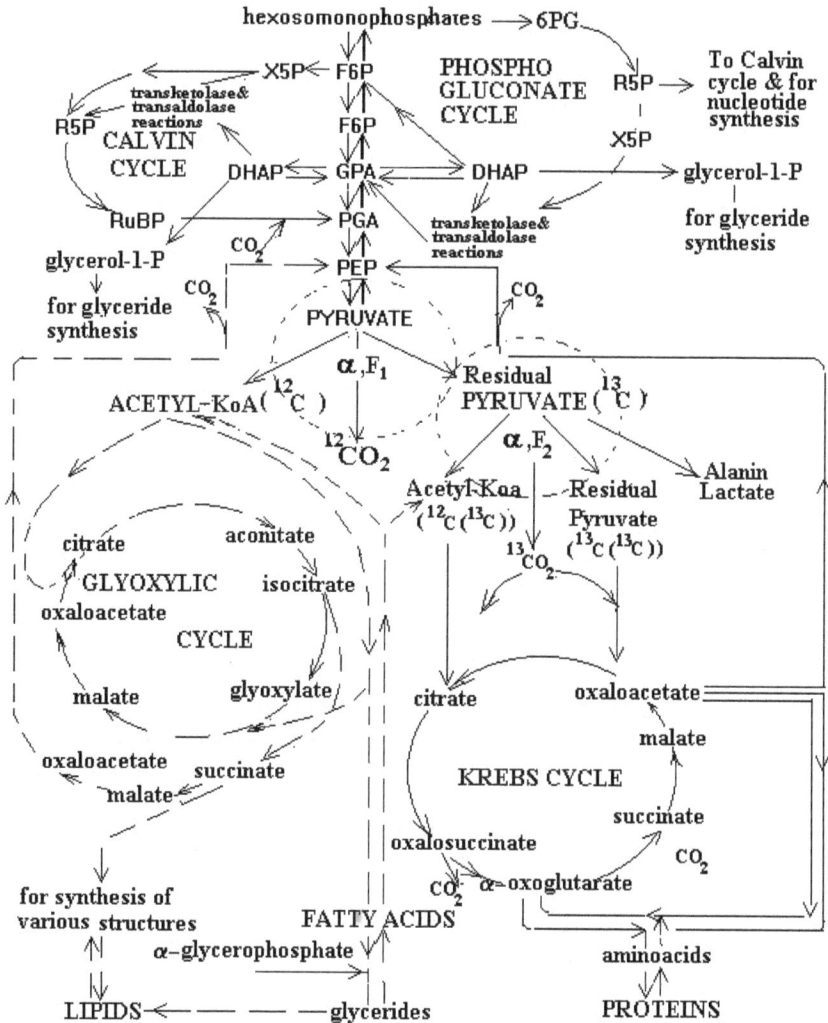

Fig. 8. The simplified diagram depicting temporal organization of secondary metabolism in glycolytic chain (see the text). Dotted lines denote the enzymatic pyruvate decarboxylase complex where carbon isotope fractionation occurs. Abbreviations: X5P, xylose-5-phosphate; R5P, ribose-5-phophate; RuBP, ribulose-1,5-bisphosphate; 6PG, 6-phosphogluconate; F6P, fructose 6-phosphate; FBP, fructose-1,5-bisphosphate; PGA, phosphoglyceric acid; DHA-P, dihydroxyacetone phosphate; PEP, phosphoenolpyruvate; α, carbon isotope fractionation coefficient; F, the extent of pyruvate pool depletion.

idea suitable for autotrophic organisms taking into consideration that heterotrophic organisms have originated first in the course of evolution while autotrophs emerged later adding pentose phosphate reductive cycle (Calvin cycle) to glycolytic chain to feed it with substrates (Ivlev, 2009).

Hence photosynthesizing organisms have inherited glycolytic chain from the precursors with all its functions, regulation and temporal organization and with carbon isotope fractionation mechanism as well. Following studies of ^{13}C distribution regularities in autotrophic and heterotrophic biomass showed they have much in common and confirmed this assumption. Unlike to photosynthetic oscillations, glycolytic ones were found to be long-term. According to Sel'kov (1975, 1978), glycolytic oscillations consist of two phases: glycolysis and gluconeogenesis. In glycolysis, which correlates with dark period of photosynthesis, carbon flux goes "down" the chain (Fig. 8). It means carbohydrates (starch) accumulated in carboxylase phase of photosynthetic oscillations transform into lipids and proteins. In gluconeogenesis, which correlates with light period of photosynthesis, carbon flux goes "up". It means that pools of lipids and proteins accumulated in the dark partly destroy and form the reverse substrate flux directed to carbohydrates. "Up" and "down" indicate only general direction of transformations, since glycolytic and gluconeogenetic pathways do not coincide entirely. The fructose-1,6-bisphosphate futile cycle, is the main regulator of glycolytic oscillations, capable to work in opposite directions (hydrolysis/phosphorilation) depending on concentration ratio of hexosomonophosphates to fructose-1,6-bisphophate (Sel'kov, 1975, 1978).

Carbon isotope fractionation occurs in phase of glycolysis and relates to pyruvate dehydrogenase complex, which is the main cross-point in the chain. Due to pyruvate decarboxylation occurring in this point the pyruvate pool is depleted followed by the Raleigh isotope effect. Glycolysis is organized in such a way that metabolites derived of C_2-fragments (fatty acids, carotenoids, steroids, etc.) referred to lipids, emerge when the extent of the pool depletion is less than a half ($F_1 < 0,5$). It causes lipids in general are enriched in ^{12}C relative to ambient carbohydrates. This piece of glycolysis phase is depicted on the Fig.8 as dotted circle with isotope characteristics α and F_1. The second period of glycolysis mainly corresponds to Krebs cycle functioning and protein synthesis. This piece is depicted on the Fig.8 as dotted circle with isotope characteristics α and F_2. The glycolysis proceeds when the extent of pyruvate pool depletion is more than a half ($F_2 > 0.5$). That is why total proteins are enriched in ^{13}C relative to lipids and to ambient carbohydrates as well. It was adopted that ^{13}C patterns of metabolites related to glycolytic chain are determined solely by isotope fractionation in pyruvate decarboxylation and by the specificity of the following enzymatic interactions since no proofs are available evidencing for carbon isotope fractionation in gluconeogenesis phase.

Now let's see carbon isotope fractionation in pyruvate decarboxylation in a more detail. The important role of the reaction is conditioned at least by two reasons. First, the reaction is located at the cross-point of central metabolic pathways. Hence carbon isotope fractionation is typical to all photosynthesizing organisms. Second, the products of the reaction are used as structural units for the synthesis practically for all secondary metabolites. Taking into account the kinetic nature of isotope effect, metabolic pathways and specificity of enzymatic interactions the intramolecular carbon isotope distributions of many metabolites can be easily predicted to be compared with the experimental data (see below). To get this objective it is necessary to find out ^{13}C distribution in the structural units produced in pyruvate decarboxylation and their dependence on the Raleigh effect.

Three structural units are produced in the above reaction. They are CO_2, evolved in decarboxylation, (C_1-fragments), acetyl-KoA (C_2-fragments), and residual pyruvate (C_3-fragments). According to the isotope effect theory (Melander & Saunders, 1983) only the atoms located at the ends of the broken bonds are subjected to kinetic isotope effect (Fig.9). It means that the effect results in heterogeneous intramolecular isotope distribution

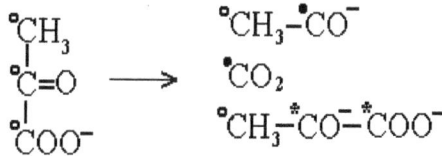

Fig. 9. Three types of carbon atoms resulting in pyruvate decarboxylation.

Empty circles denote atoms with non-changeable isotopic composition in the reaction; filled circles denote atoms getting enriched in ^{12}C relative to the respective atoms of the pyruvate molecules subjected to decarboxylation; asterisks denote atoms getting enriched in ^{13}C to the respective atoms of the initial pyruvate molecules.

Given the isotope composition of the initial pyruvate as that of G6P, derived in carboxylase phase, and taking it as reference level, it is convenient to divide all the atoms in the above fragments into three types. At that one should consider kinetic nature of pyruvate decarboxylation carbon isotope effect and Raleigh effect of pool depletion.

1. Methyl atoms of C_2-fragments and C_3-fragments. Their isotopic compositions during carboxylation remain unchanged and are inherited from the corresponding G6P carbon atoms. It was accepted as an internal standard;

2. Carbonyl carbon atoms of C_2-fragments and CO_2 disposed at the ends of the cleaved C-C bonds. Depending on the extent of pyruvate pool depletion F, their isotope composition can be both enriched in ^{12}C, if F is less than 0,5, or depleted in ^{12}C, if F is more than 0,5.

3. Carboxyl and the neighboring carbonyl atoms of C_3-fragments. Their isotope composition at any F is enriched in ^{13}C relative to atoms of the first and second type.

The ^{13}C distributions in metabolites were analyzed by means of their skeleton reconstruction with allowance of the known pathway, and the specificity of the enzymatic reactions and mixing of atoms in metabolic cycles. The comparison of the theoretically expected and experimentally observed isotope distributions gives the strong arguments in favor of the glycolytic oscillations

3.1 Some examples of ^{13}C distribution in secondary metabolites affirming the oscillatory character of glycolytic chain metabolism

Isotope distributions in lipid components, made of C_2-fragments, are the easiest objects for the isotopic pattern analysis. With allowance for the known fatty acids synthesis pathway (Strickland, 1963), namely the condensation of C_2-fragments according to the head-to-tail principle, there are only odd carbon atoms of skeleton that change their isotope ratios (atoms of the second type). The even atoms (atoms of the first type) remain their isotope composition inherited from the atoms of nutrient carbohydrate. In Table 6 carbon isotope distributions of some fatty acids isolated from lipid fraction of *E. coli*, grown on glucose with the known isotope ratio are presented.

Fatty acids	$\delta^{13}C$, ‰	Odd atoms		Even atoms	
	Total C	N	$\delta^{13}C$, ‰	N	$\delta^{13}C$, ‰
Myristic 14:0	-13.7	1	-27.1		...
Palmitic 16:0	-12.2	1	-15.2		
Palmitoleic 16:1	-13.0	1	-19.2		
		9	-16.0	10	-9.5
9,10-Methylenepalmitic 17: cycle	-13.7	1	-20.3		
Vaccenic 18:1	-12.6	1	-13.9	12	-9.5
		11	-15.8		

Note: $\delta^{13}C$ of nutrient glucose is equal to 9,96‰

Table 6. ^{13}C distribution in some fatty acids from lipid fraction of *E. coli* grown on glucose of the known carbon isotope composition (Monson & Hayes 1982).

As follows from Table 6, isotopic data completely correspond to the known fatty acid synthesis pathway. It confirms that the ^{13}C pattern is determined by isotope effect in pyruvate decarboxylation. Isotope composition of the even atoms (C-10 and C-12) is close to that of the carbon atoms of nutrient glucose, while $\delta^{13}C$ values of the odd atoms (C-1, C-9 and C-11) vary from -13,9 to -27,1. The variations in $\delta^{13}C$ of odd atoms prove they belong to the second type and indirectly evidence on the existence of Raleigh effect accompanying pyruvate pool depletion. The latter in turn indicates the existence of the oscillations. The odd atoms of the fatty acids in all cases are enriched in ^{12}C relative to nutrient glucose ($\delta^{13}C=$ -9,96). In the frame of the model, it means the fatty acids are derived at the extent of pool depletion less than 0,5.

Nutrient glucose	Fatty acids	^{13}C distribution in acetate		
		Total carbon	Carboxyl atom	Methyl atom
-9.0	-12.2	-3.3	+15.0	-8.8

Table 7. ^{13}C distribution in acetate evolved by *E. coli* into the medium in the fermentation of the microorganisms on glucose of the known isotope composition (Blair et al, 1985)

The similar results one can see from Table 7. Acetate, like fatty acids, is made of C_2 units, what is confirmed by its ^{13}C pattern. Carboxyl atom has unusual "heavy" isotope composition ($\delta^{13}C$ = + 15‰), while methyl (even) atom has carbon isotope composition close to the nutrient glucose. Such unusual the ^{13}C enrichment of carboxyl atoms supports again its relation to the Raleigh effect and evidences the acetate is formed at high level of pyruvate pool depletion. On contrary, fatty acids have "light" carbon isotope composition ($\delta^{13}C$ = - 12,2‰) evidencing that their carboxyl atom is enriched in ^{12}C relative to glucose and the fatty acid molecules are derived at extent of pool depletion less than 0,5, as in previous example.

Similar conclusions may be done from the analysis of ^{13}C distribution in quite different class of compounds, plant monoterpenes which also made of C_2 structures (Fig. 10). On the top of

Fig.10 the known synthesis pathway of monoterpenes from C_2-fragments is shown (Nicolas, 1963). As before methyl atoms denoted by empty circles (first type atoms) have approximately equal isotope composition and affirm that isotope effect in CO_2 assimilation was about -25‰ for all studied plants. Isotope ratio of carboxyl atoms (second type atoms) denoted by filled circles vary in a wide range what is expected for them and affirm that fatty acids synthesis in cell cycle have some time length.

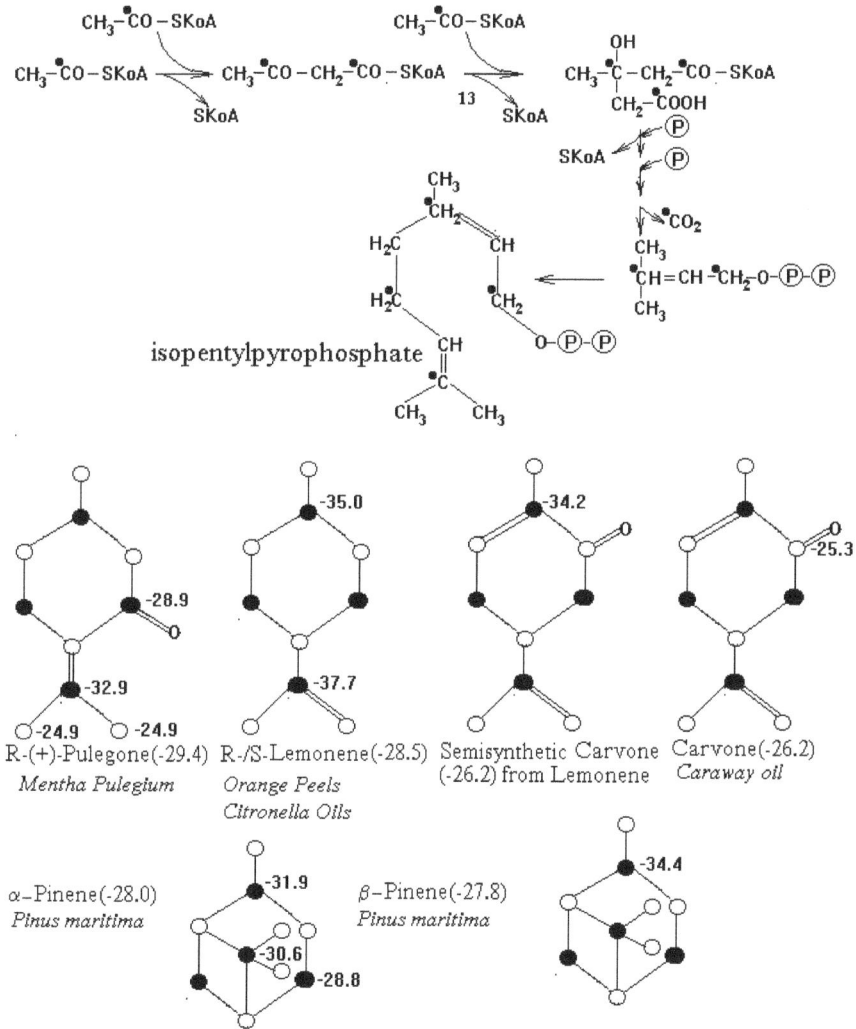

The empty and filled circles indicate carbon atoms of the first and second type

Fig. 10. Biosynthesis pathway of monoterpenes (Nicolas, 1963) and ^{13}C distribution in some plant compounds (Schmidt et al, 1995).

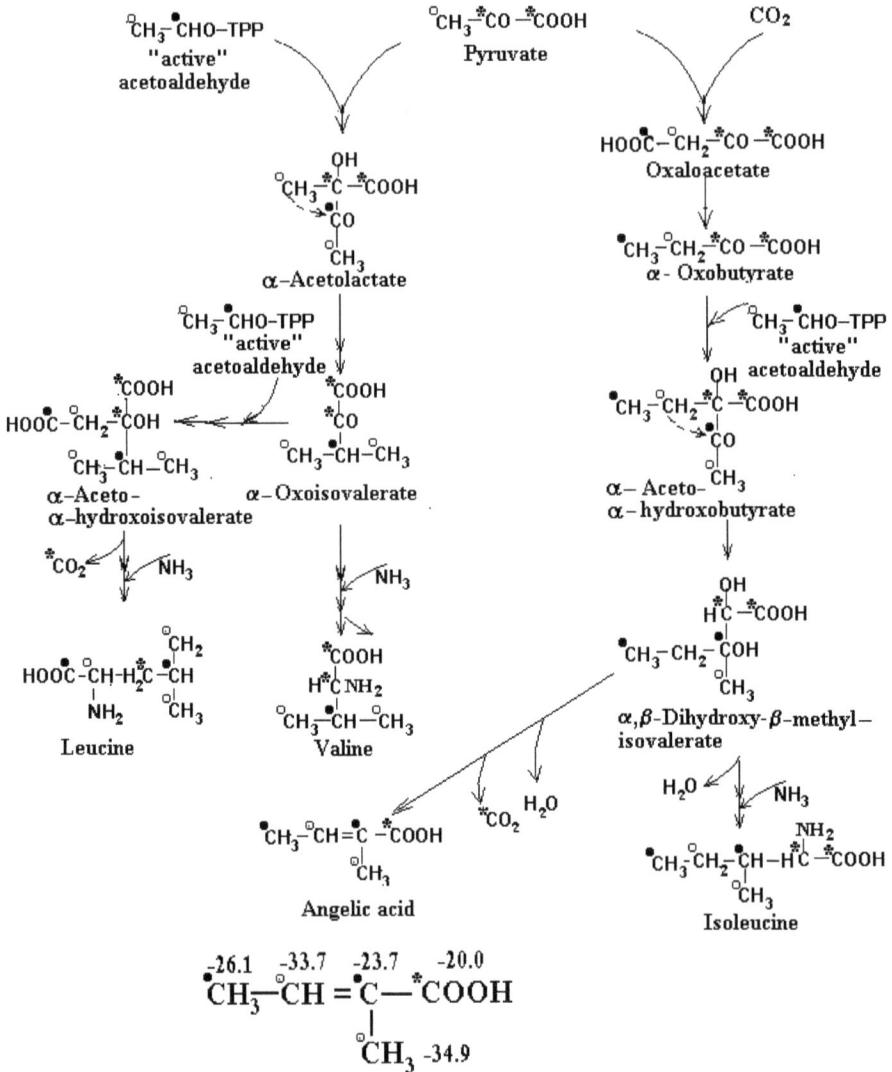

Fig. 11. Biosynthesis pathway of branched aminoacids (Metun, 1963) and ^{13}C distribution in angelic acid precursor of isoleucine, isolated from plant *Angelica Archangelica* (Schmidt et al, 1995).

Two possible synthesis pathways for branched amino acids in photosynthesizing organisms, known from the literature, are shown on Fig.11. One of them leads to leucine (left), another to isoleucine (right). C_2 and C_3 – fragments are the structural units used for their synthesis. Isotopic atoms of all three types, denoted as before, are included in the molecules. Angelic acid, the precursor of isoleucine, was experimentally studied. The entire coincidence with the expected ^{13}C distribution is observed in spite of the internal regrouping of the molecule which occurs at the step of α-aceto-α-hydroxybuterate formation.

[13]C-Distribution in sinigrin is shown on Fig. 12. It is a glucosinolate that belongs to the family of glucosides found in some plants of the *Brassicaceae* family. At the top of the figure the scheme of its biosynthesis pathway is drawn. The same very good coincidence of the predicted ^{13}C distribution with that of observed takes place.

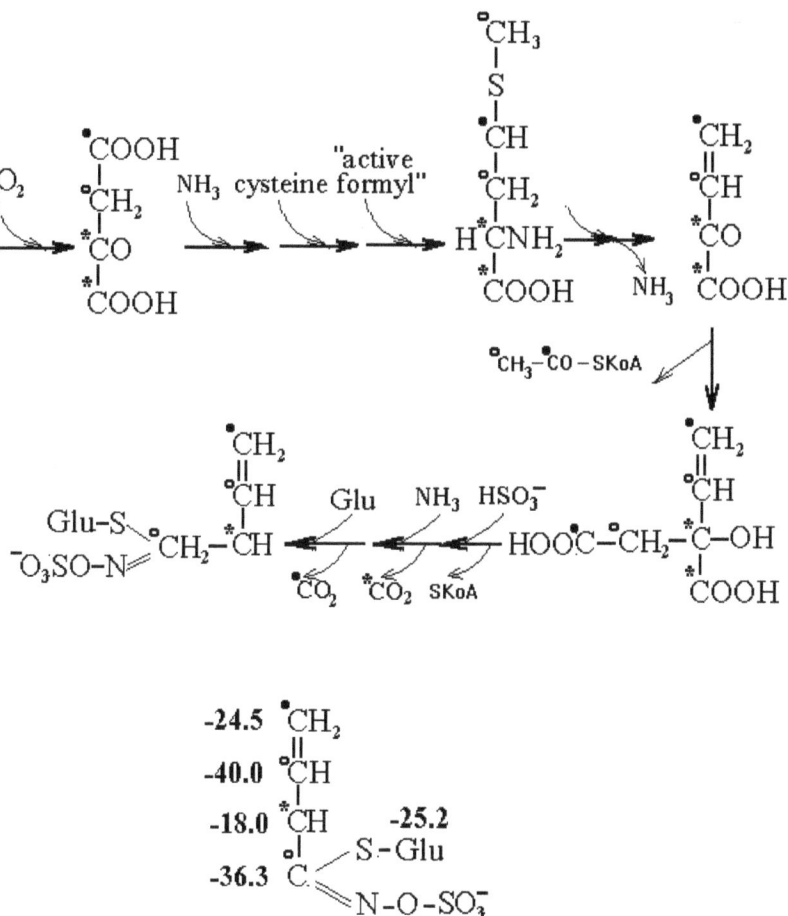

Fig. 12. Biosynthesis pathway and ^{13}C distribution of sinigrin, isolated from plant Angelica Archangelica (Schmidt et al, 1995).

According to the scheme, all three types of isotopic atoms are involved in sinigrin skeleton formation. There are two atoms of the first type which are most enriched in ^{12}C, one atom relating to the third type which was expectedly most enriched in ^{13}C, and one atom of the second type with intermediate isotope composition. The full coincidence of the observed sinigrin isotopic pattern to that expected from the above pathway and the Raliegh law gives one more strong argument in favor of oscillatory character of glycolytic metabolism. Some difference in δ^{13}C values for the first type atoms are the result of two different measurement techniques. Carbon atom of the first type adjacent to double C=C bond was measured by means of NMR technique which is less précised than mass-spectrometric technique. Nevertheless both atoms are considerably "lighter" than atoms of other types and characterize isotope effect in RuBP carboxylation (or carbon isotope ratio of carbohydrates synthesized in carboxylase phase). Notably, glucose in carbohydrate part of sinigrin (-25.2‰), as compared with atoms of the first type, is much more "heavier". It means that it is originated from the labile carbohydrates formed in oxygenase phase of photosynthetic oscillations. Some other arguments proving the oscillation concept are given in the recent publications (Ivlev, 2010, 2011)

4. Conclusions

The performed analysis of isotopic data proves that primary carbon metabolism in photosynthesis and secondary metabolism in glycolytic chain are of oscillatory character. They discover the existence of Raliegh effect that in turn evidences that cells work in filling/depletion regime and there is strict temporal organization of metabolic events. The examination of isotopic patterns of metabolites allows establishing the sequence of metabolites biosyntheses in cell cycle. It is a very important since it changes the fundamental view on the mechanisms underlying all cell processes. This means that besides metabolic pathways one should consider the parameter of temporal organization (Lloid, 2009).
The existence of regularities in ^{13}C distribution proves the following assertions. 1) cell cycles are rather stable, by other words, cell oscillations are in-phase, i.e. at given functional state conditions temporal sequence of metabolic events weakly depends on the environmental factors; 2) cycle oscillations in different cells are synchronized. This fact is in compliance with the known independence of metabolic clocks on the same factors (Shnol', 1996). It means that oscillatory characteristics are determined by the internal properties of the system itself. By the other words the stable temporal sequence of metabolites syntheses which determine isotopic regularities of ^{13}C distribution in metabolites is formed in the course of evolution (Ivlev, 2009). The changes in the environmental conditions can partly change the sequence of events in metabolic organization to better adaptation of organisms to the environments.

5. References

Abdurakhmanova, Z.N., Aliev, A.K., and Abdullaev, A., (1990) Photosynthetic carbon metabolism and conversion of glycolic acid in cotton leaf during ontogeny. *Fiziol. Rast.*, Vol. 37, pp. 675–578.

Abelson, P.H., Hoering, T.C. (1961) Carbon isotope fractionation in formation of amino acids by photosynthetic organisms. *Proc. Nat. Acad. Sci. (USA)* Vol. 47. pp. 623-629.

Barbour, M.M., Mcdowell, N.G., Tcherkez, G., Bickford, Ch.P., Hanson, D.T. (2007) A new measurement technique reveals rapid post-illumination changes in the carbon isotope composition of leaf respired CO_2. *Plant Cell Environ.* Vol.30, pp. 469 – 482.

Blair, N., Lea, A., Munez, E., Olsen, J., Kwong, E., Des –Marais, D. (1985) Carbon isotopic fractionation in heterotrophic microbial metabolism. *Appl. Environ. Microbial* Vol. 50, pp. 996-1002.

Borland, A.M., Griffiths, H., Broadmeadow, M.S., Fordham, M.C., and Maxwell, C. (1994) Carbon isotope composition of biochemical fractions and the regulation of carbon balance in leaves of the C_3-Crassulenean acid metabolism intermediate *Clusia minor* L. growing in Trinidad. *Plant Physiol.*, Vol. 105, pp. 493-501

Buchachenko, A.L. (2003) Are chemical transformations in an enzyme-substrate complex reversible? The experience of the fractionation of isotopes in enzymatic reactions. *Russ. J. Phys. Chem.*, Vol.77, pp. 1298–1302

Cernusak, L.A., Tcherkez, G., Keitel, C., Cornwell, W.K, Santiago, L.S., Knohl, A., Barbour, M.M., Williams, D.G., Reich, P.B., Ellsworth, D.S., Dawson, T.E., Griffiths, H.G., Farquhar, G.D., Wright, I.J. (2009) Why are non-photosynthetic tissues generally ^{13}C enriched compared with leaves in C_3 plants? Review and synthesis of current hypotheses. *Funct. Plant Biol.* Vol. 36, pp. 199–213.

Craig, H. (1957). Isotopic standards for carbon and oxygen and correlation factors for mass-spectrometric analysis of carbon dioxide. *Geochim. et Cosmchim. Acta* , Vol.12. pp. 133-149

Dubinsky, A.Yu. , Ivlev, A.A. (2011) Computational analysis of the possibility of the oscillatory dynamics in the processes of CO_2 assimilation and photorespiration. *Biosystems* Vol. 103, pp. 285 – 290.

Dubinsky, A.Yu. , Ivlev, A.A. Igamderdiev, A.U. (2010) Theoretical analysis of the possibility of existence of oscillations in photosynthesis. *Biophysics* Vol.55, pp 71-74.

Edwards, G., Walker, D. 1983. C3, C4 : Mechanisms, cellular, and environmental regulation of photosynthesis. Blackwell Scientific Publications, Oxford, London, Boston, Melbourne.

Farquhar, G.D., Hubick, K.T., Condon, A.G., Richards, R.A. (1989) Carbon isotope fractionation and plant water-use efficiency. In: *Stable Isotopes in Ecological Research* Eds. Rundel, Ehleringer, Nagy .Vol. 68, pp. 21-40, Springer, New York, Berlin, London, Tokyo

Farquhar, G.D., Lloid, (1993) Carbon and oxygen isotope effects in the exchange of carbon dioxide between terrestrial plants and the atmosphere. In: *Stable isotopes and plant carbon - water relations* Eds. Ehleringer, Hall, Farquhar, pp. 47-70, Academic Press, San Diego – Boston

Farquhar, G.D., O'Leary, M.H., Berry, J.A. (1982). On the relationship between carbon isotope discrimination and intercellular carbon dioxide concentration in leaves. *Aust.J.Plant Physiol.*, Vol.9. pp. 121-137

Galimov, E.M. (1985) *The biological fractionation of isotopes*. Academic Press, London.

Galimov, E.M., Kodina, L.A., Generalova, V.N., Bogachova, M.V. 1977. A study of carbon isotope distribution in biogenic compounds. In: A.V. Sidorenko, ed. *The 8th International Congress on Organic Geochemistry*. Abstracts, Vol. 2, p.156, Moscow.

Gessler, A., Tcherkez, G., Peuke, A.D., Ghashghaie, J.G., Farquhar G.D. (2008) Experimental evidence for diel variations of the carbon isotope composition in leaf, stem and phloem sap organic matter in *Ricinus communis*. *Plant, Cell and Environment*, Vol. 31, pp. 941-953.

Gillon, J.S., Griffiths, H. (1997). The Influence of (photo)respiration on carbon isotope discrimination in plants. *Plant Cell Environment*, Vol.20. pp. 1217-1230

Hayes, J.M., (2001) Fractionation of the isotopes of carbon and hydrogen in biosynthetic processes. In: Valley, Cole (eds) *Stable isotope geochemistry. (Reviews in mineralogy and geochemistry*, vol. 43) Mineralogical Society of America, Washington DC, pp 225–277

Igamderdiev, A.U. (1988) Photorespiration and biochemical evolution of plants. *Uspekhi Sovr.Biologii* (russ.), Vol.105. pp. 488-504.

Igamderdiev, A.U. (1991) Peroxisomal oxidation in plants. *Fiziol.rastenii* Vol.38, pp.773-785/\.

Igamberdiev, A.U., Ivlev, A.A., Bykova, N.V., Threlkeld, Ch., Lea, P.J., and Gardestrom, P. (2001) Decarboxylation of glycine contributes to carbon isotope fractionation in photosythetic organisms. *Photosynthesis Research.* Vol.67, pp. 177-184.

Igamberdiev, A.U. Mikkelsen, T.N., Ambus, P. ,Bauwe, H., Lea, P.J., Gardeström, P. (2004) Photorespiration contributes to stomatal regulation and carbon isotope fractionation : a study with barley, potato and *Arabidopsis* plants deficient in glycine decarboxylase. *Photosynthesis Research,* Vol.81, pp.139-152.

Ivanov, M.V., Zyakun, A.M., Gogotova, G.I., Bondar', V.A. (1978) Fractionation of carbon isotopes by photosinthesizing bacteria, grown on bicarbonate enriched in carbon – 13. *Dokl. Akad. Nauk SSSR* (russ.), Vol. 242, pp. 1417 - 1420.

Ivlev, A.A., (1989) On discreteness of CO_2 assimilation by C_3 plants in the light. *Biofizika (russ.)* Vol. 34, pp. 887-891

Ivlev, A.A., (1992) Carbon isotope effects and a coupled mechanism of photosynthesis and photorespiration. *Sov.Plant Physiol.,* Vol.39, pp. 825 - 835.

Ivlev, A.A., (1993) On the flows of "light" and "heavy" carbon during photosynthesis and photorespiration coupling . *Russ. J. Plant Physiol.,* Vol. 40, pp. 752-758.

Ivlev, A.A. Carbon isotope effect as a tool to study photosynthesis. (2003). *Chemical probes in biology /* ed. M.P. Schneider. Kluwer Academic Publishers. Netherlands pp. 269-285.

Ivlev, A.A. (2004) Intramolecular isotopic distributions in metabolites of the glycolytic chain. *Biophysics,* Vol.49. pp. 414-429.

Ivlev, A.A., (2005) An isotopic effect in the glycine dehydrogenase reaction underlies the intramolecular isotope heterogeneity of glucose carbon in starch synthesized during photorespiration. *Biophysics,* Vol. 50, pp. 1079–1086.

Ivlev, A.A. (2008) *Isotope fractionation and cell mechanisms of carbon metabolism in photosynthesizing cell.* RGAY–MSKhA of Timiryazev. Moscow.

Ivlev, A.A. (2009) *The relationship of the evolution of photosynthesis with geological history of the Earth.* RGAY–MSKhA of Timiryazev. Moscow.

Ivlev, A.A. (2010) Oscillatory character of carbon metabolism in photosynthesis. Arguments and facts. *Biol. Bull.* Vol. 37, pp.211-220.

Ivlev, A.A. (2011) Oscillatory character of carbon metabolism in photosynthesizing cells according to data on carbon isotope composition. *Uspekhi Sovr. Boilogy.* Vol. 131. pp. 178-192.

Ivlev, A.A., Bykova, N.V., and Igamberdiev, A.U. (1996) Fractionation of carbon ($^{13}C/^{12}C$) isotopes in glycine decarboxylase reaction . *FEBS Letters,* Vol. 386, pp. 174-176.

Ivlev, A.A., Dubinsky, A.Yu., Igamberdiev, A.Y. (2010) Oscillatory pattern of photosynthetic CO_2 assimilation affects ^{13}C distribution in carbohydrates. *Izvestiya of TAA* (russ.), № 7, pp. 1 – 23.

Ivlev, A.A., Dubinsky, A.Yu., (2011) On the nature light enhanced dark respiration (LEDR) of plants. *Biophysics* (in press)

Ivlev, A.A., Igamberdiev, A.U., Threlkeld, Ch., and Bykova, N.V. (1999) Carbon isotope effects in the glycine decarboxylase reaction *in vitro* on mitochondria from pea and spinach. *Russ.J. Plant Physiol.,* Vol. 46, pp. 653-660.

Ivlev, A.A., Igamberdiev, A.U, Dubinsky, A.Yu. 2004. Isotopic composition of carbon metabolites and metabolic oscillations in the course of photosynthesis. *Biophysics,* Vol.49 (Suppl. 1), pp. 3–16.

Ivlev, A.A., Lapin, A.V., and Brizanova, L.Ya., (1987) Distribution of carbon isotopes ($^{13}C/^{12}C$) in glucose of maize starch. *Fiziol. Rast.,* Vol. 34, pp. 493–498.

Ivlev, A.A., Kalinkina, L.G. (2001) Experimental evidence for the isotope effect in photorespiration. *Russ.J.Plant. Physiology,* Vol. 48, pp. 400-412.

Ivlev, A.A., Pichuzhkin, V.I., Knyazev, D.A., (1999) Developmental changes in the carbon isotope composition of wheat organs in relation to photorespiration. *Russ. .J. Plant Physiol.*, Vol. 46, pp. 443–451.

Kalinkina, L.G., Naumova N.G. (1992) The content of free amino acids in sea and freshwater Chlorella cells under saline conditions in the presence of the inhibitors of glycolate pathway. *Fiziologia rastenii* , Vol. 38, pp. 559 – 569.

Kalinkina, L.G., and Udel'nova ,T.M. (1990) Effect of photorespiration on the stable carbon isotope fractionation in marine *Chlorella. Fiziol. Rast.* (russ.) Vol. 37, pp. 96-104.

Kalinkina, L.G., and Udel'nova ,T.M. (1991) A mechanism of glycolate pathway involvement in free proline accumulation in *Chlorella* under conditions of salinity. *Fiziol. Rast.* (russ.) Vol. 38, pp. 948-958.

Kyrsanov, A.A. (1976) *Assimilate transport in plants* (russ.), Nauka, Moscow.

Lerman, J.C., Deleens, E., Nato, A., Moyse., A. (1974) Variations in the carbon isotope composition of a plant with crassulecean acid metabolism, *Plant Physiol.*, Vol. 53, pp. 581–584.

Laisk, A. Kh. (1977) *Kinetics photosynthesis and photorespiration in C_3 plants.* Nauka, Moscow.

Lloyd, D., 2009. Oscillations, synchrony and deterministic chaos. U. Lüttge et al. (eds.), Progress in Botany 70, 69-91. Springer-Verlag, Berlin.

Mauser, H., King, W.A., Gready, J.E., Andrews, T.J. (2001). CO_2 fixation by Rubisco: computational dissection of the key steps of carboxylation, hydration and C–C bond cleavage. *J. Am. Chem. Soc.* Vol.123, pp. 10821–10829.

Melander, L., Saunders, W.H. (1983) *Reaction rates of isotopic molecules.* Wiley Interscience Publ., New York

Metun, J. (1963) Biosynthesis of amino acids, in P. Bernfeld (ed.) *Biogenesis of Natural Compounds.* Pergamon Press. Oxford. pp.9-33.

Monson, K.D., Hayes, J.M. (1982). Carbon isotopic fractionation in the biosynthesis of bacterial fatty acids. Ozonolysis of unsaturated fatty acids as a means of determining the intramolecular distribution of carbon isotopes. *Geochim. et Cosmochim. Acta,* Vol.46. pp. 139 -149

Nicolas, G. (1963) Biogenesis of terpenes. In: P. Bernfeld (Ed.) *Biogenesis of Natural Compounds.* pp.549-590. Pergamon Press. Oxford.

O'Leary, M.H., Yapp, C.J. (1978). Equilibrium carbon isotope effect on a decarboxylation reaction. *Biochem. Biophys. Res. Comm.,* Vol.80, pp. 155-160

Oliver, D.J., Neuberger, H., Bourguignon, J., Douce, L. (1990) Glycine metabolism by plant mitochondria. *Physiol. Plant.* Vol.80, pp. 487-491.

Park, R., Epstein, S. (1960). Carbon isotope fractionation during photosynthesis. *Geochim et Cosmochim. Acta,,* Vol.21, pp. 110- 119

Park, R., Epstein, S. (1961). Metabolic fractionation of $^{13}C/^{12}C$ in plants. *Plant. Physiol.,* Vol.36,. pp. 133-139

Raven, J.A., Griffiths, H., Glidewell, S.M., Preston, T. (1982) The mechanism oxalate biosynthesis in higher plants: investigations with the stable isotopes oxygen-18 and carbon-13. *Proc. R. Soc. (London),* Ser.B, Vol. 216. pp. 87 - 101.

Rivera, E.R., Smith, B.N. (1979) Crystal morphology and ^{13}Carbon/^{12}Carbon composition of solid oxalate in Cacti. *Plant Physiol.,* Vol.64, pp. 966 - 970.

Rossmann, F., Butzenlechner, M., Schmidt, H.-L. (1991) Evidence for nonstatistical carbon isotope distribution in natural glucose. Plant Physiology.Vol.96. pp. 609-614.

Roussel, M.R., Ivlev, A.A., Igamberdiev, A.U. (2007) Oscillations of the internal CO_2 concentration in tobacco leaves transferred to low CO_2. *J. Plant Physiol.,* Vol. 164, pp 1182–1196.

Roussel, M.R., Igamberdiev, A.Y. Dynamics and mechanisms of oscillatory photosynthesis. *BioSystems* 2011. Vol. 103. pp. 230-238.

Sanadze, G.A., Black, C.C., Tevzadze, I.T., Tarkhnishvili, G.M. (1978) A change in the $^{13}CO_2/^{12}CO_2$ isotope ratio during photosynthesis by C_3 and C_4-plants. *Fiziol. Rast.* (russ.), Vol. 25, pp. 171 - 172.

Saranga , Y., Flash, I., Patersson, A.H.,Yakir, D. (1999) Carbon isotope ratio in cotton varies with growth stage and plant organ. Plant Science, Vol.142, pp. 47 - 56.

Sel'kov, E.E. (1975) Stabilization of energy charge, oscillations and multiplicity of stationary states in energy metabolism as a result of purely stochoimetric regulation. *Europ J Biochem* Vol. 59, pp. 151-157.

Sel'kov, E.E. (1978). Temporal organization of energy metabolism and cell clock. In: *Regulation of energy exchange and physiological state of organism.* Ed. Kondrashova, pp.15-32, Nauka, Moscow.

Schmidt, H.-L., Kexel, H., Butzenlechner, M., Schwarz, S., Gleixner, G., Thimet S., Werner, R.A., Gensler, (1995). M. 2. Non-statistical isotope distribution in natural compounds: mirror of their biosynthesis and key for their origin assignment.. E. Wada, T. Yoneyama, M. Minagawa, T. Ando, B.D. Fry (Eds.) pp.17-35, Kyoto : Kyoto University Press

Schmidt, H.-L. (2003) Fundamentals and systematics of the non-statistical distributions of isotopes in natural compounds. *Naturwissenschaften*, Vol.90, pp. 537–552

Smith, B.N., Epstein, S. (1971) Two categories of $^{13}C/^{12}C$ ratios for higher plants. *Plant Physiol.*.Vol.47, pp. 380-384

Shnol', S.T. (1996) Biological clocks: brief review of the course of studies and modern condition of the problem. Soros Educ. J. N 7, pp26 – 32.

Strickland, K.P. (1963) Biogenesis of lipids. In: P. Bernfeld (ed.) *Biogenesis of Natural Compounds.* Pergamon Press. Oxford. pp.82-131.

Saranga, Y., Flash, I., Patersson, A.H., Yakir, D. (1999) Carbon isotope ratio in cotton varies with growth stage and plant organ. *Plant Sci..*, Vol. 142, pp. 47–56.

Tapia, O., Andres, J., Safont, V.S., (1995) Transition structure in vacuo and the theory of enzyme catalysis. Rubisco's catalytic mechanism: a paradigm case? *J. Mol. Struct.* Vol.342, pp. 131–140.

Tcherkez, G., Farquhar, G., Badeck, F., Ghashghaie, J. (2004). Theoretical consideration about carbon isotope distribution in glucose of C_3-plants. *Funct. Plant Biol.* Vol. 31, pp. 857–877.

Varshavsky,Ya.M. (1988) On distribution of carbon heavy isotope (^{13}C) in biological systems. *Biofizika* (russ.) Vol.33, pp. 351-355

Vogel, J.C. (1993). Variability of carbon isotope fractionation during photosynthesis. In: *Stable isotopes and plant carbon - water relations* Eds. Ehleringer, Hall, Farquhar, pp. 29-46, Academic Press, San Diego – Boston

Voznesenskii, V.L., Glagoleva, T.A., Zubkova, E.K., Mamushina, N.S., Filippova, L.A., and Chulanovskaya, M.V. (1982) Metabolism of ^{14}C in *Chlorella* during prolonged cultivation in the presence of $^{14}CO_2$. *Fiziol. Rast.* (russ.), Vol.29, pp. 564-571.

White, J.W. (1993) Implication of carbon isotope discrimination studies for breeding common bean under water deficits, in *Stable Isotopes and Plant Carbon–Water Relations.* Eds. Ehleringer, Hall, Farquhar, pp. 387–398 Acad. Press, San Diego - Boston

4

Photosynthetic Carbon Metabolism: Plasticity and Evolution

Roghieh Hajiboland
Plant Science Department,
University of Tabriz
Iran

1. Introduction

Carbon metabolism is the important part of the photosynthetic process in that plant green cells convert physical and chemical sources of energy into carbohydrates. Over the last 50 years, knowledge and understanding of carbon metabolism has improved considerably. Photosynthetic carbon metabolism can no longer be explained by a single, invariable cycle. It is no longer restricted to just the chloroplast or even to a single cell. In addition to carbon reduction, nitrogen assimilation, sulphate reduction and other aspects of intermediary metabolism are tightly connected with this process.

Like other physiological processes, photosynthesis differs greatly among various plant species and under different environmental conditions. Over the evolutionary history of land plants, selection pressures led to evolution of variants of photosynthetic carbon metabolism namely C_4 and crassulacean acid metabolism (CAM) pathways. In this chapter we will focus on the effect of environmental factors on photosynthetic carbon metabolism in order to show the considerable plasticity of this process. In addition, the processes in the evolution of main types of photosynthetic carbon metabolisms will be discussed regarding anatomical, physiological and molecular evidences.

2. Photosynthetic carbon metabolism: A general description

The pathway by which all photosynthetic eukaryotic organisms incorporate CO_2 into carbohydrates is known Calvin cycle or photosynthetic carbon reduction (PCR) cycle. The PCR cycle can be divided into three primary stages: (1) carboxylation which fixes the CO_2 in the presence of the five-carbon acceptor molecule, ribulose bisphosphate (RuBP), and converts it into two molecules of a three-carbon acid. The carboxylation reaction is catalyzed by the enzyme ribulose-1,5-bisphospahate carboxylase-oxygenase (Rubisco). (2) reduction, which consumes the ATP and NADPH produced by photosynthetic electron transport to convert the three-carbon acid to trios phosphate, and (3) regeneration, which consumes additional ATP to convert some of the triose phosphate back into RuBP to ensure the capacity for the continuous fixation of CO_2 (Fig. 1).

The first stable intermediate of Calvin cycle is a three-carbon acid, 3-phosphoglycerate. Therefore, the PCR cycle is commonly referred to as the C_3 cycle.

2.1 Photorespiration, principles and significance

An important property of Rubisco is its ability to catalyze both the carboxylation and the oxygenation of RuBP. Oxygenation is the primary reaction in a process known as photorespiration. Photosynthesis and photorespiration work in opposite directions, photorespiration results in loss of CO_2 from cells that are simultaneously fixing CO_2 by the Calvin cycle. The C_2 glycolate cycle, also known as the photosynthetic carbon oxidation cycle, begins with the oxidation of RuBP to 3P-glycerate and P-glycolate (Fig. 2).

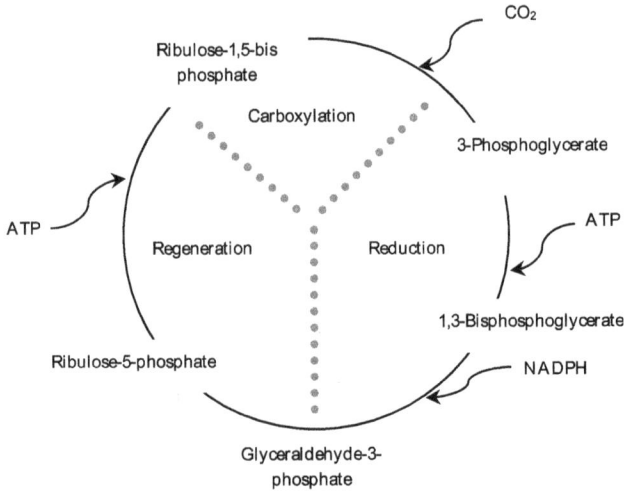

Fig. 1. The three stages of the photosynthetic carbon reduction (PCR) cycle or Calvin cycle.

In normal air (21% O_2), the rate of photorespiration in sunflower leaves is about 17% of gross photosynthesis. Every photorespired CO_2, however, requires an input of two molecules of O_2 and the true rate of oxygenation is about 34% and the ratio of carboxylation to oxygenation is about 3 to 1. The ratio of carboxylation to oxygenation depends, however, on the relative levels of O_2 and CO_2 since both gases compete for binding at the active site on Rubisco. Increase in the relative level of O_2 (or decrease in CO_2) shifts the balance in favor of oxygenation. An increase in temperature will also favor oxygenation. Increase in temperature declines the solubility of gases in water, but O_2 solubility is less affected than CO_2. Thus O_2 will inhibit photosynthesis, measured by net CO_2 reduction. There is also an energy cost associated with photorespiration and the glycolate pathway. Not only is the amount of ATP and NAD(P)H expended in the glycolate pathway following oxygenation (5ATP+3NADPH) greater than that expended for the reduction of one CO_2 in the PCR cycle (3ATP+2NADPH), but there is also a net loss of carbon. Photorespiration appears to be a costly energy and carbon acquisition. It is logical to ask why should the plants indulge in such an apparently wasteful process? Several ideas have been proposed (Hopkins and Hüner, 2004; Foyer et al., 2009; Bauwe, 2011).

2.1.1 Oxygenation is an unavoidable consequence of evolution

It has been proposed that the oxygenase function of Rubisco is an inescapable process. Rubisco evolved at a time when the atmosphere contained large amounts of CO_2 but little oxygen. Under these conditions, an inability to discriminate between the two gases would have had little significance to the survival of the organism. It is believed that oxygen began to accumulate in the atmosphere primarily due to photosynthetic activity and the atmospheric content of O_2 had increased to significant proportions during the following stages of evolution of land plants. By this view, then, the oxygenase function is an evolutionary "hangover" that has no useful role.

However, this is an oversimplified view of photorespiration since photorespiratory mutants of *Arabidopsis* proved to be lethal under certain growth conditions, indicating the essential nature of the photorespiratory pathway in C_3 plants. There is no evidence that selection pressures have caused evolution of a form of Rubsico with lower affinity to O_2.

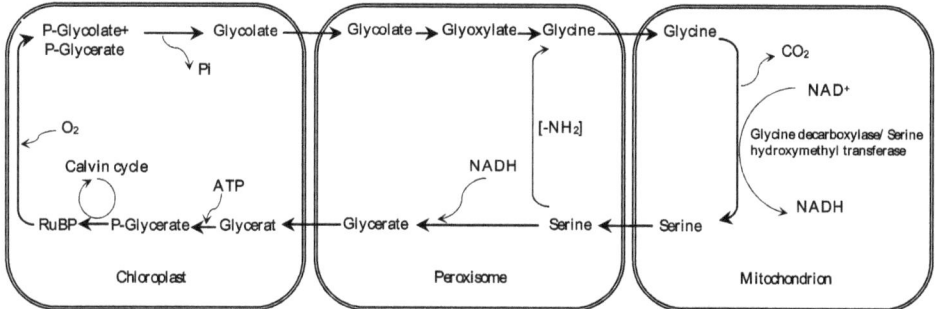

Fig. 2. The photorespiratory glycolate pathway.

2.1.2 Plants have turned this apparent evolutionary deficiency into a useful metabolic sequence

The glycolate pathway, undoubtedly serves as a scavenger function. For each two turns of the cycle, two molecules of phosphoglycolate are formed by oxygenation. Of these four carbon atoms, one is lost as CO_2 and three are returned to the chloroplast. The glycolate pathway thus recovers 75% of the carbon that would otherwise be lost as glycolate. There is also the possibility that some of the intermediates, serine and glycine, for example, are of use in other biosynthetic pathways, although this possibility is still subject of some debate.

2.1.3 Photorespiration functions as a safety valve for dissipation of excess excitation energy

A significant decline in the photosynthetic capacity of leaves irradiated in the absence of CO_2 and O_2 has been reported. Injury is prevented, however, if sufficient O_2 is present to permit photorespiration to occur. Apparently, the O_2 consumed by photorespiration is sufficient to protect the plant from photo-oxidative damage by permitting continued operation of the electron transport system. This could be of considerable ecological value under conditions of high light and limited CO_2 supply, for example, when the stomata are closed due to water stress. Photorespiratory mutants of *Arabidopsis* are more sensitive to photoinhibition than their wild type counterparts.

In order to increase crop productivity efforts have been made on the inhibition or genetically eliminating photorespiration. Effort has been expended in the search for chemicals that inhibit the glycolate pathway or selective breeding for low-photorespiratory strains through finding a Rubisco with lower affinity for oxygen. All of these efforts have been unsuccessful, presumably because the basic premise that photorespiration is detrimental to the plant and counterproductive is incorrect.

Clearly, success in increasing photosynthesis and improving productivity lies in other directions. A mechanism for concentrating CO_2 in the photosynthetic cells could be one way to suppress photorespiratory loss and improve the overall efficiency of carbon assimilation. That is exactly what has been achieved by C_4 and CAM plants. A limited extent of photorespiration in C_4 and CAM plants is a consequence of mechanisms that concentrate CO_2 in the Rubisco environment and thereby suppress the oxygenation reaction (Hopkins and Hüner, 2004; Foyer et al., 2009; Bauwe, 2011)

3. C_4 mode of carbon assimilation

C_4 plants are distinguished by the fact that the first product is a four-carbon acid oxaloacetate (OAA). The key to the C_4 cycle is the enzyme phosphoenol pyruvate carboxylase (PEPC), which catalyzes the carboxylation of PEP using the bicarbonate ion (HCO_3^-) as the substrate (rather than CO_2). C_4 plants also exhibit a number of specific anatomical, physiological and biochemical characteristics that constitute the "C_4 syndrome". One particular anatomical feature characteristic of most C_4 leaves is the presence of two distinct photosynthetic tissues. In C_4 leaves the vascular bundles are quite close together and each bundle is surrounded by a tightly fitted layer of cells called the bundle sheath. Between the vascular bundles and adjacent to the air spaces of the leaf are the more loosely arranged mesophyll cells (Fig. 3). This distinction between mesophyll and bundle sheath photosynthetic cells called Kranz anatomy plays a major role in the C_4 syndrome (Bhagwat, 2005; Edwards and Voznesenskaya, 2011).

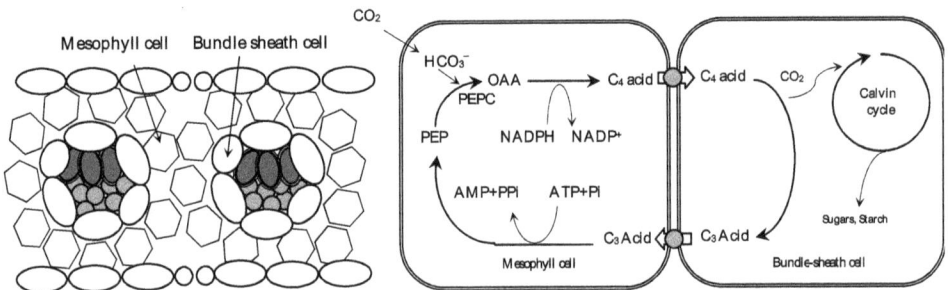

Fig. 3. Leaf anatomy of a C_4 plant (left). Note the tightly fitted bundle sheath cells (ovals) surrounded by a concentric layer of mesophyll cells (hexagons). Schematic of the C_4 photosynthesis carbon assimilation cycle (right).

There are certain similarities between the PCR cycle and C_4 metabolism. Like Rubisco, the PEPC carboxylation reaction is virtually irreversible and, consequently, energetically very favorable. Reducing potential is required at some point to remove the product and ATP is required to regenerate the acceptor molecule, PEP (Fig. 3). A very significant difference between the PCR cycle and C_4 metabolism, however, is that once in the bundle sheath cell, the C_4 acid is decarboxylated, giving up the CO_2 originally assimilated in the mesophyll cell. This decarboxylation means that, unlike the C_3-cycle, the C_4 cycle does not of itself result in any net carbon reduction. The plant relies ultimately on the operation of the PCR cycle in the bundle sheath chloroplast for the synthesis of triose phosphates.

Within the general pattern of the C_4 cycle described above there are three variations include NADP-malic enzyme type, NAD-malic enzyme and PEP carboxykinase types. Regardless of these variations, the principal effect of the C_4 cycle remains to concentrate CO_2 in the bundle-sheath cells where the enzymes of the PCR cycle are located. By shuttling the CO_2 in the form of organic acids it is possible to build much higher CO_2 concentrations in the bundle-sheath cells than would be possible relying on the diffusion of CO_2 alone. The concentration of CO_2 in bundle-sheath cells may reach 60 mM about tenfold higher than that in C_3 plants. Higher CO_2 concentrations would suppress photorespiration and support higher rates of photosynthesis. Under optimal conditions, C_4 crop species can assimilate CO_2 at rates two to three times that of C_3 species. All this productivity, however, has an energy cost to building the CO_2 concentration in the bundle-sheath cells. For every CO_2 assimilated, two ATP must be expended in the regeneration of PEP. This is in addition to the ATP and NADPH required in the PCR cycle. Thus the net energy requirement for assimilation of CO_2 by the C_4 cycle is five ATP and two NADPH (Bhagwat, 2005; Edwards and Voznesenskaya, 2011).

4. CAM mode of carbon assimilation

Another CO_2 concentrating mechanism is CAM. This specialized pattern of photosynthesis was originally studied in the family Crassulaceae. One of the most striking features of CAM plants is an inverted stomatal cycle i.e. the stomata open mainly during the nighttime hours and are usually closed during the day. This means that CO_2 uptake also occurs mainly at night. In addition, CAM plants are characterized by an accumulation of malate at night and its subsequent depletion during hours and storage carbohydrate levels. Nocturnal stomatal opening supports a carboxylation reaction producing C_4 acids that are stored in the large vacuoles. Accumulation of the organic acids leads to a marked acidification of these cells at night. The acids are subsequently decarboxylated during daylight hours and the resulting CO_2 is fixed by the PCR cycle. As in C_4 plants, the enzyme PEPC is central to CAM operation (Fig. 4). CAM species may be distinguished by the enzymes which catalyse organic acid decarboxylation, NAD-malic enzyme, NADP-malic enzyme and PEP-carboxykinase types (Bhagwat, 2005; Dodd et al., 2002; Holtum et al., 2005).

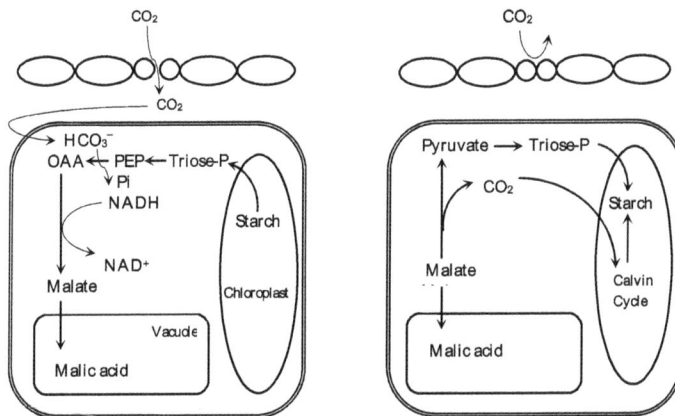

Fig. 4. Schematic of crassulacean acid metabolism (CAM).

5. Carbon isotope discrimination in carbon assimilation pathways

There are two naturally occurring stable isotopes of carbon, ^{12}C and ^{13}C. Most of the carbon is ^{12}C (98.9%), with 1.1% being ^{13}C. The overall abundance of ^{13}C relative to ^{12}C in plant tissue is commonly less than in the carbon of atmospheric carbon dioxide, indicating that carbon isotope discrimination occurs in the incorporation of CO_2 into plant biomass. Variation in the $^{13}C / ^{12}C$ ratio is the consequence of so called "isotope effects," which are expressed during the formation and destruction of bonds involving a carbon atom, or because of other processes that are affected by mass, such as gaseous diffusion (Farquhar et al., 1989). Isotope effect, denoted by α, is defined as the ratio of carbon isotope ratios in reactant and product:

$$\alpha = R_r / R_p \qquad\qquad (1)$$

R_r is the $^{13}C / ^{12}C$ molar ratio of reactant and R_p is that of the product. For plants R_a (R_r) is isotopic abundance in the air and R_p is defined isotopic abundance in the plant. For numerical convenience, instead of using the isotope effect ($\alpha = R_a / R_p$), it has been proposed to use the Δ, the deviation of α from unity, as the measure of the carbon isotope discrimination by the plant:

$$\Delta = \alpha - 1 = R_a / R_p - 1 \qquad\qquad (2)$$

Isotopic composition is another parameter is specified as $\delta^{13}C$ values, R_s is the $^{13}C / ^{12}C$ molar ratio of the standard:

$$\delta^{13} = R_p / R_s - 1 \qquad\qquad (3)$$

Isotopic composition of plants (δ^{13}) is negative, whereas the process of CO_2 diffusion and carboxylation by Rubisco have positive discrimination (i.e. against $^{13}CO_2$). Therefore, Δ values are usually positive while those of δ^{13} are usually negative. Typically, Δ and δ^{13} values are ~$10\text{-}35 \times 10^{-3}$, which is normally presented as 10-35‰ ("per mil").

Because Rubisco preferentially fixes the light ^{12}C isotope over the heavier ^{13}C, plant tissues are enriched in ^{12}C relative to the bulk atmosphere. However, C_4 plants exhibit much lower rates of Δ (and higher rates of δ^{13}) than C_3 plants. Plants exhibiting CAM have intermediate values which appear to be related to the relative proportions of C_3 and C_4 fixation by these species (Farquhar et al., 1989). The evolutionary modifications that lead to the enhancement of fixation by the C_4-cycle in C_3-C_4 intermediates are also associated with reduction in Δ values and increase in δ^{13} values from C_3 to near C_4 values (Hobbie and Werner, 2004)(See below).

6. Effect of environmental factors on photosynthetic carbon metabolism

6.1 Effect of water availability

Plant and cell water balance is determined by water lost in transpiration and water absorption from the soil. When transpiration exceeds absorption, cell turgor and relative water content (RWC) decrease, while the concentration of cellular contents increases. Under these conditions, osmotic potential and water potential fall.

RWC normalizes water content by expressing it relative to the fully turgid state and is an easily measured indicator of water status:

$$RWC\% = (\text{fresh mass} - \text{dry mass})/(\text{water saturated mass} - \text{dry mass}) \times 100 \qquad (4)$$

Low cell turgor and RWC slow growth and decrease the stomatal conductance for H_2O (g_s). Water deficiency covers the range from fully hydrated cells (100% RWC), as the control state with metabolism functioning at the potential rate to very dehydrated cells (50% RWC or less) at which the cell will not recover when rehydrated.

Progressive decrease in RWC decreases A (net CO_2 assimilation rate) of leaves. A depends on the activity of Rubisco per unit leaf, the rate of RuBP synthesis (hence on capture of photosynthetically active radiation, PAR) and on the CO_2 supply, determined by g_s and the ambient CO_2 concentration (C_a) (Lawlor, 2001). CO_2 supply to the PCR cycle in the chloroplast is determined by C_a and conductance of the pathway for diffusion between air and enzyme active sites, principally g_s in the gas phase and g_m in the liquid phase, which includes all physicochemical and biochemical factors (von Caemmerer 2000). The CO_2 concentration within the leaf, C_i, depends on A, g_s and C_a:

$$A = g_s \, (C_a - C_i) \qquad (5)$$

The CO_2 concentration at the active sites of Rubisco in the chloroplasts (C_c) is given by:

$$A = g_m \, (C_i - C_c). \qquad (6)$$

By measuring A as a function of C_i (A/C_i response curve) under standard PAR flux, the limitations to A could be assessed. The maximum rate of A under saturating CO_2 (C_a, C_i and C_c) and light in fully hydrated leaves is defined as A_{pot}. To achieve the same A_{pot} at small RWC as at large RWC, C_c must saturate Rubisco and so Ca must be sufficient to overcome the limitation of g_s. If A_{pot} at small RWC does not attain the value of A_{pot} at large RWC, despite CO_2 saturation, then metabolism is inhibited (Lawlor and Cornic, 2002).

Experimental studies on CO_2 assimilation of C_3 plants under decreasing RWC show that there are fundamental differences between species in the relative roles of g_s limitation of CO_2 supply and metabolic limitation of A_{pot}. Basically data obtained from various species fall into two groups, which have been called Type 1 and Type 2 responses (Fig. 5).

Type 1: With RWC=90% –75%, increasing C_a to 5% restores A fully to the A_{pot} of control leaves (Cornic and Massacci 1996). At RWC<75% restoration of A_{pot} to the value at 100% RWC is not achieved and the response to CO_2 becomes progressively smaller. Therefore, in Type 1 response, there are two main, relatively distinct phases with a transition between them. The stomatal limitation phase occurs at RWC between 100 and 75%, without effect on A_{pot}, so that A may be restored to A_{pot} by large concentration of CO_2. The metabolic phase of limitation is at lower RWC <75% where A_{pot} is limited by metabolism (Lawlor and Cornic, 2002).

Type 2: Elevated CO_2 increases A to A_{pot} in unstressed leaves (RWC 100 to 75%), but A is progressively less stimulated as RWC decreases, i.e. A is not restored to the unstressed A_{pot}, so the potential rate of CO_2 assimilation is decreased (Tezara *et al.* 1999). Inhibition of A with 10% or greater C_a, rather than restoration, shows that metabolism is affected by elevated CO_2, but as A is not restored to A_{pot}, photosynthetic metabolism must be impaired. The evidence is therefore of partial metabolic inhibition of A by moderate stress and substantial inhibition at more severe stress. In Type 2, the phases are not distinct but progressive, lacking the two clearly distinguished phases of Type 1. Stomatal regulation, i.e. decreased g_s, dominates at relatively large RWC, leading to a lower C_i and C_c. g_s becomes progressively less important and metabolic limitations more important as RWC falls (Lawlor and Cornic, 2002).

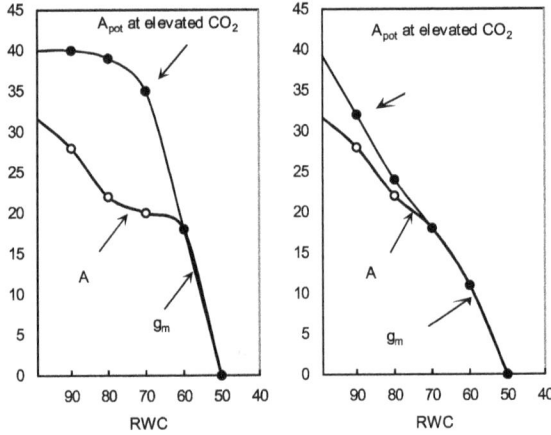

Fig. 5. There are two types of response of A_{pot} to decreasing RWC: Type 1 (left) with no decrease as RWC falls from 100 to~75% and Type 2 with progressive decrease over this range (right). Redrawn according to Lawlor and Cornic (2002) with permission.

Although effect of different experimental approaches could not be excluded in different response curve of species as described by type 1 and 2, this difference is most likely related to differences in the particular characteristics of metabolism in different species. They may reflect differences in the cell water balance and differences in cell elastic modulus. Alternatively, they could reflect differences in sensitivity of a basic process to changing cellular conditions in different species or under different conditions, e.g. if ionic concentration in chloroplasts differed and thus resulted in different rates of ATP synthesis (Lawlor and Cornic, 2002).

The potential capacity for light harvesting, energy transduction, electron transfer in reaction centers of the photosystems (PSII and PSI) and electron transport in thylakoids are unaffected by a wide range of RWC, only severe loss of RWC decreases photosystem activity and alters the structure of PSII (Giardi et al. 1996). The rate of electron transport at saturating PAR is determined by sink capacity for electrons, principally A at large RWC. Decreased sink capacity for electrons results in increased non-photochemical energy dissipation in Type 1 and 2 responses (Müller et al., 2001). However, maintenance of A_{pot} in the Type 1 response will enable greater CO_2 recycling and energy use within the tissue than if A_{pot} decreases as in Type 2. At low RWC, where A_{pot} decreases in both responses, electron transport is decreased because of biochemical, as opposed to biophysical, limitations (Lawlor and Cornic, 2002).

As A falls with decreasing RWC, the amount of assimilate available for export as triose-phosphate from chloroplast to cytosol and sucrose synthesis diminishes. Sucrose content in leaves falls in rapidly stressed leaves at RWC<80%, caused by low A and continued respiration. Thus, it is very unlikely that accumulation of assimilates would result in feedback inhibition of A or that the capacity of the triose-phosphate-P_i transporter in the chloroplast envelope is affected by low RWC. The rate of sucrose synthesis also depends on the activity of sucrose phosphate synthase, which is greatly decreased by even small loss of RWC. Sucrose phosphate synthase is subjected to complex control, including allosteric modulation by glucose 6-phosphate and phosphorylation by a protein kinase using ATP (S.C. Huber and J.L. Huber, 1996). The latter activates SPS under osmotic stress, suggesting that inactivation may be related to decreased ATP content (Lawlor and Cornic, 2002).

Water deficiency changes the proportion of different carbohydrates. Starch, glucose and fructose concentrations increase with mild drought but sucrose change little. Such changes may be adaptive i.e. osmoregulation, and is associated with increased soluble (vacuolar) acid invertase activity in C_3 species (Pelleschi et al. 1997).

6.2 Effect of temperature
6.2.1 Carbon metabolism under low temperature and during cold acclimation
Low temperature is one of the most important factors affecting plant performance and distribution (Stitt and Hurry, 2002). Photosynthetic carbon metabolism is greatly influenced by low temperature, directly via the modulation of enzymes activity and indirectly via changes in sink demand of plants experiencing low temperature stress.

Ecological and physiological studies have uncovered a strong correlation between sugar concentrations and frost resistance (Guy et al., 1992). Sugars either act as osmotica or protect specific macromolecules during dehydration. Changes in the subcellular concentration and distribution of sugars might also provide a mechanism to protect specific compartments. For example, although sucrose is largely restricted to the cytosol of leaves at high temperatures, there are reports that it accumulates in the chloroplast in cold-acclimated cabbage (Fowler et al., 2001).

In recent years, molecular studies unravel mechanisms underlying responses of photosynthetic metabolism under low temperatures. The constitutive increases in the frost tolerance was observed in an *Arabidopsis* mutant over-expressing one dehydration-responsive element gene that is correlated with increased sugar contents (Gilmour et al., 2000). A freezing sensitive mutant with impaired cold acclimation has sugar levels that are lower than those of wild type *Arabidopsis* plants (McKown et al., 1996).

Decreased temperatures lead to an acute Pi-limitation of photosynthesis. Indeed, some of the changes in photosynthetic metabolism that occur during cold acclimation are reminiscent of the response to low Pi (Nielsen et al., 1998). Evidence that changes in Pi concentration or availability to metabolism contribute to cold acclimation has been provided by studies of *pho1* and *pho2* mutants. *pho1* mutant with decreased shoot Pi concentration shows accentuation of the low-temperature-induced increase of sucrose phosphate synthase activity and of cytosolic fructose-1,6-bisphosphatase and sucrose phosphate synthase gene expression, abolishment of the decrease in the transcript levels of genes encoding for Rubsico and light harvesting chlorophyll a/b binding protein after chilling, accentuation of the cold-induced shift in carbon allocation from starch to sucrose and increase in the proline accumulation after chilling (Hurry et al., 2000). The opposite of these metabolic characteristics was observed in *pho2* mutant with higher shoot Pi concentration compared with control (Hurry et al., 2000). These results reveal that signals relating to altered Pi concentration or availability to metabolism lead to the activation and increased expression of enzymes in the sucrose synthesis pathway. They also lead to changes in the relative activities of enzymes in the Calvin cycle (Hurry et al., 2000).

Functional importance of sugar metabolism has also been demonstrated during cold acclimation. Optimal rates of photosynthesis require an appropriate balance between the rates of carbon fixation and sucrose synthesis (Stitt, 1996). Excessive sucrose synthesis depletes the phosphorylated Calvin cycle intermediates and inhibits the regeneration of ribulose-1,5-bisphosphate. Conversely, inadequate sucrose synthesis leads to accumulation of phosphorylated intermediates and depletion of Pi, resulting in inhibition of ATP synthesis, accumulation of 3-phosphoglycerate and inactivation of Rubisco. A sequence of events reverses the inhibition of sucrose synthesis and photosynthesis as the plants acclimate to low temperatures (Hurry et al., 2000). Short- and mid-term adjustments act primarily on sucrose synthesis but also stimulate photosynthesis by relieving the acute Pi-limitation. Longer-term adjustments affect photosynthesis directly. The recovery has two important functions: increased sucrose production (Strand et al., 1999) and protection against photoinhibition by allowing increased turnover of the photosynthetic electron chain (Hurry et al., 2000). The transfer of warm-grown *Arabidopsis* plants to 4°C leads to the post-translational activation of sucrose phosphate synthase within 30 min. Over the next few days, sucrose synthesis is stimulated by two further adjustments. One is a selective increase in the expression of cytosolic fructose-1,6-bisphosphatase and sucrose phosphate synthase genes, the two key regulated enzymes in the pathway of sucrose synthesis (Strand et al., 1999). The second is a shift in the subcellular distribution of Pi. In the leaves, most of the Pi is in the vacuole. Indirect evidence indicates that the Pi distribution shifts towards the cytoplasm at low temperatures allowing phosphorylated metabolites to increase without depleting the free Pi (Hurry et al., 2000).

Full acclimation occurs in leaves that develop at low temperature. They retain the selective increase in the expression of sucrose synthesis enzymes, and also have higher activities of all of the Calvin cycle enzymes on a fresh weight or leaf area basis. Two factors contribute to this increase. First, whereas transcripts for light harvesting chlorophyll a/b binding protein and Rubisco genes decrease after transfer to low temperature, they recover in leaves that develop at low temperature. This recovery occurs even though leaf sugars rise, indicating that sugar-repression of these genes is overridden at low temperature in acclimated leaves. Second, leaves that mature at low temperature have reduced water and increased protein contents due to an increase in the volume of the cytoplasm relative to that of the vacuole (Strand et al., 1997; 1999) .

6.2.2 Carbon metabolism under higher temperatures

Photosynthesis is very responsive to high temperatures (Knight and Ackerly, 2003). In semi-arid regions, temperature and precipitation are often negatively correlated, with lower rainfall in warmer environments. Therefore, studies on photosynthetic thermal tolerance of plants are complicated by the fact that a variety of environmental factors can affect photosynthesis, including plant water status, soil salinity and light levels. Photosynthetic acclimation can occur on the scale of minutes to hours in response to moderately elevated temperatures. During long-term effects of various co-existing factors, however, morphological processes may also contribute to the plastic acclimation responses of photosynthesis (Knight and Ackerly, 2003). Reduction of specific leaf area and increased expression levels of small heat shock proteins are form adaptations of plants carbon metabolism to high temperatures. Small heat shock proteins dominate protein synthesis during and after high temperature stress. Variation between species for expression levels of the chloroplast small heat shock proteins following heat stress is positively correlated with the maintenance of PSII electron transport (Preczewskiet at al., 2000).

6.3 Effect of CO_2 concentration

Human activities have caused the concentration of atmospheric CO_2 to increase continuously from about 280 parts per million (ppm) at the beginning of the 19th century to 369 ppm at the beginning of the 21st century (Prentice et al., 2001). Future projections of atmospheric CO_2 concentration range between about 450 and 600 ppm by the year 2050 and are strongly dependent on future scenarios of anthropogenic emissions.

Long-term studies on the effects of CO_2 enrichment on plants have provided a rich suite of data and understanding about a wide variety of plant responses (McLeod and Long, 1999). Initial short-term experiments demonstrated that elevated CO_2 concentrations partially alleviated the limitation of C_3 (but not C_4) photosynthesis by CO_2 supply and acted as a negative feedback on transpiration in both C_3 and C_4 species (Long, 1991). Subsequent and often longer-term experiments have shown that photosynthesis could acclimate downwards in response to CO_2 enrichment, and there is now some evidence to suggest that photosynthesis is stimulated in C_4 species in response to CO_2 enrichment (Ghannoum et al 2000). In species with the C_3 photosynthetic pathway, high irradiance can lead to photoinhibition. Field studies have now demonstrated that CO_2 enrichment can reduce the severity of photoinhibition, although this effect is dependent on rubisco activity (Hymus et al., 2000).

6.3.1 Effect on stomatal conductance and water use efficiency

Leaf thickness generally increases whereas specific leaf area decreases as a result of CO_2 enrichment. A detailed analysis of leaf development in Scots pine (*Pinus sylvestris*) after four years of exposure to CO_2 enrichment confirmed that leaf thickness was increased but also indicated reductions in stomatal density (Lin et al., 2001). This stomatal-density response confirms observations of reduced stomatal conductance in response to CO_2 enrichment.

Recently, a gene *HIC* (*HI*gh Carbon dioxide) has been identified whose disruption leads to large increases in the number of stomata initiated in response to CO_2 enrichment (Gray et al., 2000). The *HIC* gene encodes an enzyme involved in the synthesis of those long-chain fatty acids that are typically found in the cuticle of leaves. Changes in these fatty acids may influence the cell-to-cell signaling of stomatal development. The short-distance cell-to-cell signaling of stomatal development is complimented by longer distance systemic signaling of stomatal development. The systemic signal allows the development of stomata in immature leaves to be controlled after CO_2 concentration is detected by mature leaves (Lake et al., 2001).

Early experiments demonstrated significant reductions in stomatal conductance under CO_2 enrichment. Analysis of 13 long-term (i.e. duration of more than one year) field-based studies on tree species demonstrated an overall reduction of 21% in stomatal conductance (Medlyn et al., 2001). The observation of reduced stomatal conductance was much more consistent in the longer-term than in the shorter-term studies. In combination with the partial down regulation of photosynthetic rate as the plants acclimatized to elevated CO_2, the reduction in stomatal conductance led to a 40% increase in instantaneous water use efficiency (Woodward, 2002).

6.3.2 Assimilates allocation

Carbon allocation to reproduction is strongly stimulated after a long-term of CO_2 enrichment. In a long-term study, trees growing in the enriched CO_2 were twice as likely to be reproductively mature, and produced three times more seeds than control plants growing in ambient CO_2 concentrations (LaDeau and Clark 2001). This result indicates that CO_2 enrichment hastens significantly the onset of seed production, a feature that may prove to be effective in tracking climatic change. In contrast, flowering and seed set in grasslands, where species may have deterministic life cycles, were unaffected, reduced or stimulated under CO_2 enrichment. The species-specific nature of these responses indicates a strong potential for CO_2 enrichment to change the composition of plant communities (Woodward, 2002).

6.3.3 Acclimation of plants to CO_2 enrichment

Long-term field experiments indicate that leaf photosynthesis is stimulated by CO_2 enrichment in C_3 species from 7% for legume herbs to 98% for *Pinus radiata* (Long et al., 2004). However, in the longer term when photosynthesis exceeds the capacity for carbohydrate export and utilization, a down regulation process is expected. This response is exacerbated by genetic limitations, such as determinate growth patterns, and environmental limitations, such as N deficiency or low temperature (Ainsworth et al., 2004).

The causes of photosynthetic downregulation have been variously ascribed to a reduction in carbohydrate sink strength, a limited capacity to sequester carbon in a storage form, changes in nitrogen allocation and a reduction of rubisco concentration (Woodward, 2002). These responses indicate not only a decreased expression of photosynthetic genes but also a co-ordination of the carbon to nitrogen balance (Paul and Foyer 2001). For example, nitrate and ammonium uptake, and nitrate reductase activity are sensitive to CO_2. These co-ordinating activities match photosynthetic capacity with the capacities for growth and carbon storage (Walch-Liu et al., 2001).

Respiration rates have been observed to decline, or remain unchanged with CO_2 enrichment, depending on the species. Rates of dark respiration are directly correlated with leaf nitrogen content (Hamilton et al., 2001). Therefore, when CO_2 enrichment leads to a reduction in leaf nitrogen concentration, respiration also declines. Surprisingly, CO_2 enrichment increases the average number of mitochondria in each cell, even though leaf respiration rate decreases in response to elevated CO_2 across a diverse selection of plant species (Griffin et al., 2001).

At elevated CO_2 concentrations, Rubisco content was decreased by about 20%, but in contrast there was little change in capacity for Ribulose-1,5-bisphosphate regeneration and little or no effect on photosynthetic rate. In long-term CO_2 enrichments, the loss of Rubisco cannot be explained as the result of an overall decline in leaf N, but instead appears specific and accounts for most of the decrease in N per unit of leaf area. These results suggest that loss of Rubisco is more appropriately described as an acclimatory change benefiting N use efficiency rather than as down-regulation (Long et al., 2004). Both genetic and experimental modifications of source-sink balance provide results consistent with current models of carbohydrate feedback on Rubisco expression. There is no evidence of acclimation in C_4 species under long term CO_2 enrichment, and increases in photosynthesis and production are consistent with the hypothesis that this results from improved water use efficiency. The findings have important implications both for predicting the future terrestrial biosphere and understanding how crops may need to be adapted to the changed and changing atmosphere.

7. Plasticity in CAM

Photosynthetic gas exchange pattern of CAM consists of four phases (Osmond 1978). Phase I consisted of nocturnal uptake of CO_2 via open stomata, fixation by PEPC and vacuolar storage of CO_2 in the form of organic acids, mainly malic acid. Daytime remobilization of vacuolar organic acids, decarboxylation and refixation plus assimilation of CO_2 behind closed stomata in the Calvin-cycle were named phase III. Between these two phases there are transitions when stomata remain open for CO_2 uptake for a short time during the very early light period (phase II) and reopen again during the late light period for CO_2 uptake with direct assimilation to carbohydrate when vacuolar organic acid is exhausted (phase IV) (Fig. 6).

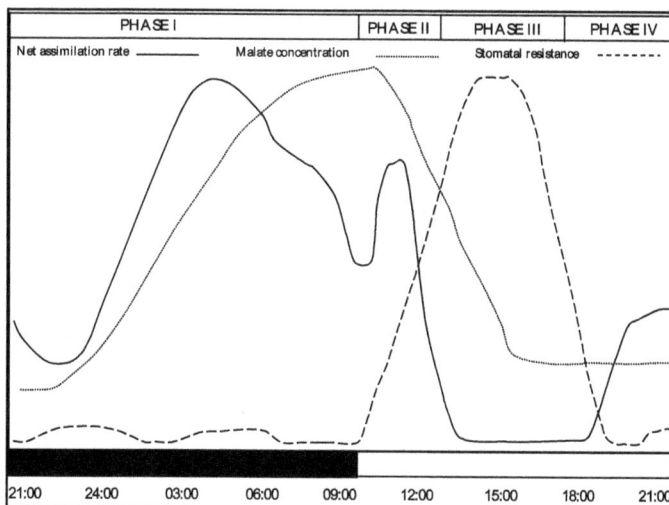

Fig. 6. Diurnal course of net CO_2 assimilation rate, malate accumulation and stomatal resistance in a CAM plant.

High internal CO_2-concentrations in phase III resulting from malate decarboxylation repress photorespiration, due to a generally low O_2/CO_2-ratio in CAM plants. The main advantage of CAM, in comparison to C_3 photosynthesis, is that it offers much higher water use efficiency, due to stomatal closure during the day. Such an adaptation is beneficial to plants living in dry and saline environments.

Plasticity in the expression of various CAM phases described above is a ubiquitous feature of the majority of CAM plants. CAM is intimately linked with the environment and can be perturbed by temperature, light level and water status. Plasticity in expression of CAM has been shown in the members of Crassulaceae. Thinner-leaved species of Crassulaceae are highly plastic in photosynthetic expression and behave like C_3 species, both in terms of the duration of diurnal atmospheric CO_2 uptake and light use efficiency. In contrast, thicker-leaved, more succulent species suffer from extreme CO_2-diffusion limitation and are more strongly bound to nocturnal CO_2 fixation for the daytime supply of carbon (Winter, 1985). Carbon isotope discrimination studies showed also that CAM photosynthesis is inducible (facultative) or constitutive (obligate) (Kluge et al., 1995). In addition, the stage of plant development affects CAM expression in plants. Several C_3/CAM intermediate species change their mode of metabolism in response to stress conditions (Dodd et al. 2002). Facultative or inducible CAM species use the C_3 pathway to maximize growth at times of sufficient water supply but switch to CAM as a means of reducing water loss while maintaining photosynthetic integrity during periods of limited water supply. This C_3/CAM intermediate pathway is predominantly found among the Aizoaceae, Crassulaceae, Portulacaceae and Vitaceae (Smith and Winter 1996). CAM expression in facultative species is primarily determined by genotype and the severity of water limitation, but may also be induced by ontogenetic and other environmental factors. Although no unique enzymes are required to facilitate the C_3–CAM transition, the imposition of water stress has a profound influence on the abundance and regulation of enzymes involved in organic acid and carbohydrate formation, turnover and intracellular transport functions (Cushman and Borland, 2002). Salt-induced C_3–CAM transition is linked with an increased activity of antioxidative enzymes. It has been suggested that, the redox status in the proximity of PSII in the C_3/CAM intermediate plants controls the expression of key genes encoding scavengers of reactive oxygen species such as superoxide dismutase, ascorbate peroxidase and activity of NADP-malic enzyme (Ślesak et al., 2002).

7.1 Effect of environmental factors on CAM

Generally, water is considered to be the most important factor and CAM to be an adaptation to water-shortage stress. However, CAM is also observed in submerged freshwater plants (Keeley, 1996). CO_2 has been considered as the central factor and most important driving force for CAM. It has been assumed that early CAM evolution occurred during geological times when atmospheric CO_2 concentration was low (Raven and Spicer, 1996). CAM is a CO_2-concentrating mechanism due to the much higher substrate affinity of PEPC for HCO_3^- than of the C_3-photosynthesis/Calvin cycle carboxylase Rubisco for CO_2. Thus, during the dark period a concentrated CO_2 pool is built up in the form of vacuolar malic acid accumulation, and during phase III its remobilization in the light leads to internal CO_2 concentrations that may be 2-60 times more than atmospheric CO_2 concentration. For aquatic plants, this CO_2-concentrating mechanism provides a benefit for CO_2 acquisition. For terrestrial plants, the benefit of CO_2-concentrating by CAM is considered to be related to water use as will be discussed below (Lüttge et al., 2004).

7.1.1 Water availability

For terrestrial plants, the greatest benefit of CAM is considered to be increased water use efficiency because stomatal opening during the dark period causes much less transpirational loss of water than opening during the light period. With this high water use efficiency, CAM plants not only inhabit arid habitats e.g. cacti, agaves and euphorbs, but also inhabit tropical rainforests. These CAM species are mainly epiphytic and subjected also to the particular problems of water supply in this habitat (Zotz and Hietz, 2001). In addition to CAM phase-dependent stomatal responses affecting WUE, CAM plants have other structural and functional ways for water storage. The large vacuolar concentrations of nocturnally accumulated organic acids are osmotically active. The increased osmotic pressure drives water uptake into the cells, which is associated with increased turgor pressure. This allows CAM plants extra acquisition of water, particularly towards the end of the dark period when vacuolar organic acid levels become rather high. It may be a particular advantage in moist, tropical forests with dew formation occurring mainly during the late dark period. During acid remobilization in phase III, osmotic and turgor pressures decline again but the water gained is available to the plants (Lüttge, 2004). CAM also occurs in some resurrection plants such as *Haberla rhodopensis* and *Ramonda serbica* (Gesneriaceae) that are desiccation-tolerant and can shift between biosis and anabiosis as they dry out and are rewatered, respectively (Markovska et al., 1997).

7.1.2 Light

Light quality and intensity affects CAM in different ways. Intensity of photosynthetically active radiation during the day (phase III) determines the rate of organic acid mobilization from the vacuole. A signaling function of light is also obvious i.e. long-day dependent induction of CAM. Phytochrome, the red-light receptor involved in photoperiodism, elicits CAM expression (Brulfert et al., 1985). In C_3/CAM intermediate species, light responses of stomata change dramatically when CAM is induced. In *Portulacaria afra*, blue-light and red-light responses of stomata in the C_3-state are lost in the CAM-state. In *M. crystallinum* after the C_3-CAM transition, the opening response of guard cells to blue and white light is lost in parallel with light-dependent xanthophyll formation. The xanthophyll zeaxanthin is involved in the signal transduction chain from light to stomatal opening (Tallman et al., 1997).

7.1.3 Salinity

One of the major effects of salinity is osmotic stress, and hence there are intimate relationships to drought stress. Therefore, considering CAM as a major photosynthetic accommodation to water stress, CAM might be expected to be a prominent trait among halophytes. Moreover, halophytes are often succulent as they sequester NaCl in large central vacuoles, which is called salt succulence (Ellenberg, 1981). However, observations do not support this expectation as, in general, halophytes are not CAM plants and CAM plants are not halophytes. Generally CAM plants, including desert succulents, are highly salt sensitive (Lüttge, 2004). CAM plants inhabiting highly saline ecosystems are either effectively functional salt excluders at the root level, such as some cacti or complete escape from the saline substrate by retreat to epiphytic niches (Lüttge, 2004). The single exception is the annual facultative halophyte and facultative CAM species *Mesembryanthemum crystallinum* (Cushman and Bohnert, 2002). This plant can grow well in the absence of NaCl but has its growth optimum at several hundred mM NaCl in the medium and can complete its life cycle at 500 mM NaCl (Lüttge, 2002).

7.2 CAM physiotypes

There are some photosynthetic physiotypes for the metabolic cycle of CAM include full CAM, CAM idling, CAM cycling, C_3/CAM and C_4/CAM (Table 1). In CAM idling stomata remain closed day and night and the day/night organic acid cycle is fed by internal recycling of nocturnally re-fixed respiratory CO_2. In CAM cycling, stomata remain closed during the dark period but some nocturnal synthesis of organic acid fed by respiratory CO_2 occurs, and stomata are open during the light period with uptake of atmospheric CO_2 and direct Calvin-cycle CO_2 reduction (C_3-photosynthesis) in addition to assimilation of CO_2 remobilized from nocturnally stored organic acid. CAM idling is considered as a form of very strong CAM, while CAM cycling is weak CAM (Sipes and Ting, 1985). In the epiphytic

Codonanthe crassifolia (Gesneriaceae), CAM cycling was observed in well-watered plants and CAM idling in drought-stressed plants. CAM cycling that scavenges respiratory CO_2 appears to be a starting point for CAM evolution (Guralnick et al., 2002). The various forms of weak and strong CAM may be restricted to different individual species or may also be expressed temporarily in one given species. For example, *Sedum telephium* has the potential to exhibit pure C_3 characteristics when well-watered and a transition to CAM when droughted, including a continuum of different stages of CAM expression which are repeatedly reversible under changing drought and watering regimes (Lee and Griffiths, 1987).

CAM physiotypes	Phase of CO_2 fixation	Phase of stomatal closure	Diel Fluctuation of malate concentration	Diel pH Fluctuation
Full CAM	I	II, III, IV	>15	High
CAM idling	---	I, II, III, IV	>15	High
CAM cycling	II, III, IV	I	>5	Low
C_3/CAM				Intermediate
C_4/CAM				Intermediate

Table 1. Various CAM physiotypes with different degrees of CAM expression.

There are true intermediate species (C_3/CAM) that can switch between full C_3 photosynthesis and full CAM. The large genus *Clusia*, comprises three photosynthetic physiotypes, i.e. C_3, C_3/CAM and CAM. There are also some C_4/CAM intermediate species, e.g. *Peperomia camptotricha*, *Portulaca oleracea* and *Portulaca grandiflora* (Guralnick et al., 2002). Only succulent C_4 dicotyledons are capable of diurnal fluctuations of organic acids, where dark-respiratory CO_2 is trapped in bundle sheaths by PEPC and the water storage tissue in the succulent leaves may also participate in the fixation of internally released CO_2. In *Portulaca*, this may be a form of CAM cycling in leaves with C_4 photosynthesis, while stems perform CAM idling (Guralnick et al., 2002). However, although C_4 photosynthesis and weak CAM occur in the same leaves, they are separated in space and do not occur in the same cells.

Compatibility of CAM and C_4 photosynthesis has been questioned (Sage, 2002a). Incompatibility of C_4 photosynthesis and CAM may be due to anatomical, biochemical and evolutionary incompatibilities. The separation of malate synthesis and decarboxylation in space in C_4 photosynthesis and in time in CAM, respectively, and the primary evolution of C_4 photosynthesis for scavenging photorespiratory CO_2 and of CAM for scavenging respiratory CO_2 (CAM cycling) may be the most important backgrounds of these incompatibilities. Although single cells may perform C_4 photosynthesis, there is intracellular compartmentation of carboxylation and decarboxylation, and these cells never perform CAM. Unlike C_3-CAM coupling, there is never C_4-CAM coupling and both pathways only occur side by side in C_4/CAM intermediate species (Sage, 2002a).

7.3 CAM evolution

CAM occurs in approximately 6% of plants, comprising monocots and dicots, encompassing 33 families and 328 genera including terrestrial and aquatic angiosperms, gymnosperms and *Welwitschia mirabilis* (Sayed, 2001). Its polyphyletic evolution was facilitated because there are no unique enzymes and metabolic reactions specifically required for CAM. CAM in the terrestrial angiosperms is thought to have diversified polyphyletically from C_3 ancestors sometime during the Miocene, possibly as a consequence of reduced atmospheric CO_2 concentration (Raven and Spicer, 1996). There is strong evidence that the evolutionary direction has been from C_3/CAM intermediates to full CAM, paralleled by specialization to and colonization of new, increasingly arid habitats (Kluge et al., 2001). A rearrangement and appropriately regulated complement of enzyme reactions present for basic functions in any green plant tissue are sufficient for performing CAM (Lüttge 2004). However, CAM-specific isoforms of key enzymes have evolved. Analysis of PEPC gene families from facultative and obligate CAM species led to the conclusion that during the induction of CAM, in addition to the existing housekeeping isoform, a CAM-specific PEPC isoform is expressed, which is responsible for primary CO_2 fixation of this photosynthetic pathway (Cushman and Bohnert 1999). A single family member of a small gene family (e.g. four to six isogenes) is recruited to fulfill the increased carbon flux demand of CAM. The recruited family member typically shows enhanced expression in CAM-performing leaves. Remaining isoforms, which presumably fulfill anapleurotic 'housekeeping' or tissue-specific functional roles, generally have lower transcript abundance and show little change in expression following water deficit. This 'gene recruitment' paradigm is likely to apply to other gene families as well (Cushman and Borland, 2002). In addition to enzymes involved in malate synthesis and mobilization, CAM induction involves large increases in carbohydrate-forming and - degrading enzymes (Häusler et al. 2000). Such activity changes are matched by corresponding changes in gene expression of at least one gene family member of glyceraldehyde-3-phosphate dehydrogenase, enolase and phosphoglyceromutase (Cushman and Borland, 2002). CAM induction causes a dramatic increase in transcripts encoding PEP-Pi and glucose-6-phosphate-Pi translocators, with expression peaking in the light period, whereas transcripts for a chloroplast glucose transporter and a triose-phosphate transporter remain largely unchanged (Häusler et al. 2000).

Duplication events appear to be the source of CAM-specific genes recruited from multigene families during CAM evolution (Cushman and Bohnert 1999). Enzyme isoforms with different subcellular locations are also thought to have evolved through gene duplication of pre-existing. Following gene duplication, modification of multipartite cis-regulatory elements within non-coding 5′☐ and 3′ flanking regions is likely to have occurred, conferring water-deficit-inducible or enhanced expression patterns for CAM-specific isogenes (Cushman and Borland, 2002).

Transcriptional activation appears to be the primary mechanism responsible for increased or enhanced expression of CAM-specific genes following water-deficit stress. Most changes in transcript abundance correlate with changes in protein amounts arising from de novo protein synthesis. Alterations in the translational efficiency of specific mRNA populations may also contribute significantly to the expression of key CAM enzymes (Cushman and Borland, 2002).

8. C_3-C_4 intermediate species

Evolution of C_4 species undoubtedly involved steps in which anatomical characteristics were between those of C_3 and C_4 species.

Evidences suggest that C_4 plants have evolved from ancestors possessing the C_3 pathway of photosynthesis and this has occurred independently over 45 times in taxonomically diverse groups (Sage, 2004). Naturally occurring species with photosynthetic characteristics intermediate between C_3 and C_4 plants have been identified in the genera *Eleucharis* (Cyperaceae), *Panicum* (Poaceae), *Neurachne* (Poaceae), *Mollugo* (Aizoaceae), *Moricandia* (Brassicaceae), *Flaveria*, (Asteraceae) *Partheniurn* (Asteraceae), Salsola (Chenopodiaceae), Heliotropium (Boraginaceae) and *Alternanthera* (Amaranthaceae) (Brown and Hattersley 1989; Rawsthorne, 1992; Voznesenskaya et al., 2001; Muhaidat, 2007). All of these genera include C_3 species and most also include C_4 species.

The intermediate nature of these species is reflected in the isotopic composition (δ^{13}), CO_2 compensation point (Γ) as well as in the differential distribution of organelles in the bundle sheath cells (Table 2).

Photosynthetic type	δ^{13}Value (‰)	Γ (μmol mol^{-1})	Organelles in bundle sheath cells (%)	
			Chloroplasts	Mitochondria +Peroxisomes
C_3	~ −30	48–62	9-11	8-19
C_3–C_4	~ −28	9–17	13-25	25-52
C_4	~ −15	3–5	28-53	30-74

Table 2. Main characteristics of C_3-C_4 species from various genera showing the intermediate nature of these species.

Intermediate species are also recognized in their CO_2 net assimilation rate as a function of intercellular CO_2 concentration and in the CO_2 compensation point as a function of O_2 concentration in the medium (Fig. 7).

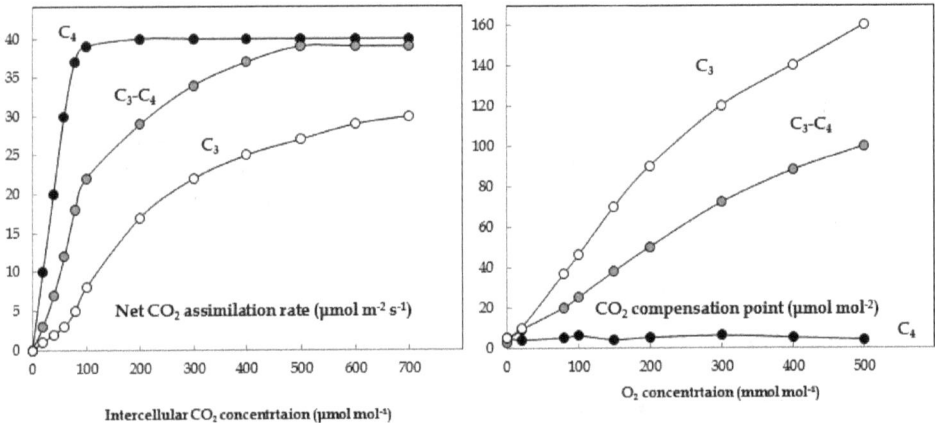

Fig. 7. Generalized curves for net assimilation rate (left) and compensation point (right) of CO_2 in C_3, C_4 and C_3-C_4 intermediate species.

8.1 Leaf anatomy

C_3-C_4 species have anatomical characteristics between those of C_3 and C_4. The vascular bundles are surrounded by chlorenchymatous bundle sheath cells reminiscent of the Kranz anatomy of leaves of C_4 plants (Fig. 8). However, the mesophyll cells are not in a concentric ring around the bundle sheath cells as in a C_4 leaf, but are arranged as in leaves of C_3 species where interveinal distances are also much greater. In all intermediate species, the bundle sheath cells contain large numbers of organelles. Numerous mitochondria, the peroxisomes and many of the chloroplasts are located centripetally in the bundle sheath cells. The mitochondria are found along the cell wall adjacent to the vascular tissue and are overlain by the chloroplasts. Quantitative studies have shown that the mitochondria and peroxisomes are four times more abundant per unit cell area than in adjacent mesophyll cells and that these mitochondria have twice the profile area of those in the mesophyll (Brown and Hattersley, 1989; McKown and Dengler, 2007, 2009).

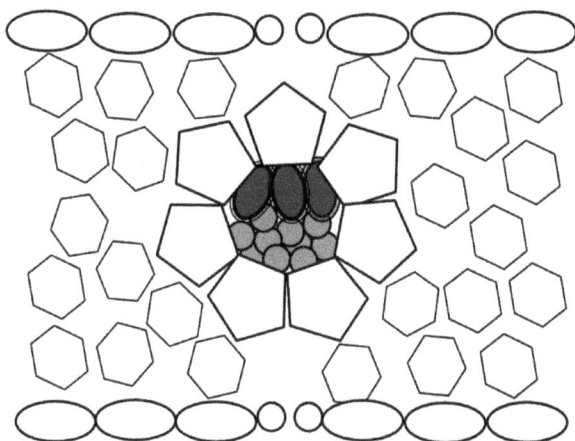

Fig. 8. Leaf anatomy in a C_3-C_4 intermediate species. Note the concentric layer of not well-developed bundle sheath cells (large hexagons) surrounded by not concentrically-arranged mesophyll cells (small hexagons).

Although some of the C_3-C_4 species, notably in *Flaveria* and *Moricandia*, do not have very well developed Kranz anatomy, they all exhibit a tendency to partition more cells to the bundle sheath and to concentrate organelles in bundle sheath cells. The tendency to partition organelles to the bundle sheath was not accomplished in a parallel way in the various C_3-C_4 species. The small bundle sheath cells in *Neurachne minor*, for example, resulted in only 5% of the total cell profile area being in the bundle sheath. But the high concentration of organelles in bundle sheath cells compensated for their small size. In other C_3-C_4 species, increased partitioning of organelles in bundle sheath cells compared to C_3 species resulted from both higher organelle concentrations and increased bundle sheath cells size and/or number relative to mesophyll cells (Brown and Hattersley, 1989; McKown and Dengler, 2007, 2009). In addition, C_3-C_4 intermediate species plasmodesmatal densities at the bundle sheath/mesophyll interface approach those of C_4 species and are much greater than those of the C_3 species studied (Brown et al, 1983).

8.2 Leaf gas exchange in C_3-C_4 intermediate species

Photosynthetic rates of C_3 and C_3-C_4 intermediate species are comparable in a range of light and atmospheric gas compositions, but the responses of gas exchange parameters which provide a measure of photorespiratory activity differ widely between these two photosynthetic groups. In contrast to C_3 plants where Γ is essentially unaffected by light intensity, Γ is strongly light-dependent in C_3-C_4 intermediate species. There is no evidence that the oxygenation reaction of Rubisco was itself being suppressed to any major extent by a C_4-like mechanism. Whereas about 50% of the photorespiratory CO_2 of a C_3 leaf is recaptured before it escapes from the leaf, it was estimated that up to 73% is recaptured in a C_3-C_4 leaf. Clearly, the improved recapture of CO_2 could account for a low Γ in C_3-C_4 species but a mechanism was required to explain how this improvement occurred (Hunt et al., 1987; Sudderth et al., 2007).

8.3 Biochemical mechanisms in C_3-C_4 intermediate species
Because of the intermediate nature of Γ and the somewhat C_4-like leaf anatomy of the C_3-C_4 species, many researchers attempted to show that these species had a partially functional C_4 cycle which accounted for their low rates of photorespiration and hence Γ. However, there is now good evidence that C_3-C_4 intermediates in the genera *Alternanthera, Moricandia, Panicum* and *Parthenium* do not have a C_4 cycle which could account for their low rates of photorespiration. Activities of PEPC and the C_4 cycle decarboxylases are far lower than in C_4 leaves, and Rubisco and PEPC are both present in mesophyll and bundle sheath cells. Label from $^{14}CO_2$ is not transferred from C_4 compounds to Calvin cycle intermediates during photosynthesis. There was clearly another explanation for low apparent photorespiration in these species. Since gas exchange measurements indicated that CO_2 was being extensively recaptured via photosynthesis, and the unusual leaf anatomy was at least in part consistent with this mechanism, the location of the photorespiratory pathway in leaves of the C_3-C_4 species has been examined (Rawsthorne, 1992).

It was shown that, the differential distribution of glycine decarboxylase is a major key to the unusual photorespiratory metabolism and Γ of C_3-C_4 intermediate species. This enzyme is abundant in the mitochondria of leaves of higher plants but is only detected at very low levels in mitochondria from other tissues. Glycine decarboxylase has four heterologous subunits (P, H, T, and L) which catalyse, in association with serine hydroxymethyltransferase, the metabolism of glycine to serine, CO_2 and ammonia. The P, H, T, and L subunits are all required for activity of gdc but the P subunit catalyses the decarboxylation of glycine. Immunocytological and in-situ hybridization studies have shown that the P subunit, is absent from the mesophyll mitochondria and the expression of the P subunit gene in the mesophyll is specifically prevented in the leaves of C_3-C_4 intermediate species. It seems likely, therefore, that the differential distribution of glycine decarboxylase must contribute to the observed reduction in apparent photorespiration in the C_3-C_4 species (Rawsthorne, 1992; Yoshimura et al., 2004).

9. Evolution of C_4 photosynthesis

C_4 photosynthesis is a series of biochemical and anatomical modifications that concentrate CO_2 around the carboxylating enzyme Rubisco. Many variations of C_4 photosynthesis exist, reflecting at least 45 independent origins in 19 families of higher plants. C_4 photosynthesis is present in about 7500 species of flowering plants, or some 3% of the estimated 250 000 land plant species. Most C_4 plants are grasses (4500 species), followed by sedges (1500 species) and dicots (1200 species). C_4 photosynthesis is an excellent model for complex trait evolution in response to environmental change (Furbank et al., 2000; Sage, 2001; Keeley and Rundel 2003; Sage, 2004; Sage et al., 2011).

Molecular phylogenies indicate that grasses were the first C_4 plants, arising about 24–34 million yr ago. Chenopods were probably the first C_4 dicots, appearing 15 –20 million yr ago. By 12-14 million yr ago, C_4 grasses were abundant enough to leave detectable fossil and isotopic signatures. By the end of the Miocene, C_4-dominated grasslands expanded across many of the low latitude regions of the globe, and temperate C_4 grasslands were present by 5 million yr ago (Cerling et al., 1999).

Rubisco and the C_3 mode of photosynthesis evolved early in the history of life and apparently were so successful that competing forms of net photosynthetic carbon fixation have gone extinct. In high CO_2 atmospheres, Rubisco operates relatively efficiently. However, the active site chemistry that carboxylates RuBP can also oxygenate i.e. photorespiration. In the current atmosphere, photorespiration can inhibit photosynthesis by over 30% at warmer temperatures ($> 30°C$). Evolving a Rubisco that is free of oxygenase activity also appears unlikely because the active site biochemistry is constrained by similarities in the oxygenase and carboxylase reactions. In the absence of further improvements to Rubisco, the other solution to the photorespiratory problem is to enhance the stromal concentration of CO_2 or to reduce O_2. Reducing O_2 is unlikely due to unfavorable energetics. Increasing CO_2 around Rubisco by 1000 ppm would nearly eliminate oxygenase activity, and under circumstances of high photorespiration could justify the additional energy costs required to operate a CO_2 pump (von Caemmerer and Furbank, 2003).

PEPC is the other major carboxylase in C_3 plants. In its current configuration, however, PEP carboxylation does not allow for net CO_2 fixation into carbohydrate, because the carbon added to PEP is lost as CO_2 in the Krebs cycle. For PEPC to evolve into a net carboxylating enzyme, fundamental rearrangements in carbon flow would also be required, while the existing role of PEPC would have to be protected or replaced in some manner (Sage, 2004).

Instead of evolving novel enzymes, CO_2 concentration requires changes in the kinetics, regulatory set points, and tissue specificity of existing enzymes. This pattern of exploiting existing biochemistry rather than inventing new enzymes is the general rule in complex trait evolution. Given these considerations, it is no surprise that the primary means of compensating for photorespiration in land plants has been the layering of C_4 metabolism over existing C_3 metabolism. All C_4 plants operate a complete C_3 cycle, so in this sense the C_4 pathway supplements, rather than replaces, C_3 photosynthesis. Because it uses existing biochemistry, the evolutionary trough that must be crossed to produce a C_4 plant is relatively shallow, and could be bridged by a modest series of incremental steps (Furbank et al., 2000; Sage, 2001; Keeley and Rundel 2003; Sage, 2004; Sage et al., 2011).

9.1 Effect of environmental factors on C_4

C_4 photosynthesis has been described as an adaptation to hot and dry environments or to CO_2 deficiency. These views, however, have been challenged in recent publications. C_4 plants do not appear to be any more drought-adapted than C_3 species from arid zones and a diverse flora of C_4 grasses occurs in the tropical wetland habitats. In addition, there is a disparity between the timing of C_4 expansion across the earth and the appearance of low atmospheric CO_2. C_4-dominated ecosystems expanded 5 and 10 million yr ago, but no obvious shift in CO_2 has been documented for this period (Cerling, 1999). Indeed, C_4

photosynthesis is not a specific drought, salinity or low-CO_2 adaptation, but it as an adaptation that compensates for high rates of photorespiration and carbon deficiency. In this context, all environmental factors that enhance photorespiration and reduce carbon balance are responsible for evolution of C_4 photosynthesis. Heat, drought, salinity and low CO_2 are the most important factors, but others, such as flooding, could also stimulate photosynthesis under certain conditions (Sage, 2004).

9.1.1 Heat. Salinity and drought

High temperature is a major environmental requirement for C_4 evolution because it directly stimulates photorespiration and dark respiration in C_3 plants. The availability of CO_2 as a substrate also declines at elevated temperature due to reduced solubility of CO_2 relative to O_2. Aridity and salinity are important because they promote stomatal closure and thus reduce intercellular CO_2 level, again stimulating photorespiration and aggravating a CO_2 substrate deficiency. Relative humidity is particularly low in hot, arid regions, which will further reduce stomatal conductance, particularly if the plant is drought stressed. The combination of drought, salinity, low humidity and high temperature produces the greatest potential for photorespiration and CO_2 deficiency (Ehleringer and Monson, 1993), so it is not surprising that these environments are where C_4 photosynthesis would most frequently arise. Many C_3-C_4 intermediates are from arid or saline zones, for example intermediate species of *Heliotropium, Salsola, Neurachne, Alternanthera* and a number of the *Flaveria* intermediates (Sage, 2004).

C_4 photosynthesis may have evolved in moist environments as well, which can be consistent with the carbon-balance hypothesis if environmental conditions are hot enough to promote photorespiration. The sedge lineages largely occur in low-latitude wetlands, indicating they may have evolved on flooded soils and the aquatic C_4 species certainly evolved in wet environments (Bowes et al., 2002). In the case of the aquatic, single-celled C_4 species, warm shallow ponds typically become depleted in CO_2 during the day when photosynthetic activity from algae and macrophytes is high. Many of the C_3-C_4 intermediates such as *Flaveria linearis, Mollugo verticillata* also occur in moist, disturbed habitats such as riverbanks, roadsides and abandoned fields indicate that disturbance is also an important factor in C_4 evolution, particularly for lineages that may have arisen in wetter locations (Monson 1989).

9.1.2 Low CO_2 concentration

In recent geological time, low CO_2 prevailed in the earth's atmosphere. For about a fifth of the period of past 400 000 yr, CO_2 was below 200 ppm. Because low CO_2 prevailed in recent geological time, discussions of C_4 evolution must consider selection pressures in atmospheres with less CO_2 than today. In low CO_2, C_3 photosynthesis is impaired by the lack of CO_2 as a substrate in addition to enhanced photorespiration (Ehleringer, 2005). As a result, water and nitrogen-use efficiencies and growth rates are low, competitive ability and fecundity is reduced and recovery from disturbance is slow (Ward, 2005). There is a strong additive effect between heat, drought and salinity and CO_2 depletion, so that, the inhibitory effects of heat, drought and salinity increase considerably in low CO_2.

Manipulation of the biosphere by human and increases in atmospheric CO_2 could halt the rise of new C_4 life forms and may lead to the reduction of existing ones (Edwards *et al.*, 2001). However, certain C_4 species are favored by other global change variables such as climate warming and deforestation. Hence, while many C_4 species may be at risk, C4 photosynthesis as a functional type should not be threatened by CO_2 rise in the near term (Sage, 2004).

9.2 Evolutionary pathways to C_4 photosynthesis

Evolution was not directed towards C_4 photosynthesis, and each step had to be stable, either by improving fitness or at a minimum by having little negative effect on survival of the genotype. The predominant mechanisms in the evolution of C_4 genes are proposed to be gene duplication followed by nonfunctionalization and neofunctionalization (Monson, 1999, 2003), and alteration of *cis*-regulatory elements in single copy genes to change expression patterns (Rosche and Westhoff, 1995). Major targets for non- and neofunctionalization are the promoter and enhancer region of genes to allow for altered expression and compartmentalization, and the coding region to alter regulatory and catalytic properties. Both non- and neofunctionalization can come about through mutations, crossover events, and insertions of mobile elements (Kloeckener-Gruissem and Freeling, 1995; Lynch & Conery, 2000). A model for C_4 evolution has been presented that recognizes seven significant phases (Sage, 2004) (Table 3).

10. Single cell C_4 photosynthesis

The term Kranz anatomy is commonly used to describe the dual-cell system associated with C_4 photosynthesis, consisting of mesophyll cells containing PEPC and initial reactions of C_4 biochemistry, and bundle sheath cells containing enzymes for generating CO_2 from C_4 acids and the C_3 carbon reduction pathway, including Rubisco. Kranz anatomy is an elegant evolutionary solution to separating the processes, and for more than three decades it was considered a requirement for the function of C_4 photosynthesis in terrestrial plants (Edwards et al., 2001).

This paradigm was broken when two species, *Borszczowia aralocaspica* and *Bienertia cycloptera*, both representing monotypic genera of the family Chenopodiaceae, were shown to have C_4 photosynthesis within a single cell without the presence of Kranz anatomy (Voznesenskaya et al., 2001; Sage, 2002b; Edwards and Voznesenskaya, 2011). *Borszczowia* grows in central Asia from northeast of the Caspian lowland east to Mongolia and western China, whereas *Bienertia* grows from east Anatolia eastward to Turkmenistan and Pakistani Baluchestan (Akhani et al., 2003).

Single-cell C_4 plants can capture CO_2 effectively from Rubisco without Kranz anatomy and the bundle sheath cell wall barrier. Photosynthesis in the single-cell systems is not inhibited by O_2, even under low atmospheric levels of CO_2, and their carbon isotope values are the same as in Kranz-type C_4 plants, whereas the values would be more negative if there were leakage of CO_2 and overcycling through the C_4 pathway (Voznesenskaya et al., 2001; Edwards and Voznesenskaya, 2011).

Borszczowia has a single layer of elongate, cylindrical chlorenchyma cells below the epidermal and hypodermal layers, which surround the veins and internal water storage tissue. The cells are tightly packed together with intercellular space restricted to the end of the cells closest to the epidermis. The anatomy of *Bienertia* leaves with respect to photosynthetic tissue is very different in that there are two to three layers of shorter chlorenchyma cells that surround the centrally located water-storage and vascular tissue in the leaf. The cells are loosely arranged, with considerable intercellular space around them (Edwards et al., 2004).

Stage	Events
General Preconditioning	**Modification of the gene copies without losing the original function:** multiplication of genes by duplication → selection and screen for adaptive functions in the short-lived annuals and perennials → reproductive barriers → genetically isolated populations.
Anatomical Preconditioning	**Decline of distance between mesophyll (MC) and bundle sheath cells (BSC) for rapid diffusion of metabolites:** reduction of interveinal distance and enhancement of BSC layer size → adaptive traits without relationship with photosynthesis: improvement of structural integrity in windy locations and enhancement of water status of the leaf in hot environments → selection. Easier reduction of MC and BSC distance in species with parallel venation (grasses) than in species with reticulate venation (dicots) → C_4 photosynthesis first arose in grasses and is prolific in this family.
Creating Metabolic Sink for Glycine Metabolism and C_4 Acids	**Increase in bundle sheath organelles:** the number of chloroplasts and mitochondria in the bundle sheath increases in order to maintain photosynthetic capacity in leaves with enlarged BSC→ increased capacity of BSC to process glycine from the mesophyll → subsequent development of a photorespiratory CO_2 pump → further increase in organelle number → greater growth and fecundity in high photorespiratory environments → maintaining incremental rise in BSC organelle content → significant reduction in CO_2 compensation points.
Glycine Shuttles and Photorespiratory CO_2 Pumps	**Changes in the glycine decarboxylase (GDC) genes:** duplication of GDC genes, production of distinct operations with separate promoters in the MC and BSC → loss of function mutation in the MC GDC → movement of glycine from MC to the BSC to prevent lethal accumulation of photorespiratory products → subsequent selection for efficient glycine shuttle.
Efficient Scavenging of CO_2 Escaping from the BSC	**Enhancement of PEPC activity in the MC**: reorganization of expression pattern of enzymes: specific expression of C_4 cycle enzymes in the MC and localization of Rubisco in BSC, increase in the activity of carboxylating enzymes: NADP-ME, NAD-ME through increasing transcriptional intensity, increased PPDK activity in the later stages.
Integration of C_3 and C_4 Cycles	**Avoidance of competition between PEPC and Rubisco in the MC for CO_2 and ATP increase in the phases of C_4 cycle:** further reorganization of the expression pattern of enzymes: reduction in the carbonic anhydrase activity in chloroplasts of BSC for preventing its conversion to bicarbonate and its diffusion out of the cell without being fixed by Rubisco, increase in the cytosol of MC to support high PEPC activity → large gradient of CO_2 between BSC and MC, reduction of MC Rubisco activity in the later stages.
Optimization and Whole-Plant Coordination	**Selection for traits that allow plants to exploit the productive potential of the C_4 pathway to the maximum:** adjustment and optimization of photosynthetic efficiency, kinetic properties and regulatory set-points of enzymes to compensate for changes in the metabolic environment: (1) Optimization of NADP-ME regulation in the earlier phases of C_4 evolution: increase in the specific activity of NADP-ME and reduction of Km for malate. (2) Optimization of PEPC in the final stages of C_4 evolution: reduction of sensitivity of PEPC to malate, increased sensitivity to the activator glucose-6-phosphate, increased affinity for bicarbonate and reduced for PEP. (3) Optimization of Rubisco: evolving into a higher catalytic capacity but lower specificity with no negative consequences. (4) Improvement of water-use efficiency: increased stomatal sensitivity to CO_2 and light→ enhancing the ability of stomata to respond to environmental variation at relatively low conductances, reduction of leaf specific hydraulic conductivity by increasing leaf area per unit of conducting tissue.

Table 3. The main evolutionary pathways towards C_4 photosynthesis (Adapted from Sage, 2004).

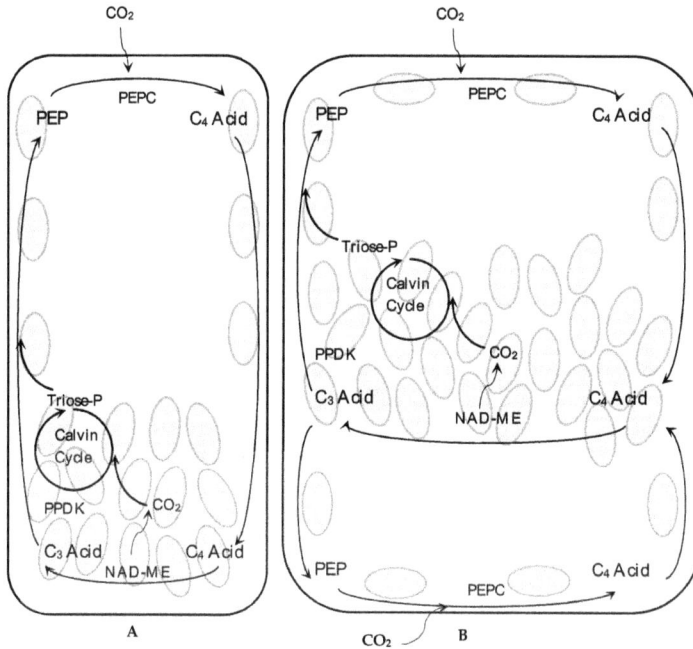

Fig. 9. Model of proposed function of C_4 photosynthesis in the two types of single cell systems in *Borszczowia* (A) and *Bienertia* (B). Note that chloroplasts are in two distinct cytoplasmic compartments.

A model has been proposed for the operation of C_4 photosynthesis in a single chlorenchyma cell in *Borszczowia* and *Bienertia* (Edwards et al., 2004; Edwards and Voznesenskaya, 2011). In *Borszczowia*, atmospheric CO_2 enters the chlorenchyma cell at the distal end, which is surrounded by intercellular air space. Here, the carboxylation phase of the C_4 pathway assimilates atmospheric CO_2 into C_4 acids. Two key enzymes in the process are pyruvate-Pi dikinase (PPDK), located in chloroplasts at the proximal part and PEPC, located in the cytosol. The C_4 acids diffuse to the proximal part of the cell through a thin, cytoplasmic space at the periphery of the middle of the cell, which is devoid of organelles. In the proximal end, the C_4 acids are decarboxylated by NAD-malic enzyme (NAD-ME) in mitochondria that appear to be localized exclusively in this part of the cell. The CO_2 is captured by Rubisco that is localized exclusively in chloroplasts surrounding the mitochondria in the proximal part of the cell (Fig. 9A).

In *Bienertia* there is a similar concept of organelle partitioning in a single cell to operate the C_4 process. However, it has a very different compartmentation scheme (Fig. 9B). Atmospheric CO_2 enters the cell around the periphery, which is exposed to considerable intercellular air space, and here the carboxylation phase of the C_4 pathway functions to convert pyruvate and CO_2 into OAA through the combined action of PPDK in the chloroplast and PEPC in the cytosol. C_4 acids diffuse to the central cytoplasmic compartment through cytoplasmic channels and are decarboxylated by NAD-ME in mitochondria, which are specifically and abundantly located there. Chloroplasts in the central cytoplasmic compartment surround the mitochondria and fix the CO_2 by Rubisco, which is only present in the chloroplasts of this compartment, through the C_3 cycle (Edwards et al., 2004; Edwards and Voznesenskaya, 2011).

Single-cell C_4 photosynthesis could simply be an alternative mechanism to Kranz type C_4 photosynthesis. Although it may be equally complex in its control of compartmentation of functions, is less complex in that it does not require the cooperative function of two cell types, nor does it require development of Kranz anatomy. Single-cell C_4 allows more flexibility in mode of photosynthesis than Kranz-type C_4 plants by, for example, shifting from C_3 to C_4 depending on environmental conditions (Edwards et al., 2004; Edwards and Voznesenskaya, 2011).

11. Conclusion

Life on earth largely depends on the photosynthetic carbon fixation using light energy. Energy-rich sugar molecules are the basis of many growth and developmental processes in plants. Reduced carbon products in the leaves, however, are used not only for synthesis of carbohydrates but also in a number of primary and secondary metabolic pathways in plants including nitrogen assimilation, fatty acid synthesis and phenolic metabolism.

Photosynthetic carbon assimilation is an investment of resources and the extent of this investment responds to the economy of the whole plant. Maintenance of energy homeostasis requires sophisticated and flexible regulatory mechanisms to account for the physiological and developmental plasticity observed in plants. It this regard, sugars not only are the prime carbon and energy sources for plants, but also play a pivotal role as a signaling molecule that control metabolism, stress response, growth, and development of plants.

Environmental factors determine the distribution and abundance of plants and evolutionary adaptation is an inevitable response to environmental change. Throughout the course of geological time, the environments in which plants grew have been changing, often radically and irreversibly. Physiological adaptation to environmental variables cannot improve without associated changes in morphology and anatomy. Evolution of C_4 plants is an excellent example of parallel evolution of leaf physiology and anatomy. Finally, any physiological evolution must be associated with changes at biochemical and molecular level. This chapter provides an introduction to theses area with a focus on plasticity in the carbon metabolism and evolution of variants of the carbon assimilation pathways.

12. References

Ainsworth, E.A.; Rogers, A.; Nelson, R. & Long, S.P. (2004). Testing the source-sink hypothesis of down-regulation of photosynthesis in elevated [CO_2] with single gene substitutions in *Glycine max*. *Agricultural and forest meteorology*, Vol. 122, pp. 85-94, ISSN 0168-1923

Akhani, H.; Ghobadnejhad, M. & Hashemi, S.M.H. (2003). Ecology, biogeography, and pollen morphology of *Bienertia cycloptera* Bunge ex Boiss. (Chenopodiaceae), an enigmatic C_4 plant without Kranz anatomy. *Plant Biology*, Vol. 5, pp. 167-78, ISSN 1435-8603

Bauwe, H. (2011). Photorespiration: The Bridge to C_4 Photosynthesis. In: *C_4 Photosynthesis and Related Concentrating Mechanisms, Advances in Photosynthesis and Respiration*, Vol.

32, Raghavendra, A. S. & Sage, R. F. (Eds.), pp. 81-108, ISBN 978-90-481-8530-6, Springer, Dordrecht, The Netherlands.

Bhagwat, A.S. (2005). Photosynthetic carbon assimilation of C_3, C_4 and CAM pathways. In: *Hand Book of Photosynthesis*, 2nd Edition, Pessarakli, M. (Ed.). pp. 376-389, ISBN 0-8247-5839-0, CRC Press, Taylor & Francis Group, Boca Raton, FL, USA.

Bowes, G.; Rao, S.K.; Estavillo, G.M. & Reiskind, J.B. (2002). C_4 mechanisms in aquatic angiosperms: comparisons with terrestrial C_4 systems. *Functional Plant Biology*, Vol. 29, pp. 379–392, ISSN 1445-4408

Brown, R.H.; Bouton, J.H.; Rigsby, L. & Rigler, M. (1983). Photosynthesis of grass species differing in carbon dioxide fixation pathways. VIII. Ultrastructural characteristics of *Panicum* species in the Laxa group. *Plant Physiology*, Vol. 71, pp. 425-431, ISSN 0032-0889

Brown, R.H. & Hattersley, P. W. (1989) Leaf anatomy of C_3-C_4 species as related to evolution of C_4 photosynthesis. *Plant Physiology*, Vol. 91, pp. 1543-1550, ISSN 0032-0889

Brulfert, J.; Vidal, J.; Keryer, E., Thomas, M.; Gadal, P. & Queiroz, O. (1985). Phytochrome control of phosphoenolpyruvate carboxylase synthesis and specific RNA level during photoperiodic induction in a CAM plant and during greening in a C_4 plant. *Physiologie Vegetale*, Vol. 23, pp. 921-928, ISSN 0570-1643

Cerling, T.E. (1999). Paleorecords of C_4 plants and ecosystems. In: C_4 *Plant Biology*, Sage, R.F. & Monson, R.K. (Eds.), pp. 445–469, ISBN 0126144400, Academic Press, San Diego, CA, USA.

Cornic, G. & Massacci, A. (1996). Leaf photosynthesis under drought stress. In: *Photosynthesis and the Environment*, Baker, N. R. (Ed.), pp. 347–366, ISBN 9780792343165, Kluwer Academic Publishers, Dordrecht, The Netherlands.

Cushman, J.C. & Bohnert, H.J. (1999). Crassulacean acid metabolism: molecular genetics. *Annual Review of Plant Physiology and Plant Molecular Biology*, Vol. 50, pp. 305-332, ISSN 1040-2519

Cushman, JC. & Bohnert, H.J. (2002). Induction of Crassulacean acid metabolism by salinity – molecular aspects. In: *Salinity: Environment – Plants – Molecules*, Läuchli, A. & Lüttge, U. (Eds.), pp 361–393, ISBN 978-90-481-5965-9, Kluwer Academic Publishers, Dordrecht, The Netherlands.

Cushman, J.C. & Borland, A.M. (2002). Induction of crassulacean acid metabolism by water limitation. *Plant, Cell and Environment*, Vol. 25, pp. 297-312, ISSN 0140-7791

Dodd, A.N.; Borland, A.M.; Haslam, R.P.; Griffith, H. & Maxwell, K. (2002). Crassulacean acid metabolism: plastic fantastic. *Journal of Experimental Botany*, Vol. 53, pp. 569-580, ISSN 0022-0957

Edwards, G.E. & Voznesenskaya, E.V. (2011). C_4 Photosynthesis: Kranz forms and single-cell C_4 in terrestrial plants. In: C_4 *Photosynthesis and Related Concentrating Mechanisms, Advances in Photosynthesis and Respiration, Vol. 32*, Raghavendra, A. S. & Sage, R. F. (Eds.), pp. 29-61, ISBN 978-90-481-8530-6, Springer, Dordrecht, The Netherlands.

Edwards, G.E.; Franceschi, V.R.; Ku, M.S.B.; Voznesenskaya, E.V.; Pyankov, V.I. & Andreo, C.S. (2001). Compartmentation of photosynthesis in cells and tissues of C_4 plants. . *Journal of Experimental Botany*, Vol. 52, pp. 577–90, ISSN 0022-0957

Edwards, G.E.; Franceschi, V.R. & Voznesenskaya, E.V. (2004). Single-cell C_4 photosynthesis versus the dual-cell (Kranz) paradigm. *Annual Review of Plant Biology*, Vol. 55, pp. 173–96, ISSN 1543-5008

Ehleringer, J.R. & Monson, R.K. (1993). Evolutionary and ecological aspects of photosynthetic pathway variation. *Annual Review of Ecology and Systematics*, Vol. 24, pp. 411–439, ISSN 0066-4162

Ehleringer, J.R. (2005). The influence of atmospheric CO_2, temperature, and water on the abundance of C_3/C_4 taxa. In: *A History of Atmospheric CO_2 and its Effects on Plants, Animals and Ecosystems*, Ehleringer, J.R., Cerling, T.E. & Dearling, D. (Eds.), pp. 214-231, ISBN 978-0-387-22069-7, Springer, Berlin, Germany.

Ellenberg, H. (1981). Ursachen des Vorkommens und Fehlens von Sukkulenten in den Trockengebieten der Erde. *Flora*, Vol. 171, pp. 114-169, ISSN 0367-2530

Farquhar, G.D.; Ehleringer, R. & Hubick, K.T. (1989). Carbon isotope discrimination and photosynthesis. *Annual Review of Plant Physiology and Plant Molecular Biology*, Vol. 40, pp. 503–37, ISSN 1040-2519

Fowler, D.B.; Breton, G.; Limin, A.E.; Mahfoozi, S. & Sarhan, F. (2001). Photoperiod and temperature interactions regulate low-temperature-induced gene expression in barley. *Plant Physiology*, Vol. 127, pp. 1676-1681, ISSN 0032-0889

Foyer, C.H.; Bloom, A.J.; Queval, G. & Noctor, G. (2009). Photorespiratory metabolism: genes, mutants, energetics, and redox signaling. *Annual Review of Plant Biology*, Vol. 60, pp. 455–484, ISSN 1543-5008

Furbank, R.T.; Hatch, M.D. & Jenkins, C.L.D. (2000). C_4 photosynthesis: mechanism and regulation. In: *Photosynthesis: Physiology and Metabolism*, Leegood, R.C., Sharkey, T.D. & von Caemmerer, S. (Eds.), pp. 435-457, ISBN 978-0-7923-6143-5, Kluwer Academic Publishers, The Netherlands.

Ghannoum, O.; Von Caemmerer, S.; Ziska, L.H. & Conroy, J.P. (2000). The growth response of C_4 plants to rising atmospheric CO_2 partial pressure: a reassessment. *Plant, Cell and Environment*, Vol. 23, pp. 931-942, ISSN 0140-7791

Giardi, M.T.; Cona, A.; Geiken, B.; Kučera, T.; Maojídek, J. & Mattoo, A.K. (1996). Long-term drought stress induces structural and functional reorganization of photosystem II. *Planta*, Vol. 199, pp. 118–125, ISSN 0032-0935

Gilmour, S.J.; Sebolt, A.M.; Salazar, M.P.; Everard, J.D. & Thomashow, M.F. (2000). Overexpression of the *Arabidopsis* CBF3 transcriptional activator mimics multiple biochemical changes associated with cold acclimation. *Plant Physiology*, Vol. 124, pp. 1854-1865, ISSN 0032-0889

Gray, J.E.; Holroyd, G.H.; Van der Lee, F.; Bahrami, A.R.; Sijmons, P.C.; Woodward, F.I.; Schuch, W. & Hetherington, A.M. (2000). The *HIC* signaling pathway links CO_2 perception to stomatal development. *Nature*, Vol. 408, pp. 713-716, ISSN 0028-0836

Griffin, K.L.; Anderson, O.R.; Gastrich, M.D.; Lewis, J.D.; Lin, G., Schuster, W.; Seemann, J.R.; Tissue, D.T.; Turnbull, M.H. & Whitehead, D. (2001). Plant growth in elevated CO_2 alters mitochondrial number and chloroplast fine structure. *Proceedings of the National Academy of Sciences USA*, Vol. 98, pp. 2473-2478, ISSN 0027-8424

Guralnick, L.J.; Edwards, G.; Ku, M.S.B.; Hockema, B. & Franceschi, V.R. (2002). Photosynthetic and anatomical characteristics in the C_4-crassulacean acid metabolism-cycling plant, *Portulaca grandiflora*. *Functional Plant Biology*, Vol. 29, pp. 763-773, ISSN 1445-4408

Guy, C.L.; Huber, J.L.A. & Huber, S.C. (1992). Sucrose phosphate synthase and sucrose accumulation at low-temperature. *Plant Physiology*, Vol. 100, pp. 502-508, ISSN 0032-0889

Hamilton, J.G.; Thomas, R.B. & DeLucia, E.H. (2001). Direct and indirect effects of elevated CO_2 on leaf respiration in a forest ecosystem. *Plant, Cell and Environment*, Vol. 24, pp. 975-982, ISSN 0140-7791

Häusler, R.E.; Baur, B., Scharte, J.; Teichmann,T.; Eicks, M.; Fischer, K.L.; Flügge, U-I.; Schubert, S.; Weber, A. & Fischer, K. (2000). Plastidic metabolite transporters and their physiological functions in the induc ible crassulacean acid metabolism plant *Mesembryanthemum crystallinum*. *The Plant Journal*, Vol. 24, pp. 285-296, ISSN 0960-7412

Hobbie, E.A. & Werner, R.A. (2004). Intramolecular, compound-specific and bulk carbon isotope patterns in C_3 and C_4 plants: a review and synthesis. *New Phytologist*, Vol. 161, pp. 371-385, ISSN 0028-646X

Holtum, J.A.M.; Smith, J.A.C. & Neuhaus, H.E. (2005). Intracellular transport and pathways of carbon flow in plants with crassulacean acid metabolism. *Functional Plant Biology*, Vol. 32, pp. 429-449, ISSN 1445-4408

Hopkins, W.G. & Hüner, N.P.A. (2004). Introduction to Plant Physiology. ISBN 0-471-38915-3, John Wiley & Sons, Inc., USA.

Huber, S.C. & Huber, J.L. (1996). Role and Regulation of sucrosephosphate synthase in higher plants. *Annals of Review of Plant Physiology and Plant Molecular Biology*, Vol. 47, pp. 431-444, ISSN

Hunt, S.; Smith, A.M. & Woolhouse, H.W. (1987). Evidence for a lightdependent system for reassimilation of photorespiratory CO_2, which does not include a C_4 cycle, in the C_3-C_4 intermediate species *Moricandia arvensis*. *Planta*, Vol. 171, pp. 227-234, ISSN 0032-0935

Hurry, V.; Strand, Å.; Furbank, R. & Stitt, M. (2000). The role of inorganic phosphate in the development of freezing tolerance and the acclimatization of photosynthesis to low temperature is revealed by the *pho* mutants of *Arabidopsis thaliana*. *The Plant Journal*, Vol. 24, pp. 383-396, ISSN 0960-7412

Hymus, G.J.; Dijkstra, P.; Baker, N.R.; Drake, B.G. & Long, S.P. (2000). Will rising CO_2 protect plants from the midday sun? A study of photoinhibition of *Quercus myrtifolia* in a scrub-oak community in two seasons. *Plant, Cell and Environment, Vol.* 24, pp. 1361-1368, ISSN 0140-7791

Keeley, J.E. & Rundel, P.W. (2003). Evolution of CAM and C_4 carbon-concentrating mechanisms. *International Journal of Plant Sciences*, Vol. 164, pp. S55-S77, ISSN 1058-5893

Keeley, J.E. (1996). Aquatic CAM photosynthesis. In: *Crassulacean Acid Metabolism. Biochemistry, Ecophysiology and Evolution*, Winter, K. & Smith, J.A.C. (Eds.), pp. 281-295, ISBN 3540581049, Springer, Berlin, Germany.

Kloeckener-Gruissem, B. & Freeling, M. (1995). Transposon-induced promoter scrambling: a mechanism for the evolution of new alleles. *Proceedings of the National Academy of Sciences USA*, Vol. 92, pp. 1836-1840, ISSN

Kluge, M.; Brulfert, J.; Rauh, W.; Ravelomanana, D. & Ziegler, H. (1995). Ecophysiological studies on the vegetation of Madagascar: a $\delta^{13}C$ and δD survey for incidence of Crassulacean acid metabolism (CAM) among orchids from montane forests and succulents from the xerophytic thorn-bush. *Isotopes in Environmental and Health Studies*, Vol. 31, pp. 191-210, ISSN 1025-6016

Kluge, M.; Razanoelisoa, B. & Brulfert, J. (2001). Implications of genotypic diversity and phenotypic plasticity in the ecophysiological success of CAM plants, examined by studies on the vegetation of Madagascar. *Plant Biology*, Vol. 3, pp. 214-222, ISSN 1435-8603

Knight, C.A. & Ackerly, D.D. (2003). Evolution and plasticity of photosynthetic thermal tolerance, specific leaf area and leaf size: congeneric species from desert and coastal environments. *New Phytologist*, Vol. 160, pp. 337-347, ISSN 0028-646X

LaDeau,. SL. & Clark, J.S. (2001). Rising CO_2 levels and the fecundity of forest trees. *Science*, Vol. 292, pp. 95-98, ISSN 0036-8075

Lake, J.A.; Quick, W.P.; Beerling, D.J. & Woodward, F.I. (2001). Signals from mature to new leaves. *Nature*, Vol. 411, p. 154, ISSN 0028-0836

Lawlor, D.W. & Cornic, G. (2002). Photosynthetic carbon assimilation and associated metabolism in relation to water deficits in higher plants. *Plant, Cell and Environment*, Vol. 25, pp. 275–294, ISSN 0140-7791

Lee, H.S.J & Griffiths, H. (1987). Induction and repression of CAM in *Sedum telephium* L. in response to photoperiod and water stress. *Journal of Experimental Botany*, Vol. 38, pp. 834-841, ISSN 0022-0957

Lin, J.; Jach, M.E. & Ceulemans, R. (2001). Stomatal density and needle anatomy of Scots pine (*Pinus sylvestris*) are affected by elevated CO_2. *New Phytologist*, Vol. 150, pp. 665-674, ISSN 0028-646X

Long, S.P. (1991). Modification of the response of photosynthetic productivity to rising temperature by atmospheric CO_2 concentrations: Has its importance been underestimated? *Plant, Cell and Environment*, Vol. 14, pp. 729-739, ISSN 0140-7791

Long, S.P.; Ainsworth, E.A.; Rogers, A. & Ort, D.R. (2004). Rising atmosphere carbon dioxide: Plants FACE the Future. *Annual Review of Plant Biology*, Vol. 55, pp. 591-628, ISSN 1543-5008

Lüttge, U. (2004). Ecophysiology of crassulacean acid metabolism (CAM). *Annals of Botany*, Vol. 93, pp. 629-652, ISSN 0305-7364

Lüttge, U. (2002). Performance of plants with C_4-carboxylation modes of photosynthesis under salinity. In: *Salinity: Environment – Plants – Molecules*, Läuchli, A. & Lüttge, U. (Eds.), pp 113-135, ISBN 978-90-481-5965-9, Kluwer Academic Publishers, Dordrecht, The Netherlands.

Lynch, M. & Conery, J.S. (2000). The evolutionary fate and consequences of duplicate genes. *Science*, Vol. 290, pp. 1151-1155, ISSN

Markovska, Y.; Tsonev, T. & Kimenov, G. (1997). Regulation of CAM and respiratory recycling by water supply in higher poikilohydric plants *Haberlea rhodopensis* Friv.

and *Ramonda serbica* Pancic at transition from biosis to anabiosis an vice versa. *Botanica Acta*, Vol. 110, pp. 18-24, ISSN 0932-8629

McKown, A.D. & Dengler, N.G. (2007). Key innovations in the evolution of Kranz anatomy and C_4 vein pattern in *Flaveria* (Asteraceae). *American Journal of Botany*, Vol. 94, pp. 382-399, ISSN 0002-9122

McKown, A.D. & Dengler, N.G. (2009). Shifts in leaf vein density through accelerated vein formation in C_4 *Flaveria* (Asteraceae). *Annals of Botany*, Vol. 104, pp. 1085-1098, ISSN 0305-7364

McKown, R.; Kuroki, G. &Warren, G. (1996). Cold responses of *Arabidopsis* mutants impaired in freezing tolerance. *Journal of Experimental Botany*, Vol. 47, pp. 1919-1925, ISSN 0022-0957

McLeod, A.R. & Long, S.P. (1999). Free-air carbon dioxide enrichment (FACE) in Global Change Research: A review. *Advances in Ecological Research*, Vol. 28, pp. 1–55, ISSN 0065-2504

Medlyn, B.E.; Barton, C.V.M.; Broadmeadow, M.S.J.; Ceulemans, R.; De Angelis, P.; Forstreuter, M.; Freeman, M.; Jackson, S.B.; Kellomäki, S.; Laitat, Rey, A.; Robertnz, P.; Sigurdsson, B.D.; Strassemeyer, J.; Wang, K.; Curtis, P.S. & Jarvis, P.J. (2001). Stomatal conductance of forest species after long-term exposure to elevated CO_2 concentration: a synthesis. *New Phytologist*, Vol. 149, pp. 247-264, ISSN 0028-646X

Monson, R.K. (1989). On the evolutionary pathways resulting in C_4 photosynthesis and crassulacean acid metabolism (CAM). *Advances in Ecological Research*, Vol. 19, pp. 57-101, ISSN 0065-2504

Monson, R.K. (1999). The origins of C_4 genes and evolutionary pattern in the C_4 metabolic phenotype. In: *C_4 Plant Biology*, Sage, R.F. & Monson, R.K. (Eds.), pp. 377–410, ISBN 0126144400, Academic Press, San Diego, CA, USA.

Monson, R.K. (2003). Gene duplication, neofunctionalization, and the evolution of C_4 photosynthesis. *International Journal of Plant Science*, Vol. 164, pp. S43–S54, ISSN 1058-5893

Muhaidat, R. (2007). Diversification of C_4 Photosynthesis in the Eudicots: Anatomical, Biochemical and Physiological Perspectives. PhD Thesis, University of Toronto, Toronto, Canada.

Müller, P.; Li, X.P. & Niyogi, K.K. (2001). Non-photochemical quenching. A response to excess light energy. *Plant Physiology*, Vol. 125, pp. 1558-1566, ISSN 0032-0889

Nielsen, T.H.; Krapp, A.; Roper-Schwarz, U. & Stitt, M. (1998). The sugar-mediated regulation of genes encoding the small subunit of Rubisco and the regulatory subunit of ADP glucose pyrophosphorylase is modified by phosphate and nitrogen. *Plant, Cell and Environment*, Vol. 21, pp. 443-454, ISSN 0140-7791

Osmond, C.B. (1978). Crassulacean acid metabolism: a curiosity in context. *Annual Review of Plant Physiology*, Vol. 29, pp. 379-414, ISSN 0066-4294

Paul, M.J. & Foyer, C.H. (2001). Sink regulation of photosynthesis. *Journal of Experimental Botany*, Vol. 52, pp. 1383-1400, ISSN 0022-0957

Pelleschi, S.; Rocher, J.P. & Prioul, J.-L. (1997). Effect of water restriction on carbohydrate metabolism and photosynthesis in mature leaves. *Plant, Cell and Environment*, Vol. 20, pp. 493-503, ISSN 0140-7791

Preczewski, P.J.; Heckathorn, S.A.; Downs, C.A. & Coleman, J.S. (2000). Photosynthetic thermotolerance is quantitatively and positively associated with production of specific heat-shock proteins among nine genotypes of *Lycopersicon* (tomato). *Photosynthetica*, Vol. 38, pp. 127-134, ISSN 0300-3604

Prentice, C.; Farquhar, G.; Fasham, M.; Goulden, M.; Heimann, M.; Jaramillo, V.; Kheshgi, H.; Le Quéré, C.; Scholes, R. & Wallace, D. (2001). The carbon-cycle and atmospheric carbon dioxide. In: *Climate Change 2001: The Scientific Basis*, Houghton, J.T.; Ding, Y., Griggs, D.J., Noguer, M. & van der Linden P.J. (Eds.), pp. 183–237, ISBN 0521014956, Cambridge University Press, Cambridge, UK.

Raven, J.A. & Spicer, R.A. (1996). The evolution of Crassulacean acid metabolism. In: *Crassulacean Acid Metabolism. Biochemistry, Ecophysiology and Evolution*, Winter, K. & Smith, J.A.C. (Eds.), pp. 360-385, ISBN 3540581049, Springer, Berlin, Germany.

Rawsthorne, S. (1992). C_3-C_4 intermediate photosynthesis: linking physiology to gene expression. *The Plant Journal*, Vol. 2, pp. 267-274, ISSN 0960-7412

Rosche, E. & Westhoff, P. (1995). Genomic structure and expression of the pyruvate, orthophosphate dikinase gene of the dicotyledonous C_4 plant *Flaveria trinervia* (Asteraceae). *Plant Molecular Biology*, Vol. 29, pp. 663-678, ISSN 0167-4412

Sage, R.F. (2004). The evolution of C_4 photosynthesis. *New Phytologist*, Vol. 161, pp. 341-370, ISSN 0028-646X

Sage, R.F. (2001). Environmental and evolutionary preconditions for the origin and diversification of the C_4 photosynthetic syndrome. *Plant Biology*, Vol. 3, pp. 202–213, ISSN 1435-8603

Sage, R.F. (2002)a. Are crassulacean acid metabolism and C_4 photosynthesis incompatible? *Functional Plant Biology*, Vol. 29, pp. 775-785, ISSN 1445-4408

Sage, R.F. (2002)b. C_4 photosynthesis in terrestrial plants does not require Kranz anatomy. *Trends in Plant Science*, Vol. 7, pp. 283-85, ISSN 1360-1385

Sayed, O.H. (2001) Crassulacean acid metabolism 1975-2000 a check list. *Photosynthtica*, Vol. 39, pp. 339-352, ISSN 0300-3604

Sage, R.F.; Kocacinar, F. & Kubien, D. S. (2011). C_4 photosynthesis and temperature. In: *C_4 Photosynthesis and Related Concentrating Mechanisms*, Raghavendra, A. S. & Sage, R. F. (Eds.), pp. 161-195, ISBN 978-90-481-8530-6, Springer, Dordrecht, The Netherlands.

Sipes, D.L. & Ting, I.P. (1985). Crassulacean acid metabolism and crassulacean acid metabolism modifications in *Peperomia camptotricha*. *Plant Physiology*, Vol. 77, pp. 59-63, ISSN 0032-0889

Ślesak, I.; Miszalski, Z.; Karpinska, B.; Niewiadomska, E.; Ratajczak, R. & Karpinski, S. (2002). Redox control of oxidative stress responses in the C_3/CAM intermediate plant *Mesembryanthemum crystallinum*. *Plant Physiology and Biochemistry*, Vol. 40, pp. 669-677, ISSN 0981-9428

Smith, J.A.C. & Winter, K. (1996). Taxonomic distribution of Crassulacean acid metabolism. In: *Crassulacean Acid Metabolism. Biochemistry, Ecophysiology and Evolution*, Winter, K. & Smith, J.A.C. (Eds.), pp. 427–436, ISBN 3540581049, Springer, Berlin, Germany.

Stitt, M. (1996). Metabolic regulation of photosynthesis. In: *Photosynthesis and the Environment*, Baker N.R. (Ed.), pp. 151-190, ISBN 9780792343165, Kluwer Academic Publishers, Dordrecht, The Netherlands.

Strand, Å.; Hurry, V., Gustafsson, P. & Gardeström, P. (1997). Development of *Arabidopsis thaliana* leaves at low temperatures releases the suppression of photosynthesis and photosynthetic gene expression despite the accumulation of soluble carbohydrates. *The Plant Journal*, Vol. 12, pp. 605-614, ISSN 0960-7412

Strand, Å.; Hurry, V.; Henkes, S.; Huner, N.; Gustafsson, P.; Gardeström, P. & Stitt, M. (1999). Acclimation of *Arabidopsis* leaves developing at low temperatures. Increasing cytoplasmic volume accompanies increased activities of enzymes in the Calvin cycle and in the sucrose-biosynthesis pathway. *Plant Physiology*, Vol. 119, pp. 1387-1397, ISSN 0032-0889

Sudderth, E.A.; Muhaidat, R.; McKown, A.D.; Kocacinar, F. & Sage, R.F. (2007). Leaf anatomy, gas exchange and photosynthetic enzyme activity in *Flaveria kochiana*. *Functional Plant Biology*, Vol. 34, pp. 118-129, ISSN 1445-4408

Tallman, G.; Zhu, J.; Mawson, B.T.; Amodeo, G.; Nouki, Z.; Levy, K. & Zeiger, E. (1997). Induction of CAM in *Mesembryanthemum crystallinum* abolishes the stomatal response to blue light and light-dependent zeaxanthin formation in guard cell chloroplasts. *Plant and Cell Physiology*, Vol. 38, pp. 236-242, ISSN 0032-0781

Tezara, W.; Mitchell, V.J.; Driscoll, S.P. & Lawlor, D.W. (1999). Water stress inhibits plant photosynthesis by decreasing coupling factor and ATP. *Nature*, Vol. 401, pp. 914–917, ISSN 0028-0836

von Caemmerer, S. & Furbank, R.T. (2003). The C_4 pathway: an efficient CO_2 pump. *Photosynthesis Research*, Vol. 77, pp. 191-207, ISSN 0166-8595

von Caemmerer, S. (2000). Biochemical Models of Leaf Photosynthesis. ISBN 9780643063792, CSIRO Publishing, Collingwood, Australia.

Voznesenskaya, E.V.; Artyusheva, E.G.; Franceschi, V.R.; Pyankov, V.I.; Kiirats, O.; Ku, M.S.B. & Edwards, G.E. (2001). *Salsola arbusculiformis*, a C_3-C_4 intermediate in Salsoleae (Chenopodiaceae). *Annals of Botany*, Vol. 88, pp. 337-348, ISSN 0305-7364

Voznesenskaya, E.V.; Franceschi, V.R.; Kiirats, O.; Freitag, H .& Edwards, G.E. (2001). Kranz anatomy is not essential for terrestrial C_4 plant photosynthesis. *Nature*, Vol. 414, pp. 543–46, ISSN 0028-0836

Walch-Liu, P.; Neumann, G. & Engels, C. (2001). Elevated atmospheric CO_2 concentration favors nitrogen partitioning into roots of tobacco plants under nitrogen deficiency by decreasing nitrogen demand of the shoot. *Journal of Plant Nutrition*, Vol. 24, pp. 835-854, ISSN 0190-4167

Ward, J. (2005). Evolution and growth of plants in a low CO_2 world. In: *A History of Atmospheric CO2 and its Effects on Plants, Animals and Ecosystems*, Ehleringer, J.R., Cerling, T.E. & Dearling, D. (Eds.), pp. 232-257, ISBN 978-0-387-22069-7, Springer, Berlin, Germany.

Winter, K. (1985). Crassulacean acid metabolism. In: *Photosynthetic Mechanisms and the Environment*, Barber, J. & Baker, N.R. (Eds.), pp. 329-387, Elsevier, Amsterdam, The Netherlands, ISBN 0444806741

Woodward, F.I. (2002). Potential impacts of global elevated CO_2 concentrations on plants. *Current Opinion in Plant Biology*, Vol. 5, pp. 207-211, ISSN 1369-5266

Yoshimura, Y., Kubota, F. & Ueno, O. (2004). Structural and biochemical bases of photorespiration in C_4 plants: quantification of organelles and glycine decarboxylase. *Planta*, Vol. 220, pp. 307-317, ISSN 0032-0935

Zotz, G. & Hietz, P. (2001). The physiological ecology of vascular epiphytes: current knowledge, open questions. *Journal of Experimental Botany*, Vol. 52, pp. 2067-2078, ISSN 0022-0957

5

Photosynthesis and Quantum Yield of Oil Palm Seedlings to Elevated Carbon Dioxide

H.Z.E. Jaafar and Mohd Hafiz Ibrahim
Universiti Putra Malaysia (UPM)
Malaysia

1. Introduction

Photosynthesis is a metabolic process through which green plants synthesize organic compounds from inorganic raw materials in the presence of sunlight. This process can be regarded as a procedure of converting radiant energy of the sun into chemical energy of plant tissues in the form of organic molecules. Photosynthesis increases the total free energy available to organism and provides energy to the world, directly or indirectly, necessary for sustaining all forms of life on earth. Farming is basically a system of exploiting solar energy to synthesize organic matter through photosynthesis. The yield of crop plants ultimately depends on the size and efficiency of their photosynthetic system (Anderson, 2000). The most important factors of biomass production of any crop are the amount of radiation intercepted by the crop and the effectiveness of using the radiation in dry matter production. All organisms on earth need energy for growth and maintenance. As a result, higher plants, algae and certain types of bacteria capture this energy directly from the sunlight and utilise it for the biosynthesis of essential food materials for dry matter increase. The plant photosynthetic apparatus contain the necessary pigments in leaf able to absorb light and channel the energy of the excited pigment molecules into a series of photochemical and enzymatic reactions. Light energy is absorbed by protein-bound chlorophylls that are located in light-harvesting complexes and the energy migration to photosynthetic reaction centres results in electron excitation and transfer to other components of the electron transfer chain (Hall and Rao, 1999).

Carbon dioxide is a trace gas in the atmosphere, presently accounting for about 0.037%, or 370 parts per million (ppm), of air. The partial pressure of ambient CO_2 (Ca) varies with atmospheric pressure and is approximately 36 pascals (Pa) at sea level. The current atmospheric concentration of CO_2 is almost twice the concentration that has prevailed during most of the last 160,00 years, as measured from air bubbles trapped in glacial ice in Antarctica. For the last 200 years, CO_2 concentrations during the recent geological past have been low, fluctuating between 180 and 260 ppm. These low concentrations were typical of times extending back to the Cretaceous, when Earth was much warmer and the CO_2 concentration may have been as high as 1200 to 2800 ppm (Ehleringer et al., 1991). However, with the rapid increases in world population and economic activity, a doubling of the present atmospheric [CO_2], assuming a mean annual increase rate of 1.5 ppm, which was observed over the past decade 1984–1993 (Stoskoptf, 1981), could be expected before the end of the 21st century (Baker and Ort, 1992). Rising atmospheric $[CO_2]$ could benefit many economically important crops, especially the C3; however, gains may or may not be realized in long-term growth because of the interaction of various environmental factors that complicate the issue (Farquhar and Sharkley, 1982). The current CO_2 concentration of the atmosphere is increasing by about 1 ppm each year, primarily because of the burning of fossil fuels. Since 1958, when systematic measurements of CO_2 began at Mauna Loa, Hawaii, atmospheric CO_2 concentrations have increased by more than 17% (Keeling et al. 1995), and by 2020 the atmospheric CO_2 concentration could reach up to 600 ppm.

With the increase in $[CO_2]$, many crops may be affected either positively or negatively. Oil palm is an industrial perennial plant widely cultivated in Southeast Asia where it plays a major role in the economics of the regional income. Claimed to be the most productive oil bearing plant as compared to coconut, olive, rapeseed and soybean, the crop has contributed about 8.2% of Malaysian gross domestic products (GDP) and the second largest economic contributor after exported goods and petroleum. In 2009, large area of about 4.6 million ha had been cultivated with oil palm. As the concentration of CO_2 is expected to increase to 600 ppm by 2020, the productivity of oil palm could also be increased. However, the research on oil palm acclimation to increased CO_2 issue is still lacking especially in the leaf gas exchange aspects.

Net photosynthesis and quantum yield are good indicators of plant acclimation to elevated CO_2. The notion of photosynthetic efficiency in the literature involves some different terms including photosynthetic rate, quantum yield of carbon assimilation and photochemical efficiency of PSII, which is often expressed as a ratio of variable to maximal fluorescence (Xu and Shen, 2000). These terms are different but they linked to each other. Both photosynthetic rate and quantum yield are related to characteristics of the leaf, cell, and chloroplast itself and the environmental conditions. Photosynthetic rate is often expressed as number of molecules of CO_2 fixed or O_2 evolved per unit leaf area per unit time while quantum yield is expressed as number of molecules of CO_2 fixed or O_2 evolved per photon absorbed. The efficiency of photosynthesis of the whole plant is crucial to agriculture, forestry, ecology, etc. when it comes to analyzing productivity for food and fuels and many other product users. The quality and quantity of photosynthetic incident light (or photosynthetic active radiation, PAR), temperature and water availability, mineral nutrients availability and utilization, photorespiratory losses, presence of pollutants in the atmosphere and in the soil (heavy metals), etc., are some of the factors that affect plant productivity. How these factors interact with the changing environment is now the subject of much practical and basic research.

This chapter focuses on the photosynthetic responses, particularly net photosynthesis and apparent quantum yield, of oil palm to enhanced growth $[CO_2]$. The leaf gas exchange and apparent quantum yield response of oil palm seedling to elevated $[CO_2]$ will be discussed. The net photosynthesis and apparent quantum yield data are directly collected from LICOR 6400 using light response curve analysis. As the photosynthetic mechanism of a plant species is the major determinant of how it will respond to rising atmospheric $[CO_2]$, understanding the mechanisms of photosynthesis acclimation to rising $[CO_2]$ could potentially be translated into a basic framework for improving the efficiency of crop production in a future climate-changed world.

2. Factors limiting photosynthetic rates

A number of external environmental factors can influence the rate of photosynthesis, leading to up-regulation or down-regulation of photosynthetic capacity. These factors might be low or high temperature, deficiency or over supply of water or nutrient, low CO_2 or high O_2 concentration, and low light intensity. In the meantime, many plant internal factors including developmental hormones and respiration, etc. may also have a significant effect on net photosynthetic rate. The main limitation site of net photosynthetic rate in C3 plants, however, is often in the reaction centre catalyzed by the enzyme ribulose-1,5-bisphosphate (RuBP) carboxylase/oxygenase (Rubisco). Therefore, reducing or eliminating its oxygenase function or photorespiration or increasing the affinity of the enzyme for CO_2 is a long-term goal to increase productivity of the plant (Xu and Shen, 2000).

3. Quantum yield

The quantum yield of a process in which molecules give up their excitation energy (known as "decay") is the fraction of excited molecules that decay via that pathway (Wells et al., 1982; Taiz and Zieger, 1991). The quantum yield of a process, such as photochemistry, has been defined (Clayton 1971; 1980) mathematically as the yield of photochemical products divided by total number of quanta absorbed. For a particular process, the value of quantum yield can range between 0 (when the process does not involve any decay of the excited state) and 1.0 (when the process involves deactivation of the excited state). Taiz and Zieger (1991) explained that all possible processes would ultimately contribute to a sum 1.0 of the quantum yields. Basically, in functional chloroplast that is kept in dim light, the quantum yield of photochemistry is approximately 0.95, the quantum yield of fluorescence is 0.05 or lower, and the quantum yields of other processes are negligible. The vast majority of excited chlorophyll molecules, therefore, lead to photochemistry.

The reciprocal of the quantum yield is called the quantum requirement. For a high yielding crop, it is not only a high photosynthetic rate in strong light is important but a high quantum yield in weak light is also crucial. Therefore, Ort and Baker (1988) believed that the improvement of future crop production should aim at increasing their quantum yield. It is well known that in photosynthesis of the Calvin-Benson cycle, the assimilation of one molecule of CO_2 into carbohydrate requires 2NADPH and 3ATP. The production of 2NADPH is the result of transporting four electrons from $2H_2O$ to 2NADP along an electron transport chain. Because the chain includes two photosystems in series, two photons are needed for one electron transport. Thus, at least eight photons are required for the production of 2NADPH. Therefore, the maximal or theoretical quantum yield for photosynthetic carbon assimilation is 0.125 mole CO_2/mole photons. In field studies Xu (1988) found that the apparent quantum yield of photosynthetic carbon assimilation often displayed a significant midday decline in many C3 plants such as soybean and wheat but not in C4 plants such as maize and sorghum. It was deduced that photoinhibition may be a cause of the midday decline of the photosynthetic efficiency (Xu et al. 1990). The molecular mechanism of photoinhibition, however, is still not fully understood. For more than a decade photoinhibition has been considered almost synonymous with photodamage to the photosynthetic apparatus (Nigoyi, 1999). In addition to photoinhibition, enhanced photorespiration is another cause of the midday decline in the photosynthetic efficiency of C3 plants (Guo et al., 1994). For a long time photorespiration has been considered a wasteful process. Many efforts have been made to eliminate or reduce photorespiration but no success has yet been reported. Extensive screening programs involving several species (wheat, barley, oats, soybean, potato, tall fescue) failed to identify genotypes with low CO2 compensation point (Hay and Walker, 1989). Attempts to select C3 plants with low rates of photorespiration and high rates of net photosynthesis also have had little success (Xu and Shen, 2000).

3.1 Factors affecting quantum yield
3.1.1 External factors

Emerson and Lewis (1943) showed that the values of quantum yield were related to the quality of light. A high quantum yield was measured at red light around 680 nm. The quantum yields of sun and shade leaves grown under different light intensities were similar, although there was a significant difference in light saturated photosynthetic rate between them (Oquist and Hallgren, 1982). At 21% O_2 and a temperature range of 15–35°C the quantum yield decreased gradually with temperature increase in C3 plants but not in C4 plants (Ku and Edwards, 1978; Xu and Shen, 2000). Water deficiency and excessive water or flooding could lead to a decline in quantum yield (Mohanty and Boyer, 1976). After several rainy days, the photosynthetic quantum efficiency became lower in spinach leaves (Li et al. 1991). The reason may be that the reduction of NADP is severely hindered in swollen chloroplasts under hypotonic conditions (Ye et al., 1995). Decreasing O_2 concentration or increasing CO_2 concentration in air could increase quantum yield in C3 plants but not in C4 plants (Monson et al., 1982). This may be due to decreased excitation energy transport from antenna pigments to PSII reaction centers and enhanced excitation energy dissipation as heat under phosphate deficiency conditions (Jacob, 1995; Xu and Shen, 2000).

3.1.2 Internal factors

Among all internal factors, photorespiration has the most significant effect on quantum yield. The effects of air temperature and CO_2 or O_2 concentration on quantum yield mentioned earlier, in fact, are related to the changes in photorespiratory rate caused by these factors. In normal air and at 20–25°C, the quantum yields of C3 and C4 plants were similar. However, when the air temperature was over 30°C, the quantum yield in C4 plants was slightly higher than that in C3 plants (Ehleringer and Pearcy, 1983). When photorespiration was inhibited by high CO_2 and/or low O_2, C4 plants had about 30% lower quantum yields than C3 plants because they used two additional ATP molecules in the C4 pathway for fixation of one molecule of CO_2 to form carbohydrate (Osmond et al. 1980; Xu and Shen, 2000).

4. C$_3$ species response to elevated CO$_2$

The present atmospheric [CO_2] limits the photosynthetic capability, growth, and yield of many agricultural crop plants, among which the C3 species show the greatest potential for response to rising [CO_2] (Xu et al., 1984). Current atmospheric CO_2 and O_2 levels and C3 Rubisco specificity factors translate into photorespiratory losses of 25% or more for C3 species (Farquhar and Sharkey, 1982). The projection that a rise in atmospheric [CO_2] will reduce the deleterious effect of O_2 on C3 photosynthesis but that it has a negligible effect on C4 photosynthesis is indeed supported by experimental growth data. Exposure of C3 plants to elevated [CO_2] generally results in stimulated photosynthesis and enhanced growth and yield (Sharkawy et al., 1990).

A compilation of the existing data available from the literature for C3 agricultural crops, including agronomic, horticultural, and forest tree species, shows an average enhancement in net CO_2 exchange rates up to 63% and growth up to 58% with a doubling of the present atmospheric $[CO_2]$ (Brinkman and Frey, 1978). Long-term exposure to elevated $[CO_2]$ leads to a variety of acclimation effects, which include changes in the photosynthetic biochemistry and stomatal physiology and alterations in the morphology, anatomy, branching, tillering, biomass, and timing of developmental events as well as life cycle completion (Evan and Dunstone, 1970). A greater number of mesophyll cells and chloroplasts have been reported for plants grown under elevated $[CO_2]$ (Poskuta and Nelson, 1986). In terms of dark respiration, the exposure of plants to elevated $[CO_2]$ usually results in lowering the dark respiration rate, which can be explained by both direct and indirect effects (Pettigrew and Meredith, 1994). The mechanism for the direct effect appears to be an inhibition of the enzymes in the mitochondrial electron transport system, and for the indirect (acclimation) effect of elevated $[CO_2]$ on dark respiration may be related to changes in tissue composition (Yin et al., 1956; Xu and Shen, 2000).

Many C3 species grown for long periods at elevated $[CO_2]$ show a down-regulation of leaf photosynthesis (Zelith, 1982); and carbohydrate source-sink balance is believed to have a major role in the regulation of photosynthesis through the feedback inhibition (Wells et al., 1986). Source-sink imbalances may occur during exposure to elevated $[CO_2]$ when photosynthetic rate exceeds the export capacity or the capacity of sinks to use photosynthates for growth resulting in an accumulation of carbohydrates in photosynthetically active source leaves (Dong, 1991). Although growth is enhanced under elevated $[CO_2]$ the extent to which starch and soluble sugars accumulate depends largely on the species differences. In many plants, the increase in starch also seems to be greater than that of soluble sugars. More frequently observed is the correlation between starch accumulation and inhibition of leaf photosynthesis (Wells et al., 1986) implying that high starch content may be responsible for down regulation of photosynthesis under elevated $[CO_2]$. For many plant species, the long exposure to elevated $[CO_2]$ has also resulted in a down-regulation of Rubisco (Mohanty and Boyer, 1976). Zhang et al. (1992) observed the down-regulation in cotton, cucumber, parsley, pea, radish, soybean, spinach, tobacco and wheat exposed to elevated $[CO_2]$ due to increased leaf acid invertase activities, an indication of starch accumulation in the leaf; conversely, an up-regulation of photosynthesis in bean, plantain and sunflower was also detected suggesting variations in responses by species differences to elevated $[CO_2]$.

Levels of soluble sugars in plant cells have been shown to influence the regulation of expression of several genes coding for key photosynthetic enzymes (Osmond et al., 1980; Xu and Shen, 2000). The buildup in carbohydrates may signal the repression but does not directly inhibit the expression of Rubisco and other proteins that are required for photosynthesis (Oquist et al., 1982). Although the signal transduction pathway for regulation of the sugar-sensing genes may involve phosphorylation of hexoses, derived from sucrose hydrolysis by acid invertase via hexokinase (Guo et al., 1996), unknown gaps still exist between hexose metabolism and repression of gene expression at elevated growth $[CO_2]$ (Hong and Xu, 1998; Xu and Shen, 2000).

5. Photosynthetic and quantum yield up-regulation under elevated CO_2

Carbon dioxide is the substrate that through the light and dark reactions of photosynthesis, are combined into dry mass (Pinkard et al., 2010). Thus $[CO_2]$ can be a major factor limiting photosynthesis (Hall and Rao, 1992). Stomata regulate the diffusion of CO_2 into leaves;

stomata can respond sensitively to $[CO_2]$ as part of a proportionate response to the CO_2 requirement for photosynthesis; increasing concentrations are, therefore, associated with a closing response and vice-versa (Pinkard et al., 2010). Hence, elevated $[CO_2]$ is anticipated to increase or up-regulate photosynthesis, decrease stomatal conductance and increase intrinsic water-use efficiency i.e. the ratio of leaf photosynthesis to stomatal conductance (Long et al., 2004). Many factors other than $[CO_2]$ determine photosynthetic rate, and the law of limiting factors (von Liebig, 1840) will ultimately determine photosynthetic responses to $[CO_2]$ where more than one limiting factors may be involved. Sala and Hoch (2009) suggested that elevated CO_2 would improve carbon balance in light-limited as well as high-light environments through the CO_2 enhancement of quantum yield. The present $CO_2:O_2$ ratio of the air constrains photosynthesis by 30-40% because of O_2 inhibition of carboxylation and associated photorespiration (Booth and Jayanovic, 2005). As CO_2 concentration increases, quantum yield is increased because the ratio of carboxylation to oxygenation by Rubisco increases and photorespiration decreases (Pinkard et al., 2010). Several studies have shown that CO_2 enrichment enhances photosynthesis and growth under limiting irradiance condition, and in some cases the relative enhancement was greater at low rather than at high irradiances (Gifford et al., 1981). In C3 plants, elevated $[CO_2]$ increases the quantum yield of photosynthesis by reducing photorespiration caused by the oxygenase activity of Rubisco. Maximum, single-leaf quantum yield was increased from 0 065 to 0 080 (Long and Drake, 1992).

Up-regulation of photosynthesis refers to a significant increase in the light-saturated rate of photosynthesis (A_{sat}), the rate of photosynthesis under ambient light (A), and/or diurnal photosynthesis (A). Elevated $[CO_2]$ up-regulates photosynthesis by increasing the carboxylation rate (Vc) of ribulose bisphosphate carboxylase (Rubisco) and competitively inhibiting the oxygenation of ribulose bisphosphate (RuBP), thereby reducing photorespiration (Luo and Reynold, 1999). Elevated $[CO_2]$ is also associated with the expression of several other changes that affect photosynthesis. The common observation of reduced stomatal conductance, gs, will tend to dampen the extent to which any up-regulation is expressed at a leaf-scale, but may conserve water such that stand-scale responses are positive (Ainsworth and Rogers, 2007).

Photosynthetic acclimation refers to longer-term adaptive changes in the photosynthetic responses to external stimuli that reduce the net level of the initial response; acclimation is also referred to as down-regulation (Pinkard et al., 2010). Acclimation is commonly observed and arises from the plant's need to balance all resources that are allocated for photosynthetic processes, including the external $[CO_2]$ (Gunderson and Wullschleger, 1994). For elevated $[CO_2]$, acclimation is mechanistically linked to decreased maximum apparent carboxylation velocity (Vc_{max}) and reduced investment in Rubisco (Rogers and Humphries, 2000). It is also associated with reduction in N content. These changes are linked to a decrease in control of A_{max} by Vc_{max} but an increase in the regeneration of RuBP, J_{max} (Long and Drake, 1992). There is also an increase in starch and sugar content.

In the short term, rising $[CO_2]$ increases photosynthesis in many of the woody species as have been studied by Ainsworth and Long (2005). These species have the potential to yield significantly with increases in the rates of biomass accumulation. The allocation of dry mass to the above-ground parts in forest free air carbon dioxide enrichment (FACE) experiment was also found to increase by 28% on the average; this includes a greater allocation to woody components. In general, larger responses in growth, biomass production and leaf area index to elevated [CO_2] have been observed in trees than other functional types (Ainsworth and Long, 2005). Nevertheless, there is often a poor correlation between photosynthetic capacity measured as A_{max} and total biomass production under elevated [CO_2] enrichment (Oren et al., 2001).

6. Environmental factors determining the response of photosynthesis to elevated [CO_2]

6.1 Nitrogen supply

At elevated [CO_2] condition, when plant photosynthesis becomes RubP limited, Rubisco will be in excess of requirements (Ainsworth and Rogers, 2007). The excess capacity for carboxylation could be reduced through a reduction in the activation state of Rubisco (Cen and Sage, 2005). Alternatively, because less Rubisco is required by these plants at elevated [CO_2], redistribution of the excess N invested in Rubisco could further increase N use efficiency at elevated [CO_2] without negatively impacting potential C acquisition (Parry et al., 2003; Ainsworth and Rogers, 2007). However, there is only benefit in reducing the amount of N invested in Rubisco at elevated [CO_2] when the resources invested in it can be usefully deployed elsewhere (Parry et al. 2003). Ainsworth and Long (2005) reported that the stimulation in A_{sat} at elevated [CO_2] was 23% lower in plants grown with a low N supply. Meanwhile, under elevated [CO_2] the Vc_{max} was decreased at both high and low N, with greater reduction of 85% in low N condition. This result is in agreement with summaries of earlier studies conducted in controlled environments and field enclosures by Ainsworth and Rogers (2007), Drake et al. (1997), Moore et al. (1999) and Stitt and Krapp (1999), and is consistent with current understanding of the mechanism underlying acclimation under elevated [CO_2]. When plants are N limited, sink development is restricted, C supply is in excess of demand, and the sugar feedback mechanism as outlined earlier can operate to reduce Rubisco content and increase N use efficiency. As N supply increases, the limitation imposed by sink capacity decreases and the sugar linked signal for down-regulating Rubisco content is reduced (Ainsworth and Rogers, 2007; Drake et al. 1997; Rogers et al. 1998; Long et al., 2004).

6.2 Sink strength

Defined here as the capacity to utilize photosynthate, sink strength can be a major constraint on carbon acquisition (Ainsworth and Rogers, 2007). A reduced or insufficient sink capacity may be the result of many potentially limiting processes e.g. N supply (Rogers et al., 1998), genetic constraints (Ainsworth et al., 2004), temperature (Ainsworth et al., 2003b) or developmental changes (Bernacchi et al., 2005; Rogers and Ainsworth, 2006). However, the net result is the same, i.e. the appearance of a carbohydrate-derived signal that can lead to the subsequent down-regulation (acclimation) of photosynthetic machinery, principally Rubisco (Stitt and Krapp, 1999; Long et al., 2004). Davey et al. (2006) showed that poplar grown at elevated [CO_2] had a large sink capacity. Poplar was able to export >90% of its photosynthate during the day; it also had a large capacity for the temporary storage of overflow photosynthate as starch (Ainsworth and Rogers, 2007; Stitt and Quick 1989; Davey et al., 2006). These two traits enabled poplar to maintain high photosynthetic rates at elevated [CO_2] and avoid a major source–sink imbalance that could lead to a reduction in the potential for C acquisition (Ainsworth and Rogers, 2007). In contrast, *L. perenne* can become extremely sink limited at elevated [CO_2] (Rogers and Ainsworth, 2006). As reported, large accumulations of carbohydrate which build up in grasses over several days and weeks are common (Fischer et al., 1997; Isopp et al., 2000; Rogers & Ainsworth 2006). The most likely explanation for the sink limitation observed in L. perenne is an insufficient N supply (Fischer et al., 1997; Rogers et al., 1998).

The excess of C and shortage of N may explain why grasses reduced their Rubisco content at elevated [CO_2] despite the negative impact on potential carbon gain (Rogers et al., 1998). Therefore, a reduction in carboxylation capacity would be expected. However, legumes can trade photosynthate for reduced forms of N with their bacterial symbionts (Rogers et al., 2006b). The benefit of an increase in N use efficiency resulting from the reduction of Rubisco content and the sugar-derived signal required for a reduction in carboxylation capacity would not be expected (Ainsworth & Rogers, 2007). It follows that acclimation in legumes is likely to occur through reduction in Rubisco activity rather than through a loss of Rubisco protein content. This occurs in order to maintain the balance between the supply and demand for the products of the light reactions (Ainsworth & Rogers, 2007). Alternatively, other nutrient limitations may also impact N-fixation and sink capacity at elevated [CO_2] (Almeida et al., 2000; Hungate et al., 2004).

7. Oil palm responses to elevated CO₂

An experiment was carried out using three levels of [CO_2] (400, control; 800 and 1200 μmol mol^{-1} CO_2) to demonstrate the responses of oil palm seedlings on photosynthesis and quantum yield to elevated [CO_2]. Photosynthetic light response curves of oil palm were measured at growth CO_2 concentrations on with an open flow infrared gas analyzer with an attached red LED light source (LI-6400, Li-Cor, Inc., Lincoln, NE). Measurements began with approximately 5 minutes of saturating light (1500 μmol m^{-2}s^{-1}) followed by nine incremental reductions until the irradiance was 0 μmol m^{-2}s^{-1}. Decreasing light was used rather than increasing light to reduce the equilibrium time required for stomatal opening and photosynthetic induction (Kubiske & Pregitzer, 1996). Preliminary trials indicated that photosynthetic rates reached steady state within 2 minutes following each incremental decrease in light. Measurements were made on fully expanded leaves. Gas exchange measurements were restricted to the hours between 0800 and 1200 hours on sunny days to minimize diurnal effects on photosynthesis. Leaf temperatures averaged 28.89 ± 0.89°C within each measurement period. The differences in light response curves due to CO_2 concentration were examined by calculating and statistically comparing light-saturated photosynthesis (A_{sat}), dark respiration (Rd), light compensation point (Γ) (where A = Rd), and apparent quantum yield (φ). Apparent quantum yield (φ) and Rd were estimated from the measured data. Values of φ were calculated as the slope of photosynthesis (A) versus the incident irradiance. Light compensation points (Γ) were estimated by extrapolating between measured data. The shape of the average light response curve in each CO_2 concentration and canopy position was modeled by fitting data to a non-rectangular hyperbola (Leverenz, 1987; Leverenz, 1995) by means of a nonlinear least squares curve-fitting program (JMP, SAS Institute, Inc., Cary, NC)

Photosynthetic light response curves of oil palm are shown in Figure 1. Light-saturated net photosynthesis (A_{sat}) was greater in elevated treatment than in ambient CO_2. The elevated [CO_2] exposure of oil palm seedling resulted in higher rates of A_{sat} (P < 0.001; Figure 2a). The enhancement of A_{sat} by [CO_2] enrichment was significantly greater for 1200 μmol mol^{-1} CO_2

followed by 800 μmol mol^{-1} and 400 μmol mol^{-1} CO_2. During the experiment, elevated CO_2 had caused the A_{sat} of oil palm to increase by 52 to 78% compared to the ambient. Apparent quantum yield, calculated from the initial slope of the light response curves, was slightly lower in ambient treatment compared to the elevated CO_2 treatments (800 and 1200 μmol mol^{-1} CO_2; Figure 2b). During measurements, the elevated treatment exhibited a higher quantum yield than the ambient leaves ($P < 0.05$). Quantum yield was enhanced by 2 fold and 3 fold respectively in the 800 and 1200 μmol mol^{-1} CO_2 treatments compared to 400 μmol mol^{-1} CO_2 one.

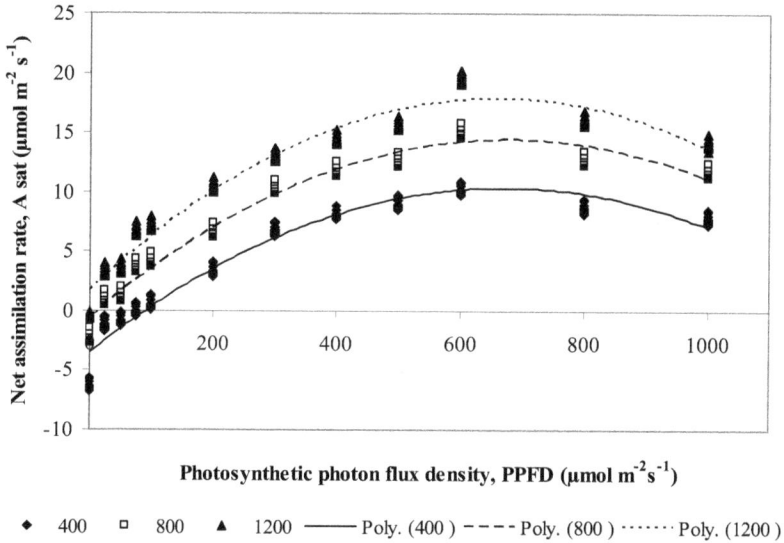

Fig. 1. Light response curve as affected by different CO_2 levels in oil palm seedlings

Elevated CO_2 leaves had a lower light compensation point (Γ) and dark respiration rate (R_d) than the ambient leaves (Figure 2c; 2d). Estimated from the photosynthetic light response curves, it was also demonstrated a significant effect of elevated CO_2 on Γ or Rd ($P \leq 0.01$). Using the light response curves, the light compensation point for ambient CO_2 was recorded at 78.21 followed by 66.21 for 400 and 800 μmol mol^{-1} CO_2 respectively and 30.24 for 1200 μmol mol^{-1} CO_2 treatment. It was also observed that the dark respiration was reduced by 1.72 to 3.21 μmol m^{-2}s^{-1} compared to ambient levels that recorded 5.71 μmol m^{-2}s^{-1}.

Many studies have suggested that the enriched CO_2 leaves respond to atmospheric CO_2 enrichment to a greater extent than the ambient- CO_2 leaves as a result of increased quantum yields (Hanstein & Felle, 2002). A small increase in quantum yield may increase daily carbon gain under low light conditions (Kiirats et al., 2002). In our study, elevated CO_2 increased apparent quantum yields in the 800 and 1200 μmol mol^{-1} CO_2 treatments. Light response curve analysis of oil palm seedling had showed that CO_2 enriched seedlings had reduced their dark respiration rate by 43 to 70% through enhancement of their A_{sat} and apparent quantum yield (φ) by 52 to 78% and 15 to 62%, respectively. The enhancement of Γ and φ signify direct inhibition of the activity of key respiratory enzymes under elevated CO_2 (Drake et al., 1997). This result has been supported by Henson and Haniff (2005) who reported productivity or dry matter production of plants would increase if respiration could be minimized without affecting gross assimilation, or if gross assimilation could be increased without increasing respiration. This shows that increase in CO_2 would enhance gross assimilation and reduce respiration by compensating respiration rate with high carbon gain. Usually, compensation irradiance is reduced while quantum efficiency is increased in plant under elevated CO_2 (Vavin et al., 1995). The same result was also observed by Kubiske and Preigitzer (1996) with red oak seedlings grown at elevated CO_2 in shaded open top chamber.

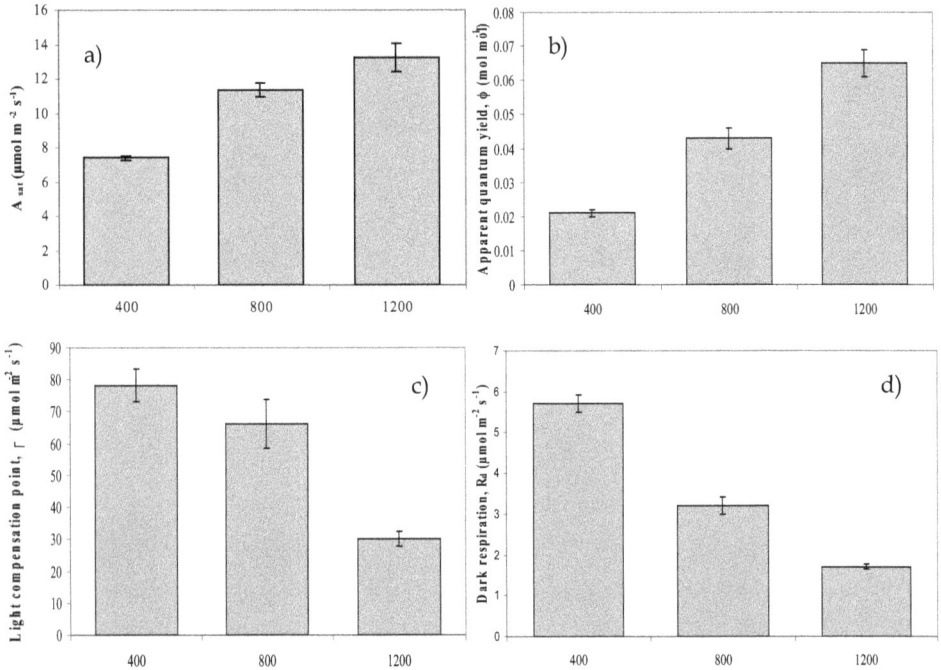

Fig. 2. The maximum net assimilation rate Asat (a), apparent quantum yield (b), Light compensation point (c) and dark respiration rate (d) as affected by different CO2 levels in oil palm seedlings

In oil palm, it was found that enrichment with high levels of [CO_2] have enhanced the leaf gas exchange of oil palm seedlings. As CO_2 levels increased from 400 to 800 and 1200 μmol mol^{-1} CO_2 the net photosynthesis (Figure 3) and water use efficiency (Figure 4) were also improved. Net photosynthesis (A) and water use efficiency (WUE) were been enhanced by respective 211 to 278% and 158 to 224% when enriched with [CO_2] (800 and 1200 μmol mol^{-1} CO_2). As CO_2 levels increased, it was observed that the intercellular CO_2 (Ci) increased higher in oil palm seedling treated with high levels of [CO_2] (Figure 5). The Ci for 1200 μmol mol^{-1} CO_2 recorded the highest (361.11 μmol mol^{-1} CO_2) value followed by that of 800 μmol mol^{-1} CO_2 (311.11 μmol mol^{-1} CO_2) with the lowest at 400 μmol mol^{-1} CO_2 that recorded 289.12 μmol mol^{-1}. Up-regulation of A may as represented by increases in leaf intercellular CO_2 concentration (C_i) that could also be related to increase in the thickness of the leaves (high SLA) achieved under elevated [CO_2] that contains high photosynthetic protein especially Rubisco (Ramachandra & Das, 1986). The latter might also up–regulate several enzymes related to carbon metabolism which simultaneously increase the C_i (Anderson et al., 2001). This data imply that high A under elevated CO_2 could be due to more efficient net assimilation resulting from extra carbon fixation as exhibited by high C_i per unit area which is related to increased thickness of mesophyll layer, mainly due to increased palisade layer (Lawson et al., 2002). Up-regulation of A may as represented by increases in leaf intercellular CO_2 concentration (C_i) may also be related to increase in the thickness of the leaves under elevated CO_2 that contain high photosynthetic protein especially Rubisco (Ramachandra and Das, 1986). The latter might up–regulate several enzyme related to carbon metabolism that simultaneously increase the C_i (Anderson et al., 2001). This data implied that high A under elevated CO_2 could be due to more efficient net assimilation due to extra carbon fixation exhibited by high C_i per unit area which is related to increased thickness of mesophyll layer, mainly due to increased palisade layer (Lawson et al., 2002).

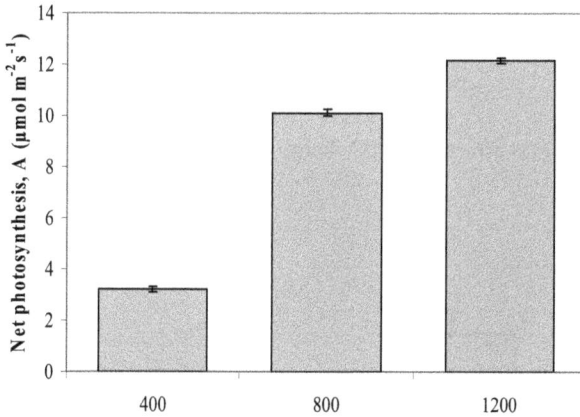

Fig. 3. Net photosynthesis as affected by CO2 levels in oil palm seedlings

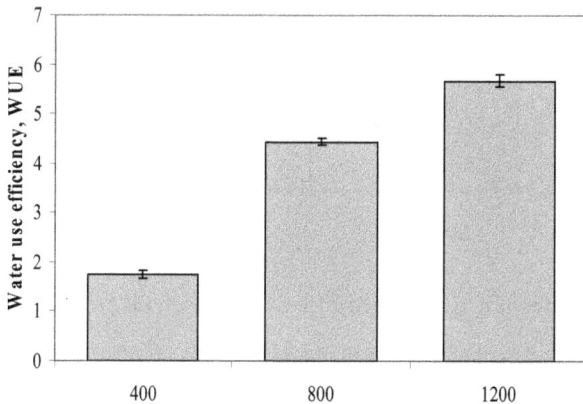

Fig. 4. Water use efficiency as affected by CO_2 levels in oil palm seedlings

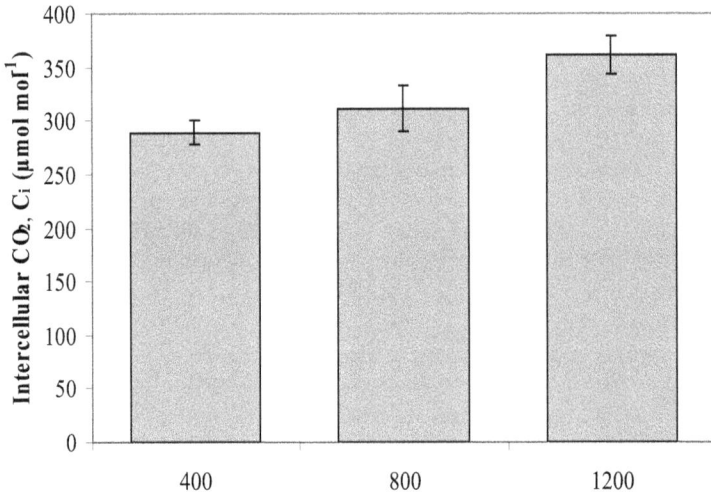

Fig. 5. Intercelluar CO_2 as affected by CO_2 levels in oil palm seedlings

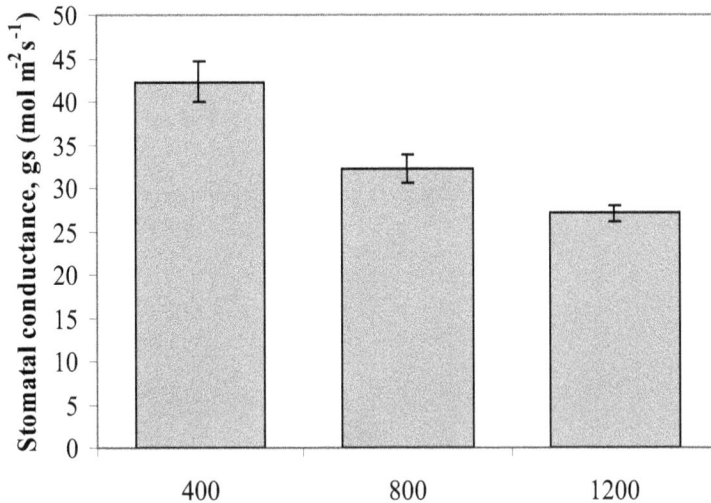

Fig. 6. Stomatal conductance, gs as affected by CO_2 levels in oil palm seedlings

Statistically, higher A was observed over the controlled plants as the levels of enrichment increased. Similar result was obtained by Downtown (1990) and Van and Megonigal (2001). Theoretically, exposure to higher [CO_2] would increase A by increasing the availability of the substrate (CO_2; Downtown, 1990). In the present study, the increase in A might be justified by reduced light compensation point (Γ) and dark respiration rate (R_d), with the plant enriched with high CO_2 having enhanced apparent quantum yield and net assimilation rates (Kubiske & Preigitzer, 1996).

Despite increases in A and WUE, stomatal conductance of oil palm seedlings enriched with high levels of CO_2 decreased as levels of CO_2 increases (Figure 6). In 400 µmol mol^{-1} CO_2, stomatal conductance recorded a value at 42.3 mmol m^{-2} s^{-1}; with increasing [CO_2] to 800 and 1200 µmol mol^{-1} the stomatal conductance documented lower values (27.1 to 32.3 mmol m^{-2} s^{-1}). Further enhancing the plants to 800 and 1200 µmol mol^{-1} had shown to reduce stomatal conductance (g_s) of the CO_2-enriched seedlings versus the ambient CO_2-treated plants with lowest g_s. The decreased g_s simultaneously reduced the transpiration rate (E) of plant under elevated CO_2. This phenomenon is usually reported in plant treated with high than ambient CO_2 (Rashke, 1986; Lodge et al., 2001; Lawson et al., 2002). It was believed that reduced g_s might contribute to plant acclimation to high intercellular CO_2 (C_i) (Morrison & Jarvis, 1980).

It was also found that nitrogen levels were influenced by CO_2 levels applied to the oil palm seedlings. From Figure 7 it is observed that the nitrogen levels were highest in leaves followed by stems and lowest in the roots. As [CO_2] levels increased from 400 to 1200 µmol mol^{-1} CO_2 the nitrogen content decreased highly in 1200 followed by 800 and lowest in 400 µmol mol^{-1} CO_2. This implies that plant enriched with high levels of [CO_2] have high dilution of nitrogen content in the plant tissues. Nitrogen content was influenced by the application of CO_2 levels to the seedlings. As the levels of CO_2 increased from 400 to 1200 µmol mol^{-1} CO_2, nitrogen content were found to be reduced. The decrease in nitrogen content with increasing CO_2 levels has been reported by Porteus et al. (2009). Several researchers attributed this phenomenon to decreasing uptake of nitrogen as transpiration rate (E) was decreased due to reduction in stomata conductance (g_s) under elevated CO_2 level (Conroy and Hawking, 1993).

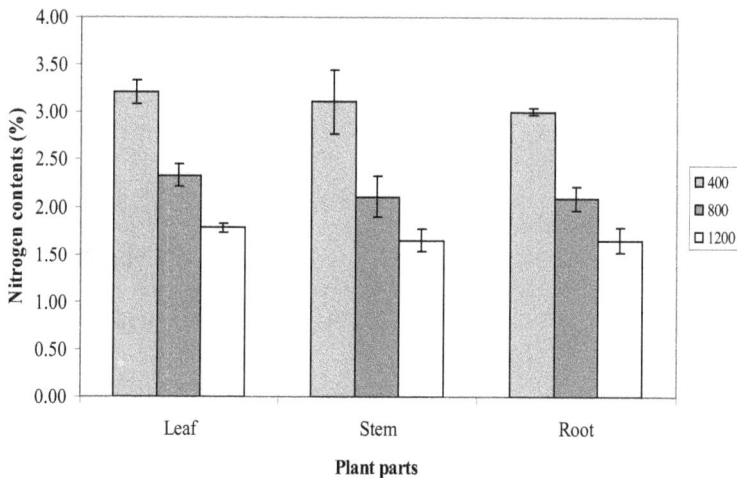

Fig. 7. Nitrogen levels as affected by CO_2 levels in different part of oil palm seedlings

8. Conclusion

The results demonstrated by the oil palm seedlings indicated a positive response of oil palm seedling to elevated CO_2 enrichment in term of enhanced photosynthesis rate and quantum yield as compared with ambient CO2 condition. The positive impact of oil palm seedlings to CO2 enrichment was shown by enhancement of the leaf gas exchange characteristics of oil palm. The positive responses have been shown to cause increases in net photosynthesis, Asat, apparent quantum yield and reduction of dark respiration rate and light compensation point. The findings suggest that in the next 22nd century, it would be expected that oil palm to benefit from changes in the climate as long as temperature does not increase beyond the palm optimum level. Further research in the future needs to be conducted to confirm these effects especially involving many environmental conditions under elevated $[CO_2]$. Producing crops under climate change conditions, then, would be a growing challenge in the new agriculture of the world.

9. Acknowledgment

The authors are grateful to the Ministry of Higher Education Malaysia and the Research Management Centre of Universiti Putra Malaysia for supporting this work under the Graduate Research Assistant Grant Scheme.

10. References

Ainsworth E.A., Rogers A., Blum H., Nösberger J. & Long S.P. (2003b) Variation in acclimation of photosynthesis in Trifolium repens after eight years of exposure to free air CO2 enrichment (FACE). *Journal of Experimental Botany* Vol 54, pp. 2769–2774

Ainsworth, E.A. & Long, S.P. (2005).What havewe learned from15 years of free-air CO2 enrichment (FACE)? Ameta-analytic reviewof the responses of photosynthesis canopy properties and plant production to rising CO2. *New Phytologist* Vol165, pp 351–372.

Ainsworth, E.A. & Rogers, A., 2007. The response of photosynthesis and stomatal conductance to rising [CO2]: mechanisms and environmental interactions. *Plant, Cell & Environment* Vol 30, pp 258–270.

Almeida J.P.F., Hartwig U.A., Frehner M., Nösberger J. & Löscher A. (2000) Evidence that P deficiency induces N feedback regulation of symbiotic N2 fixation in white clover (Trifolium repens L.). *Journal of Experimental Botany* Vol 51, pp. 1289–1297.

Anderson J.M. (2000). "Strategies of Photosynthetic Adaptations and Acclimation." In Probing Photosynthesis: Mechanisms, Regulation and Adaptation. M. Yunus, U. Pathre, P. Mohanty, eds. London

Baker, N.R. & Ort, D.R. (1992). Light and crop photosynthesis performance. In: NR Baker, H Thomas, eds. Crop Photosynthesis: Spatial and Temporal Determinants. Amsterdam: Elsevier Science Publishers, 1992, pp 289–312.

Bernacchi C.J., Portis A.R.,Nakano H., von Caemmerer S. & Long S.P. (2002) Temperature response of mesophyll conductance. Implications for the determination of rubisco enzyme kinetics and for limitations to photosynthesis in vivo. *Plant Physiology* Vol 130, pp 1992–1998.

Booth, T.H.& Jovanovic, T. (2005). Tree Species Selection and Climate Change in Australia. Australian Greenhouse Office, Canberra.

Brinkman, M.A. & Frey, K.J.(1978). Flag leaf physiological analysis of oat isolines that differ in grain yield from their recurrent parents. *Crop Science* Vol 18, pp.69–73

Cen Y.P. & Sage R.F. (2005) The regulation of rubisco activity in response to variation in temperature and atmospheric CO2 partial pressure in sweet potato. *Plant Physiology* Vol 139 pp 979–990

Clayton, R.K. (1971) Light and living matter: A guide to the study of photobiology. New York, USA: McGraw-Hill.

Clayton, R.K. (1980) Photosynthesis: Physical mechanism and chemical patterns. Cambridge, England: Cambridge University Press

Conroy, J. and Hocking, P. (1993). Nitrogen nutritions of C_3 plants at elevated carbon dioxide concentration. *Plant physiology* Vol 89: 570 – 576.

Davey P.A., Olcer H., Zakhleniuk O., Bernacchi C.J., Calfapietra C., Long S.P. & Raines C.A. (2006) Can fast-growing plantation trees escape biochemical down-regulation of photosynthesis when grown throughout their complete production cycle in the open air under elevated carbon dioxide? *Plant, Cell & Environment* Vol 29, pp. 1235–1244.

Dong, S.T. (1991). Studies on the relationship between canopy apparent photosynthesis and grain yield in high-yielding winter wheat. *Acta Agronomy Singapore* Vol 17, pp 461–469

Downton, W.J.S., Grant, W.J.R. and Chacko, E.K. (1990). Effect of elevated carbon dioxide on the photosynthesis and early growth of mangosteen (*Garcinia mangostana L*). *Scientia Horticulturae* Vol 44, pp 215 – 225

Drake B.G., Gonzalez-Meler M.A. & Long S.P. (1997) More efficient plants: a consequence of rising atmospheric CO2? *Annual Review of Plant Physiology and Plant Molecular Biology* Vol 48, pp. 609–639.

Drake, B.G., Gonzalez, M.M.A., and Long, S.P. (1997). More efficient plants : a consequence of rising atmospheric CO_2. *Annual Review Plant Physiology Plant Molecular Biology* Vol 48 pp 609 – 639

Ehleringer, J. & Pearcy, R.W.(1983). Variation in quantum yield for CO2 uptake among C3 and C4 plants. *Plant Physiology* Vol 73, pp. 555–559

Ehleringer, J. R.; Sage, R. F.; Flanagan, L. B. & Pearcy, R. W. (1991). Climate change and the evolution of C4 photosynthesis. *Trends in Ecology Evolution*. Vol 6, pp. 95–99

Emerson, R. & Lewis, C.M. (1943). The dependence of the quantum yield of chlorella photosynthesis on wave length of light. *American Journal of Botany* Vol 30, pp.165–178

Evans, L.T.& Dunstone, R.L.(1970). Some physiological aspects of evolution in wheat. *Australian Journal Biological Science* Vol 23 pp.725–741

Farquhar, G.D. & Sharkey, T.D. (1982). Stomatal conductance and photosynthesis. *Annual Review Plant Physiology* Vol33, pp. 317–345

Gifford, R.M.& Evans, L.T. (1981). Photosynthesis, carbon partitioning, and yield. *Annual Review of Plant Physiology* Vol 32, pp 485–509

Gunderson, C.A. & Wullschleger, S.D. (1994). Photosynthetic acclimation in trees to

Guo, L.W.; Xu,D.O.; & Shen, Y.K.(1996). Photoinhibition of photosynthesis without net loss of D1 protein in wheat leaves under field conditions. *Acta Botany Singapore* Vol 38, pp. 196–202

Guo, L.W; Xu, D.Q & Shen, Y.K. (1994). The causes of midday decline of photosynthetic efficiency in cotton leaves under field conditions. *Acta Phytophysiol Singapore* Vol 20, pp.360–366

Hall, D.O. & Rao, K.K. (1999). Photosynthesis. Cambridge: Cambridge University Press

Hanstein S.M. & Felle H.H. (2002) CO2-triggered chloride release from guard cells in intact fava bean leaves. Kinetics of the onse of stomatal closure. *Plant Physiology* Vol 130, pp. 940–950.

Hay, R.K.M & Walker, A.K.(1989). An Introduction to the Physiology of Crop Yield. New York: Longman Scientific & Technical copublished in the United States with John Wiley & Sons, pp 31–86.

Henson,I.E. and Haniff, M.H., (2006). Carbon dioxide enrichment in oil palm canopies and its possible influence on photosynthesis. *Oil Palm Bulletin*. Vol 51, pp 1–10

Hong, S.S.& Xu, D.Q. (1998).Light-induced increase in initial chlorophyll fluorescence Fo level and its possible mechanism in soybean leaves. In: G Garab, ed. Photosynthesis: Mechanisms and Effects. Vol III. Dordrecht, The Netherlands: Kluwer Academic Publishers, 1998, pp 2179–2182.

Hungate B.A., Stiling P.D., Dijkstra P., Johnson D.W., Ketterer M.E., Hymus G.J., Hinkle C.R. & Drake B.G. (2004) CO2 elicits long-term decline in nitrogen fixation. *Science* Vol 304, pp. 1291.

Jacob, J.(1995). Phosphate deficency increases the rate constant of thermal dissipation of excitation energy by photosystem II in intact leaves of sunflower and maize. *Australian Journal of Plant Physiology* Vol 22, pp. 417–424

K.V.R., McCarthy, H., Hendrey, G., McNulty, S.G.& Katul, G.G. (2001). Soil fertility limits carbon sequestration by forest ecosystems in a CO2-enriched atmosphere. *Nature* Vol 411, pp 469–472.

Keeling, C. D.; Whorf, T. P. Wahlen, M. & Van der Plicht, J. (1995) Interannual extremes in the rate of rise of atmospheric carbon dioxide since 1980. *Nature* Vol 375, pp. 666–670.

Kiirats O.,Lea P.F., Franceschi V.R. & Edwards G.E. (2002) Bundle sheath diffusive resistance to CO2 and effectiveness of C4 photosynthesis and refixation of photorespired CO2 inaC4 cycle mutant and wild-type Amaranthus edulis. *Plant Physiology* Vol 130,pp. 964–976.

Ku, S.B; & Edwards, G.E. (1978). Oxygen inhibition of photosynthesis. III. Temperature dependence of quantum yield and relation to O2/CO2 solubility ratio. *Planta* Vol 140 pp.1–6.

Kubiske, M.E. and Pregitzer, K.S. (1996). Effects of elevated CO_2 and light availability on the photosynthetic response of trees of contrasting shade tolerance. *Tree Physiology* Vol 16, pp 351-358.

Lawson, T., Carigon, J., Black, C.R., Colls, J.J., Landon, G. and Wayers, J.D.B. (2002). Impact of elevated CO_2 and O_3 on gas exchange parameters and epidermal characteristics of potato (*Solanum tuberosum* L.). *Journal of Experimental Botany*. Vol 53, No 369, pp 737-746

Li, D.Y.; Ye, J.Y.& Shen, Y.K.(1991). Effect of rainy weather on the photosynthetic efficiency in spinach. *Plant Physiology Community Singapore* Vol 27, pp.413–415

Lodge, R.J., Dijkstra, P., Drake, B.G. and Morrison, J.I.L. (2001). Stomatal acclimation to increased level of carbon dioxide in a Florida scrub oak species *Quercus myritifolia*. *Plant Cell. Environment* Vol 14, pp 729 – 739

Long S.P., Ainsworth E.A., Rogers A. & Ort D.R. (2004) Rising atmospheric carbon dioxide: plants FACE the future. *Annual Review of Plant Biology* Vol 55, pp 591–628.

Long, S.P. & Drake, B.G. (1992). Photosynthetic CO2 assimilation and rising atmospheric CO2 concentration. In: Baker, N.R., Thomas, H. (Eds.), Topics in Photosynthesis Research. Elsevier, pp. 69–104.

Long, S.P. &Drake, B.G. (1992). Photosynthetic CO2 assimilation and rising atmospheric CO2 concentration. In: Baker, N.R., Thomas, H. (Eds.), Topics in Photosynthesis Research. Elsevier, pp. 69–104.

Long, S.P., Ainsworth, E.A., Rogers, A.& Ort, D.R., (2004). Rising atmospheric carbon dioxide: Plants face the future. *Annual Review of Plant Biology* Vol 55, pp 591– 628

Luo, Y. & Reynolds, J. (1999). Validity of extrapolating field CO2 experiments to predict carbon sequestration in natural ecosystems. *Ecology* Vol 80, pp 1568–1583

Mohanty, P. & Boyer, J.S. (1976). Chloroplast response to low leaf water potentials. IV. Quantum yield is reduced. *Plant Physiology* Vol 57, pp 704–709

Mohanty,P. & Boyer, J.S. (1976). Chloroplast response to low leaf water potentials. IV. Quantum yield is reduced. *Plant Physiology* vol 57 pp.704–709

Monson, R.K; Littlejohn, J.L.& Williams, G.J. (1982). The quantum yield for CO2 uptake in C3 and C4 grasses. *Photosynthesis Research* Vol 3, pp.153–159

Moore B.D., Cheng S.H., Sims D. & Seemann J.R. (1999) The biochemical and molecular basis for photosynthetic acclimation to elevated atmospheric CO2. *Plant, Cell & Environment* Vol 22, pp 567–582.

Morison, J.I.L. (1987). Intercellular carbon dioxide concentration and stomatal responses to carbon dioxide. In *Stomatal Function*, ed. Zeiger, E., Farquhar, G.D. and Cowan, I.R., pp. 229 - 251. Stanford, California: Stanford University Press.

Niyogi, K.K. (1999). Photoprotection revisited: genetic and molecular approaches. Annual Revision Plant Physiology Plant Molecular Biology Vol 50, pp .333–359

Oquist, G.; Brunes, L. & Hallgren, G.E. (1982). Photosynthetic efficiency of Betula pendula acclimated to different quantum flux densities. *Plant Cell Environment* Vol 5, pp 9–15

Oquist,G.; Brunes, L.& Hallgren, G.(1982). Photosynthetic efficiency during ontogenesis of leaves of Betula pendula.*Plant Cell Environment* Vol 5, pp. 17–21

Oren, R., Ellsworth, D.S., Johnsen, K.H., Phillips, N., Ewers, B.E., Maier, C., Schafer,K.V.R., McCarthy, H., Hendrey, G., McNulty, S.G.& Katul, G.G. (2001). Soil fertility limits carbon sequestration by forest ecosystems in a CO2-enriched atmosphere. *Nature* Vol 411, pp 469–472.

Ort, D.R. & Baker, N.R. (1988). Consideration of photosynthetic efficiency at low light as a major determinant of crop photosynthetic performance. *Plant Physiology Biochemistry* Vol 26, pp.555-565.

Osmond, C.B.; Bjorkman, O & Anderson, D.J. (1980). Physiological Processes in Plant Ecology. Berlin: Springer-Verlag, pp. 291-377

Parry M.A.J., Andralojc P.J., Mitchell R.A.C., Madgwick P.J. & Keys A.J. (2003) Manipulation of Rubisco: the amount, activity, function and regulation. *Journal of Experimental Botany* Vol 54 pp 1321-1333.

Pettigrew, W.T.; & Meredith, B.R. (1994). Leaf gas exchange parameters vary among cotton genotypes. *Crop Science* Vol 34, pp700-705

Pinkard, E.A.; Beadle, C.L.; Mendham, D.S; Carter, J.& Glen, M. 2010. Determining photosynthetic responses of forest species to elevated [CO2]: Alternatives to FACE. *Forest ecology and Management* Vol 260, pp 1251 - 1261

Porteaus,F., Hill, J., Ball, A.S., Pinter, P.J., Kimbal, B.A., Wall, G.W. and Ademsen, F.J. and Morris, C.F.(2009). Effects of free air carbon dioxide enrichment (FACE) on the chemical composition and nutritive value of wheat grain straw. *Animal Feed Science Technology* Vol 149, pp 322 - 332.

Poskuta, J.W. & Nelson, C.J.(1986). Role of photosynthesis and photorespiration and of leaf area in determining yield of tall fescue genotypes. *Photosynthetica* Vol 20, pp.94-101

Ramachandra, A.R. and Das, V.S.R. (1986). Correlation between biomass production and net photosynthetic rates and kinetic properties of RuBP carcoxylase in certain C_3 plants. *Biomass* Vol 10, pp 157 - 164

Raschke, K. (1986). The influence of carbon dioxide content of the ambient air on stomatal conductance and the carbon dioxide concentration in leaves. In Enoch, H.Z. and Kimball, B.A. [eds] *Carbon dioxide enrichment of greenhouse crops*, Volume 2, ed. pp. 87 -102, Boca Raton: CRC Press.

Rogers, A. and Ainsworth, E.A. (2006) The response of foliar carbohydrates to elevated carbon dioxide concentration. In Managed Ecosystems and CO2. Case Studies, Processes and Perspectives (eds J. Nösberger, S.P. Long, R.J. Norby,M. Stitt, G.R. Hendrey & H. Blum), pp. 293-308. Springer-Verlag, Heidelberg, Germany.

Rogers A., Fischer B.U., Bryant J., Frehner M., Blum H., Raines C.A. & Long S.P. (1998) Acclimation of photosynthesis to elevated CO2 under low-nitrogen nutrition is affected by the capacity for assimilate utilization. Perennial ryegrass under free air CO2 enrichment. *Plant Physiology* Vol 118, pp 683-689

Rogers A.,Allen D.J.,Davey P.A., et al. (2004) Leaf photosynthesis and carbohydrate dynamics of soybeans grown throughout their life-cycle under Free-Air Carbon dioxide enrichment. *Plant,Cell & Environment* Vol 27,pp 449-458

Rogers, A.& Humphries, S.W. (2000). Amechanistic evaluation of photosynthetic acclimation at elevated CO2. *Global Change Biology* Vol 6, pp. 1005-1011.

Rogers A.,Gibon Y., Stitt M.,Morgan P.B.,Bernacchi C.J.,OrtD.R. & Long S.P. (2006b) Increased C availability at elevated carbon dioxide concentration improves N assimilation in a legume.*Plant, Cell & Environment* Vol 29, pp. 1651-1658

Sala, A.& Hoch, G. (2009). Height-related growth declines in ponderosa pine are not due to carbon limitation. *Plant, Cell and Environment* Vol32, pp 22–30

Schimel, D. (2006). Rising CO2 levels not as good for crops as thought. *Science* Vol 312,

Sharkawy, M.A.; Cock, J.H.; Lynam, J.K; Hernandez, A.P.& Cadavid, L.L.F. (1990). Relationships between biomass, root yield and single-leaf photosynthesis in field-grown cassava. *Field Crop Research* Vol 25, pp. 183–201

Stitt M. & Krapp A. (1999) The interaction between elevatedbcarbon dioxide and nitrogen nutrition: the physiological and molecular background. *Plant, Cell & Environment* Vol 22, pp 583–621.

Stitt M. & Quick P. (1989) Photosynthetic carbon partitioning: its regulation and possibilities for manipulation. *Physiologia Plantarum* Vol 77, pp 663–641

Stoskopf, N.C.(1981).Understanding Crop Production. Reston, VA: Reston Publishing Company, pp 1–12.

Taiz, L. & Zieger, E. (1991). *Plant Physiology.* The Benjamin/Cummings Publishing Company, Inc., ISBN 0-8053-0153-4, California, USA

Van, C.D. and Megonigal, J.P. (2002). Productivity of Acer rubrum and taxodium distichum seedlings to elevated carbon dioxide and flooding. *Environmental pollution* Vol 116, pp 31 – 36

Vivin, .P, Gross, P., Aussenac, G. and Guehl, J.M., (1995). Whole plant CO_2 exchange, carbon partitioning and growth in *Quercus robur* seedlings exposed to elevated CO_2. *Plant Physiology Biochemistry* Vol 33, pp 201 – 211

Von Liebig, J. (1840). Die organische Chemie in Ihrer Anwendung auf Agricultur und Physiologie. Friedrich Vieweg und Sohn Braunschweig, Germany

Wells, R., Schulze, L.L., Ashley, D.D., Boerma, H.R. & Brown, R.H. (1982). Cultivar differences in canopy apparent photosynthesis and their relationship to seed yield in soybeans. *Crop Science*, Vol 22, pp. 886–890

Wells, R; Meredith, W.R & Williford, J.F. (1986). Canopy photosynthesis and its relationship to plant productivity in near-isogenic cotton lines differing in leaf morphology. *Plant Physiology* Vol 82, pp. 635–640

Xu, D.O. & Shen, Y.K. (2000). Photosynthetic efficiency and crop yield. In handbook of plant and crop physiology Revised and updated. pp 821 – 830.

Xu, D.Q. (1988). Photosynthetic efficiency. *Plant Physiology Community Singapore* Vol 24, pp.1–6

Xu, D.Q; Xu, B.J &.Shen, G.Y.(1990). Diurnal variation of photosynthetic efficiency in C3 plants. *Acta Phytophysiology Singapore* Vol 16, pp 1–5

Xu,D.Y.; Li, D.Y.; Shen,Y.G & Liang, G.A. (1984). On midday depression of photosynthesis of wheat leaf under field conditions. *Acta Phytophysiol Singapore* vol 10 pp.269–276.

Ye, J.Y.; Li, D.Y & Shen, Y.G. (1995). Effect of hypotonic swelling on photosynthesis in spinach intact chloroplasts. *Acta Phytophysiol Singapore* Vol 21, pp.73–79.

Yin, H.C.; Shen, Y.C.; Chen, Y; Yu, C.H. & Li, P.C. 1956. Accumulation and distribution of dry matter in rice after fllowering. *Acta Botany Singapore* Vol 5, pp 177–184

Zelitch, I. (1982). The close relationship between net photosynthesis and crop yield. *Bioscience* Vol 32 pp796–802

Zhang, S.Y.; Lu, G.Y. ;Wu, H.; Shen, Z.X.; Zhong, H.M.; Shen,Y.G.; Xu, D.Y.; Ding, H,G. &, Hu, W.X. (1992). Photosynthesis of major C3 plants on Qinghai plateau. Acta Botany Singapore Vol 34, pp 176–184

6

The Role of C to N Balance in the Regulation of Photosynthetic Function

Vladimir I. Chikov and Svetlana N. Batasheva

Kazan Institute of Biochemistry and Biophysics of the Russian Academy of Sciences
Russia

1. Introduction

Among the numerous factors affecting plant growth and development one of the most important ones is mineral nutrition. Here, phosphorous and potassium sustain energetics and metabolite transport in the cell. The efficiency of their using can be elevated by repeated circulation of the atoms in conjugated processes. But nitrogen is expended in building plant cell mass and the growth is impossible without continues inflow of new portions of the element. That is why it has a special impact on all physiological processes, including photosynthesis and assimilate transport.

Data accumulated so far about the interaction between carbon and nitrogen metabolisms in plants indicate its key role in regulation of plant vital functions. C to N balance in a whole plant was shown to take part in regulation of photosynthesis, germination, senescence, morphogenesis (Malamy and Ryan, 2001; Martin et al., 2002; Paul and Foyer, 2001; Paul and Pellny, 2003). Nevertheless, mechanisms underlying this regulation are still elusive.

In order to find out the regulatory mechanisms it is necessary to consider all points of contact between carbon and nitrogen metabolisms in plant, and one point that is often overlooked is a significant influence of nitrogen on photo-assimilate transport in plants.

2. The influence of nitrogen nutrition level on photosynthesis and photo-assimilate partitioning

2.1 The influence of nitrogen nutrition on photoassimilate transport

At the first glance, the literature data on the action of nitrogen nutrition on assimilate transport is controversial. For instance, in some publications it is noted that additional nitrogen nutrition delays assimilate export from leaves (Kudryavtsev & Roktanen, 1965; Marty, 1969; Vaklinova et al., 1958; Zav'yalova, 1976), while in others the opposite effect is noted (Anisimov, 1959; Grinenko, 1964; Hartt, 1970; Pristupa & Kursanov, 1957).

Very important here is a period of the plant ontogenetic development, in which plant nitrogen nutrition level changes. In starving for the element juvenile plants, in which the sink-source system consists only of leaves and roots, nitrogen fertilization leads to intensified inflow of photo-assimilates to roots for active metabolization of mineral nitrogen, and these results will probably be interpreted as activation of transport processes by nitrogen. On the other hand, nitrogen supply of older plants with formed sink-source relationships usually results in relative inhibition of assimilate export from source leaves to sink organs (Table 1).

Treatment	Radioactivity of a plant part, %	
	Leaf	Ear
Control	36.1 ± 4.1	51.8 ± 6.2
Nitrogen	50.6 ± 3.8	28.2 ± 1.7
% to control	140	54

Table 1. The influence of pre-planting nitrogen fertilization of soft wheat cv. Saratovskaya 29 on ^{14}C-photoassimilate export from leaves and their inflow into ears (% radioactivity of above ground plant part) in the milky stage of grain development (Tarchevsky et al., 1973)

The characteristic feature of photosynthetic carbon metabolism in plants grown on increased nitrogen background is a lowered ratio of labeled sucrose to hexoses (Table 2). Using our own method of extraction of the labeled photosynthetic products from the apoplast (Chikov et al., 2001), it was established that at increasing nitrate nutrition level the ratio of labeled sucrose to hexoses decreases in the apoplast rather than in mesophyll cells (Table 2). Thus, the enhanced sucrose hydrolysis is a property of the apoplastic compartment.

Treatment	Upper part		^{14}C-donor part	
	leaves	apoplast	leaves	apoplast
Control (Non-fertilized)	15.0 ± 0.15	149.5 ± 25.0	16.9 ± 7.0	128.0 ± 20.0
NO$_3$-fertilized	11.6 ± 3.4	36.7 ± 6.7	13.6 ± 0.7	38.1 ± 3.8
control/ NO$_3$	1.29	4.07	1.24	3.36

Table 2. The influence of nitrate nutrition on the ratio of labeled sucrose to hexoses in the leaves of flax plants (Chikov et al., 2001)

Delayed assimilate export from leaves of plants fertilized with nitrogen is believed to be linked with intensified synthesis of nitrogen containing compounds and diverting to the process carbon fixed in photosynthesis with lesser formation of transport photosynthetic products, i.e. sugars (Champigny & Foyer, 1992). The data on increased sucrose hydrolysis in the apoplast which is an intermediate in the sucrose transfer to the phloem suggest that the reason of lowered export lies not in the shortage of sugars but in the mechanism itself of their transport from leaves.

2.2 The influence of nitrogen nutrition on plant photosynthetic carbon metabolism (PCM)

As a rule, plants grown at various levels of nitrogen differ dramatically in their morphological features such as sizes and densities of leaf blades, photosynthetic pigment contents, the ratios of above-ground part to root weights, etc.; and all this confuses interpretation of data on PCM. That is why to identify differences between the influence of different nitrogen forms it is desirable to assess physiological and biochemical characteristics before pronounced visual changes of plants become obvious. Bearing all this in mind, we compared the influence of nitrate on PCM to that of urea (as a reduced nitrogen form) the next day after plant fertilization. The experiments were performed with the upper leaf of wheat plants, grown at moderate level of full mineral nutrition till the stage of kariopsides formation. At this time the export function of the upper leaf is most expressed. On the eve of the experiment, plants were watered with solutions of calcium nitrate or urea with concentrations calculated so to be equal to 2 grams of N per pot. Next day from 10 to 12 a.m. the flag leaf was exposed to $^{14}CO_2$ for 2 min and fixed to study PCM. To establish specialities of the influence of nitrogen fertilization on photorespiratory glycolate pathway PCM was investigated under two CO_2 concentrations (0.03 and 0.3%) and two O_2 concentration (21% and 1%). Notably, the gas composition in the leaf photosynthetic chamber was altered only for the period of $^{14}CO_2$ assimilation by the leaf.

The experiment showed significant differences in the action of oxidized and reduced N on PCM (Table 3). The influence of urea and nitrates on PCM had some common features as well as distinct ones. Irrespective of the form of N used, the introduction of ^{14}C into phosphorous esters of sugars decreased and into malate, aspartate and alanine increased, that implied diminished phosphoglyceric acid (PGA) reduction to sugar phosphates and its enhanced non-reductive metabolism. Additionally, in plants fertilized with nitrates, the formation of glycolate pathway products (serine, glycine, glycolate) increased.

Treatment	Photosynthesis intensity ($\mu g\ CO_2 m^{-2}s^{-1}$)	Free sugars	Phosphorus esters of sugars	Serine, glycine, glycolate	Alanine, malate, aspartate
21% O_2; 0.03% CO_2					
Control	680	42.2 ± 1.0	27.6 ± 1.1	11.8 ± 0.1	7.7 ± 0.4
Urea	740	44.5 ± 0.8	24.1 ± 0.9	11.3 ± 1.0	12.2 ± 1.1
Nitrate	260	40.1 ± 1.5	16.3 ± 2.6	18.9 ± 2.2	11.8 ± 0.4
1% O_2; 0.3% CO_2					
Control	990	44.9 ± 1.0	28.9 ± 1.4	4.0 ± 0.6	13.2 ± 0.6
Urea	1430	47.3 ± 1.1	14.1 ± 2.5	2.4 ± 0.4	22.0 ± 0.8
Nitrate	1030	46.8 ± 2.0	14.2 ± 1.6	9.6 ± 0.5	22.5 ± 1.0

Table 3. The influence of different CO_2 and O_2 concentrations and N forms on ^{14}C distribution among some labeled products after 2 min $^{14}CO_2$ assimilation (% radioactivity of water-ethanol soluble fraction) in wheat (Chikov & Bakirova, 1999)

According to common knowledge, glycolate is synthesized from ribulose-1,5-bisphosphate in the RuBP-oxygenase reaction of photosynthesis which requires oxygen, and the oxygenase reaction competes with carboxylase one for RuBP. All this occurs in the joint active centre of the Rubisco enzyme. Thus, O_2 and CO_2 compete for the binding of RuBP in the reaction center, and to lower the activity of oxygenase reaction of Rubisco one needs to decrease the concentration of O_2 and increase the concentration of CO_2. Such a situation was created in the experiment: in the period of $^{14}CO_2$ assimilation a gas mixture of oxygen (1%) and carbon dioxide (0.3%) was delivered into the photosynthetic leaf chamber containing a treated leaf at the same concentration of $^{14}CO_2$.

As a result, ^{14}C incorporation into the products of glycolate metabolism relatively (%) reduced in all plants; however, in nitrate plants it was the least expressed. If one calculates the formation of glycolate pathway products in unit mass of fixed carbon dioxide ($\mu g\ CO_2$

m^{-2} s^{-1}) he will find that in control and urea fed plants the formation of these products lessened twofold while in nitrate plants it augmented twofold.

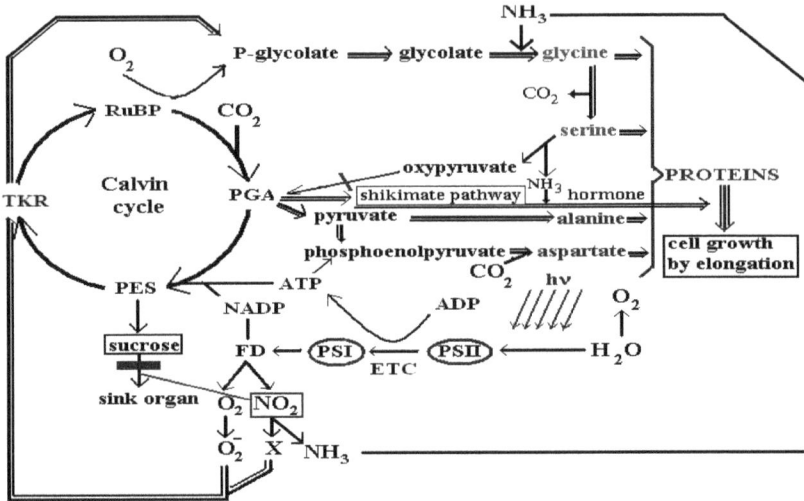

Fig. 1. The scheme of photosynthetic carbon metabolism regulation. ETC – electron transport chain, FD – ferredoxin, PES – phosphorous esters of sugars, PGA – phosphoglyceric acid, PSI – photosystem I, PSII – photosystem II, RuBP – ribulose 1,5-bisphosphate, TKR – transketolase reaction, X – unknown oxidizer, double line – intensification of a process

Based on this data the following conclusions can be made. Firstly, the suppression of RuBP-oxigenase activity by low O_2 and elevated CO_2 occurs only without nitrates. When nitrates are present, the formation of glycolate and its metabolites even enhances. Secondly, as the amount of CO_2 fixation product (PGA) increases at the saturable CO_2 concentration its non-reductive metabolism with appearance of alanine, malate and aspartate rises. Apparently, the latter is partly associated with rising CO_2 fixation in dark type-reaction catalyzed by phosphoenolpyruvate (PEP) carboxylaze (see the scheme in fig. 1).

The active production of glycolate under conditions suppressing RuBP-oxydenase reaction suggests that in the presence of nitrates glycolate is formed from other (not RuBP) phosphorous esters of sugars (PES) in transketolase reactions. Transketolase reaction requires superoxide radical (Asami & Akasava, 1977), produced in the Mehler reaction (Takabe et al., 1980).

The enhanced oxidation of PES may account for their decreased radioactivity and overall elevated non-carbohydrate tendency of photosynthesis in nitrate fed plants. The possibility of such a mechanism was declared long ago (Asami, Akasava, 1977). One of the probable mechanisms of transketolaze reaction (TKR) activation and glycolate formation from sugar di-phosphates upstream of RuBP might be an inhibited activity of phosphatases of fructose- and sedoheptulose-diphosphates (Heldt et al., 1978).

For fully grown leaves, which are typical exporters of assimilates, glycolate metabolism mainly terminates with formation of PGA (see the scheme in Fig. 1), that returns back to the Calvin cycle and is reduced to sugars. In the case of hindered sugar export the glycolate pathway becomes less closed. Glycine and serine, amino acids derived from glycolate, and also, alanine and aspartate, resulting from non-reducing PGA metabolism, can be used for the protein synthesis when the leaf recommences its growth through expansion. In our experiments with the removal of fruit elements from a cotton plant a leaf which was a source of assimilates could augment its size as many as 1.5-2 times in 10-15 days after the exposure (Chikov, 1987).

The return of the carbon of glycolate into the Calvin cycle is carried along the chain 2 glycolate $\rightarrow$ 2 glyoxylate + NH_2 $\rightarrow$ 2 glycine $\rightarrow$ serine + NH_2 $\rightarrow$ oxyglycerate + NH_2 $\rightarrow$ glycerate + ATP $\rightarrow$ PGA.

If the products of the glycolate pathway do not return to the Calvin cycle but accumulate and used in synthetic processes then the carbon will come to glycerate in smaller quantities. In this case the ratio of radioactivities of glycolate+glycine+serine to that of glycerte must increase.

This conclusion was confirmed in studies of the kinetics of ^{14}C introduction into those compounds. In the fertilized plants labeled carbon entered such compounds as glycolate, glycine, serine and alanine to a greater extent than in the controls, while glycerate, conversely, to a lesser extent (Fig. 2). As a result, the ratio of glycolate+glycine+serine/glycerte increased several-fold. The kinetic curves show that these differences need certain time (not less than 30 s) to reveal oneselves, that supports the metabolism of these substances in the direction of glycolate $\rightarrow$ glycine $\rightarrow$ serine $\rightarrow$ glycerate.

It is interesting that under conditions of hindered assimilate export from leaves after removal of sink organs a similar kinetics of ^{14}C introduction from $^{14}CO_2$ into glycolate and glycine was found at ambient CO_2 concentration (Chikov, 1987). The results were in good agreement with the idea of transketolase mechanism of glycolate formation, because in experimental plants labeled carbon from $^{14}CO_2$ appeared in glycolate earlier than that occurred in control ones. Characteristically, at saturable CO_2 concentration the kinetics curves for control and experimental plants were the same and both resembled the curve for experimental plants under ambient CO_2 (Chikov, 1987). The data indicate again that under the circumstances of slowed assimilate export and saturable CO_2 concentration glycolate and the products of its metabolism are derived from sugar phosphates that are predecessors of RuBP.

Thus, inhibition of assimilate export from the leaf and an increase of the Warburg effect in leaf photosynthetic gas-exchange, observed in some cases (Chikov, 1987), are accompanied by enhanced CO_2 metabolism through the glycolate pathway, but with increasing potion of glycolate formed in reactions not related with RuBP-oxidase one, most likely, in transketolase reaction of the Calvin cycle. This mechanism probably works also at delay of assimilate export from leaves under elevated level of plant nitrate nutrition. In the case of nitrate presence in the leaf, an oxidizer required to perform the transketolase reaction could appear from NO_2^- reduction in the chloroplast ETC (see Fig. 1).

In both cases growth processes become enhanced in the source leaf, for which early photosynthetic products in the form of amino acids alanine, serine, glycine and aspartate are used. It should be mentioned that these four amino acids (of 20 proteinogenic ones) represent over 30% (by number, not by weight) of fraction 1 protein (the main chloroplastic protein).

Furthermore, the suppression of sugar export can enhance the metabolism of phosphoerythroses through the shikimate pathway with the formation of aromatic (see Fig. 1) amino acids (tyrosine, phenylalanine, tryptophan) and then hormonal substances (auxins), creating an additional substrate base for metabolism rearrangement in the leaf (and whole plant). So, a key factor for triggering metabolism readjustment in the leaf (the plant) is inhibition of sugar outflow from leaves.

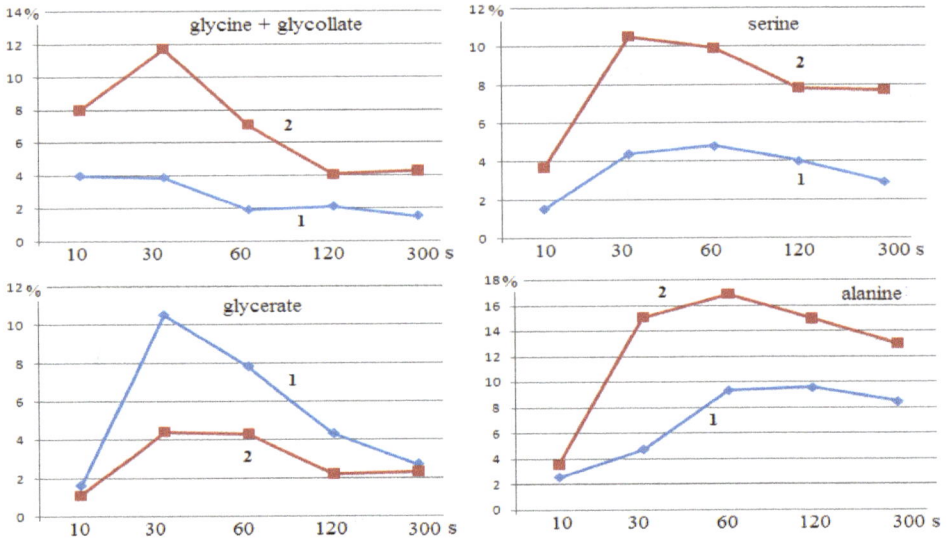

Fig. 2. The kinetics of ^{14}C introduction into some products of photosynthesis (% radioactivity of water-ethanol soluble fraction) in leaves of wheat plants cv. Moskovskaya-35. 1 – control; 2 - fertilized with 2 g of nitrogen as $Ca(NO_3)_2$

2.3 The influence of nitrogen supply on the dynamics of post-photosynthetic conversion of ^{14}C-products of photosynthesis

The actions of oxidized and reduced N vary not only in primary photosynthetic products formation but in their subsequent metabolisation to the end transport compound, sucrose.

After short exposure of leaves to $^{14}CO_2$, ^{14}C content in sucrose was reduced under N treatments, irrespectively of the N form, which seemed to be due to intensive formation of non-carbohydrate substances (organic and amino acids). Then in the next 30 min ^{14}C accumulated in the end photosynthetic product, sucrose, in all plant variants (Fig. 3). However in plants fed with nitrogen this accumulation occurred at higher rates than in control, and as a result, ^{14}C content in sucrose in both N fed variants almost equaled and approached the control level.

From this moment sucrose radioactivity began to decrease in all variants. During 1.5 h it reduced by 75% in control and by 45% in plants fed with urea. In the next 20 h sucrose radioactivity in the plants was running down steadily to the level of 4-5% of the maximal value. Unlike these two variants, nitrate plants exhibited even descent of sucrose radioactivity from 2 h point and during the next 20 h not reaching 30% of the maximal level. All this evidenced significant distinction of the sucrose transport in nitrate plants from those both in control and urea fed ones.

In summary, at water or urea feeding labeled primary photosynthetic products are quite successfully converted into sucrose with its subsequent export in the post-photosynthetic period, while in nitrate fed plants labeled assimilates are piled up as sucrose which remains in the source leaves for a long time period.

To reveal a mechanism of assimilate export delay from leaves special model experiments were performed, where nitrate or urea solutions were injected into isolated plant shoots.

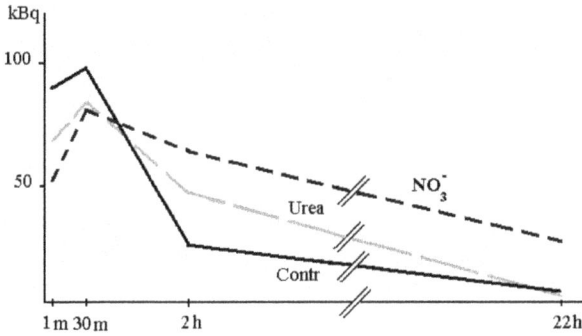

Fig. 3. The influence of nitrate or urea fertilization of wheat plants on the dynamics of ^{14}C content in sucrose after 1 min $^{14}CO_2$ assimilation by leaves

3. The influence of different N forms in the transpiration water stream on photosynthetic carbon metabolism and assimilate transport from leaves

3.1 Methodical peculiarities of the procedure

To study the immediate influence of increased nitrogen nutrition level on photosynthesis and assimilate transport we have designed a special device (Fig. 4) for introduction of solutions or water into a shoot under the pressure of 10^4 Pa, corresponding to a normal root pressure. Preliminary studies had shown that a shoot could survive for several days under conditions of direct sunlight without demonstrating any visible lesions.

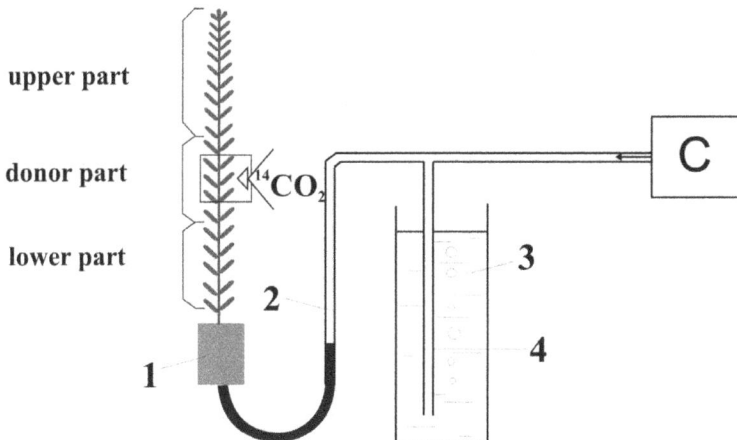

1 – a mandrel for shoot fixation; 2 – a silicone tube with a solution fed into the apoplast; 3 – a bath with water and a monostat (4) drowned in it to the depth of 1 m to imitate root pressure; C –a compressor

Fig. 4. The scheme of solution introduction into an isolated flax shoot

For photosynthesis investigations a central shoot part was exposed to $^{14}CO_2$ using a photosynthetic chamber. This allowed not only to determine ^{14}C distribution among the labeled products of 2-3 min photosynthesis, but also to assess the character of allocation of labeled photosynthetic products along the shoot to different shoot parts (^{14}C-source leaves, the apex, parts above and below the ^{14}C-source part) in the post-photosynthetic period.

There are two main types of plants differing in phloem loading strategy and transportable photosynthetic products (Lohaus & Fischer, 2002). These are plants with the apoplastic phloem loading strategy in which photoassimilates move outside the plasma membrane of parenchyma cells and are loaded from the apoplast into the companion cells or sieve elements using energy-dependent transporters. In other plants photoassimilates move from assimilating to conducting cells along the system of plasmodesmata without membrane crossing. For our experiments we have chosen flax (*Linum usitatissimum* L.) as a plant with apoplastic phloem loading and a willow-herb (*Chamerion angustifolium* (L.) Holub.) as a symplastic plant.

3.2 The introduction of nitrate solution into an apoplastic plant

The study of [14]CO_2 distribution among the labeled products of photosynthesis in flax leaves has shown that nitrate injection into the apoplast had the same effect on metabolism as fertilization of plants with nitrogen via soil. The introduction of nitrates resulted in decreased ^{14}C incorporation into sucrose and the ratio of labeled sucrose to hexoses, and increased ^{14}C distribution into the glycolate pathway products (Table 4).

Compounds	Water	KNO₃ (0.5%)	KNO₃ (1.5%)	Ca(NO₃)₂ (0.5%)	Urea (2.5%)
Sucrose	59.2 ± 1.6	52.6 ± 2.5	48.1 ± 2.8	47.0 ± 1.8	56.4 ± 0.9
Phosphorous esters of sugars	3.2 ± 0.3	2.1 ± 0.8	3.5 ± 0.3	0.3 ± 0.0	2.9 ± 0.8
Hexoses	4.3 ± 0.6	4.0 ± 0.5	6.0 ± 0.5	5.0 ± 0.8	3.5 ± 0.5
Sucrose/hexoses	13.8	13.2	8.0	9.4	16.1
Amino acids	21.1 ± 1.1	29.2 ± 3.1	32.1 ± 3.2	29.4 ± 3.3	24.5 ± 0.8
including: glycine	1.6 ± 0.2	2.1 ± 0.5	2.6 ± 0.7	3.7 ± 0.4	1.8 ± 0.1
serine	5.7 ± 0.3	14.9 ± 2.5	15.3 ± 1.6	11.8 ± 2.3	5.3 ± 0.3
alanine	11.7 ± 0.7	9.4 ± 0.3	10.8 ± 1.0	10.1 ± 1.0	13.5 ± 0.7
Organic acids	3.6 ± 0.3	6.5 ± 0.9	4.3 ± 0.4	5.9 ± 0.7	6.4 ± 0.2
including: glycerate	0.8 ± 0.1	4.5 ± 0.9	1.6 ± 0.1	1.7 ± 0.2	0.7 ± 0.1
malate	1.8 ± 0.1	0.9 ± 0.1	1.5 ± 0.3	3.1 ± 0.5	4.8 ± 0.1
Pigments	2.7 ± 0.2	1.2 ± 0.1	2.0 ± 0.2	3.0 ± 0.4	1.5 ± 0.1
Others	5.9	4.4	4	9.4	4.8

Table 4. The influence of nitrate and urea feeding into the apoplast on ^{14}C distribution among labeled products in flax source leaves immediately after exposure to [14]CO_2 (% radioactivity of water-soluble fraction)

As it was indicated above, increased label incorporation into the glycolate pathway products is also characteristic of enhanced soil nitrogen nutrition, however at soil fertilization ^{14}C was mostly incorporated into glycine and glycolate while at direct nitrate feeding into the shoot the radioactivity of serine was higher (Table 4). Watering of plants grown in the soil with $Ca(NO_3)_2$ solution on the eve of the day of experiment apparently leaves the plants enough time for activation of protein synthesizing systems, utilizing serine. The protein synthesizing systems of plants supplied with nitrates through the transpiration water stream were probably not ready for utilization of ample amounts of newly-formed amino acids and this resulted in ^{14}C accumulation in amino acids (first of all, serine). The speed of nitrate reduction exceeds the flow through the GOGAT-pathway approximately by 25%, leading to reduced nitrogen piling up in the immediate products such as ammonium and glutamine as well as photorespiration metabolites, glycine and serine (Stitt et al., 2002).

Why does ^{14}C distribution into sucrose immediately after $^{14}CO_2$ assimilation decrease under nitrate nutrition of plants? The common explanation of reduction in sucrose synthesis after nitrogen nutrition of plants is the following: at enhanced inflow of nitrogen into the plant photosynthetic products and energy are diverted from sucrose synthesis to the formation of nitrogen-containing compounds (Champigny & Foyer, 1992). However, this explanation does not accord with different actions of oxidized and reduced nitrogen on sucrose production (Batasheva et al., 2007). Undoubtedly, the reduction of nitrates requires more energy spending, but the electron transport in chloroplast is known to have significant plasticity and different types of mitochondrial electron transport are probably take part in the maintenance of necessary $ATP/NAD(P)H$ ratio in the cell (Noctor & Foyer, 1998; 2000).

Another reason of decreased sucrose synthesis under nitrate nutrition may be a feedback regulation of photosynthesis resulted from slowdown of sucrose transport from leaves.

Injection of nitrate solution into a flax shoot resulted in pronounced response of photosynthetic rate, photosynthetic carbon metabolism and assimilate transport measured in 3 h after $^{14}CO_2$ assimilation (Batasheva et al., 2007). That is why we decided to study those changes in more detail.

Studies of the dynamics of ^{14}C distribution along the shoot have revealed that feeding nitrates to the apoplast resulted in inhibition of assimilate export. In 30 min after $^{14}CO_2$ assimilation in shoots fed with KNO_3, relative content and distribution of ^{14}C outside the ^{14}C-donor part did not virtually differ from those in control (Table 5). In additional 2.5 h of post-photosynthesis, in control plants, ^{14}C relative contents outside the donor part increased with significant rising of ^{14}C content in the lower part (Table 5). In shoots fed with nitrates, ^{14}C content outside the donor part in 2.5 h also increased but in lesser degree than in control, and the overall ^{14}C relative content in the lower shoot part became smaller whereas in the upper part – greater compared to control.

In our previous experiments it was found that feeding urea solution (0.15% (w/v)) containing the same amount of nitrogen as potassium nitrate solution (0.5% (w/v)), led to almost the same distribution of ^{14}C through the shoot as feeding water (Batasheva et al., 2007). NH_4NO_3 solution feeding led to lesser inhibition of assimilate export and slightly changed the pattern of ^{14}C partitioning along the shoot compared to those under KNO_3 solution feeding (Batasheva et al., 2007).

In flax, most ^{14}C-sucrose after being loaded into the phloem terminals in source leaves moves downwards within stem phloem vessels, and ^{14}C content in the plant lower parts gradually raises (Chikov & Bakirova, 2004). During their movement along the stem assimilates can partly escape into the stem apoplast, and long distant transport is controlled by retention and retrieval mechanisms in the phloem (Ayre et al., 2003). The assimilates lost to the apoplast can either be re-loaded back into the phloem (Kühn et al., 1997) or be transported upwards with the transpiration water stream. If assimilates leaked in the apoplast cannot be quickly loaded back into the phloem due to some reason then the portion of the assimilates inflowing into the upper plant part with the transpiration water stream would increase. It seems likely that this is the reason of increased ^{14}C content in the upper parts of the shoots fed with nitrates. Enhancement of sucrose hydrolysis in the apoplast in the presence of nitrates must result in appearing large amounts of labeled hexoses, which can not be loaded into the phloem (Turkina et al., 1999) and are easily carried away upwards with the transpiration stream.

Shoot part	H_2O		KNO_3	
	30 min	3 h	30 min	3 h
Above ^{14}C-donor part	3.5 ± 0.6	5.2 ± 0.9	3.1 ± 0.4	9.8 ± 2.0
including: top	0.2 ± 0.0	1.0 ± 0.3	0.2 ± 0.0	3.3 ± 1.2
leaves	$2.0 \pm 0.$	1.7 ± 0.4	$1.3 \pm 0,5$	1.4 ± 0.3
cortex	0.9 ± 0.0	1.5 ± 0.4	$1.1 \pm 0,1$	3.1 ± 0.6
wood	0.4 ± 0.05	1.0 ± 0.30	$0.5 \pm 0,1$	2.0 ± 0.5
^{14}C-donor part	82.2 ± 2.8	67.5 ± 0.9	82.0 ± 1.3	76.3 ± 2.5
Below ^{14}C-donor part	14.3 ± 0.9	27.3 ± 1.1	14.9 ± 1.3	13.9 ± 3.8
including: leaves	9.9 ± 1.1	13.5 ± 1.5	7.3 ± 2.1	9.0 ± 1.4
cortex	3.0 ± 0.3	6.7 ± 0.8	5.8 ± 1.7	2.2 ± 0.6
wood	1.4 ± 0.2	7.1 ± 1.6	1.8 ± 0.7	2.7 ± 0.8
Above/below	4.1	5.25	4.8	1.4

Table 5. The influence of nitrate feeding through the transpiration water stream on ^{14}C distribution among the organs of flax in 30 min and 3 h after $^{14}CO_2$ assimilation by the middle shoot part (% of whole shoot radioactivity)

As we have shown previously (Batasheva et al., 2007), increased ^{14}C content in the lower part of plants fed with water is not connected with label accumulation due to synthesis of any high-molecular weight substances in this period.

In 30 min after $^{14}CO_2$ assimilation relative content of ^{14}C-sucrose in source leaves increased both in control and in nitrate fed plants (Table 6). In the following 2.5 h in water fed shoots ^{14}C-sucrose content decreased to lower values compared to those observed immediately after $^{14}CO_2$ fixation whereas in nitrate fed shoots it continued growing. Thus, in 3 h after $^{14}CO_2$ assimilation relative ^{14}C content in sucrose in source leaves of the shoots fed with KNO_3 rose significantly up to 75%. Relative radioactivity of hexoses practically did not change which resulted in increase of the labeled sucrose/hexoses. The similar picture of ^{14}C-sucrose dynamics was described above for wheat plants fertilized with nitrates (Fig. 3).

Localization of labeled sucrose in leaves was determined by autoradioagraphy of whole leaves taken from ^{14}C-donor part in 30 min and 3 h after $^{14}CO_2$ assimilation. It turned out that in 30 min after $^{14}CO_2$ assimilation in water fed shoots the label was concentrated mainly in leaf large veins whereas in nitrate fed shoots – outside them (Fig. 6). It indicates that upon nitrate feeding the accumulation of the labeled assimilates occurred either in mesophyll cells or in minor vein cells from which they were not transported to large veins.

Labelled compounds	H₂O		KNO₃	
	30 min	3 h	30 min	3 h
Sucrose	71.6 ± 1.7	50.7 ± 1.6	67.7 ± 0.5	75.0 ± 1.6
Glucose	9.3 ± 1.7	17.2 ± 1.7	7.6 ± 0.3	4.8 ± 0.6
Fructose	4.9 ± 0.5	15.0 ± 1.9	3.3 ± 0.6	4.2 ± 0.4
Glycine	0.9 ± 0.5	1.9 ± 0.6	2.4 ± 0.3	1.0 ± 0.3
Serine	1.7 ± 1.3	2.2 ± 0.2	5.3 ± 1.3	1.4 ± 0.5
Aspartate	0.2 ± 0.0	0.2 ± 0.1	0.2 ± 0.1	0.2 ± 0.1
Glutamate	0.9 ± 0.1	0.5 ± 0.1	0.9 ± 0.2	0.5 ± 0.0
Malate	1.3 ± 0.1	2.5 ± 0.5	1.4 ± 0.3	1.0 ± 0.2
Alanine	1.3 ± 0.1	1.0 ± 0.3	1.2 ± 0.1	1.4 ± 0.1
Pigments	1.9 ± 0.3	2.2 ± 0.5	1.5 ± 0.1	1.6 ± 0.1
Others	6.0	6.6	8.5	8.9

Table 6. The incorporation of ^{14}C into the products of photosynthesis in 30 min and in 3 h after 2.5 min $^{14}CO_2$ assimilation

In 3 h of post-photosynthesis in control leaves ^{14}C mostly disappeared from large bundles. In nitrate fed leaves the differences between ^{14}C contents inside and outsides the large bundles became even more contrasty. It seemed likely that in control shoots ^{14}C-sucrose export exceeded its synthesis while in nitrate fed shoots ^{14}C-sucrose synthesis was not compensated by its removing from the leaf. Earlier it was shown that feeding urea (0.15% (w/v)) into the apoplast led to the same picture of ^{14}C-sucrose dynamics as feeding water (Batasheva et al., 2007).

Thus, nitrate feeding resulted in graduate accumulation of sucrose in source leaves, which was observed on the background of assimilate export suppression. It is interesting that similar dynamics of sucrose radioactivity changes in mature leaves was observed by Möller and Beck (1992), who studied metabolism of labeled sucrose when unlabelled sucrose was constantly inflowing into the apoplast. In the work high sucrose content in the apoplast led to labeled sucrose accumulation within cells. A question arises why labelled sucrose was not hydrolyzed especially under the conditions of inhibited export appearing when sucrose is fed into the apoplast or observed in our experiments upon nitrate feeding.

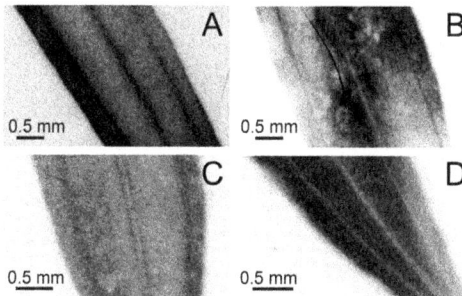

Fig. 6. The influence of water or KNO₃ solution (0.5%) feeding on ^{14}C distribution within the donor leaf in 30 min (A – water; B – nitrate) and 3 h (C – water; D – nitrate) after exposure to $^{14}CO_2$. Darker places correspond to greater ^{14}C contents

It is probable that labeled sucrose was in the leaf conducting system where sucrose hydrolyzing activity is negligible (Dubinina et al., 1984). On the other side, we can not rule out the possibility that at least a part of labeled sucrose was accumulated in mesophyll cells. For instance, in barley shoots, suppression of assimilate export resulted in sucrose accumulation in bundle sheath parenchyma cells and in mesophyll cells and probably in the bundles themselves. In the mesophyll and bundle sheath cells an increase in glucose contents was also observed (Koroleva et al., 1998; Pollock et al., 2003).

Thus, it turned out that nitrates fed into the cut shoots with the transpiration water stream exerted the same action on assimilate outflow and ^{14}C distribution among the products of photosynthesis as nitrate fertilization.

The similar action of nitrate fertilization (Chikov et al., 2001) and artificial nitrate infusion into the apoplast on ^{14}C distribution among the products of photosynthesis, primarily, the ratio of labeled sucrose to hexoses, allowed us to conclude that enhanced sucrose hydrolysis under increased nitrate fertilization of plants was in some way connected with the presence of nitrate anion in the apoplast. Urea did not exert such an effect (Batasheva et al., 2007).

Thus, a possible reason of changes in metabolism at soil fertilization of plants with nitrates may be an enhancement of sucrose hydrolysis in the apoplast at nitrate income therein. Because monoses formed in the process of sucrose hydrolysis cannot be loaded into the phloem terminals and, consequently, take part in assimilate transport, assimilate outflow from leaves becomes lowered. The monoses have to return back to mesophyll cells and then, decreased photosynthesis rate and its non-carbohydrate tendency can be the consequences of feedback inhibition of photosynthesis and sucrose formation.

The possible influence of nitrates through activation of sucrose hydrolysis by cell wall invertase is supported by numerous data on the similar action of nitrates and sugars on certain gene expressions, primarily those connected with nitrogen metabolism (Stitt et al., 2002). Furthermore, it was shown that one of nitrate carrier genes is induced by NO_3^- and decrease of its transcript abundance in the dark could be prevented by addition of sucrose. The gene could also be induced in the absence of external NO_3^- by a sharp decline of medium pH from 6.5 to 5.5 (Forde, 2000), which could cause an activation of the apoplastic invertase, because its pH optimum lies in the acidic pH range (Brovchenko, 1970).

In plants with the apoplastic type of phloem loading inhibition of photoassimilate transport by putting a cold collar on the petiole was associated with appearance of large central vacuoles in phloem companion cells (Gamalei & Pakhomova, 2000). So, it could be possible that in our experiments in the post-photosynthetic period sucrose accumulated not in the transport path itself but in vacuoles formed in companion cells.

Significant changes in minor vein cell structure as a response to increase in nitrate concentration in the apoplast were found (Fig. 7). In control leaves, companion cells were characterized by well developed system of cell wall invaginations and very slight vacuolization of their protoplasts (Fig. 7A). In 30 min after beginning of KNO_3 (0.5%) solution feeding sieve elements filled up with electron-transparent vesicles, and a large central vacuole was formed in companion cells (Fig. 7B). Some indication of endocytosis allows to guess that the vacuolation was a result of seizure of extracellular milieu containing sugar in high concentration. This could be the way by which companion cells protected themselves from osmotic stress. The vacuole was growing in size for the next 2 h (Abdrakhimov et al., 2008).

When the dynamics of changes in ultrastructure of leaves in the presence of nitrate in the apoplast was studied it turned out that in mesophyll, bundle sheath and phloem parenchyma cells the mitochondria matrices clarified and dictyosomes curled in the first 30 min, but repaired their structure by 1 h after beginning of nitrate feeding (Abdrakhimov et al., 2008). We supposed that such a two-phase response could consist of a quick direct action of nitric oxide, formed from nitrate, on leaf ultrastructure and slower changes related with inhibited photoassimilate export (Abdrakhimov et al., 2008)

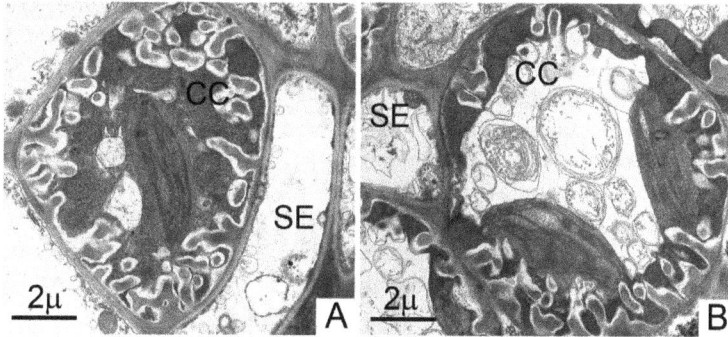

Fig. 7. Ultrastructure of terminal vein cells in the leaves of common flax 1 h after supplying water (a) or 0.5% KNO_3 (b) to the apoplast. CC - companion cell; SE - sieve element (Batasheva et al., 2007)

Thus, analysis of flax leaf terminal vein ultrastructure revealed vacuole formation in companion cells in response to nitrate feeding into the apoplast. One can guess that, similarly to putting a cold collar on the petiole, nitrate feeding into the apoplast initially creates hindrances to assimilate transport within the sieve elements or assimilate transport from companion cells to sieve elements. Augmentation of callose synthesis is known to happen very quickly in response to low temperature, within several minutes (Kursanov, 1976). It is tempting to speculate that when nitrates are excessive NO is generated, which, as a stress signal, can trigger synthesis of callose.

3.3 The influence of nitrates fed into the apoplast of a symplastic plant

The introduction of nitrate solution into the plant apoplast of willow-herb shoots (control – water) caused an inhibition of $^{14}CO_2$ fixation by leaves (Khamidullina et al., 2011). Analysis of distribution of labeled photosynthetic products among the plant organs in 3 h after $^{14}CO_2$ assimilation has shown that the intensity of assimilate export from willow-herb leaves (both in control and treated plants) was lower than that from flax leaves. In willow-herb plants it was 8.1 % in control and 4.5% in nitrate plants, while under the same conditions in flax plants more than 40% of ^{14}C-assimilates formed in leaves were exported (Khamidullina et al., 2011). Such a difference in the export rate in plants with various types of phloem loading could be linked with synthesis of transport products represented by tri- and tetra-saccharides in symplastic plants. Probably it is due to variable mechanisms of transport, longer time required for oligosaccharide synthesis, and different diurnal dynamics of assimilate export (Gamalei, 2004) the transfer of labeled photosynthetic products was slower in willow-herb than in flax.

As in the flax plants, in the willow-herb plants nitrates stimulated the transfer of labeled assimilates to the upper shoot part. The difference will become larger if the content of labeled assimilates in the upper shoot part is assessed relatively to that exported from the ^{14}C-donor leaf. As a result the ratio of exported labeled assimilates in the upper part to that in the lower part was 4.8 in control and 2.4 in nitrate plants. Because the labeled assimilates are likely to get into the upper shoot part with the transpiration water stream after their leakage from the phloem into the stem apoplast the data evidences the enhancement by nitrates of the permeability of phloem tubes to transport photosynthetic products. The latter could be due to the slowdown of phloem transport and increased concentration of products in the stem phloem or reduced return of sugars from the apoplast back into the phloem (Kühn, et al., 1999).

The analysis of ^{14}C distribution among the labeled products of low molecular weight fraction revealed both common traits and differences in the action of nitrates on photosynthetic metabolism in flax and willow herb. In both cases ^{14}C introduction into sucrose decreased under nitrates, that indicates the similarity of nitrate influence on photosynthetic carbon metabolism in symplastic and apoplastic plants. The influence on photosynthetic carbon metabolism is not connected with the presence of potassium cation, because the same changers were also observed under calcium nitrate nutrition of plants (Chikov et al., 1998).

After comparing the dynamics of ^{14}C acquirement by sucrose in flax and willow herb the similarity was also found. After short expositions to $^{14}CO_2$ mirroring the intensity of sucrose synthesis a decrease in sucrose radioactivity was observed. Later in treated plants the label piled up in sucrose as a result of sucrose export arrest. Thus, the characters of post-photosynthetic changes of sucrose radioactivity in both plant types were resemblant.

As it is known in symplastic plants a significant fraction of exported sugars is represented by oligosaccharides (Pristupa, 1959). Probably that is why the dynamics of changes in ^{14}C content in this group of compounds is similar to that of sucrose. After analysis of the data on oligosaccharide and sucrose content in the willow-herb it was found that in nitrate plants ^{14}C was gradually accumulated with time in sucrose and oligosaccharides. In control, this parameter changed insignificantly (within the error).

Thus, nitrate anion hinders assimilate export irregardless of the way of sugar transfer from leaf mesophyll cells to the phloem, suggesting that the mechanisms of nitrate action on assimilate transport in both plant types have much in common.

The radioactivity of phosphorous esters of sugars (PES) in nitrate fed willow-herb plants immediately after 3 min exposition to $^{14}CO_2$ was increased apparently indicating their hampered conversion to export sugars (Khamidullina et al., 2011). The general elevation of radioactivity in the glycolate pathway products (serine, glycine and glycolate) suggests that photooxidation processes become activated under nitrate feeding. As it was shown by Chikov and Bakirova (1999) the enhanced production of glycolate pathway products under increased nitrate nutrition may be connected not with atmospheric oxygen but with reactive oxygen species, generated in the course of nitrate reduction in the chloroplast electron transfer chain, because their formation only slightly declined after a decrease in oxygen concentration down to 1%. A possible existence of other sources of glycolate and its metabolites under delayed assimilate export from the leaf was demonstrated previously in the kinetic experiments (Chikov et al., 1985).

In 3 h after beginning of nitrate feeding into the apoplast the most evident changes of ultrastructure were observed in the mitochondrion and vacuolar system of cells of conducting system (Khamidullina et al., 2011). The electron density of mitochondrial matrix increased and christa lumens swelled indicating increased osmotic pressure in the cytosol of cells in conducting system. This was paralleled by accumulation of fibrillar inclusions represented by polymer substance that generated a homogenous network evenly distributed throughout the organelle interior on the cross-sections (Fig. 8).

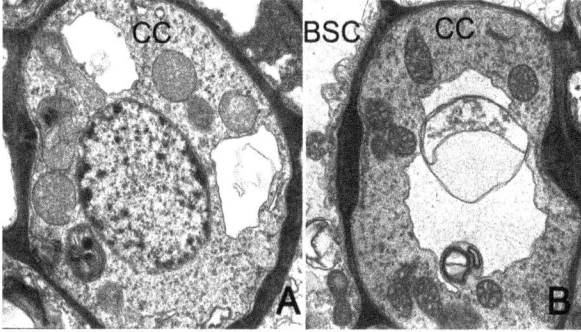

Fig. 8. Companion cells in the leaves of fireweed plants in 3 h after water or potassium nitrate (50 mM) feeding into the apoplast. BSC – bundle sheath cells, CC – companion cells (Khamidullina et al., 2011)

The nature of the inclusions is unknown, but, bearing in mind the abundance of the polymer and the fact that analogous formations were revealed also in the sieve elements we have guessed that the observed structures is either photosynthetic transport product itself or the result of its polymerization. But irregardless of the way of formation its accumulation must also express a blockage of transfer of carbohydrate transport product in willow-herb plants. The organelles covered with semi-permeable membranes, mitochondria and plastids are known to be ideal intracellular osmometers (Gamalei, 2004). Obviously, hindered transport of low molecular weight compounds from leaf blades must result in their piling up in the places of phloem loading and osmotic contraction of the organelles. That is just what was observed in our experiments (Fig. 8). Similar changes we have earlier seen in an apoplastic plant, flax, in 3 h after beginning of nitrate feeding through the transpiration stream (see above). However, oligosaccharides rather than sucrose are accumulated in companion cell vacuoles, diagnosable by fibrillar inclusions.

4. The influence of NO donor, sodium nitroprusside on photosynthetic carbon metabolism

As it was shown above in model experiments, under direct feeding of salts, containing nitrate anion, in the plant apoplast, decreased $^{14}CO_2$ assimilation rate and photoassimilate export from leaves with changes in photosynthetic carbon metabolism and ultrastructure of phloem companion cells were observed. Elevation of nitrate content in a plant is a condition advantageous for nitric oxide generation from nitrite by both enzymatic and non-enzymatic ways (Neill et al., 2003). From whence a participation of NO signal system in triggering a mechanism, arresting assimilate export from leaves under increased nitrate nutrition level may be proposed. To test this proposition we investigated the influence of NO donor, sodium nitroprusside, on photosynthetic metabolism and ultrastructure of cells of flax leaf blades.

As in experiments with injection of potassium nitrate solution into the shoot, sodium nitroprusside solutions (SNP) with concentrations of 50 µM, 100 µM and 1mM were introduced into cut flax shoots. In 30 min a middle shoot part containing 8-10 leaves was placed into a photosynthetic chamber, where $^{14}CO_2$ (0.03%) was delivered at natural sunlight (^{14}C-donor part of a shoot).

Feeding SNP solution into the apoplast resulted in reduced $^{14}CO_2$ assimilation, with the inhibition of $^{14}CO_2$ fixation by SNP (50 µM, 100 µM) being almost equal to that by potassium nitrate (50 mM). Further elevation of nitroprusside concentration to 1 mM enhanced the inhibition of photosynthesis up to 75% (Table 7, Batasheva et al., 2010). Such a sharp drop of carbon dioxide uptake after increasing SNP concentration up to 1 mM can partially be explained by NO participating in stomatal closure (Mata & Lamattina, 2001). NO involved in stomatal movements can be generated from nitrate anion by nitrate reductase (Desikan et al., 2002).

SNP introduction into the shoot affected ^{14}C-assimilate distribution throughout the plant similarly to nitrate feeding. In the treated plants assimilate outflow from the donor part slowed down and their transfer into the upper part relatively increased (Table 7). Whereas in control the main part of ^{14}C-products of photosynthesis exported from the donor-part appeared in the lower shoot part, after feeding SNP the relatively higher content of exported ^{14}C-assimilates was found in the upper shoot part (Batasheva et al., 2010).

Shoot parts	Control (water)	SNP (50 µM)	SNP (100 µM)	SNP (1mM)
Top	2.5 ± 0.5	3.0 ± 1.4	6.0 ± 0.8	1.8 ± 0.2
Upper shoot part	5.3 ± 1.3	6.3 ± 2.3	9.1 ± 0.5	11.5 ± 6.2
Above ^{14}C-source part	7.8	9.8	15.1	13.3
^{14}C-source leaves	42.3 ± 4,5	49.2 ± 4,4	46.0 ± 1,8	21.4 ± 2.8
^{14}C-stem	24.2 ± 3.4	31.5 ± 8.1	28.5 ± 4.2	57.1 ± 13.2
^{14}C-source part summarized	66.5	80.6	74.5	78.5
Below ^{14}C-source part	25.7 ± 4.8	9.5 ± 3.1	10.4 ± 2.0	8.0 ± 4.4
Above/Below part	3.29	0.97	0.69	0.60
Uptake of $^{14}CO_2$ (% control)	100	88	86	25

Table 7. The influence of sodium nitroprusside solution introduced into the flax shoot with the transpiration water stream on ^{14}C distribution among different plant parts in 3 h after $^{14}CO_2$ assimilation by the middle shoot part (% total shoot radioactivity) (Batasheva et al., 2010)

SNP treatment resulted in altered photosynthetic carbon metabolism. Analysis of ^{14}C distribution among the labeled products of photosynthesis has shown that the largest relative changes occurred in the glycolate pathway compounds (serine, glycine, glycolate) and sugars (Table 8). As in the case of nitrate feeding into the apoplast (Batasheva et al., 2007) the portion of ^{14}C in sucrose decreased, leading to the lowered ratio of labelled sucrose to hexoses. Elevated monosaccharide availability relatively enhanced production of other storage sugars (oligosaccharides). As a result the ratio of sucrose to oligosaccharides reduced almost three times.

Compounds	Control (H_2O)	Sodium nitroprusside (1 mM)
Sucrose	48.0 ± 3.9	33.6 ± 1.1
Hexoses	5.3 ± 0.9	10.1 ± 1.9
Serine+glycine+glycolate	0.7 ± 0.2	4.1 ± 0.4
Amino acids	22.0 ±1.7	23.3 ± 1.8
Oligosaccharides	3.3 ± 0.9	6.2 ± 1.2
Pigments	2.6 ± 0.5	3.1 ± 0.7
Others	20.7	22.7

Table 8. The influence of sodium nitroprusside solution (1 mM) infused into the flax shoot with the transpiration water stream on ^{14}C distribution among the labeled compounds of 2.5 min $^{14}CO_2$ assimilation by leaves (% water-ethanol soluble fraction) (Batasheva et al., 2010)

On the background of lower ^{14}C content in sucrose the formation of the glycolate pathway products substantially increased, which is also a characteristic of metabolism under increased nitrate concentration in the plant. However, in distinction from nitrate treatment under nitroprusside injection ^{14}C income into amino acids did not increase compared to control. This was probably related with the absent of additional nitrogen as a substrate for amino acid synthesis.

Infiltration of plants with NO donor lead to appreciable alterations in the organization of both assimilating and conducting system cells (Fig. 9). In 30 min after beginning of feeding sodium nitroprusside into the apoplast the structural changes became obvious (Fig. 9 B, C). They were expressed in vacuolization of companion cells with appearing of a large central vacuole (Fig. 9B). The peaks of the crests of cell wall labyrinth became osmophilic and often reached the vacuole interior. Structural changes of the assimilating cell-bundle sheath cell-phloem parenchyma domain expressed in clarification of mitochondrion matrices (a place of glycine decarboxylation) and dictyosome curling into ring-shaped structures (Batasheva et al., 2010).

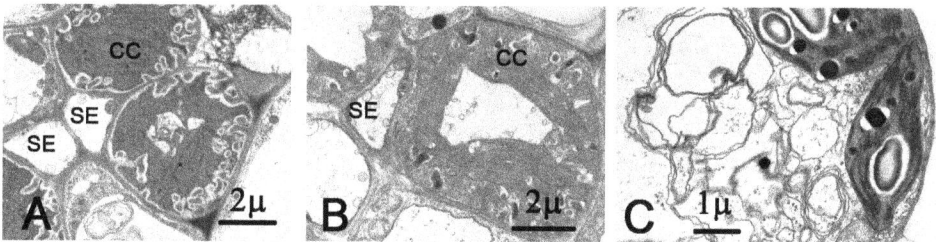

Fig. 9. The influence of sodium nitroprusside solution (50 µM) on ultrastructure of flax leaf companion cells. A – control, B, C –after 30 min of introduction of sodium nitroprusside with the transpiration water stream; CC – companion cell, SE – sieve element (Batasheva et al., 2010)

NO induced formation of numerous multimembrane and multivesicular structures in assimilating cells (Fig. 9C) in many ways analogous to those observed in 30 min after beginning of nitrate salts feeding into the apoplast (Abdrakhimov et al., 2008).

Because the effects of NO and nitrates on cell ultrastructure were similar, one can suppose that a product of incomplete reduction of nitrate, NO, is capable of inhibiting assimilate transport directly in phloem cells. This suggestion is supported by the observation that NO is able to elevate callose content in the leaf, β-1,3-glucane, taking part, inter alia, in plugging sieve plate pores (Paris et al., 2007). Exhibitive of this mechanism is also data on increased NO synthesis (Zottini et al., 2007) and inhibited callose destruction (Serova et al., 2006) in the presence of salicylic acid, implying that the suppression of callose breakdown may be mediated by nitric oxide.

Appearing of vesicles inside the vacuole owing to endocytosis can be indicative of rising osmolarity of cell environment, probably as a consequence of sugar accumulation in the leaf apoplast due to slowdown of sugar export along the phloem.

The discovered facts allow us to propose that a likely reason of assimilate transport suppression by nitrate is generation of nitric oxide under increased nitrate concentration in cells and point to involvement of NO signal system in regulation of assimilate movement in the whole plant system.

5. A possible regulatory interaction of nitrates and sugars at alterations of other external factors

In the plant there is a regulatory link between the two main mass flows – sugars of photosynthetic origin and nitrates. The flows are heading towards each other and their interaction reacts to a change in sink-source relationships between assimilating and photosynthate consuming organs.

Because an influx of nitrates into a plant is determined by root system activity, this assimilate consumer holds a special place compared to other sink organs, competing with the latters for photosynthates. Different sink organs are uneven in their functions. Unlike other acceptors, such a sink as the root system, could provide feed-back to photosynthetic apparatus not only through sugar consumption but also through export of mineral nutrients (primarily nitrates) and triggering NO signal system (Fig. 10).

The results of the works (Batasheva et al., 2007; Abdrakhimov et al., 2008) on ultrastructure and radioautography of leaves after NO_3^- intrusion into the apoplast evidence a block to the sugar flow at the level of long-distance transport, most likely, either on the stage of transition of primary sieve tubes to the vascular bundles or when sucrose is moving along the phloem. NO donor, sodium nitroprusside, is known to induce callose accumulation (Paris et al., 2007), involved in clogging the pores in the sieve elements of phloem.

The degree of nitrate reduction is directly related to the sugar availability (Stitt & Krapp, 1999). The process of nitrate uptake has long been known to require photosynthetic energy. At low doses of nitrates and intense photosynthesis they are almost completely reduced in the roots with nitrogen coming to aboveground organs in the form of amides and amino acids. As nitrate concentration in the roots rises, in substances exported from roots to leaves a fraction of amino acids increases at first, and then also NO_3^-.

A disproportion of the two main mass flows (nitrate and sugars) is likely to be of a great importance. Here, two modes of events are possible:

1) a sudden elevation of nitrate income into the plant relatively to the existing (established) small amount of synthesized photosynthetic products (sugars);

2) sugar availability increases at unchanged (or decreased) nitrate influx into the plant.

The first may occur after additional input of mineral (nitrate) fertilizers, shortage of irradiance or partial loss of plant leaf blades, while the second, conversely, under raise of lighting, damage or cutting of root system by machinery in the course of planting treatment.

Depending on the character of such a disequilibrium the meeting of the two disturbed mass flows may occur either in the aboveground plant part or in the root system. According to the place of the meeting, generation of NO and triggering of NO signal system may take place in the root area or in the shoot. Both are also influenced by the intensity of the transpiration, that carries mineral nutrients from roots to leaves with the water stream.

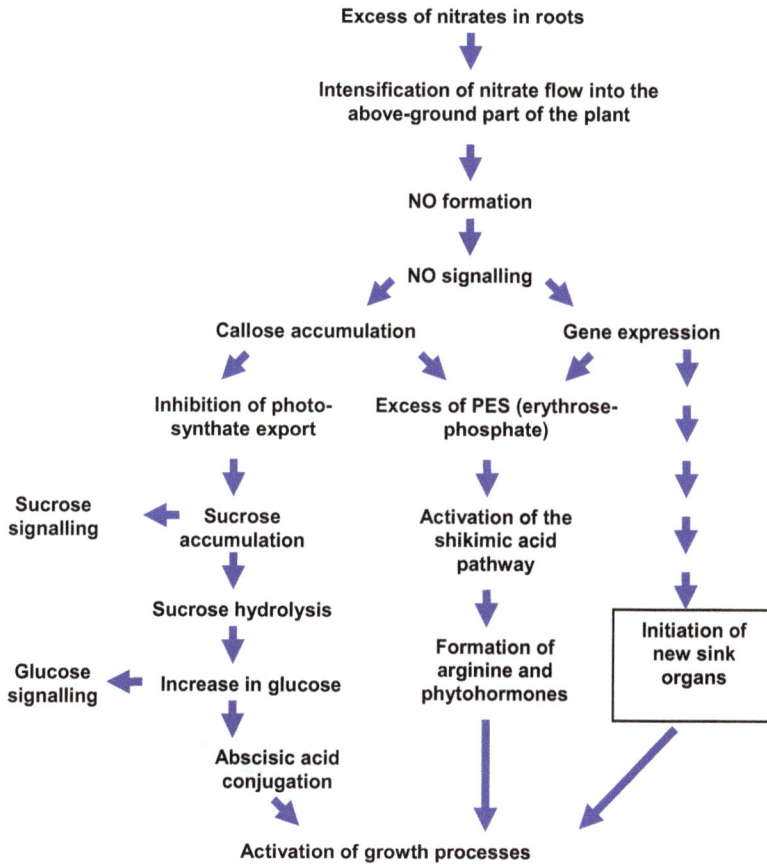

Fig. 10. Scheme of NO-signal system participation in the rearrangement of metabolism in the whole plant under increased nitrate

When excessive nitrates appear and the available nitrate reducing enzymes and their activity are not enough to utilize this massive flow of nitrates, the latters will surge into the upper plant part without having been reduced, where they will interact with the sugar flow, and NO signal system will become triggered. Oppositely, when sugars are in excess, they will be transported to roots and interact with nitrates therein.

In the first case, growth processes will be initiated in the above-ground plant part, and gradual nitrate utilization in leaves (shoots) will lead to a step-down in root nitrates, and augmented photosynthetic apparatus will send extensively increased sugar flow to the roots. In the second case, interaction of nitrates with sugars and NO signal system triggering will occur in roots and the process of new secondary root formation will become activated.

This thesis is well illustrated by experiments of L. B. Vysotskaya (Vysotskaya, 2001). Removal of the largest part of roots from seven-days old wheat seedlings suppressed shoot growth and activated biomass growth of remaining roots in as soon as 2 h. In the remaining roots auxins and cytokinins were accumulated while in the growing part of the shoot a rapid decline of auxin content compared to intact plants was found. This suggests that excessive sugar flow to the reduced root system creates prerequisites for interaction of the changed nitrate to sugar ratio and NO signal system triggering (alike an analogue of apical dominance alleviation) and synthesis of cytokinines that activate new root formation. Initially, the prerequisites are most probably the immediate fueling of the nitrate uptake process by better sugar supply of roots. Additional nitrates, in their turn, will trigger NO signal system.

The proposed concept on the role of NO signaling in the regulation of plant metabolism is supported by split-root experiments where plant roots were exposed to culture mediums of different concentrations (Trapeznikov et al., 1999). By placing one part of roots of an individual potato plant into a medium of high concentration and the other part into low-salt one, the authors have found that in the concentrated medium a massive formation of small (absorbing) roots occurred while in the low-salt one numerous tubers appeared (Fig. 11).

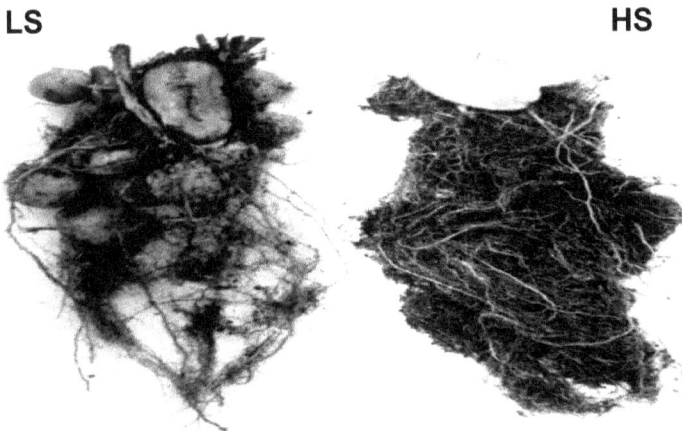

Fig. 11. Root system of an individual potato plant at local nutrition. HS – high-salt culture medium, LS – low-salt culture medium (Trapeznikov et al., 1999)

6. Conclusion

Nitrate has been shown to act as a signal molecule, inducing expression of genes, primarily related with nitrogen metabolism and organic acid synthesis. However, low sugar level in the plant inhibits nitrate assimilation, overriding signals from nitrogen metabolism (Stitt et al., 2002). In this regard a concept has appeared that for regulation of various processes in the plant not sugar and nitrate concentrations are important but a certain ratio between them which was called a C/N-balance (Coruzzi & Bush, 2001).

We believe that the link between nitrates and sugars is to be sought not at the molecular level, i.e. at the level of their metabolism in the cell or their influence on gene expression, but at a higher level - at the level of transport of these substances within the plant. This view is supported by observation that information on the nitrogen and carbon status of the plant is transmitted over long distances, revealed by the well known effect of root nitrate on the metabolism of the above-ground plant part and on shoot to root weight ratio (Scheible et al., 1997). In this connection, there is now a large group of studies devoted to the search of a "signal" coordinating shoot and root responses to nitrogen availability (Walch-Liu et al., 2005).

Activation of the hydrolysis of sucrose in the apoplast in the presence of nitrates is in good agreement with a similar effect of nitrate and sugars on the expression of several genes (Stitt et al., 2002), as well as with discerned differences in systemic and local effect of nitrate on the morphogenesis of the roots (Zhang et al., 2007). And the systemic action of nitrate is associated with its negative influence on the flow of assimilates to roots (Scheible et al., 1997).

Currently, signaling functions are ascribed not only to nitrate but also to products of nitrate reduction. Depending on the ratio of available carbon and nitrogen in the plant the ratio of oxidized and reduced nitrogen will vary. The influence of the products of nitrate reduction was noted to be opposite to nitrate influence, though the mechanism of their action is also as yet unknown, but supposed to involve glutamine content or glutamine/2-oxoglutarate ratio (Foyer and Noctor, 2002; Stitt et al., 2002).

Since an increase in amount of nitrates in the plant creates conditions favorable for the generation of nitric oxide from nitrite in both enzymatic and non-enzymatic ways (Neill et al., 2003), we can assume that the signaling effects of nitrate are partially realized through the formation of nitric oxide and triggering of NO-signaling system. This is confirmed by found similarities in actions of nitrate and nitric oxide generator, sodium nitroprusside, on assimilate transport and metabolism. However, in contrast to nitrate, nitric oxide preferably activates genes involved in plant defense (Grün et al., 2006).

Actually, the difference of nitrate and nitric oxide actions may be due to differences in the activity of amino acid synthesis, that, as was mentioned above, can also perform signaling roles. Study of the dynamics of gene expression activation under the influence of nitrate showed that many genes induced by nitrate in the first 0-5 and 5-10 minutes are subjected to negative regulation by as early as 20 minutes (Castaings et al., 2011).

Thus, there remains a lot to be elucidated in the signaling mechanism of nitrate and the study of mechanisms of nitrate influence on the transport of sugars can be very promising, not only for this area of research, but also to discovering how the different processes in the plant are interrelated.

7. References

Abdrakhimov, F.A., Batasheva, S.N, Bakirova, G.G. & Chikov, V.I. (2008). Dynamics of ultrastructural changes in common flax leaf blades during assimilate transport inhibition with nitrate anion. *Tsitologiya*, Vol. 50, pp. 700-710, ISSN 0041-3771

Anisimov, A.A. (1959). Movement of assimilates in wheat seedlings associated with the conditions of root nutrition. *Soviet Journal of Plant Physiology*, Vol. 6, No. 2, pp. 138-143

Asami, S. & Akasava, T. (1977). Enzimicformation of glycolate in chromatium: Role superoxide radical in transketolase – type mechanism. *Biochemistry*, Vol. 16, pp. 2202-2209, ISSN 0006-2960

Ayre, B.G., Keller, F. & Turgeon, R. (2003). Symplastic continuity between companion cells and the translocation stream: long-distance transport is controlled by retention and retrieval mechanisms in the phloem. *Plant Physiology*, Vol. 131, pp. 1518-1528, ISSN 0032-0889

Batasheva, S.N., Abdrakhimov, F.A., Bakirova, G.G. & Chikov, V.I. (2007) Effect of nitrates supplied with the transpiration flow on assimilate translocation. *Russian Journal of Plant Physiology*, Vol.54, pp. 373-380, ISSN 1021-4437

Batasheva, S.N., Abdrakhimov, F.A., Bakirova, G.G., Isaeva, E.V. & Chikov, V.I. (2010). Effects of sodium nitroprusside, the nitric oxide donor, on photosynthesis and ultrastructure of common flax leaf blades. *Russian Journal of Plant Physiology*, Vol.57, pp. 376-381, ISSN 1021-4437

Brovchenko, M.I. (1970). Sucrose hydrolysis in the free space of leaf tissues and invertase localization. *Soviet Journal of Plant Physiology*, Vol. 17, No. 1, pp. 31-39.

Castaings, L., Marchive, C., Meyer, C. & Krapp, A. (2011). Nitrogen signaling in Arabidopsis: how to obtain insights into a complex signaling network. *J. Exp. Bot.*, Vol. 62, pp. 1391-1397, ISSN 0022-0957

Champigny, M.-L. & Foyer, C.H. (1992). Nitrate activation of cytosolic protein kinases diverts photosynthetic carbon from sucrose to amino acid biosynthesis. Basis for a new concept. *Plant Physiol.*, Vol. 100, pp. 7-12, ISSN 0032-0889

Chikov, V.I. (1987). *Fotosintez i transport assimilyatov (Photosynthesis and assimilate transport)*, Nauka, Moscow

Chikov, V. & Bakirova, G. (1999). Relationship between Carbon and Nitrogen Metabolism in Photosynthesis. The Role of Photooxidation Processes. *Photosynthetica*, Vol. 37, pp. 519-527, ISSN 0300-3604

Chikov, V.I. & Bakirova, G.G. (2004). Role of the apoplast in the control of assimilate transport, photosynthesis, and plant productivity. *Russian Journal of Plant Physiology*, Vol. 51, pp. 420- 431, ISSN 1021-4437

Chikov, V.I. Bakirova, G.G., Ivanova, N.P., Nesterova, T.N. & Chemikosova, S.B. (1998). A change of photosynthetic carbon metabolism in wheat flag leaf under fertilization with ammonia and nitrate. *Physiology and Biochemistry of Cultivated Plants*, Vol. 30, pp. 333-341, ISSN 0522-9310

Chikov, V.I., Avvakumova, N.I., Bakirova, G.G., Belova, L.A. & Zaripova, L.M. (2001). Apoplastic transport of ^{14}C-photosynthates measured under drought and nitrogen supply. *Biologia Plantarum*, Vol. 44, pp. 517-521, ISSN 0006-3134

Chikov, V.I., Bulka, M.E. & Yargunov, V.G. (1985). Effect of removal of reproductive organs on photosynthetic $^{14}CO_2$ metabolism in cotton leaves. *Soviet Journal of Plant Physiology*, Vol. 32, pp. 1055-1063.

Coruzzi, G, & Bush, DR. (2001). Nitrogen and carbon nutrient and metabolite signaling in plants. *Plant Physiology*, Vol. 125, pp. 61-64, ISSN 0032-0889

Desikan, R., Griffiths, R., Hancock, J. & Neill, S. (2002). A new role for an old enzyme: Nitrate reductase-mediated nitric oxide generation is required for abscisic acid-induced stomatal closure in Arabidopsis thaliana. *Proc. Natl. Acad. Sci.*, Vol. 99, pp. 16314-16318, ISSN 0027-8424

Dubinina, I.M., Burakhanova, E.A. & Kudryavtseva, L.F. (1984). Suppression of invertase activity in sugar beet conducting bundles as an essential prerequisite for sucrose transport. *Soviet Journal of Plant Physiology*, Vol. 31, pp. 153-161

Forde, B.G. (2000). Nitrate transporters in plants: structure, function and regulation. *Biochim. Biophys. Acta*, Vol. 1465, pp. 219-235, ISSN 0006-3002

Foyer, C.H. & Noctor, G. (2002) Photosynthetic nitrogen assimilation: inter-pathway control and signaling, In: *Photosynthetic nitrogen assimilation and associated carbon and respiratory metabolism*, C.H. Foyer & G. Noctor (Eds), pp. 1-22, Kluwer Academic Publishers, ISBN 0-7923-6336-1, the Netherlands

Gamalei, Y.V. & Pakhomova, M.V. (2000). The time course of carbohydrate transport and storage in the leaves of the plant species with symplastic and apoplastic phloem loaded under the normal and experimentally modified conditions. *Russian Journal of Plant Physiology*, Vol. 47, pp. 109-128, ISSN 1021-4437

Gamalei, Yu. V. (2004) *Transportnaya sistema sosudistykh rastenii (Transport System of Vascular Plants)*, St. Petersburg Univ, ISBN 5-288-03343-9, St. Petersburg

Grinenko, V.V. (1964) Metabolism of cotton plants under conditions of disturbed ratio of mineral nutrients. In *Role of mineral elements in plant metabolism and productivity*, pp. 113-120, Nauka Moscow

Grün, S., Lindermayr, C., Sell, S. & Durner, J. (2006). Nitric oxide and gene regulation in plants. *J. Exp. Bot.*, Vol. 57, pp. 507-516, ISSN 0041-3771

Hartt, C. E. (1970). Effect of nitrogen deficiency upon translocation of ^{14}C in sugar-cane. *Plant Physiol.*, Vol. 46, pp. 419–423, ISSN 0032-0889

Heldt, H.W., Chon C.J., Lilley R. McC. & Portis A.R. (1978). The role of fructose- and sedoheptulosebisphosphatase in the control of CO_2 fixation: Evidence from the effects of Mg^{++} concentration, pH and H_2O_2 in Proceedings of the 4th International Congress on Photosynthesis, pp. 469–478

Khamidullina, L.A., Abdrakhimov, F.A., Batasheva, S.N., Frolov, D.A. & Chikov, V.I. (2011). Effect of Nitrate Infusion into the Shoot Apoplast on Photosynthesis and Assimilate Transport in Symplastic and Apoplastic Plants. *Russian Journal of Plant Physiology*, Vol. 58, No. 3, pp. 484-490, ISSN 1021-4437

Koroleva, O.A., Farrar, J.F., Tomos, A.D. & Pollock, C.J. (1998). Carbohydrates in individual cells of epidermis, mesophyll, and bundle sheath in barley leaves with changed export or photosynthetic rate. *Plant Physiology*, Vol. 118, pp. 1525-1532, ISSN 0032-0889

Kühn, C., Barker, L., Bürkle, L. & Frommer, W.-B. (1999). Update on sucrose transport in higher plants. *J. Exp. Bot.*, Vol. 50, pp. 935-953, ISSN 0022-0957

Kudryavtsev, V. A. & Roktanen, G.-L. (1965). The influence of mineral nutrition regime on the formation of generative organs and some metabolic indices of tomato under different lighting conditions. Agrochemistry, Vol. 6, No. 1, pp. 88-93

Kühn, C., Franceschi, V.R., Schulz, A., Lemoine, R. & Frommer W.B. (1997). Localization and turnover of sucrose transporters in enucleate sieve elements indicate macromolecular trafficking. Science, Vol. 275, pp. 1298-1300, ISSN 0036-8075

Kursanov, A.L. (1976). Transport assimilyatov v rastenii, Nauka Moscow. Translated under the title (1984) Assimilate transport in plants, Elsevier, Amsterdam

Lohaus, G. & Fischer, K. (2002). Intracellular and Intercellular Transport of nitrogen and carbon, In: Photosynthetic nitrogen assimilation and associated carbon and respiratory metabolism, C.H. Foyer & G. Noctor (Eds), pp. 239-263, Kluwer Academic Publishers, ISBN 0-7923-6336-1, the Netherlands.

Malamy, J.E. & Ryan, K.S. (2001). Environmental regulation of lateral root initiation in Arabidopsis. Plant Physiology, Vol. 127, pp. 899-909, ISSN 0032-0889

Martin, T., Oswald, O. & Graham, I.A. (2002). Arabidopsis seedling growth, storage mobilization, and photosynthetic gene expression are regulated by carbon:nitrogen availability. Plant Physiology, Vol. 128, pp. 472-481, ISSN 0032-0889

Marty, K. S. (1969). Effect of topdressing'nitrogen at heating time on carbon assimilation of rice plant during the ripening period. Indian J. Plant Physiol., Vol. 12, pp. 202–210.

Mata, C.G. & Lamattina, L. (2001). Nitric oxide induces stomatal closure and enhances the adaptive plant responses against drought stress. Plant Physiol., Vol. 126, pp. 1196-1204, ISSN 0032-0889

Möller, I. & Beck, E. (1992). The fate of apoplastic sucrose in sink and source leaves of Urtica dioica. Physiologia Plantarum, Vol. 85, pp. 618-624, ISSN 0031-9317

Neill, S.J., Desikan, R. & Hancock, J.T. (2003). Nitric oxide signalling in plants. New Phytologist, Vol. 159, pp. 11–35, ISSN 0028-646X

Noctor, G. & Foyer, C.H. (1998). A re-evaluation of the ATP : NADPH budget during C3 photosynthesis: a contribution from nitrate assimilation ad its associated respiratory activity? J. Exp. Bot., Vol. 49, pp. 1895-1908, ISSN 0022-0957

Noctor, G. & Foyer, C.H. (2000). Homeostasis of adenylate status during photosynthesis in a fluctuating environment. J. Exp. Bot. ,Vol. 51, pp. 347-356, ISSN 0022-0957

París, R., Lamattina, L. & Casalongué, C.A. (2007). Nitric Oxide Promotes the Wound-Healing Response of Potato Leaflets. Plant Physiol. Biochem., Vol. 45, pp. 80-86, ISSN 0981-9428

Paul, M.J. & Foyer, C.H. (2001). Sink regulation of photosynthesis. J. Exp. Bot., Vol. 52, pp. 1383–1400, ISSN 0022-0957

Paul, M.J. & Pellny, T.K. (2003). Carbon metabolite feedback regulation of leaf photosynthesis and development. J. Exp. Bot., Vol. 54, pp. 539-547, ISSN 0022-0957

Pollock, C., Farrar, J., Tomos, D., Gallagher. J, Lu, C. & Koroleva, O. (2003). Balancing supply and demand: the spatial regulation of carbon metabolism in grass and cereal leaves. J. Exp. Bot., Vol. 54, pp. 489-494, ISSN 0022-0957

Pristupa, N. A. & Kursanov, A. L. (1957). Downward flow of assimilates and its relationship with uptake by the root. *Agrochemistry*, Vol. 4, pp. 417-424

Pristupa, N. A. (1959). About transport form of carbohydrates in pumpkin plants. *Soviet Journal of Plant Physiology*, Vol. 6, pp. 30-38

Scheible, W.R., Lauerer, M., Schulze, E.D., Caboche, M. & Stitt, M. (1997). Accumulation of nitrate in the shoot acts as a signal to regulate shoot-root allocation in tobacco. *The Plant Journal*, Vol. 11(4), pp. 671-691, ISSN 0960-7412

Serova, V.V., Raldugina, G.N. & Krasavina, M.S. (2006) Salycylic acid inhibits callose hydrolysis and disrupts transport of tobacco mosaic virus. *Dokl. Akad. Nauk*, Vol. 406, pp. 705-708, ISSN 0869-5652

Stitt, M. & Krapp, A. (1999). The interaction between elevated carbon dioxide and nitrogen nutrition: the physiological and molecular background. *Plant, Cell & Environ.*, Vol. 22, pp. 583-621, ISSN 0140-7791

Stitt, M., Müller, C., Matt, P., Gibon, Y., Carillo, P., Morcuende, R., Scheible, W.-R., Krapp, A. (2002). Steps towards an integrated view of nitrogen metabolism. *J. Exp. Bot.*, Vol. 53, pp. 959-970, ISSN 0022-0957

Takabe, T., Asami, S. & Akazawa, T. (1980). Glycolate formation catalyzed by spinach leaf transketolase utilizing the superoxide radical. *Biochemistry*, Vol. 19, No. 17, pp. 3985−3989, ISSN 0006-2960

Tarchevsky, I.A., Ivanova, A.P. & Biktemirov, U.A. (1973). Effect of mineral nutrition on assimilate movement in wheat, *Proceedings of Biology-Soil Institute*, Vol. 20, pp 174-178

Trapeznikov, V.K., Ivanov, I. I. & Tal'vinskaya, N. G. (1999). *Local nutrition of plants*, Gilem, Ufa

Turkina, M.V., Pavlinova, O.A. & Kursanov, A.L. (1999). Advances in the study of the nature of phloem transport: the activity of conducting elements. *Russian Journal of Plant Physiology*, Vol. 46, pp. 709-720, ISSN 0015-3303

Vaklinova, S.G., Doman, N.,G., & Rubin, B.,A. (1958). The influence of different nitrogen forms on assimilation products of leaves and their distribution among above-ground and underground organs in maize seedlings. *Soviet Journal of Plant Physiology*, Vol. 5, No. 6, pp. 516-523

Vysotskaya, L.B., Timergalina, L.N., Simonyan, M.V., Veselov, S.Yu. & Kudoyarova, G.R. (2001). Growth rate, IAA and cytokinin content of wheat seedling after root pruning. *Plant Growth Regul.*, Vol. 33, pp. 51-57, ISSN 0167-6903

Walch-Liu, P., Filleur, S., Gan, Y. & Forde, B.G. (2005). Signaling mechanisms integrating root and shoot responses to changes in the nitrogen supply. *Photosynthesis Research*, Vol. 83, pp. 239-250, ISSN 0166-8595

Zav'yalova, T.F. (1976) The influence of rising dozes of nitrogen and phosphorous fertilizers on photosynthetic phosphorylation and productivity in barley. *Bulletin of Soviet Union Research Institute of fertilizers and soil science*, No. 29, pp. 37-41

Zhang, H., Rong, H. & Pilbeam, D. (2007). Signalling mechanisms underlying the morphological responses of the root system to nitrogen in Arabidopsis thaliana. *J. Exp. Bot.*, Vol. 58, pp. 2329-2338, ISSN 0022-0957

Zottini, M., Costa, A., De Michele, R., Ruzzene, M., Carimi, F. & Lo Schiavo, F. (2007). Salicylic acid activates nitric oxide synthesis in Arabidopsis, *J. Exp. Bot.*, Vol. 58, pp. 1397-1405, ISSN 0022-0957

Part 2

Special Topics in Photosynthesis

7

A Review: Polyamines and Photosynthesis

Sheng Shu, Shi-Rong Guo and Ling-Yun Yuan
College of Horticulture, Nanjing Agricultural University, Key Laboratory of Southern Vegetable Crop Genetic Improvement, Ministry of Agriculture, Nanjing China

1. Introduction

Polyamines (PAs) are low molecular weight ubiquitous nitrogenous compounds found in all living organisms (Kaur-Sawhney et al., 2003). In higher plants, the most common polyamines are spermidine (Spd), spermine (Spm) and their diamine obligate precursor putrescine (Put). They are formed by aliphatic hydrocarbons substituted with two or more amino groups (Figure.1). Because of the polycationic nature at physiological pH, PAs are present in the free form or as conjugates bound to phenolic acids and other low molecular weight compounds or to proteins and nucleic acids (Childs et al., 2003). Like hormones, PAs displaying high biological activity are involved in a wide array of fundamental processes in plants, such as replication and gene expression, growth and development, senescence, membrane stabilization, enzyme activity modulation and adaptation to abiotic and biotic stresses (Galston et al., 1997; Bais and Ravishankar, 2002; Zapata et al., 2008). Although, according to these reports, PAs seem to be important growth regulators, their precise physiological function and mechanism of action still remain unclear.

It has been shown that chloroplasts and photosynthetic subcomplexes including thylakoids, LHCII complex and PSII membranes are enriched with three major polyamines, while PSII core and the reaction center of PSII are exclusively rich in Spm (Kotzabasis et al., 1993; Navakoudis et al., 2003). The potential role of polyamines in maintaining the photochemical efficiency of plants has become a research focus. These studies mainly focused on the effect PAs exert a positive role in the photosynthesis of plants in response to various environmental stresses. In green alga, it was shown that the bound Put content of the thylakoid membrane was increased in environments with high CO_2 concentrations, which caused an increase in reaction center density and led to an increased photosynthetic rate (Logothetis et al., 2004). An increase in conjugated Put content can stabilize the thylakoid membrane, thus enhancing resistance of tobacco plants to ozone pollution (Navakoudis et al., 2003) and UV-B radiation (Lütz et al., 2005). Low temperature stress reduced the content of Put as well as the Put/Spm ratio in thylakoids and the light-harvesting complexes LHCII in *Phaseolus vulgaris* L., leading to a decrease in photosynthetic electron transport rate and inactivation of the PSII reaction center (Sfakianaki et al., 2006). Put is also involved in the induction of a photosynthetic apparatus owning high concentration of reaction center with a small functional antenna that leads to enhance photochemical quenching of the absorbed light energy (Kotzabasis et al., 1999). PAs biosynthesis is controlled by light and the Spm/Put ratio is correlated to the structure and function of the photosynthetic apparatus during photoadaptation. These studies indicated that changes of endogenous polyamines might be involved in an important protective role in the photosynthetic apparatus.

A lot of researches started to pay attention to the effects of application of exogenous polyamines on photosynthesis under various stresses. It has been demonstrated that exogenously applied polyamines can rapidly enter the intact chloroplast (He et al., 2002) and play a role in protecting the photosynthetic apparatus from adverse effects of environmental stresses (Navakoudis et al., 2003). However, the effect of polyamines on the photosynthetic efficiency of stressed plants depends on the stress level and the type of exogenous polyamines. Exogenous polyamines improved the photosynthetic capacity of salt-stressed cucumber plants by increasing the level of the photochemical efficiency of PSII (Zhang et al., 2009). In green alga *Scenedesmus obliquus* cultures, exogenously added Put was used to adjust the increase in the functional size of the antenna and the reduction in the density of active photosystem II reaction centers, so that to confer some kind of tolerance to the photosynthetic apparatus against enhanced NaCl-salinity and permit cell growth even in NaCl concentrations that under natural conditions would be toxic (Demetriou et al., 2007). Investigations into restoration of the maximum photochemical efficiency (Fv/Fm) by adding Put, Spd and Spm to low salt thylakoid showed that Spd are the most efficient ones in Fv/Fm restoration, but higher amounts of Spm and/or Spd reverse the effect and lead to a decline of the Fv/Fm (Ioannidis and Kotzabasis, 2007). When *Physcia semipinnata* was exposed to UV-A radiation, it was also found that exogenously Spd added samples had higher Chl a content and photosystem II activity than Spm and Put added samples (Unal et al., 2008). In addition, analysis of PSII particles isolated from leaf fragments floated in the presence of Put, Spd and Spm solutions under the dark conditions was conducted. It was observed that Spd could interact directly with thylakoid membranes, which was effective in the retardation of the loss of LHCII observed in water-treated detached leaves, so that they become more stable to degradation during senescence (Legocka and Zajchert, 1999).

Several studies have shown that chloroplasts contain high activities of polyamine biosynthetic enzymes and transglutaminase (TGase) catalyzing the covalent binding of polyamines to proteins (Del Duca et al., 1994; Andreadakis and Kotzabasis, 1996; Della Mea et al., 2004) (Table 1). These enzymes are also involved in regulation of photosynthesis in response to stress conditions (Wang et al., 2010). Arginine decarboxylase (ADC) has been shown to be mainly localised in the chloroplasts of leaves and nuclei of roots (Borrell et al., 1995). It was established that spinach ADC was associated with LHC of photosystem II (Legocka and Zaichert, 1999). PAs synthesised in chloroplasts evidently stabilized photosynthetic complexes of thylakoid membranes under stress conditions (Borrell et al., 1995). An evidence is supported by salt treatment induced a decreased chlorophyll content and photosynthetic efficiency in the lower arginine decarboxylase activity of mutant plants, which leads to reduced salt tolerance in Arabidopsis thaliana (Kasinathan and Wingler, 2004). TGase is present in the chloroplasts of higher plants, where its activity is modulated by the presence of light. Its substrates are Rubisco and some antenna complexes of thylakoids, such as LHCII, CP29, CP26 and CP24 (Del Duca et al., 1994). In *Dunaliella salina* whole cells, TGase seems to play a role in the acclimimation to high salt concentrations under light condition, and the content of chlorophylla and b of chloroplast were enhanced, the amount of 68kD and 55kDa polypeptides was particularly high in algae already acclimated cells (Dondini et al., 2001). Recently, Ortigosa et al. (2009) showed that the over-expression of maize plastidial transglutaminas (chlTGZ) in the young leaves of tobacco chloroplasts seemed to induce an imbalance between capture and utilization of light in photosynthesis. Although these changes were accompanied by thylakoid scattering, membrane degradation and reduction of thylakoid interconnections, transplastomic plants could be maintained and reproduced in vitro.

At present, the roles of PAs in the structure and functions of the photosynthetic apparatus are widely investigated. Most of researchers consider the PAs and their related-metabolic enzymes be positive regulators of plant photosynthesis in response to various environmental stresses. However, the specific mechanism of polyamine on the protection of photochemical efficiency of stressed-plants remains until today largely unknown. Thus, we need use advanced molecular biology and proteomic approaches to further understanding the role of PAs in the regulation of photosynthetic processes.

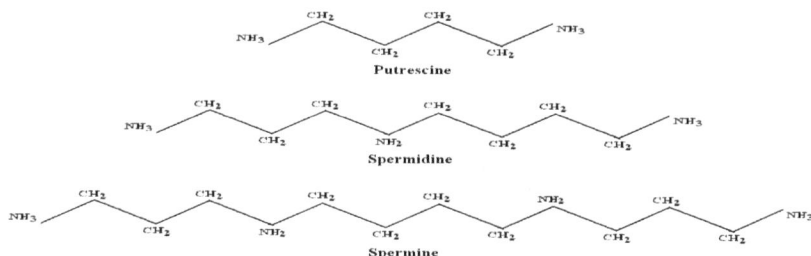

Fig. 1. The chemical structure of three major polyamines

2. Effect of polyamines on photosynthesis

2.1 Stomatal opening/ closure

Stomatal is defined by two guard cells and is responsible for gas exchange between plants and the atmosphere (Mansfield et al., 1990). Stomatal plays a key role in signal transduction, sensing and adaptive responses to abiotic stresses like drought, heat, chilling and high salinity (Hetherington and Woodward, 2003). Polyamines regulated stomatal closely correlated with the improvement of photosynthesis, which could be caused by the greater amount of CO_2 available for its fixation by photosynthetic enzymes. However, most of the roles of polyamines in regulating stomatal research are focused on environmental stress conditions, and the specific mechanisms are still unclear.

All natural polyamines, including cadaverine (Cad) and putrescine (Put), spermidine (Spd), spermine (Spm) at a given concentration, strongly inhibited opening of stomata. Liu et al. (2000) found that 1 mM Spd and Spm completely prevented light-induced stomatal opening, whereas Cad and Put inhibited this opening by 88% and 63%, respectively. Although all polyamines significantly reduced the stomatal aperture, Spd and Spm appeared to be more effective than Put at 1 mM. Çavuşoğlu et al. (2007) showed that polyamine-pretreaments could decrease stomata number and length in the upper surface under saline conditions by reducing the transpiration. In the Spm-pretreated Citrus Reticulata samples were observed in smaller stomatal aperture size than the control at a given time point or during the whole experiment (Shi et al., 2010).

Many environmental factors regulated stomatal aperture through modulation of ion channel activity in guard cells (MacRobbie, 1997). Changes in guard cell turgor that instigate stomatal movements are controlled by a number of ion channels and pumps (Raschke et al.,

1988; Ward et al., 1995). As an important player in stomatal regulation, the I Kin is an indirect target of polyamine action. A number of studies have shown that I Kin-inhibiting processes or factors often inhibit stomatal opening (Assmann, 1993). Liu et al. (2000) using patch-clamp analysis demonstrated that intracellular application of polyamines inhibited the inward K1 current across the plasma membrane of Vicia faba guard cells and modulated stomatal movement. Changes of free Ca^{2+} in the cytoplasm of guard cells are involved in stomatal aperture/closure. In the Spm-deficient mutant Arabidopsis seems to be impaired in Ca^{2+} homeostasis, which affects the stomatal movement, and that this inhibited effect was restored by application of exogenous Spd in the mutant plants (Yamaguchi et al., 2007). In addition, it has been reported that polyamine-induced ROS scavenging is an essential effect and stimulated stomatal closure (lower water loss) upon dehydration, which may function collectively to enhance dehydration tolerance (Pei et al., 2000; Bright et al., 2006). These findings suggest that polyamines target KAT1-like inward K1 channels in guard cells and modulate stomatal movements, providing a link between stress conditions, polyamine levels, and stomatal regulation.

2.2 Photosynthetic pigment

Chlorophyll (Chl) is a molecule substance that plays an important role in photosynthesis for the plant growth process, such as light absorption, and combination with protein complex, transfer the energy into carbohydrate (Meskauskiene et al., 2001). A variety of reports indicate that changes in chlorophyll levels of plants may decrease in response to environmental factors or leaf senescence (Sfichi-Duke et al., 2008; Munzi et al., 2009). Aliphatic polyamines (PAs) are involved in the delay loss of chlorophyll and lead to an increased efficiency of light capture resulting in the improvement of net photosynthetic rate, but the molecular mechanism is not clarified.

Positive effects of exogenous supplied polyamines on the content of Chla and total Chl in leaves were observed for various stresses, but there were distinct differences in the effect of three main polyamines. Unal et al. (2008) found that Chla content was significantly increased in *Physcia semipinnata* by exogenously added polyamines during exposure to UV-A radiation, and exogenously Spd added samples had higher Chla content than Spm and Put added samples. Spd delayed the loss of Chl more than Spd or Put in detached wheat leaves during dark incubation implying the importance of valency of organic cations (Subhan and Murthy, 2001). The result is in agreement with those of Aldesuquy et al. (2000) who reported that using detached wheat leaves infected with the yellow rust *Puccinia striiformis*. Among of exogenous supplied polyamines, in contrast with Put and Spd, Spm has been shown to regulate the in vivo amount of protochlorophyllide (PChlide) and Chl both in darkness and in light (Beigbeder and Kotzabasis, 1994). Beigbeder et al. (1995) suggested that the intracellular level of Put was decreased by the use of 1, 4-diamino-2-butanone inhibitor (1,4DB) dramatically increasing the PChlide levels with parallel reduction of chlorophyll. Many workers have reported retention of chlorophyll induced by exogenously supplied polyamines during the normal developmental senescence of leaves. Cheng and Kao (1983) demonstrated that Spd and Spm were effective in retarding loss of chlorophyll from detached leaves of rice, wheat and soybean. They describe the effect of locally applied polyamines as being similar to that of cytokinins. In plants of *Heliotropium* sp., leaf senescence is associated with low endogenous concentrations of polyamines (Birecka et al., 1984). One of the mechanisms by which PAs modulate chlorophyll stabilisation could be due to their modification of chlorophyll-bound proteins, catalysed by TGase. In senescing leaves, foliar spray with 0.2 mM Spm treatment prevented degradation of Chla and Chlb, and increased TGase activity, producing more PA-protein conjugates. Spm was translocated to chloroplasts and bound mainly onto fractions enriched in PSII, whose light-harvesting complexes (LHC) sub-fractions contained TGase (Serafini-Fracassini et al., 2010).

2.3 CO_2 assimilation

CO_2 assimilation is the process of carbohydrates formation which utilized the ATP and NADPH produced by light. The capacity of CO_2 assimilation is connected with the plant growth, biomass and productivity. However, the effect of polyamines on CO_2 assimilation has been investigated very rarely and only a few information concerning the relationship between PAs and carbon assimilation.

Iqbal and Ashraf (2005) analyzed the influence of pre-sowing seed treatment with PAs on growth and photosynthetic capacity in two spring wheat (*Triticum aestivum* L.) cultivars MH-97(salt intolerant) and Inqlab-91(salt tolerant). The results showed that different priming agent did not affect the net CO_2 assimilation rate. The role of three PAs [putrescine (Put), spermidine (Spd) and spermine (Spm)] in improving drought tolerance in fine grain aromatic rice (*Oryza Sativa* L.) has been appraised by Farooq et al. (2009). Three of them were used each at 10 µM as seed priming (by soaking seeds in solution) and foliar spray. Drought stress significantly reduced maximum leaf CO_2 assimilation rate, while application of PAs significantly improved the leaf CO_2 assimilation rate but decreased Gs and transpiration rate (Tr) under drought stress. The resultes suggested that PAs enhanced drought tolerance in rice was due to improved CO_2 assimilation by Rubisco in producing photosyntheate and their partitioning in dry matter yield. Huang et al. (2010) found 0.1 mM CA treatment decreased P_N, but it did not affect the photosynthetic apparatus, which suggest that the decline is at least partially attributed to a lowered RuBPC activity. However, the exogenous application of 1 mM Spd partially restored RuBPC activity in leaves, the key enzyme of carbon reduction cycle (dark reaction), and thus improving the photosynthetic rate. Chen et al. (2010) reported that exogenous Spd application decreased the carbohydrate accumulation in leaves and total sugar, sucrose content in roots under salt stress, thus reduced negative feedback inhibition to photosynthesis caused by carbohydrate accumulation. Exogenous application of Spd can alleviate the damage caused by hypoxia stress and enhance the conversion and use of carbohydrate in roots, which can promote the formation of new metabolic balance seedlings attributed to tolerance of hypoxia stress (Zhou et al., 2007)

2.4 Chloroplast ultrastructure

It is necessary to maintain structural integrity and orderliness of chloroplast that plays a role in conversion of light energy for photosynthesis (Li et al., 2009). The chloroplasts are usually 5-10 micrometer long and consist of circular DNA molecules. In higher plants, the photosynthetic machinery is mainly localized in thylakoid membranes of the chloroplasts (Kirchhoff et al., 2007). The membrane has no homogenous structure, but is subdivided into two domains: the strictly stacked grana thylakoids and the unstacked stroma lamellae (Kirchhoff et al., 2003). The structure of thylakoids is a major factor that affects functionality and performance of the photosynthetic apparatus (Ioannidis et al., 2009). It was reported that some stresses led to the decrease in the photochemical efficiency and electron transport activity might be associated with the changes of the structure of photosynthetic apparatus (Parida et al., 2003).

Several studies have showed that polyamines are involved in stabilization the structure and function of photosynthetic apparatus in response to unfavourable environment factors (Demetriou et al., 2007). Under Nacl stress, Put as organic cations dramatically enhanced lipid accumulation in the chloroplasts and prevented the membrane degradation in the granal and stromal thylakoids by interacting with the negatively charged membrane sites (Tiburcio et al., 1994). The increase in the number of plastoglobules in the chloroplasts affected by Put can result from the redirection of cell metabolism towards the products of higher reduction potential (Paramonova et al., 2003). In our study, we observed that Put can alleviate the degradation of thylakoid membrane proteins induced by salt stress, and thus making a normal stacking order in the adjacent grana thylakoids (data not published). In addition, exogenous Spd have been reported to protect the structure of chloroplasts by keeping an orderly arrangement of the thylakoids membrane and also have an ability to maintain a higher photosynthetic efficiency of *Nymphoides peltatum* under Cadmium stress (Li et al., 2009).

However, few earlier studies showed that high concentrations of polyamines may destroy the structure of chloroplasts which depended on different light conditions. Spd treatment with chloroplast for about 10 s displayed the envelope and the typical dense network of the thylakoid lamellae interspersed with numerous areas of stacked grana (Pjon et al., 1990). After 72 h, the chloroplast envelopes of spermidine-treated leaf disks incubated in the dark and under light conditions, the chloroplast envelope was destroyed by polyamine treatment, but there were distinct differences in the appearance of the chloroplast ultrastructures of dark and light-incubated leaves (Cohen et al., 1979).

2.5 Thylakoid membrane protein complexes

Photosynthetic apparatus in higher plants is a membrane bound protein complex composed of chlorophyll and carotenoid pigments that function in the conversion of light energy to chemical energy. It has been suggested that a large number of these proteins are related to photosynthesis. The thylakoid membranes within the chloroplast are the subcompartment in which the primary reactions of photosynthesis occur. These reactions are organized in the four major multisubunit protein complexes, photosystem I (PSI), PSII, the ATP-synthase complex and cytochrome b6/f complex (Hippler et al., 2001).

Polyamines such as putrescine, spermidine, spermine, and methylamine interact with protein (H-bonding) through polypeptide C=O, C-N and N-H groups with major perturbations of protein secondary structure as the concentration of amines was raised. It has been shown the chloroplasts and various photosynthetic subcomplexes including thylakoids, LHCII complex and PSII membranes are enriched with polyamines, especially are exclusively rich in PSII core and the reaction center of PSII (Kotzabasis et al. 1993; Navakoudis et al. 2003). Several studies have reported the interaction of polyamines with proteins of photosynthetic apparatus under various environmental factors. However, the action site of polyamines at photosynthetic proteins may vary with polyamine concentration and stress levels. In the alga *S. obliquus* the bound polyamines were found to be associated with both the oligomers and the monomers of LHCII, as well as with the CPs (Kotzabasis et al., 1993). However, the distribution does not reflect a constant pattern: in LHCII subcomplexes Put and Spm levels fluctuated depending on the light adaptational status of the photosynthetic apparatus. Put and Spm are bound to the photosynthetic complexes, mainly to the LHCII oligomeric and monomeric forms (Navakoudis et al., 2007). It is well documented that the LHCII protein is abundant in thylakoids and its surface is negatively charged (Standfuss et al., 2005). Kirchhoff et al. (2007) have recently shown that incubation of thylakoids under unstacking conditions leads to intermixing and randomization of the protein complexes, accompanied by disconnection of LHCII trimers from PSII and a decreased connectivity between PSIIα centers. Interestingly, exogenously added polyamines can reverse those damaging effects. Thus, there was strong indication that polyamines possessed a pivotal role in photosynthesis.

Chloroplasts also contain high activities of several polyamine biosynthetic enzymes (Borrell et al.1995; Andreadakis and Kotzabasis 1996) and transglutaminase (TGase) catalyzing the covalent binding of polyamines to proteins (Del-Duca et al., 1994; Dondini et al., 2001; Della-Mea et al., 2004). Polyamines can covalently bind with thylakoid membrane proteins under the TGase form of protein- glutamyl-polyamine or protein-glutamyl-polyamines-glutamyl-protein (Dondini et al., 2003), such as D1, D2 protein and cytochrome b6/f, together with the involvement of polyamine, could be important for stabilisation of molecular complexes in the thylakoid membranes of osmotically stressed oat leaves (Besford et al., 1993). Immunodetection of TGase in thylakoid fraction revealed that formation of covalent bonds between PAs and proteins by TGase is involved in regulating the process of chloroplast senescence (Sobieszczuk-Nowicka et al., 2009).

However, high concentration of polyamines added to submembrane fractions of photosynthetic apparatus causes a strong inhibition of PSII activity (Hamdani et al., 2010). FTIR spectroscopy analysis Spd and Spm at higher cation concentrations (5 and 10 mM), the result showed that the polyamine significant alterations of the thylakoid protein secondary structure with a decrease of the α-helical domains from 47% (uncomplexed PSII) up to 37% (cation complexes) and an increase in the β-sheet structure from 18% up to 29% (Bograh et al., 1997). So far, this specific inhibition mechanism of polyamines is not clear, but it is likely that the proteins were affected by these polycations re either extrinsic polypeptides associated with the oxygen evolving complex or portions of integral polypeptides protruding at the surface of the PSII membranes (Beauchemin et al., 2007).

3. Polyamines and stress photosynthesis

3.1 Salinity

Salt stress causes an initial water-deficit and ion-specific stresses resulting from changes in K^+/Na^+ ratios. Thus, it leads to an increased Na^+ and Cl^- concentrations that decrease plant growth and productivity by disrupting physiological processes, especially photosynthesis (Shu et al., 2010). Salt stress affects photosynthetic efficiency of plant through stomatal limitation and non-stomatal limitations, such as stomatal closure (Meloni et al., 2003), chlorophyll content loss (Sudhir and Murthy, 2004), inhibition of Rubisco activity (Brugnoli and Björkman, 1992; Ziska et al., 1990), and degradation of membrane proteins in photosynthetic apparatus (Khan and Ungar, 1997).

It has been suggested that exogenous application of polyamines can to some extent alleviate salinity-induced decline in photosynthetic efficiency, but this effect strongly depended both on PAs concentrations or types and stress levels (Duan et al., 2008). The maximum quantum efficiency of PSII (Fv/Fm) measured in leaves of salt-stressed cucumber seedlings was not

Species	Types of polyamine-related enzyme	Stress reponse	Measured photosynthetic parameters	Author
Helianthus tuberosus	TGase	Ca and light stimulate	Chlorophyll-a/b antenna complex (LHCII, CP24, CP26 and CP29);large subunit of ribulose-1,5-bisphosphate, carboxylase/oxygenase.	Del-Duca et al. (1994)
Avena sativa	ADC	Osmotic stress	ADC activity in the thylakoids membrane	Borrell et al. (1995)
Dunaliella salina	TGase	Salt stress	Thylakoid photosynthetic complexes and Rubisco	Dondini et al. (2001)
Spinacia oleracea	SAMDC	Chilling	Photoinhibition	He et al.(2002)
Helianthus tuberosus	TGase	Light condition	Light-harvesting complex II thylakoid activity	Della-Mea et al. (2004)
Arabidopsis thaliana	ADC	Salt stress	Chl contents photosynthetic efficiency	Kasinathan, and Wingler (2004)
Barley	TGase	Senescence	Thylakoid fraction	Sobieszczuk-Nowicka et al. (2009)
Zea mays	TGase	Natural condition	Chloroplast transformation; Grana stacking; PSII and PSI	Ioannidis et al. (2009)
Arabidopsis	ADC	Drought	Transpiration rate stomata conductance and aperture	Alcázar et al. (2010)
Nicotiana tabacum	TGase	Oxidative stress	Thylakoid remodelling chloroplast ultratructure	Ortigosa et al. (2010)
Poncirus trifoliata	ADC	Multiple stresses	Chl contents stomatal density	Wang et al. (2011)

Table 1. High activities of arginine decarboxylase (ADC), adenosylmethionine decarboxylase (SAMDC) and trantaminase (TGase) are involed in regulation of photosynthesis in response to environmental stress. SAMDC is an enzyme that catalyzes the conversion of S-adenosyl methionine to S-adenosylmethioninamine. ADC is an important enzyme responsible for putrscine synthesis. TGase is present in the chloroplasts of higher plants and its activity is modulated by the presence of light.

much influenced by 1mM Spd application, although Spd could ameliorate plant growth and increase net photosynthetic rate (P_N), stomatal conductance (Gs), intercellular CO_2 concentration (Ci), actual efficiency of photosystem II ($\Phi PS\,II$) and the coefficient of photochemical quenching (qP) of cucumber seedlings subjected to salinity (Li et al., 2007). 10

mM Put also alleviated the reduction of salt stress on P_N. However, Put had no effect on Gs and transpiration rate (Tr), and aggravated the reduction of salt stress on Ci. The result suggested that Put strongly affects photosynthetic apparatus involving in enhancement of photochemical quenching rather than regulation of stomatal closure or opening (Zhang et al., 2009). It was also found that exogenous Put improved the photoadaptability of Scenedesmus by altering the content of LHCII monomers and oligomers and PSI and PSII core proteins in the thylakoid membranes (Navakoudis et al., 2007). In green alga Scenedesmus under high salinity, exogenously applied Put effectively decreased the functional size of the antenna and increased the density of active PSII reaction centers, thereby reducing the salt-induced increase in excitation pressure that may cause oxidative damage to the photosynthetic apparatus (Demetriou et al., 2007). Application of Put, Spd and Spm through the root was also effective in alleviating the salt damage to PSI and PSII activities (Chattopadhayay et al., 2002). However, the effects of polyamines on the salt damage to the structure and functions of the photosynthetic apparatus of higher plants remain until today largely unknown.

Several publications have reported that changes of endogenous polyamine level and forms are involved in regulating the photochemical efficiency of salt-stressed plants, and polyamines metabolism-related enzymes are closely correlated with photosynthesis. Exogenous polyamines increased bound Spd contents in chloroplasts to enhance the photosynthetic capacities of corn suffering salt stress (Liu et al., 2006). Exogenously supplied Put in the salt stressed of green alga cultures was shown to increase the Put/Spm ratio in thylakoids and lead to a decrease of the functional antenna size, both by decreasing the size of LHCII and increasing the quantum yield of PSII primary photochemistry, so that the damaging effects induced by salinity are diminished (Demetriou et al., 2007). In our study, applications of 8 mM Put to salt-stressed plant leaves increased content of endogenous polyamines in the thylakoid membranes and overcame the damaging effects of salt stress on the structure and function of the photosynthetic apparatus, which was associated with an improvement in the actual PS II efficiency (data unpublished).

3.2 Drought stress

Drought is the major abiotic stress factor limiting crop productivity in the worldwide (Wang et al., 2003; Sharp et al., 2004). Under drought stress, plants rapidly close stomata with decrease in leaf water potential, thus leading to a significant inhibition of photosynthesis (Zlatev & Yordanov, 2004). To cope with drought, plants initiate a reprogramming of transcriptional, post-transcriptional and metabolite processes that restricts water loss. Several studies have reported that exogenous PAs application is involved in improving drought tolerance against the perturbation of biochemical processes (Yang et al., 2007; Alcázar et al., 2010), but mechanisms of their action during exogenous application in modulating physiological phenomena especially in photosynthesis is not fully understood (Bae et al., 2008). Both net photosynthetic net (P_N) and water use efficiency (WUE) in leaves of rice subjected to drought stress for 7 days were significantly improved by spraying of plants with 10 µM Put, Spd and Spm solutions, while amongst the PAs, Spm was the most effective (Farooq et al., 2009). Drought resistance of two different tomato cultivars, application of 0.1mM exogenous Spd increased the P_N, Gs and Tr in the tomato seedling leaves of the two tomato cultivars and decreased Ci by preventing stoma closure and stimulating CO_2 uptake during the later period under drought stress (Zhang et al., 2010).

They also observed that the mitigative effects of exogenous Spd on photosynthesis in drought-sensitive cultivars were greater than those in high drought-resistant cultivars. In addition, exogenous Spm to pines under drought caused a decline in transpiration rates, enhanced photosynthesis and promoted osmotic adjustment, which would help to maintain turgor (Anisul et al., 2003; Pang et al., 2007). Base on the above studies, we proposed a model describing the role of PAs on photosynthesis during drought-stress: PAs may modulate the activities of certain ion channels at the plasma membrane and stimulate stomatal closure, which would help to enhance photosynthetic efficiency under drought.

3.3 Temperature stress

Temperature is an important ecological factor for plant growth and development, when the temperature is too high or too low, plant growth will stop (Berry and Raison, 1981). It has been demonstrated that high temperature or low temperature adversely affects plant growth and survival, but the impact of temperature stress on the photosynthetic apparatus is considered to be of particular significance because photosynthesis is often inhibited before other cell functions are impaired (Haldimann and Feller, 2004). Extensive research over the last years has focused on the organization and structure of photosynthetic complexes in response to high temperature or low temperature stress (Santamaria and Vierssen, 1997). There are at least three major stress-sensitive sites in the photosynthetic machinery, the PSII, the ATPase and the carbon assimilation process (Allakhverdiev et al., 2008; Mathur et al., 2011).

Heat-shock often induced changes in polyamine content which can be ascribed as protective responses aimed at structural integrity of membrane and cell walls (Edreva et al., 1998), and tolerant plants are able to increase total spermidine and spermine pools under heat stress (Bouchereau et al., 1999). Increased contents of polyamines can stabilize the structure of thylakoid membranes and prevent chlorophyll loss, playing an important role in the protective response of plants to heat stress. Because of the polycationic nature at a physiological pH, polyamines can bind strongly to cellular constituents such as nucleic acids, proteins and membranes (Childs et al., 2003). Several reports have indicated the involvement of polyamines in regulating heat stress-induced the inhibition of photosynthetic efficiency. Exogenous application of 4 mM Spd improved the plant heat-resistance in two tomato cultivars, and especially in tolerant cultivars have higher ability to hardening and higher resistance to thermal damage of the pigment-protein complexes structure and the activity of PSII than sensitive cultivars (Murkowski, 2001). At filling stage of rice, high temperature stress caused a decline in photosynthetic capacity, chlorophyll content and RuBPc activity of two different resistance cultivars. However this effect was more severe in heat-sensitive variety than in heat-resistant variety, because it is closely related to the heat-resistant cultivars have a high content of endogenous polyamines, especially Put accumulation (Huang et al., 1999). In vivo and in vitro experiment showed that exogenous Spm was effective in alleviating heat damage to the photosynthetic apparatus of cucumber, suggesting that protein complexes in thylakoids were made more stable to heat due to their binding to Spm (Li et al., 2003).

In addition, published data showed that PAs levels increased the tolerance degree of the photosynthetic apparatus to low temperatures in different plant species (Hummel et al., 2004), comparison within species has revealed that cold-tolerant varieties/lines show higher endogenous PA contents in response to low temperature than non-tolerant ones (Zhang et al., 2009). He et al. (2002) reported that Spd-pretreated cucumber plants had a high Spd content in both leaves and thylakoid membranes during chilling. They also found Spd pretreatment no effect of stomatal conductance in chilled leaves, but alleviated the decline of the maximum efficiency of PSII photochemistry (Fv/Fm), photosynthetic electron transport activity of thylakoids and activity of enzymes in carbon metabolism. Szalai et al. (1997) observed a marked increase of Put and Spd contents together with a significant decline of Fv/Fm in maize leaves during chilling at 5°C. Based on these results, we conclude that the increase of polyamine contents might be important in plant defense to the photosynthetic apparatus against the low-temperature photoinhibition. Biochemical and physico-chemical measurements showed that the response of the photosynthetic apparatus to low temperature is affected by the changes occurring in the pattern of LHCII-associated Put and Spm which adjust the size of LHCII. The decrease of Put/Spm ratio, mainly due to the reduction in the quantity of LHCII-associated Put led to an increase of the LHCII, especially of the oligomeric forms (Sfakianaki et al., 2006), which is consistent with a data obtained for spinach showed that the low temperature induced a decrease in the Put associated to the thylakoid membranes (He et al., 2002). These results suggest the Put/Spm ratio in the structure of the photosynthetic apparatus associated with the photosynthetic efficiency and the maximal photosynthetic rate, although the mechanism by which polyamines contribute to the increasing tolerance to low temperatures is not yet understood.

3.4 UV radiation stress

An important consequence of stratospheric ozone depletion is the increased transmission of solar ultraviolet radiation (UV) to the Earth's lower atmosphere and surface. Effects of enhanced UV radiation (200-320 nm) on plants have been studied on plant morphology, growth and development, and physiology aspects, which have potential adverse effects on agricultural production and natural plant ecosystems (Caldwell et al., 1998; Madronich et al., 1998; An et al., 2000; Zhang et al., 2003; Bjorn et al., 1996). UV radiation may also induce the formation of reactive oxygen species (ROS) in plants, leading to the damage of photosynthetic apparatus, lipid peroxidation and the changes of PAs levels. The potential role of polyamines in maintaining the photochemical efficiency of plants in response to UV stress has become a research focus.

In the leaves of six genotypes of silver birch (Betula pendula Roth) seedlings, PAs analyses showed that the concentrations of Put were increased and Spd decreased with elevated temperature. Therefore, there was a change in polyamine accumulation towards more conjugated forms. It has been suggested that the conformation of antenna proteins is regulated by polyamines, affecting the efficiency of light harvesting. The changes in Put and Spm metabolism implied that the moderately elevated temperature increased photosynthetic antenna size in silve birch leaves (Tegelberg et al., 2008). The functional and biochemiacal aspects of the photosynthetic apparatus in response to UV-B radiation were examined in unicellular oxygenic algae Scenedesmus obliquus, which is the first comparative characterization of the photosynthetic responses exhibited by the Wt (wild type) and Wt-lhc (a chlorophyll b-less mutant) mutant to UV-B irradiation with emphasis on the response of polyamines and xanthophylls (Sfichi-Duke et al., 2008). The results showed that light stress led to an 204% increased of Spm (which could positively regulate antenna size) and a threefold decrease in Put/Spm ratio. The attachment of Spm to thylakoids could be of structural or functional importance. Along with xanthophylls, polyamines at the LHCII level are more sensitive to UV-B stress than xanthophylls and their responsiveness is abolished in Chl b-less mutants. The tobacco cultivars Bel B and Bel W3 were employed to describe possible protective functions of polyamines against UV-B radiation in sun light simulators (GSF/Munich) with natural diurnal fluctuations of simulated UV-B (Lütz et al., 2005). The results indicated that an increase of PAs, especially of Put level, in thylakoid membranes upon elevated UV-B exposure comprises one of the primary protective mechanisms in the photosynthetic apparatus of the tobacco variety Bel B against UV-B radiation. In addition, the sensitivity to UV-B of Bel W3 (sensitive to ozone) is attributed to its incapability to enhance Put level in thylakoid membranes. After prolongation of UV-B exposure, when endogenous plant balances are being gradually restored, due to secondary responses, (e.g., biosynthesis of carotenoids and of additional flavonoids) and the plant is adapting to the altered environmental conditions, the the PAs level is being reduced. Unal et al.(2007) repoted that under exposure to UV-A for 24 and 48h, the photosynthetic quantum yield ratio of Physcia semipinnata decreased, while that of 1mM polyamine treated were not influenced. It was also found that exogenously added Spd had higher Chla content than Spm and Put added samples. By exposure of the samples with polyamines before the UV-A treatment for 48h, MDA contents was found lower than control group and other groups. These data supported that polyamines, especially Spd and Spm, would play a role in protecting Chl a content, protein content and decreasing lipid peioxidation. This is in agreement with the report of Kramer et al. (1991) whose found membrane lipid may be a target of UV-B damage and polyamine accumulation in response to UV-B radiation stress is consistent with similar reaponses to other environmental stressors. Smith et al. (2001) investigated the influence of UV-B radiation on the UV-B sensitive legume Phaseolus vulgaris L. "Top Crop" over a two-week period. Total free polyamines showed marked decreases in response to UV-B radiation, primarily due to a decrease in Put, which was correlated with UV-B induced chlorophyll loss. Kramer et al. (1992) reported that photosynthetically active radiation (PAR) had a large effect on polyamine levels in leaves, with higher levels of Put and Spd observed at 600 than at 300 μmol m^{-2} s^{-1} in both cultivars (UV-B-sensitive and–insensitive cultivars). The results indicated that the inhibition of UV-B stress by high PAR may involve polyamine accumulation.

The sensitivity of the photosynthetic apparatus to UV-B irradiation was studied in cultures of unicellular green alga Scenedesmus obliquus incubated in low light (LL) and high light (HL) conditions, treated or not with exogenous polyamines. UV-B radiation induces a decrease in the thylakoid-associated Put and an increase in Spm, so that the reduction of Put/Spm ratio leads to the increase of light-harvesting complex II (LHCII) size per active reaction center and, consequently, the amplification of UV-B effects of the photosynthetic apparatus. The separation of oligomeric and monomeric froms of LHCII from isolated thylakoids showed that UV-B induces an increase in the oligomeric forms of LHCII, which was more intense in LL than in HL. By manipulating the LHCII size with exogenous polyamines, the sensitivity degree of the photosynthetic apparatus to UV-B changed significantly. Specifically, the addition of Put decreased highly the sensitivity of LL culture to UV-B because of the inhibitory effect of Put on the LHCII size increasing, wheres the addition of Spm enhanced the UV-B injury induced in HL culture because of the increasing of LHCII size (Sfichi et al., 2004). Zacchini et al. (2004) reported that the put content in tobacco was enhanced, compared with control, in upper layers of calli 6 h after UV-C high dose stress and decreased 24 and 48 h after irradiation, though remained statistically higher than control.

No differences between control and UV-irradiated calli were detected in lower layers. Spermidine and spermine were not affected by UV treatment.

3.5 Hypoxia stress

In natural conditions, flooding, excessive irrigation and soil compaction extremely lead to oxygen deficiency in the root-zone of plant, thus causing to hypoxia stress (Drew, 1997). Hypoxia stress is one of an important environmental factor inhibits plant growth and yield (Avijie et al., 2002). Previous several studies have been shown that suitable concentration of exogenous polyamines could alleviate hypoxia stress-induced physiological damage and improve photochemical efficiency of plants (Vigne and Frelin, 2008). Zhou et al. (2006) demonstrated application of exogenous polyamines to some extent increased net photosynthetic rate (P_N) and water use efficiency (WUE) in cucumber leaves. Shi et al. (2009) also observed that exogenous Put alleviated the reduction of P_N, Gs, Ci of cucumber subjected to root-zone hypoxia through enhancing the actual and maximal nitrate reductase activities. Application of exogenous 0.05 mM Spd added to the hypoxia nutrient solutions significantly suppressed the accumulation and insoluble bound Put in roots and leaves of cucumber seedlings, which was associated with a decreased in dissipation energy (NPQ) and xanthophylls de-epoxidation state (DEPS) during hypoxia stress, while Spd enhanced maximal photochemical efficiency (Fv/Fm), PSII photochemistry rate (ΦPSII) and P_N (Jia, 2009). These results implied that exogenous polyamines increased photosynthetic capacity might be involved in regulating changes of endogenous polyamine contents in the chloroplasts, thus enhancement of root-zone hypoxia tolerance.

3.6 Oxidative stress

Oxidative stress has been postulated, years ago, to be a causal factor in responses to environmental stresses (Velikova et al., 2000; Patil et al., 2007). The basic tenet of this hypothesis is that the stress-associated decline in the functional capacity of biological systems is primarily due to the accumulation of irreparable oxidative molecular damage (Sohal et al., 1995). Adverse environmental stresses induced a decrease in photosynthetic activity which often associated with the oxidative stress (Krause, 1994). It has been shown that plant could overcome the effect of the oxidative stress and sustain photosynthetic efficiency that may be related to the scavenging of stress-induced toxic oxygen species, such as hydrogen peroxide (H_2O_2), hydroxyl radical ($OH^{\cdot}$) and superoxide radical ($O_2^{\cdot-}$) (Sopory et al., 1990).

Polyamines (PAs) are regarded as a new class of growth substances and are also well known for their positive effects on the photosynthetic efficiency under various stress conditions due to their acid neutralizing and antioxidant properties, as well as to their membrane and cell wall stabilizing abilities (He et al., 2002; Mapelli et al., 2008; Zhao and Yang, 2008). Exogenous application of spermidine (Spd) could alleviate salt-induced membrane injury of chloroplast by increasing the active oxygen scavenging ability. The activities of superoxide dismutase (SOD), ascorbate peroxidase (APX) and glutathione reductase (GR), the contents of ascorbic acid (AsA), Car and glutathione reduced form (GSH), and the ratios of GSH/GSSG in chloroplasts were increased, which increased the salinity tolerance of the photosynthetic apparatus in cucumber seedlings (Duan et al., 2008a; Duan et al., 2009b). Pretreatment with Spd markedly reduced lipid peroxidation and membrane relative permeability in wheat leaves under water stress (Duan et al., 2006). They found that Spd also improved the transcription of PSII genes and translation of the corresponding proteins, which sustains a higher activity in PSII during water stress. Yiu et al. (2009) showed that 2 mM putrescine (Put) pre-treatment confers flooding tolerance to the photosynthetic efficiency of welsh onion plants, probably through inducing the activities of various anti-oxidative systems. It has been shown that lipid peroxidation levels were significantly decreased by PAs treated with Physcia semipinnata during the exposure to UV-A (Unal et al., 2008). Among the three polyamines, Spm-treated samples had lower concentration of MDA and higher in the amount of chlorophyll a levels than spd-and put-treated samples. These results indicate that polyamines may protect photosystem II from oxidative stress.

The mechanistic details of PAs effects have not been clarified, although a proposed mechanism is based on PAs neutralizing harmful ROS in tissues or cells and inducing the reorganization of the photosynthetic apparatus. Several results have confirmed that the increase in intra-cellular polyamine content or polyamine metabolism related enzyme activity played an important role in photosynthesis of plants and in oxidative stress resistance (Dondini et al., 2001; Cochón et al., 2007; Demetriou et al., 2007). Polyamines and especially the thylakoid-associated polyamines play a decisive role in protecting photosynthetic apparatus against oxidative stress. He et al. (2008) found that Put induced the changes of endogenous polyamines in the photosynthetic apparatus to some extent, might be involved in the reduction of H_2O_2 contents and membrane lipid peroxidation under salt stress. This result is supported by our experiments that Put reduced the number of plastoglobuli resulting from gradual thylakoid membrane degradation induced by salinity (Unpublished data). It has been suggested that transglutaminase (chlTGZ) plays an important functional role in the formation of grana stacks, probably due to antenna protein polyaminylation (Ioannidis et al., 2009). Recently, in the transplastomic tobacco young leaves, oxidative stress induced the over-expression of chlTGZ in the tobacco chloroplasts seems to be related to polyaminylation of antenna proteins and thylakoid remodelling, the extended effect of this over-expression seems to induce an imbalance between capture and utilization of light in photosynthesis, changes in the photosynthetic electron transport chain and increasing oxidative risk (Ortigosa et al., 2010).

Species	PAs type	PAs dosage /conce-ntration	Organ	Mode of PAs treatment	Stress	Measured photosynthetic parameters	Author
Oryza sativa Triticumae stivum	Spd Spm	10 mM 10 mM	Leaves	Incubation	Normal conditions	Chl content	Cheng and Kao (1983)
Dicotyledo-nous komatsuna	Spd	1 mM	Leaf discs	Incubation	Dark and light conditions	Chl content photosynthetic activity chloroplast ultrastructure	Pjon et al. (1990)
Scenedesmus obliquus	Spm	320 µM	Leaves	Foliar spraying	Natural condition	Protochlorophyllide chlorophyll	Beigbeder and Kotzabasis. (1994)

Species	PAs type	PAs dosage /concentration	Organ	Mode of PAs treatment	Stress	Measured photosynthetic parameters	Author
Spinacea oleracea	Spd Spm	10 µM 0.1mM 1 mM 5 mM 10 mM	PSII	Incubation	Natural condition	FTIR, secondary structure of PSII	Bograh et al.(1997)
Triticum aestivum	Put Spd Spm	10 µM 0.1 mM 0.5 mM	Leaves	Foliar spraying	Yellow rust	Chl contents, chloroplast ultrastructure	Aldesuquy et al.(2000)
Vicia faba	Spd	0.1, 0.5, 1.0, 3.0, 6.0 mM	Leaves	Incubation	Normal conditions	Stomatal movements	Liu et al. (2000)
Lycopersicon esculentum	Spd	4 mM	Roots	Addition to Hoagland nutrient solution	Heat	Fv/Fm, Amax, Rfd	Murkowski (2001)
Triticum aestivum	Put Spd Spm	20 µM 20 µM 20 µM	Leaves	Incubation	Dark conditions	Chl contents PSI, PSII activities	Subhan et al. (2001)
Oryza sativa	Spd Spm	1.0 mM 1.0 mM	Leaves	Root medium	Salt	Total Chl content, PSI activity, PSII activity	Chattopa-dhayay et al. (2002)
Cucumis sativus	Spd	0.5 mM	Leaves	Incubation in the solution (pre-treatment)	Chilling	Chl contents Fv/Fm, qP, qN	He et al. (2002)
Pinus strobus	Spm	10ug/L	Roots	Soaking	Water	P_N, WUE, Tr	Iqbal et al. (2003)
Nicotiana tabacum	Put Spd Spm	1 mM 1 mM 1 mM	Plants	Whole plant spraying	Ozone	P_N, Fv/Fm, Chl contents, chloroplast structure	Navako-udis et al. (2003)
Mesembryan themum crystallinum	Put	1 mM	Plants	Spraying	Salt	Ultrastructure of leaf mesophyll	Paramo-nova et al. (2003)
Mesembryan themum crystallinum	Put	1 mM	Plants	Spraying	Salt	Ultrastructure of chloroplasts	Paramo-nova et al. (2004)

Species	PAs type	PAs dosage /conce- ntration	Organ	Mode of PAs treatment	Stress	Measured photosynthetic parameters	Author
Triticum aestivum	Put Spd Spm	2.5 mM 5 mM 2.5 mM	Seed	Soaking (pre- sowing)	Salt	P_N, Ci, Gs, Tr	Iqbal and Ashraf (2005)
Triticum aestivum	Spd	0.2 mM	PSII	Irrigating or spraying	Water stress	Chlcontents, PSII phothochemical activity	Duan et al. (2006)
Zea mays	Cad, Put Spd, Spm	1 mM 1 mM 1 mM 1 mM	Leaves	Foliar spraying	Salt	P_N, Fv/Fm	Liu et al. (2006)
Cucumis sativus	Put Spd Spm	1 mM 0.5 mM 0.5 mM	Leaves	Foliar spraying	Hypoxia	P_N, Gs, Tr, Ci	Zhou et al. (2006)
Spinacea oleracea	Spd Spm	Millim olar range	PSII	Incubation	Natural condition	Oxygen evolution activity, Fv/Fo,Chl fluorescence decay kinetics Thermoluminescen ce, OJIP	Beauche- mie et al. (2007)
Berley	Cad, Put Spd, Spm	10 µM 10 µM 10 µM	The whole plant	Spraying Whole plants	Salt	The stomata number, stomata index, stomata length, the epidermis cell number, stomata width	Çavuso- glu et al. (2007)
Scenedesmus obliquus	Put	100 mM	Isolated thyla- koids	Medium	Salt	Fluorescence induction, oxygen evolving activity, chlorophyll content	Demetriou et al. (2007)
Nicotiana tabacum	Put Spd Spm	1 mM 0.1 mM 0.05 mM	Leaf discs	Floated in the solutions	Normal conditions	Fv/Fm, PSIIα and PSIIβ photopho- sphorylation oxygen evolution	Ioannidis and Kotzabasis (2007)
Cucumis sativus	Spd	1 mM	Leaves	Foliar spraying	Salt	P_N, Ci, Gs Fv/Fm, qP,ΦPS II	Li et al. (2007)
Green alga	Put	1 mM	Cell	Incubation	Normal conditions	JIP-test parameters LHCIIoligomeric monomeric forms	Navako- udis et al. (2007)

Species	PAs type	PAs dosage /conce- ntration	Organ	Mode of PAs treatment	Stress	Measured photosynthetic parameters	Author
Arabidopsis	Put Spd Spm	1 Mm 1 mM 1 mM	Plants	Addition to wet filter papers	Drought	Chl contents Stomatal status	Yamaguchi et al. (2007)
Cucumis sativus	Spd	0.5mM	Leaves	Foliar spraying	Hypoxia	P_N, Gs, Tr, Ci	Zhou (2007)
Cucumis sativus	Spd	0.1 mM	Roots	Addition to Hoagland nutrient solution	Salt	P_N, Fv/Fm	Duan et al. (2008)
Physcia semipinnata	Put Spd Spm	1 Mm 1 mM 1 mM	Lichen thalli	Incubation	UV-A	Chla content, Fv/Fm	Unal et al. (2008)
Cucumis sativus	Spd	0.1 mM	Roots	Addition to nutrient solution	Salt	P_N	Duan JJ (2009)
Cucumis sativus	Spd	0.1 mM	Roots	Addition to nutrient solution	Salt	P_N	Duan JJ (2009)
Cucumis sativus	Spd	0.1 mM	Roots	Addition to Hoagland nutrient solution	Salt	P_N	Duan JJ (2009)
Oryza sativa	Put Spd Spm	10 µM 10 µM 10 µM	Whole plants	Seed soaking	Drought	P_N,Gs, Tr, WUE,	Farooq et al. (2009)
Cucumis sativus	Spd	0.05 mM	Roots	Addition to Hoagland nutrient solution	Hypoxia	P_N, Ci, Gs, Tr Fv/Fm, qP, qN ΦPSII, AQY	Jia (2009)
Cucumis sativus	Put	0.5 mM	Roots	Addition to nutrient solution	Hypoxia	P_N, Ci, Gs, Tr Vcyt, Valt	Shi et al. (2009)

Species	PAs type	PAs dosage /conce-ntration	Organ	Mode of PAs treatment	Stress	Measured photosynthetic parameters	Author
Citrus	Spm	1 mM	Leaves	Foliar spraying (pre-treatment)	Dehydra-tion	Stomatal closure/ opening	Shi et al. (2010)
Welsh onion	Put	1,2, 3mM	Plants	Applied to the substrate surface	Flood	Chl contents Fv/Fm	Yiu et al. (2009)
Cucumis sativus	Put	10mM	Leaves	Foliar spraying	Salt	Chl contents, P_N ,Gs, Tr, Ci, qP Fv/Fm, ΦPSII, NPQ	Zhang et al. (2009)
Cowpea cultivar	Spd	1 mM	Leaves	Foliar spraying	Cinnamic acid	Fv/Fm, ΦPS II Rubisco activity	Huang and Bie (2010)
Lactuca sativus	Spm	0.2~2 mM	Leaves /Leaf discs	Spraying/ Incubation	Sene-scence	Chl contents light-harvesting complexes	Serafini-Fracassini et al. (2010)
Cucumis sativus	Put	8 mM	Leaves	Foliar spraying	Salt	Fv/Fm, qP, NPQ ΦPSII, rETR	Shu et al. (2010)
Lycopersicon esculentum	Spd	0.1mM	Leaves	Foliar spraying	Drought	P_N, Gs, Tr, Ci	Zhang et al.(2010)
Cucumis Sativus	Spd	0.1mM	Roots	Addition to Hoagland nutrient solution	Salt	P_N, Gs, Ci, Tr, total soluble sugar, sucrose, starch	Chen (2011)

Table 2. Effects of exogenous polyamines (Cad, Put, Spd, Spm) on various photosynthetic parameters of plants in response to stress and non-stress conditions. Chl, chlorophyll; P_N, net photosynthetic rate; Ci, intercellular CO_2; Gs, stomatal conductance; Tr, transpiration rate; WUE, water use efficiency; Fv/Fm, maximum quantum efficiency of photosystem II; Amax, area above the fluorescence induction curve; Rfd, informs about the interaction and equilibrium between primary photosynthetic reactions and dark enzymatic reactions; qP, photochemical quenching; qN, non-photochemical quenching; PSI, photosystem I; PSII, photosystem II; ΦPSII, actual photochemical efficiency; AQY, apparent quantum efficiency; Vcyt, cytochrome; Valt, alternative respiration; Rubisco, ribulose-1,5-bisphospate carboxylase/oxygenase

4. Conclusion

From the published literature, it can be deduced that PAs play an important role in wide spectrum of physiological processes such as cell division, dormancy breaking of tubers and germination of seeds, stimulation, and development of flower buds and fruits, somatic embryogenesis, differentiation and plant morphogenesis, signal transduction and in the protection of the photosynthetic apparatus. Although considerable evidence indicates that PAs exhibits various positive effects on photosynthetic processes of plant in response to abiotic stresses (Table 2), their precise role in these specific processes is still far from being complete. So far, it seems that one of the modes of PAs action in the regulation of photosynthesis under environmental stresses is probably by reducing the production of free radicals, scavenging the free radicals and/or involved in activation of expression of genes encoding antioxidant enzymes. On the other hand, we can also speculate that conjugated PAs play an important role in the protection of related proteins in the photosynthetic apparatus. The mechanism of their action probably involves direct binding of PAs to the extrinsic proteins and the hydrophilic portions of intrinsic polypeptides of PSII through electrostatic interaction due to their poly-cationic properties, and the poly-cationic selectivity effect in decreasing order is $Spm^{4+}>Spd^{3+}>Put^{2+}$. This electrostatic interaction could provide some stability to the conformation of thylakoid proteins against various stresses and consequently help in maintaining the photosynthetic activity. In addition, the biological functions of PAs on photosynthesis may be also as a signal component to cascade with other growth regulator or hormone. In order to better elucidate the role of PAs in the photosynthesis of plants response to abiotic stress, application of advanced molecular biology and proteomic approaches will help elucidate the mechanisms of PAs in particular improvement of photochemical efficiency in plant processes involved in stress tolerance.

5. Acknowledgement

This work was funded by National Basic Research Program of China (973 Program, No.2009CB119000) and National Natural Science Foundation of China (No. 30900995; No. 31071831; No. 30871736) and A Project Fund by the Priority Academic Program Development of Jiangsu Higher Education Institutions and Supported by the China earmarked fund for Modern Agro-industry Technology Research System (CARS-25-C-03). I may not fully understand all reseachers working in field of polyamines and photosynthesis under stress conditions during writing this chapter, here I express my most sincere apologies.

6. References

Alcázar, R.; Planas, J. Saxena, T. Zarza, X. Bortolotti, C. Cuevas, J. Bitrián, M. Tiburcio, AF. & Altabella, T. (2010). Putrescine accumulation confers drought tolerance in transgenic arabidopsis plants over-expressing the homologous *Arginine decarboxylase* 2 gene. *Plant Physiology and Biochemistry*, 48: 547–552.

Aldesuquy, HS.; Abdel-Fattah, GM. & Baka, ZA. (2000). Changes in chlorophyll, polyamines and chloroplast ultrastructure of puccinia striiformis induced 'green islands' on detached leaves of *Triticum aestivum*. *Plant Physiologyand Biochemistry*, 38: 613–620.

Allakhverdiev, S.; Kreslavski, V. Klimov, V. Los, D. Carpentier, R. & Mohanty, P. (2008). Heat stress: an overview of molecular responses in photosynthesis, *Photosynthesis Research*, 98: 541–550.

Andreadakis, A. & Kotzabasis, K. (1996). Changes in the biosynthesis and catabolism of polyamines in isolated plastids during chloroplast photodevelopment. *Journal of Photochemistry and Photobiology B: Biology*, 33: 163−170.

Anisul, IM.; Blake, TJ. Kocacinar, F. & Lada, R. (2003). Ambiol, spermine, and aminoethoxyvinylglycine prevent water stress and protect membranes in *Pius strobes* L.under droght. *Tresss*, 17: 278-284.

Assmann, SM. (1993). Signal transduction in guard cells. *Annual Review of Cell and Developmental Bioogyl*, 9: 345-375.

Avijie, D.; Uehimiya, H. & Das, A. (2002). Oxygen stress and adaptation of semi-aquatic plant: Rice (*Oryza sariva*). *Journal of Plant Research*, 115: 315-320.

Bae, HH.; Kim, SH. Kim, MS. Sicher, RC. Lary, D. Strem, MD. Natarajan, S. & Bailey, BA. (2008). The drought response of *Theobroma cacao* (cacao) and the regulation of genes involved in polyamine biosynthesis by drought and other stresses. *Plant Physiology and Biochemistry*, 46: 174–188.

Beauchemin, R.; Gauthier, A. Harnois, J. Boisvert, S. Govindachary, S. & Carpentier, R. (2007). Spermine and spermidine inhibition of photosystem II: Disassembly of the oxygen evolving complex and consequent perturbation in electron donation from Tyr_Z to P680+ and the quinone acceptors Q_A^- to Q_B. *Biochimica et Biophysica Acta*, 1767: 905– 912.

Beigbeder, A. & Kotzabasis, K. (1994). The influence of exogenously supplied spermine on protochlorophyllide and chlorophyll biosynthesis. *Journal of Photochemistry and Photobiology B: Biology*, 23: 201–206.

Beigbeder, A.; Vavadakis, M. Navakoudis, E. & Kotzabasis, K. (1995). Influence of polyamine inhibitors on light-independent and light-dependent chlorophyll biosynthesis and on the photosynthetic rate. *Journal of Photochemistry and Photobiology B: Biology*, 28: 235-242.

Berry, JA. & Raison, JK. (1981). Responses of macrophytes to temperature. In: Lange, OL.; Nobcl, PS. Osmond, CB. Ziegler, H. (Eds.), Physiological Plant Ecology I. Responses to the Physical Environment. Springer-Verlag, New York, pp. 277–337.

Birecka, H.; Dinolfo, TE. Martin, WB. & Frohlich, MW. (1984). Polyamines and leaf senescence in pyrrolizidine-bearing heliotropium plants. *Phytochemistry*, 23: 991–997.

Bograh, A.; Gingras, Y. Tajmir-Riahi, HA. & Carpentier, R. (1997). The effects of spermine and spermidine on the structure of photosystem II proteins in relation to inhibition of electron transport. *Federation of European Biochemical Societies*, 402: 41–44.

Borrell, A.; Culianez-Macia, FA. Altabella, T. Besford, RT. Flores, D. & Tiburcio, AF. (1995). Arginine decarboxylase is localized in chloroplasts. *Plant Physiology*, 109: 771-776.

Bouchereau, A.; Azis, A. Larher, F. & Martin-Tanguy, J. (1999). Polyamines and environmental challenges: recent development. *Plant Science*, 103-125.

Bright, J.; Desikan, R. Hancock, JT. Weir, LS. & Neill, SJ. (2006). ABA-induced NO generation and stomatal closure in Arabidopsis are dependent on H_2O_2 synthesis. *The Plant Journal*, 45: 113–122.

Brugnoli, E. Björkman, O. (1992). Growth of cotton under continuous salinity stress: influence on allocation pattern, stomatal and non-stomatal components of photosynthesis and dissipation of excess light energy. *Planta*, 187: 335–347.

Çavuşoğlu, K.; Kılıç, S. & Kabar K. (2007). Effects of pretreatments of some growth regulators on the stomata movements of barley seedlings grown under saline (NaCl) conditions. *Plant, Soil and Environment*, 53: 524–528.

Chattopadhayay, MK.; Tiwari, BS. Chattopadhayay, G. Bose, A. Sengupta, DN. & Ghosh, B. (2002). Protective role of exogenous polyamines on salinity-stressed rice (*Oryza sativa*) plants. *Physiologia Plantarum*, 116: 192–199.

Chen, LF.; Lu, W. Sun, J. Guo, SR. Zhang, ZX. & Yang, YJ. (2011). Effects of exogenous spermidine on photosynthesis and carbohydrate accumulation in roots and leaves of cucumber (*Cucumis sativus* L.) seedlings under salt stress. *J Nanjing Agricul University*, 34(3): 31–36.

Cheng, SH. & Kao, CH. (1983). Localized effect of polyamines on chlorophyll loss. *Plant and Cell Physiology*, 24: 1465–1467.

Childs, AC.; Mehta, DJ. & Gerner, EW. (2003). Polyamine-dependent gene expression. *Cellular and Molecular Life Sciences*, 60: 1394–1406.

Cochòn, AC.; Della Penna, AB. Kristoff, G. Piol, MN. San Martín de Viale, LC. & Guerrero, NR. (2007). Differential effects of paraquat on oxidative stress parameters and polyamine levels in two freshwater invertebrates. *Ecotoxicology and Environmental Safety*, 68: 286–292.

Del-Duca, S.; Tidu, V. Bassi, R. Esposito, C. & Serafini-Fracassini, D. (1994). Identification of chlorophyll-a/b proteins as substrates of transglutaminase activity in isolated chloroplasts of Helianthus tuberosus L. *Planta*, 193: 283–289.

Della-Mea, M.; Di-Sandro, A. Dondini, L. Del-Duca, S. Vantini, F. Bergamini, C. Bassi, R. & Serafini-Fracassini, D. (2004). A zea mays 39 kDa thylakoid transglutaminase catalyses the modification by polyamines of light harvesting complex II by polyamines in a light-dependent way. *Planta*, 219: 754–764.

Demetriou, G.; Neonaki, C. Navakoudis, E. & Kotzabasis, K. (2007). Salt stress impact on the molecular structure and function of the photosynthetic apparatus-The protective role of polyamines. *Biochimica et Biophysica Acta*, 1767: 272–280.

Dondini, L.; Bonazzi, S. Del-Duca, S. Bregoli, AM. & Serafini-Fracassini, D. (2001). Acclimation of chloroplast transglutaminase to high NaCl concentration in a polyamine-deficient variant strain of *Dunaliella salina* and its wild type. *Journal of Plant Physiology*, 158: 185–197.

Drew, M. (1997). Oxygen deficiency and root metabolism: Injury and acclimation under hypoxia and anoxia. *Annual. Review of Plant Physiology and Plant Molecular Biology*, 48: 223–250.

Duan, HG.; Shu, Y. Liu, DH. Qin, DH. Liang, HG. & Lin, HH. (2006). Effects of Exogenous spermidine on photosystem II of wheat seedlings under water stress. *Joumal of Integrative Plant Biology*, 48: 920–927.

Duan, JJ.; Guo, SR. Kang, YY. Zhou, GX. & Liu, XE. (2009). Effects of exogenous spermidine on active oxygen scavenging system and bound polyamine contents in chloroplasts of cucumber under salt stress. *Acta Ecologica Sinica*, 29: 0653–0661.

Duan, JJ.; Li, J. Guo, SR. & Kang, YY. (2008). Exogenous spermidine affects polyamine metabolism in salinity-stressed *Cucumis sativus* roots and enhances short-term salinity tolerance. *Journal of Plant Physiology*, 165: 1620–1635.

Edreva, A.; Yordanov, I. Kardijeva, R. & Gesheva, E. Heat shock responses of bean plants: involvement of free radicals, antioxidants and free radical/active oxygen scavenging systems. *Biology of Plants*, 41: 185–191.

Farooq, M.; Wahid, A. & Lee, DJ. (2009). Exogenously applied polyamines increase drought tolerance of rice by improving leaf water status, photosynthesis and membrane properties. *Acta Physiologiae Plantarum*, 31:937–945.

Haldimann, P. & Feller, U. (2004). Inhibition of photosynthesis by high temperature in oak (*Quercus pubescens*) leaves grown under natural conditions closely correlates with a reversible heat-dependent reduction of the activation state of ribulose-1, 5-bisphosphate carboxylase/oxygenase. *Plant, Cell and Environment,* 27: 1169–1183.

Hamdani, S.; Yaakoubi, H. & Carpentier, R. (2010). Polyamines interaction with thylakoid proteins during stress. *Journal of Photochemistry and Photobiology B: Biology,* 11:1011–1344.

He, LX.; Ban, Y. Inoue, H. Matsuda, N. Liu, JH. & Moriguchi, T. (2008). Enhancement of spermidine content and antioxidant capacity in transgenic pear shoots overexpressing apple *spermidine synthase* in response to salinity and hyperosmosis. *Phytochemistry,* 69: 2133–2141.

He, LX.; Nada, K. Kasukabe, Y. & Tachibana, S. (2002). Enhanced susceptibility of photosynthesis to low-temperature photoinhibition due to interruption of chill-induced increase of S-adenosylmethionice decarboxylase activity in leaves of spinach (*Spinacia oleracea* L.). *Plant Cell Physiology,* 43: 196–206.

Hetherington, AM. & Woodward, FI. (2003). The role of stomata in sensing and driving environmental change. *Nature,* 424:21.

Hippler, M.; Klein, J. Fink, A. Allinger, T. & Hoerth, P. (2001). Towards functional proteomics of membrane protein complexes: analysis of thylakoid membranes from *chlamydomonas reinhardtii. The Plant Journal,* 28: 595–606.

Huang, XX. & Bie, ZL. (2010).Cinnamic acid-inhibited ribulose-1,5-bisphosphate carboxylase activity is mediated through decreased spermine and changes in the ratio of polyamines in cowpea. *Journal of Plant Physiology,* 167:47–53.

Huang, YJ.; Luo, YF. Huang, XZ. Rao, ZM. & Liu, YB. (1999). Varietal difference of heat tolerance at grain filling stage and its relationship to photosynthetic characteristics and endogenous polyamine of flag leaf in rice. *Chinese Journal of Rice Science,* 13: 205–210.

Hummel, A.; Amrani, EI. Gouesbet, G. Hennion, F. & Couée, I. (2004). Involvement of polyamines in the interacting effects of low temperature and mineral supply on *pringlea antiscorbutica* (kerguelen cabbage) seedlings. *Journal of Experimental Botany,* 55: 1125–1134.

Ioannidis, NE. & Kotzabasis, K. (2007). Effects of polyamines on the functionality of photosynthetic membrane in vivo and in vitro. *Biochimica et Biophysica Acta,* 1767: 1372–1382.

Ioannidis, NE.; Ortigosa, SM. Veramendi, J. Pintó-Marijuan, M. Fleck, I. Carvajal, P. Kotzabasis K. Santos, M. & Torné, JM. (2009). Remodeling of tobacco thylakoids by over-expression of maize plastidial transglutaminase. *Biochimica et Biophysica Acta-Bioenergetics,* 1787: 1215–1222.

Iqbal, M. & Ashraf, M. (2005). Changes in growth, photosynthetic capacity and ionic relation in spring wheat (*Triticum aestivum* L.) due to pre-sowing seed treatment with polyamines. *Plant Growth Regulation,* 46:19–30.

Jia, YX. (2009). Studies on physiological regulation function of exogenous spermidine on cucumber seedlings tolerance to hypoxia. [Degree]. Nanjing: *Nanjing Agricultural University,* 1–125.

Kasinathan, V. & Wingler, A. (2004). Effect of reduced arginine decarboxylase activity on salt tolerance and on polyamine formation during salt stress in Arabidopsis thaliana. *Physiologia Plantarum,* 121: 101–107.

Kaur-Sawhney, R.; Tiburcio, AF. Altabella, T. & Galston, AW. (2003). Polyamines in plants: An overview. *Journal of Cell and Molecular Biology,* 2: 1–12.

Khan, MA. Ungar, IA. (1997). Effect of thermoperiod on recovery of seed germination of halophytes from saline conditions. *America. Jourbal of Botany*, 84: 279–283.

Kirchhoff, H.; Haase, W. Haferkamp, S. Schoot, T. Borinski, M. Kubitscheck, U. & Rögner, M. (2007). Structural and functional self-organization of photosystem II in grana thylakoids. *Biochimica et Biophysica Acta-Bioenergetics*, 1767: 1180–1188.

Kirchhoff, H.; Hinz, HJ. & Rösgen, J. (2003). Aggregation and fluorescence quenching of chlorophyll a of the light-harvesting complex II from spinach in vitro. *Biochimica et Biophysica Acta-Bioenergetics*, 1606:105–116.

Kotzabasis, K.; Fotinou, C. Roubelakis-Angelaki, KA. & Ghanotakis, D. (1993). Polyamines in the photosynthetic apparatus. *Photosynthesis Research*, 38: 83–88.

Krause, GH. (1994). The role of oxygen in photoinhibition of photosynthesis. In C.H. Foyer and P.M. Mullineaux, (eds), Causes of photooxidative stress and amelioraton of defense systems in plants, CRC Press, Boca Raton, pp: 43–76.

Li, J.; Gao, XH. Guo, SR. Zhang, RH. & Wang, X. (2007). Effects of exogenous spermidine on photosynthesis of salt-stressed *cuellmis sativus* seedlings. *Chinese Journal of Ecology*, 26(10): 1595–1599.

Li, Y.; Shi, GX. Wang, HX. Zhao, J. & Yuan, QH. (2009). Exogenous spermidine can mitigate the poison of cadmium in nymphoides peltatum. *Chinese Bulletin of Botany*, 44: 571–577.

Li, ZJ.; Nada, K. & Tachibana, S. (2003). High-temperature-induced alteration of ABA and polyamine contents in leaves and its implication in thermal acclimation of photosynthesis in cucumber (Cucumis sativas L.). *Journal of the Japanese Society for Horticultural Science*, 72: 393–401.

Liu, J.; Zhou, YF. Zhang, WH. & Liu, YL. (2006). Effects of exogenous polyamines on chloroplast-bound polymine content and photosynthesis of corn suffering salt stress. *Acta Bot. Boreal.- Occident. Sin*, 26(2): 0254–0258.

Liu, K.; Fu, HH. Bei, QX & Luan, S. (2000). Inward potassium channel in guard cells as a target for polyamine regulation of stomatal movements. *Plant Physiology*, 124:1315–1325.

MacRobbie, EAC. (1997). Signaling in guard cells and regulation of ion channel activity. *Journal of Experimental. Botany*, 48: 515–528.

Mansfield, TA.; Hetherington, AM. & Atkinson, CJ. (1990). Some current aspects of stomatal physiology. *Annual Reviewof Plant Physiology and Plant Molecular Biology*, 41: 55–75

Mapelli, S.; Brambilla, IM. Radyukina, NL. Ivanov, YuV. Kartashov, AV. Reggiani, R. & Kuznetsov, VlV. (2008). Free and bound polyamines changes in different plants as a consequence of UV-B light irradiation. *General and Applied Plant physiology*, 34: 55–66.

Mathur, S.; Allakhverdiev, SI. & Jajoo, A. (2011). Analysis of high temperature stress on the dynamics of antenna size and reducing side heterogeneity of Photosystem II in wheat leaves (Triticum aestivum). *Biochimica et Biophysica Acta*, 1807: 22–29.

Meloni, DA. Oliva, MA. Martinez, CA. Cambraia, J. (2003). Photosynthesis and activity of superoxide dismutase, peroxidase and glutathione reductase in cotton under salt stress. *Environmental and Experimental Botany*, 49: 69–76.

Meskauskiene, R.; Nater, M. Goslings, D. Kessler, F. Camp, R. & Apel, K. (2001). A negative regulator of chlorophyll biosynthesis in arabidopsis thaliana. *PNAS*, 98:12826–12831.

Munzi, S.; Pirintsos, SA. & Loppi, S. (2009). Chlorophyll degradation and inhibition of polyamine biosynthesis in the lichen *xanthoria parietina* under nitrogen stress. *Ecotoxicology and Environmental Safety*, 72: 281–285.

Murkowski, A. (2001). Heat stress and spermidine: effect on chlorophyll fluorescence in tomato plants. *Biologia plantarum*, 44: 53–57.

Navakoudis, E.; Lütz, C. Langebartels, C. Lütz-Meindl, U. & Kotzabasis, K. (2003). Ozone impact on the photosynthetic apparatus and the protective role of polyamines. *Biochimica et Biophysica Acta-Bioenergetics*, 1621: 160–169.

Navakoudis, E.; Vrentzou, K. & Kotzabasis, K. (2007). A polyamine-and LHCII protease activity-based mechanism regulates the plasticity and adaptation status of the photosynthetic apparatus. *Biochimica et Biophysica Acta-Bioenergetics*, 1767: 261– 271.

Ortigosa, SM.; Díaz-Vivancos, P. Clemente-Moreno, MJ. Pintò-Marijuan, M. Fleck, I. Veramendi, J. Santos, M. Hernandez, J. & Torné, JM. (2010). Oxidative stress induced in tobacco leaves by chloroplast over-expression of maize plastidial transglutaminase. *Planta*, 232: 593–605.

Pang, XM.; Zhang, ZY. Wen, XP. Ban, Y. & Moriguchi, T. (2007). Polyamine, all-purpose players in response to environment stresses in plants. *Plant Stress*, 1(2): 173–188.

Paramonova, NV.; Shevyakova, NI. Shorina, MV. Stetsenko, LA. Rakitin, VY. & Kuznetsov, VIV. (2003). The effect of putrescine on the apoplast ultrastructure in the leaf mesophyll of mesembryanthemum crystallinum L. under salinity stress. *Russian Journal of Plant Physiology*, 50: 661–673.

Paramonova.; NV. Shevyakova, N I. & Kuznetsov, VIV. (2004). Ultrastructure of chloroplasts and their storage inclusions in the primary leaves of Mesembryanthemum crystallinum affected by putrescine and NaCl. *Russian Journal of Plant Physiology*, 51: 86–96.

Parida, AK.; Das, AB. & Mittra, B. (2003). Effects of NaCl stress on the structure, pigment complex composition, and photosynthetic activity of mangrove Bruguiera parviflora chloroplasts. *Photosynthetica*, 41: 191–200.

Patil, SB.; Kodliwadmath, MV. & Kodliwadmath, SM. (2007). Study of oxidative stress and enzymatic antioxidants in normal pregnancy. *Indian Journal of Clinical Biochemistry*, 22: 135–137.

Pei, ZM.; Kuchitsu, K. Ward, JM. Schwarz, M. & Schroeder, L. (1997). Differnential abscisic acid regulation of guard cell slow anion channels in arabidopsis wild-type and abi1 and abi2 mutants. *The Plant Cell*, 9:409–423.

Pjon, CJ.; Kim, SD. & Pak, JY. (1990). Effects of spermidine on chlorophyll content, photosynthetic activity and chloroplast ultrastructure in the dark and under light. *Bot. Mag. Tokyo*, 103: 43–48.

Raschke, K.; Hedrich, R. Reckmann, U. & Schroeder, JI. (1988). Exploring biophysical and biochemical components of the osmotic motor that drives stomatal movements. *Botanica Acta*, 101: 283–294.

Santamaria, L. & Vierssen, WV. (1997). Photosynthetic temperature responses of fresh-and brackish-water macrophytes: a review. *Aquatic Botany*, 58: 135–150.

Serafini-Fracassini, D.; Di Sandro, A. & Del Duca, S. (2010). Spermine delays leaf senescence in Lactuca sativa and prevents the decay of chloroplast photosystems. *Plant Physiology and Biochemistry*, 48:602–611.

Sfakianaki, M.; Sfichi, S. & Kotzabasis, K. (2006). The involvement of LHCII-associated polyamines in the response of the photosynthetic apparatus to low temperature. *Journal of Photochemistry and Photobiology B: Biology*, 84: 181–188.

Sfichi-Duke, L.; Ioannidis, NE. & Kotzabasis, K. (2008). Fast and reversible response of thylakoid-associated polyamines during and after UV-B stress: a comparative study of the wild type and a mutant lacking chlorophyll *b* of unicellular green alga *scenedesmus obliquus*. *Planta*, 228:341–353.

Sharp, RE.; Poroyko, V. Hejlek, LG. Spollen, WG. Springer, GK. Bohnert, HJ. & Nguyen, HT. (2004). Root growth maintenance during water deficits: physiology to functional genomics. *Journal of Experimental Botany*, 55: 2343–2351.

Shi, J.; Fu, XZ. Peng, T. Huang, XS. Fan, QJ & Liu, JH. (2010). Spermine pretreatment confers dehydration tolerance of citrus in vitro plants via modulation of antioxidative capacity and stomatal response. *Tree Physiology*, 30(7): 914–922.

Shi, K.; Gu, M. Yu, HJ. Jiang, YP. Zhou, YH. & Yu, JQ. (2009). Physiological mechanism of putrescine enhancement of root-zone hypoxia tolerance in cucumber plants. *Scientia Agricultura Sinica*, 42:1854–1858.

Shoal, RS.; Agarwal, A. Agarwal, S. & Orr, WC. (1995). Simultaneous overexpression of copper-and zinccontaining superoxide dismutase and catalase retards age-related oxidative damage and increases metabolic potential in *Drosophila melanogaster*. The *Journal of Biological Chemistry*, 270:15671–15674.

Shu, S.; Sun, J. Guo, SR. Li, J. Liu, CJ. Wang, CY. & Du, CX. (2010). Effects of exogenous putrescine on PSII photochemistry and ion distribution of cucumber seedlings under salt stress. *Acta Horticulturae Sinica*, 37: 1065–1072.

Sobieszczuk-Nowicka, E.; Wieczorek, P. & Legocka □J. (2009). Kinetin affects the level of chloroplast polyamines and transglutaminase activity during senescence of barley leaves. *Acta Biochimica Polonica*, 56: 255−259.

Sopory, SK.; Greenberg, BM. Mehta, RA. Edelman, M. Mattoo, AK. (1990). Free radical scavengers inhibit light dependent degradation of the 32-kDa photosystem II reaction center protein. *Z. Naturforsch. C*, 45: 412–417.

Standfuss, J.; Terwisschavan, AC. Lamborghini, M. & Kuehlbrandt, W. (2005). Mechanisms of photoprotection and nonphotochemical quenching in pea light-harvesting complex at 2.5Å resolution. *EMBO J*, 24: 919–928.

Subhan, D. & Murthy, SDS. (2001). Effect of polyamines on chlorophyll and protein contents, photochemical activity, and energy transfer in detached wheat leaves during dark incubation. *Biologia Plantarum*, 44: 529–533.

Sudhir, P. Murthy, SDS. (2004). Effects of salt stress on basic processes of photosynthesis. *Photosynthetica*, 42: 481–486.

Szalai, G.; Janda, T. Bartók, T. & Páldi, E. (1997). Role of light in changes in free amino acid and polyamine contents at chilling temperature in maize (*Zea mays*). *Plant Physiology*, 101: 434–438.

Tiburcio.; AF. Besford, RT. Capell, T. Borrell, A. Tes-tillano, PS. & Risueno, MC. (1994). Mechanisms of polyamine action during senescence responses induced by osmotic stress. *Journal of Experimental Botany*, 45: 1789–1800.

Unal, D.; Tuney, I. & Sukatar, A. (2008). The role of external polyamines on photosynthetic responses, lipid peroxidation, protein and chlorophyll a content under the UV-A (352nm) stress in *physcia semipinnata*. *Journal of Photochemistry and Photobiology B: Biology*, 90:64–68.

Velikova, V.; Yordanov, I. & Edreva, A. (2000). Oxidative stress and some antioxidant systems in acid rain-treated bean plants-protective role of exogenous polyamines. *Plant Science*, 151: 59–66.

Vigne, P. & Frelin, C. (2008). The role of polyamines in protein-dependent hypoxic tolerance of drosophila. *BMC Physiology*, 1–14.

Wang, W.; Vinocur, B. & Altman, A. (2003). Plant responses to drought, salinity and extreme temperatures: towards genetic engineering for stress tolerance. *Planta*, 218: 1–14.

Wang, J.; Sun, PP. Chen, CL. Wang, Y. Fu, XZ. & Liu, JH. (2011). An arginine decarboxylase gene *PtADC* from *Poncirus trifoliata* confers abiotic stress tolerance and promotes primary root growth in Arabidopsis. *Journal of Experimental Botany,* doi:10.1093/jxb/erq463.

Ward, JM.; Pei, ZM. & Schroeder, JI. (1995). Roles of ion channels in initiation of signal transduction in higher plants. *Plant Cell,* 7: 833–844.

Yamaguchi, K.; Takahashi, KY. Berberich, T. Imai, A. Takahashi, T. Michael, AJ. & Kusano, T. (2007). A protective role for the polyamine spermine against drought stress in arabidopsis. *Biochemical and Biophysical Research Communications,* 352: 486–490.

Yang, JC.; Zhang, JH. Liu, K. Wang, ZQ. Liu, LJ. (2007). Involvement of polyamines in the drought resistance of rice. *Journal of Experimental Botany,* 58: 1545–1555.

Yiu, JC.; Juang, LD. Fang, DYT. Liu, CW. & Wu, SJ. (2009). Exogenous putrescine reduces flooding-induced oxidative damage by increasing the antioxidant properties of welsh onion. *Scientia Horticulturae,* 120: 306–314.

Zhang, CM.; Zou, ZR. Huang, Z. & Zhang, ZX. (2010). Effects of exogenous spermidine on photosynthesis of tomato seedlings under drought stress. *Agricultural Research in the Arid Areas,* 3: 182–187.

Zhang, RH.; Li, J. Guo, SR. & Tezuka, T. (2009). Effects of exogenous putrescine on gas-exchange characteristics and chlorophyll fluorescence of NaCl-stressed cucumber seedlings. *Photosynth Research,* 100: 155–162.

Zhang, WP.; Jiang, B. Li, WG. Song, H. Yu, YS. & Chen, JF. (2009). Polyamines enhance chilling tolerance of cucumber (*Cucumis sativus* L.) through modulating antioxidative system. *Scientia Horticulturae,* 122: 200–208.

Zhao, HZ. & Yang, HQ. (2008). Exogenous polyamines alleviate the lipid peroxidation induced by cadmium chloride stress in *Malus hupehensis* Rehd. *Scientia Horticulturae,* 116: 442–447.

Zhou, GX. (2007). Studies on effects of exogenous spermidine on photosynthesis and carbohydrate metabolism in cucumber seedlings under hypoxia stress.[Degree]. Nanjing: *Nanjing Agriculture University,* pp: 1–98.

Zhou, GX.; Guo, SR. & Wang, SP. (2006). Effects of exogenous polyamines on photosynthetic characteristics and membrane lipid peroxidation of cucumis sativas seedlings under hypoxia stress. *Chinese Bulletin of Botany,* 23: 341–347.

Ziska, LH.; Seemann, JR. & DeJong, TM. (1990). Salinity induced limitations on photosynthesis in *Prunus salicina,* a deciduous tree species. *Plant Physiology,* 93: 864–870.

Zlatev, ZS. & Yordanov, IT. (2004). Effects of soil drought on photosynthesis and chlorophyll fluorescence in bean plants. *Bulg. Journal of Plant Physiol,* 30: 3–18.

8

An Overview of Plant Photosynthesis Modulation by Pathogen Attacks

Kumarakurubaran Selvaraj[1,2] and Bourlaye Fofana[1]
*[1]Crops and Livestock Research Centre, Agriculture and Agri-Food Canada,
Charlottetown, PE,
[2]Department of Biology, University of Prince Edward Island, Charlottetown, PE,
Canada*

1. Introduction

In 1893, Charles Barnes (1858-1910) proposed to designate the biological process for "synthesis of complex carbon compounds out of carbonic acid, in the presence of chlorophyll, under the influence of light" as "photosyntax" or "photosynthesis". His preference went for the word "photosyntax", but "photosynthesis" came into common usage as the term of choice (Gest, 2002). Although this definition is still widely used today, the Oxford English Dictionary (OED, 1989) has defined the biological photosynthesis as "the process by which carbon dioxide is converted into organic matter in the presence of the chlorophyll of plants under the influence of light, which in all plants except some bacteria involves the production of oxygen from water". By considering the photosynthetic bacteria that do not produce oxygen and do not necessarily require carbonic acid as carbon source, Gest (Gest, 1993) has refined these definitions as followed: "photosynthesis is a series of processes in which electromagnetic energy is converted to chemical energy used for biosynthesis of organic cell materials; a photosynthetic organism is one in which a major fraction of the energy required for cellular syntheses is supplied by light". Photosynthesis is therefore a unique source for carbon sequestration and allows the aerobic organisms to survive based upon not only on its released oxygen but also its synthesized organic compounds. As such, plants are an ideal host for pathogenic microorganisms such as fungi, bacteria and viruses and also the only food source for herbivores. Theoretically, the photosynthetic capacity of plants is unlimited when the optimal growing conditions are met. Unfortunately, terrestrial plants are constantly challenged by abiotic (UV, water, salinity, temperature) (Baker et al., 1988, Baldry et al., 1966, Barhoumi et al., 2007, Barrow and Cockburn, 1982, Bassham, 1977, Batista-Santos et al., 2011, Bauerle et al., 2007, Berry, 1975, Bischof et al., 2000, Ripley et al., 2008, Ripley et al., 2007, Roberntz and Stockfors, 1998) and biotic (pathogens, pests, animal and human) stresses that reduce their productivity and even threaten their survival (Bilgin et al., 2010, Bonfig et al., 2006, Erickson and Hawkins, 1980, Garavaglia et al., 2010, Kocal et al., 2008, Tang et al., 2009). While the regulation of plant defence responses has been extensively investigated, the effects of pathogen infection on primary metabolism, including photosynthesis, are however less known. Currently, interest in this research area is growing and some aspects of photosynthesis, assimilate partitioning, and source–sink regulation in different types of plant–pathogen interactions have been investigated. Berger et al (Berger et al., 2007) have recently reviewed how plant physiology meets phytopathology. The reader can also get more detail on this topic in previous study (Trotta et al., 2011, Essmann et al., 2008, Scharte et al., 2009). Here, we will focus our review on current knowledge on the process of higher plant photosynthesis, its outcome for both plants and fungal pathogens, the roles for some of the metabolites and transduction pathways that are implicated in this twined inter-relationship as well as the potential targets as future strategy.

2. Photosynthesis and pathogen

Plants and pathogens have developed dynamic interactions. Whereas plants tend to survive through different mechanisms following pathogen attack, the later looks for maximizing feed intake to insure its reproduction and dissemination (Korves and Bergelson, 2003, Berger et al., 2007). In this context, the photosynthetate – the energy source for both the plant and pathogens – synthesis and its availability is the focus of a struggle to death. The next section will discuss how photosynthesis proceeds in these challenging conditions.

2.1 Photosynthesis

Chloroplast is the factory for photosynthesis in higher plants. However, new evidences suggest a contribution for mitochondrial functions in the maintenance of efficient photosynthesis (Nunes-Nesi et al., 2008). The general process of photosynthesis, its outcome and limiting factors will be briefly described in the next sections.

2.1.1 Process and outcome

2.1.1.1 Process

In higher plants, the photosynthetic CO_2 fixation occurs in the green leaves, considered as source organs, with the absorption of light by chlorophyll, much of which is located in the light-harvesting complexes (LHCs) of PSII and PSI within the thylakoid membrane of chloroplasts (Murchie and Niyogi, 2011). The mesophyll cell of higher plants, due to its higher chloroplast content, is the most active photosynthetic tissue. In general photosynthesis proceeds through 2 major phases: a) a light phase that produces ATP and NADPH in the chloroplast thylakoids and released in the stroma and b) the CO_2 reduction phase in presence of water in the stroma and that consumes the ATP and NADPH generated in phase a) to produce a triose phosphate through the Calvin-Benson cycle which comprises three stages (Figure 1).

Briefly, the carboxylation of 3 molecules of ribulose 1,5 biphosphate fixes 3 molecules of CO_2 and H_2O under the ribulose 1,5 biphosphate carboxylase/oxygenase (rubisco) catalysis in the Calvin-Benson cycle and leads to 6 molecules of 3-phosphoglycerate. The 3-phosphoglycerates are then phosphorylated in presence of ATP produced during the light reaction by the catalytic action of 3-phosphoglycerate kinase into 1,3 bisphosphoglycerate which is further reduced by NADPH and NADP-glyceraldehyde 3-phosphate dehydrogenase into 6 molecules of glyceraldehyde 3-phosphate (6 triose phosphates). Of these six triose phosphates, one represents the net synthesis from CO_2 fixation and, 9 ATP and 6 NADPH are utilized. The remaining five triose phosphates are used to regenerate the ribulose 1,5 biphosphate to insure continuous CO_2 fixation (Taiz and Zeiger, 2010).

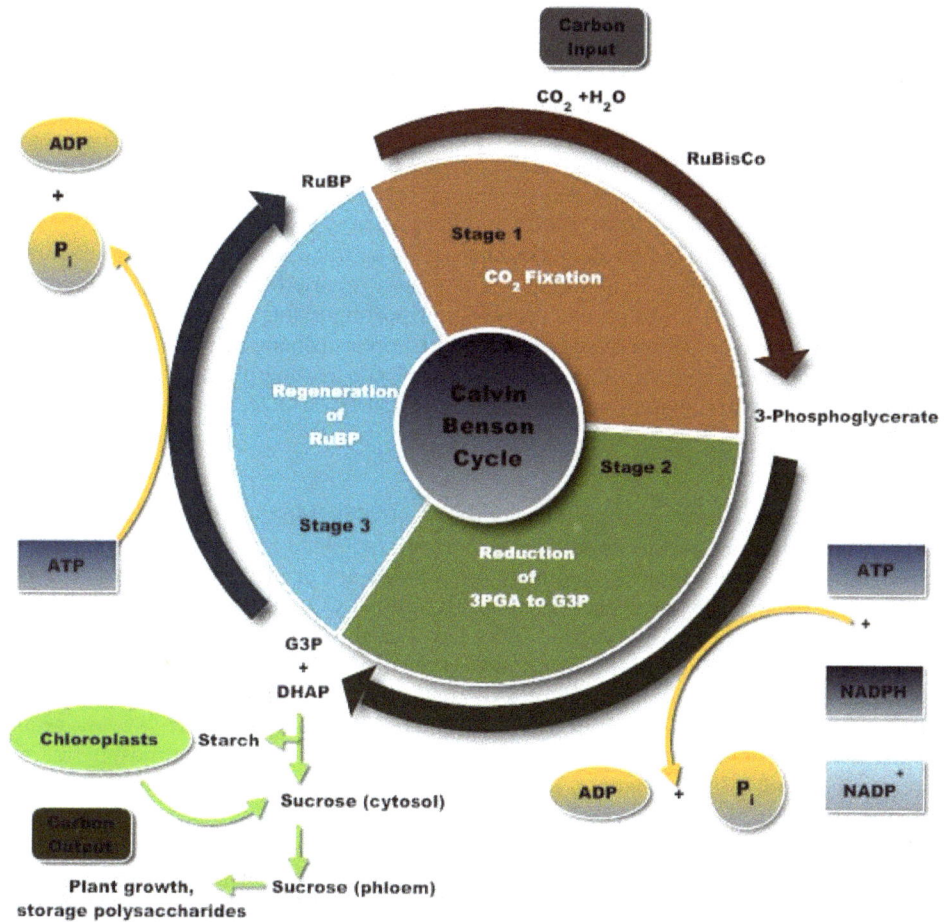

Fig. 1. Calvin-Benson cycle adapted from Taiz and Zeiger (Taiz and Zeiger, 2010)

2.1.1.2 Outcome

The outcome of CO_2 fixation by higher photosynthetic plants is the production of carbohydrates. As a result of photosynthetic CO_2 reduction during the day, starch granules accumulate in the chloroplast while an excess of assimilates are continuously allocated, mostly in the form of sucrose, to sink tissues such as developing leaves, roots, meristems, fruits, and flowers, that are unable to produce sufficient amounts of assimilates by themselves and therefore require their net import via the phloem (Kocal et al., 2008). Sucrose is loaded into the phloem in the minor veins of leaves before export (Zhang and Turgeon, 2009). Recently, Zhang and Turgeon (Zhang and Turgeon, 2009) have proposed two active, species-specific loading mechanisms. One involves transporter-mediated sucrose transfer from the apoplast into the sieve element-companion cell complex, so-called apoplastic loading. In the second putative mechanism, sucrose follows an entirely symplastic pathway, and the solute concentration is elevated by the synthesis of raffinose and stachyose in the phloem, not by transporter activity. Thus, a coordinated sequence of assimilate production, allocation, and utilization is essential for normal plant growth and development (Kocal et al., 2008). Indeed, carbohydrate accumulation in the leaves can lead to decreased expression of photosynthetic genes and accelerated leaf senescence when there is an imbalance between source and sink at the whole plant level (Paul and Foyer, 2001). Generally, when sink activity is decreased by removing active sinks or introducing nutrient deficiency, carbohydrates accumulate in leaves and photosynthesis becomes inhibited (Ainsworth and Bush, 2011, Paul and Pellny, 2003). Similarly, when sucrose export from source leaves is restricted, for example by cold girdling of petioles or down-regulation of sucrose transporter abundance, photosynthesis is inhibited. It has also been reported that a downregulation of sucrose transporter 1 (SUT1) in several sucrose-transporting plants, shown to be apoplastic loaders, led to an accumulation of sugars and leaf chlorosis (Zhang and Turgeon, 2009). In contrast, no such phenotype developed when a symplastic loading plant such as *Verbascum phoeniceum* was downregulated, emphasizing the importance of either active or passive assimilates exports.

2.1.1.3 Limiting factors

Although the Calvin-Benson cycling capacity seems unlimited in the presence of light, CO_2 and H_2O, many limiting factors counteract this highly regulated biological process. First, if the presence of CO_2 and light are required, their levels are of great importance. As light (photon flux and intensity) increases (in constant optimal CO_2), the rate of photosynthesis rises until it is saturated. An excess of light can lead to photoinhibition (Bertamini and Nedunchezhian, 2004, Murchie and Niyogi, 2011). A correlation between the *in vivo* rates of net CO_2 assimilation and the atmospheric CO_2 concentrations was observed when intact C_3 and C_4 plants were exposed to different atmospheric CO_2 concentrations (Aguera et al., 2006, Ainsworth and Rogers, 2007, Bhatt et al., 2010). In general, current atmospheric CO_2 concentration is adequate for both C_3 and C_4 plants. However, with global warming and its related rising temperature and CO_2 level, a higher biomass production and a change in C_3 and C_4 plants distribution are expected (depending on rainfall) because of their differential photosynthetic and water use efficiency. Nontheless, many temperate plant species may not adapt as rate of photosynthesis was found to decline at moderately high temperature in a temperate species such as *Arabidopsis thaliana* (Kumar et al., 2009, Kurek et al., 2007). If CO_2 generally favours photosynthesis, other environmental clues such as UV-B (Albert et al., 2008), heat shock and water deficit (Abrams et al., 1990, Ackerson and Hebert, 1981, Allakhverdiev et al., 2008, ZhangWollenweber et al., 2008), Cold (Batista-Santos et al., 2011, Bilska and Sowinski, 2010), herbivore and pathogen attacks have detrimental effects on photosynthesis (Halitschke et al., 2011, Horst et al., 2008, King and Caylor, 2010, Nabity et al., 2009). The primary targets of thermal damage in plants are the oxygen evolving complex along with the associated cofactors in photosystem II (PSII), carbon fixation by Rubisco and the ATP generating system (Allakhverdiev et al., 2008). Recent studies on the combined effect of moderate light intensity and heat stress suggest that moderately high temperatures do not cause serious PSII damage but inhibit the repair of PSII. Repair of PSII involves *de novo* synthesis of proteins, particularly the D1 protein of the photosynthetic machinery that is damaged due to generation of reactive oxygen species (ROS). Attacks by ROS during moderate heat stress principally affects the repair system of PSII, resulting in the reduction of carbon fixation and oxygen evolution, as well as the disruption of the linear electron flow (Allakhverdiev et al., 2008).

2.2 Pathogens in plants

Plants pathogens are of diverse nature and include pathogenic fungi, virus, and bacteria. Whereas all of these pathogens are of great interests for plant biologists, the sake of this contribution will be limited to the pathogenic fungi with some reference to bacteria.

2.2.1 Invasion process

Plant infection by pathogenic fungi and bacteria can occur in multiple ways. It could be passive, i.e. accidental, by suction into the plant through natural plant openings such as stomata, hydathodes or lenticels, entrance through abrasions or wounds on leaves, stems or roots (Hu and Rijkenberg, 1998b, Vidaver and Lambrecht, 2004). Infection of host plants by biotrophic plant pathogens generally involves the sequential development of specialized host-parasite interfaces, exemplified by those of haustoria, which are maintained over an extended period of time without causing significant cytological damage to host tissue in the infected region (Tariq and Jeffries, 1984, Tariq and Jeffries, 1986). Using scanning electron microscopy, Hu and Rijkenberg (Hu and Rijkenberg, 1998b) identified key time points in the formation of infection structures by *P. triticinia* on susceptible and resistant lines of hexaploid wheat. Six hours after infection, the fungus forms appressoria over stomata openings. After 12 hours, the fungus has successfully penetrated into the stroma, formed substomatal vesicles (SSV), and primary infection hyphae are visible. After SSV formation, the primary infection hypha grows and attaches to a mesophyll or epidermal cell. At 24 hours postinoculation, a septum appears separating the haustorial mother cell from the infection hypha after which the fungus forms haustorium and penetrates the cell (Hu and Rijkenberg, 1998b). Recently, Garg et al (Garg et al., 2010) described the infection process in susceptible and resistant genotypes of *Brassica napus* against *Sclerotinia sclerotiorum*. They demonstrated at the cellular level that resistance to *S. sclerotiorum* in *B. napus* is a result of retardation of pathogen development, both on the plant surface and within host tissues. There are some indications that the infection process is dependent on the nutritional status of the inoculums (Garg et al., 2010). Indeed, previous studies suggested that the presence of nutrients is essential for hyphal development, penetration and for subsequent establishment of a successful invasion of a susceptible host by the pathogen (Tariq and Jeffries, 1984, Tariq and Jeffries, 1986, Garg et al., 2010).

2.2.1.1 Outcomes of pathogen invasion

2.2.1.1.1 For the pathogen

Plant pathogens like viruses, fungi, oomycetes, and bacteria are known to interfere with the source-sink balance (Berger et al., 2007, Biemelt and Sonnewald, 2006, Seo et al., 2007), and in the case of a successful interaction, pathogens are believed to reprogram a plant's metabolism to their own benefit (Biemelt and Sonnewald, 2006). This comprises the suppression of plant defence responses and the reallocation of photoassimilates to sufficiently supply the pathogen with nutrients (Kocal et al., 2008). In accordance with this, the infected leaf is assumed to undergo a source to sink transition or retains its sink character. For example, infection of maize leaves with *Ustilago maydis* prevents establishment of C_4 photosynthesis because *U. maydis*-induced leaf galls exhibited carbon dioxide response curves, CO_2 compensation points and enzymatic activities that are characteristic of C_3 photosynthesis (Horst et al., 2008). An indication for this is provided by a stimulation of cell wall-bound invertase (cw-Inv) that mobilizes hexoses at the infection site and a decreased rate of photosynthesis (Kocal et al., 2008).

2.2.1.1.2 For the plant

Pathogen attacks result in the development of symptoms that include leaf and fruit wilt, stem and root rot (Rekah et al., 1999), coverage of leaf surface with pustule, chlorosis and necrosis (Fofana et al., 2007, Kocal et al., 2008) (Figure 2), a decreased rate of plant photosynthesis (Kocal et al., 2008), and as a consequence plant death or yield loss ensues (Berger et al., 2007).

Fig. 2. Symptoms of wheat leaf rust (*Pucinia triticina*) on wheat near isogenic RL6003 line inoculated with a) avirulent race 1 (BBB), and b) virulent race 7-2 (TJB) of *Pucinia triticina*. An incompatible interaction showing ; 1- infection types with very small pustules, no sporulation and hypersensitive reaction (BBB) and compatible interaction showing a 3+4- infection type with large pustules, abundant sporulation and chlorotic reaction around sporulations (TJB).

2.3 Photosynthesis and pathogens invasions

2.3.1 Photosynthesis efficiency under pathogen attacks

Pathogen attacks result in a decreased rate of plant photosynthesis (Kocal et al., 2008), and as a consequence yield loss (Berger et al., 2007). Pathogen infection often leads to plant death, the development of chlorotic and necrotic (Kim et al., 2010) lesions and to a decrease in photosynthetic assimilate production. Using chlorophyll fluorescence imaging, it has been reported that the changes in photosynthesis upon infection are local. In *Arabidopsis* leaves infected with *A. candida* and in tomato plants infected with *B. cinerea*, a ring of enhanced photosynthesis was detectable surrounding the area with decreased photosynthesis at the infection site. At present, it is not clear if this stimulation of photosynthesis is due to the defence strategy of the plant (Berger et al., 2007). A decrease in photosynthesis has also been reported in incompatible interactions (Bonfig et al., 2006). The decrease in photosynthesis was detectable earlier with the avirulent strain than with the virulent strain. It is suggested that plants switch off photosynthesis and other assimilatory metabolism to initiate respiration and other processes required for defence (Berger et al., 2007). Recently, Petit et al (Petit et al., 2006) characterized the photosynthetic apparatus of grape leaves infected with esca disease. Foliar symptoms were associated with stomatal closure and alteration of the photosynthetic apparatus. A decrease in CO_2 assimilation, transpiration, a significant increase in intercellular CO_2 concentration, a strong drop in the maximum fluorescence yield and the effective Photosystem II quantum yields, and a reduction of total chlorophyll but a stable carotenoid content were reported (Petit et al., 2006).

2.3.2 Mechanistic alteration of the photosynthetic capacity

Several mechanisms have been described to explain the suppression of plant defence responses and the reprogramming of the plant's metabolism to the pathogen own benefit (Garavaglia et al., 2010). The pathogens *Stagonospora nodorum* and *Pyrenophora tritici-repentis*, the causal agents of *Stagonospora nodorum* blotch (SNB) and tan spot, respectively, produce multiple effectors (Ptr ToxA, Ptr ToxB, and Ptr ToxC), also known as host-selective toxins (HSTs), that interact with corresponding host sensitivity genes in an inverse gene-for-gene manner to cause the diseases in wheat. A compatible interaction requires both the effector (HST) and the host gene and results in susceptibility as opposed to host resistance (R) genes. R-genes lead to a resistance response known as effector-triggered immunity (ETI) which includes localized programmed cell death (PCD) or hypersensitive response (HR), to restrict pathogen growth. The absence of either the effector or the host gene results in an incompatible interaction (Zhang et al., 2011, Faris et al., 2010). *Pyrenophora tritici-repentis* produces oval or diamond-shaped to elongated irregular spots that enlarge and turn tan with a yellow border and a small dark brown spot near the center causing necrotic and/or chlorotic lesions on infected leaves, which can significantly reduce total photosynthetic area and yield loss (Kim et al., 2010). The development of chlorosis in response to Ptr ToxB results from an inhibition of photosynthesis in the host, leading to the photooxidation of chlorophyll molecules as illuminated thylakoid membranes become unable to dissipate excess excitation energy (Strelkov et al., 1998). Recently, Kim et al (Kim et al., 2010) showed that treatment of wheat leaves with Ptr ToxB results in significant changes in the abundance of more than 100 proteins, including proteins involved in the light reactions of photosynthesis, the Calvin cycle, and the stress/defence response. These authors also examined the direct effect of Ptr ToxB on photosynthesis and found a net decline of photosynthesis within 12 h of toxin-treatment, long before chlorosis develops at 48–72 h. A role for ROS generation and disruptions of the photosynthetic electron transport shortly after pathogen attack or toxin treatment have been suggested as potential mechanism (Kim et al., 2010). Similar mechanism has been proposed by Allakhverdiev et al (Allakhverdiev et al., 2008) following heat stress that targets the oxygen evolving complex along with the associated cofactors in photosystem II (PSII), carbon fixation by rubisco and the ATP generating system. In another system, Kocal et al (Kocal et al., 2008) studied the role of cell wall invertase (cw-Inv) in transgenic tomato (*Solanum lycopersicum*) plants silenced for the major leaf cw-Inv isoforms during normal growth and during the compatible interaction with *Xanthomonas campestris pv vesicatoria*. Cw-Inv expression was found to be induced upon microbial infection and was most likely associated with an apoplastic hexose accumulation during the infection process. The hexoses formed are thought to aid the pathogen's nutrition (Berger et al., 2007, Biemelt and Sonnewald, 2006, Seo et al., 2007). Fungal pathogens also produce their own invertases to ensure their nutritional supply (Chou et al., 2000, Voegele et al., 2006). One of the most sophisticated mechanisms that divert plant metabolites to pathogen is the role of auxin in host-pathogen interactions (Fu et al., 2011, Navarro et al., 2006). Indole-3-acetic acid (IAA) is the major form of auxin in most plants and induces the loosening of plant cell wall, the natural protective barrier to invaders. *X. oryzae pv oryzae*, *X. oryzae pv oryzicola*, and *M. grisea* secrete IAA, which, in turn, may induce rice to synthesize its own IAA at the infection site. IAA induces the production of expansins, the cell wall-loosening proteins, and makes rice vulnerable to pathogens (Fu et al., 2011). Similarly, Garavaglia et al (Garavaglia et al., 2010) have reported an eukaryotic-acquired gene by a biotrophic phytopathogen that allows its prolonged survival on the host by counteracting the shut-down of plant photosynthesis.

3. Plant reactions to pathogen attack

During plant-pathogen interactions, the host develops a variety of defence reactions. Non-host resistance against a biotrophic fungal pathogen is often manifested as the ability of the attacked plant to prevent fungal penetration, or the ability to terminate the development and/or functioning of the fungal feeding structure such as the intracellular hypha or the haustorium before it extracts enough nutrition from the plant cells (Wen et al., 2011). These reactions may involved development of physical barriers such as exocytosis, cell wall modifications and de novo metabolites synthesis (IshiharaHashimotoMiyagawa et al., 2008).

3.1 Primary metabolites and secondary metabolite production
3.1.1 Primary metabolites
Cell wall strengthening by callosic (Wen et al., 2011) and papillae formation, cell wall apposition (Fofana et al., 2005, Wurms et al., 1999), lignin deposition (Hammerschmidt and Kuc, 1982) as well as hydrolytic PR proteins (Hu and Rijkenberg, 1998a) have been reported as first line of defence mechanism developed by plants. Components of these cell wall makeups are of primary metabolite origin. It is worth noting that papillae were at times observed as an initial response to fungal penetration. Generally, papillae relate to powdery mildew fungal penetration in two ways: in some instances, penetration fails when papillae are present and alternatively, penetration may succeed and the papilla becomes a collar for the haustorial neck (Hammerschmidt and Yang-Cashman, 1995). Using transmission electron microscopy, Fofana et al (Fofana et al., 2005) observed both outcomes for elicited and nonelicited cucumber plants, strongly suggesting that papillae formation alone may not be sufficient to explain the level of

induced resistance observed for elicited plants. Moreover, chitin labelling revealed that the walls and lobes of fungal haustoria within both treatments were undisturbed, suggesting that PR proteins, such as chitinases and β-1,3-glucanases, may not play a major role in the early events of induced resistance for cucumber (Fofana et al., 2005). However, Gonzalez-Teuber et al. (Gonzalez-Teuber et al., 2010) recently reported glucanases and chitinases as causal agents in the protection of *acacia* extrafloral nectar from infestation by phytopathogens. Nectars are rich in primary metabolites and as protective strategy, floral nectar of ornamental tobacco (*Nicotiana langsdorffii x Nicotiana sanderae*) contains "nectarins," proteins producing reactive oxygen species such as hydrogen peroxide. By contrast, *Acacia* extrafloral nectar contains pathogenesis-related (PR) proteins. This nectar is secreted in the context of defensive reactions. Gonzalez-Teuber et al (Gonzalez-Teuber et al., 2010) showed that PR proteins causally underlie the protection of *Acacia* extrafloral nectar from microorganisms and that acidic and basic glucanases likely represent the most important prerequisite in this defensive function. Salicylic acid (SA) and Jasmonic acid have long been considered as signal molecules in disease resistance (Ward et al., 1991). Recently, *Arabidopsis* GH3-type proteins functioning in auxin signaling, in association with a salicylic acid (SA)-dependent pathway, was reported to positively regulate resistance to *Pseudomonas syringae* (Jagadeeswaran et al., 2007). Accordingly, Fu et al (Fu et al., 2011) suggested that GH3-2 encodes an IAA-amido synthetase and positively regulates rice disease resistance by suppressing pathogen-induced accumulation of IAA in rice. Activation of GH3-2 confers to rice a broad spectrum and partial resistance against *Xanthomonas oryzae pv oryzae and Xanthomonas oryzae pv oryzicola* and the fungal *Magnaporthe grisea* in rice.

3.1.2 Secondary metabolites

The role for secondary metabolites in the plant's interaction with its environment is widely recognized (Rhodes, 1994). The primary metabolites deriving from photosynthesis are channeled into different metabolite pathways for the synthesis, storage, and modification (hydroxylation, glycosylation, acetylation, etc) of myriads of compounds, and for use to cope abiotic and biotic clues. Within each of the major groups of secondary metabolites such as alkaloids, phenylpropanoids and terpenoids, several thousand individual compounds accumulating in plants have been characterised and their role in plant-pathogen interactions studied (Ishihara et al., 2011). For example, induction of phenolic compounds, flavonoid phytoalexins, (Daayf et al., 1997, Fawe et al., 1998, Fofana et al., 2002, McNally et al., 2003b, McNally et al., 2003) was reported in cucumber plants following pathogen attacks and elicitor treatments. Synthesis of phytoalexins involves the rapid transcriptional activation of genes encoding a number of key biosynthetic enzymes that include anthranilate synthase (AS) (IshiharaHashimotoTanaka et al., 2008) phenylalanine ammonia-lyase (PAL), chalcone synthase (CHS) which is the early committed key enzyme of the flavonoid/isoflavonoid pathway, chalcone isomerase (CHI) and isoflavone reductase (IFR) (Dixon et al., 1995, Baldridge et al., 1998, Fofana et al., 2002, Fofana et al., 2005). The chemical nature of some of these compounds is now well elucidated (Ibanez et al., 2010, Ishihara et al., 2011, McNally et al., 2003). McNally et al (McNally et al., 2003b, McNally et al., 2003) reported the synthesis of complex C-glycosyl flavonoid phytoalexins, referred to as vitexin-6-(4-hydroxy-1-ethylbenzene) (cucumerin A) and isovitexin-8-(4-hydroxy-1-ethylbenzene) (cucumerin B), as a site-specific response to fungal penetration in cucumber. In a recent study, Ishihara et al (IshiharaHashimotoTanaka et al., 2008) reported on an induced accumulation of Trpderived secondary metabolites, including tryptamine, serotonin, and hydroxycinnamic acid amides of serotonin in rice leaves by infection with *Bipolaris oryzae*. Using enantiomers of α-(fluoromethyl)tryptophan (αFMT – R- and S- αFMT), S-αFMT but not R-αFMT effectively inhibited tryptophan decarboxylase activity extracted from rice leaves infected by *Bipolaris oryzae*, suppressed accumulation of serotonin, tryptamine, and hydroxycinnamic acid amides of serotonin in a dose-dependent manner, and lead to were severely damaged leaves showing lesions that lacked deposition of brown materials, compared to control without S-αFMT. Administration of tryptamine to S-αFMT-treated leaves restored accumulation of tryptophan-derived secondary metabolites as well as deposition of brown material and reduced damage caused by fungal infection (Ishihara et al., 2011).

3.2 Transduction pathways
Plants have developed a sophisticated innate immune surveillance system to recognize pathogens (Dodds and Rathjen, 2010, Liu et al., 2011). This surveillance system consists of an integral plasma membrane proteins with extracellular receptor domains to perceive conserved pathogen associated molecular patterns (PAMPs) presented by pathogens during infection, and an intra- cellular Resistance (R) proteins to recognize the presence of specific pathogen effector proteins in host cells (Elmore et al., 2011). Two recognition models have been reported for non-host resistance (non-race specific elicitor as signal) and gene-for-gene resistance (race-specific elicitor/ avirulence gene products as signal) interactions, with a receptor and a resistance gene product as signal perception, respectively (Romeis, 2001). Upon perception, takes place a signal transduction cascade involving protein kinases and cellular responses to the intruders ensue. Either the disease develops or resistance phenotypes are observed. For review, please see more details in previous reports (Romeis et al., 2001, Romeis, 2001, Elmore et al., 2011, Elmore and Coaker, 2011a, Liu et al., 2011, Elmore and Coaker, 2011b). The plant reaction to the outcome of signalling has a dramatic consequence on the plant's ability to photosynthesize even in incompatible interactions where HR responses lead to a localized cell death at infection sites and restrict the pathogen progression. In these conditions, with patchy leaf area (no chlorophyll for light inception), the photosynthetic capacity is reduced compared to non-infected plants.

4. New insights from the genomics and proteomics era

The genomics and proteomics era, with its high-throughput capability, has enabled the expression profiling analysis of thousands of genes and proteins simultaneously (Lee et al., 2004). Hence, the global analysis of many plant processes, including the response to pathogen attack, their interlinked regulatory networks and signalling pathways have been made possible (Duggan et al., 1999, Eulgem, 2005, GuldenerSeong et al., 2006, Schenk et al., 2000).

4.1 Gene and protein networks in plant-pathogen interactions
4.1.1 In the pathogen
One of the challenges faced by biologists in plant-pathogen interactions was their ability to differentiate plant genes from the pathogen genes. This has become feasible with the release of the genome sequences for several fungus and plant species. Exploiting these genomic resources it has been possible to design and perform the wide-genome microarray transcription profiling of the plant pathogenic fungus, *Fusarium graminearum*, grown in

culture media under different nutritional regimes and in comparison with fungal growth in infected barley (GuldenerSeong et al., 2006). Guldener et al. (GuldenerSeong et al., 2006) were able to detect the fungal gene expression during plant infection, test for sensitivity limits for detecting fungal RNA *in planta* and the potential for cross-hybridization between fungal probe sets and plant RNAs. A total of 11,994 of a possible 17,809 *Fusarium* probe sets (67.35%) were detected under various conditions of the fungus grown in culture and a total of 7132 probe sets (40.05%) were detected from the fungus during infection of barley. As of July 5, 2011 however, only 96 pathogenic genes from *F. graminearum* curated for lab experimental, molecular and biological information on genes proven to affect the outcome pathogen-host interactions in cereals (oat, wheat, barley, rye, maize), *Arabidopsis*, and tomato were reported in PHI-base database (http://www.phi-base.org/query.php) and 4000 proteins that have been annotated in MIPS *F.graminearum* Genome Database (FGDB), (GuldenerMannhaupt et al., 2006) which is far from complete. To speed up the process, a computational network approach to predict pathogenic genes for *Fusarium graminearum* was proposed by Liu et al (Liu et al., 2010). With a small number of known pathogenic genes as seed genes, the authors were able to identify a subnetwork that consists of potential pathogenic genes from the protein-protein interaction network (PPIN) of *F. graminearum*, where the genes in the subnetwork generally share similar functions and are involved in similar biological processes. The genes that interact with at least two seed genes can be identified because these genes are more likely to be pathogenic genes due to their tight interactions with the seed genes. The protein-protein interactions that connect networks with each other are thought to be the signalling pathways between biological processes. On the basis of our current understanding of pathogenicity of model pathogens, *F. graminearum* is thought to organize a complex network of proteins and other molecules, including those that might be secreted into host cells, to adapt the life inside its host plant. Hence, Zhao et al (Zhao et al., 2009) showed that *F. graminearum* protein-protein interactions (FPPI) contains 223,166 interactions among 7406 proteins which represent about 52 % of the whole *F. graminearum* proteome. Although these computational predictions are fascinating, system biology based on experiment data is also making considerable progress. Song et al (Song et al., 2011) reported the first proteome of infection structures from parasitized wheat leaves, enriched for *Puccinia triticina* (Pt) haustoria using 2-D PAGE MS/MS and gel-based LC-MS (GeLC-MS) to separate proteins. They compared the generated spectra with a partial proteome predicted from a preliminary *Pt* genome and ESTs, with a comprehensive genome-predicted protein complement from the related wheat stem rust fungus, *Puccinia graminis* f. sp. *tritici (Pgt),* and with various plant resources. The authors identified over 260 fungal proteins, 16 of which matched peptides from *Pgt*. Based on bioinformatic analyses and/or the presence of a signal peptide, at least 50 proteins were predicted to be secreted. Among those, six had effector protein signatures, some were related and the respective genes of several seem to belong to clusters. Many ribosomal structural proteins, proteins involved in energy, general metabolism and transport were detected. By measuring the gene expression over several life cycle stages of ten representative candidates using quantitative RT-PCR, all tested genes were shown to be strongly upregulated and of which four were expressed solely upon infection (Song et al., 2011). Similarly, El-Bebany et al (El-Bebany et al., 2010) identified potential pathogenicity factors including isochorismate hydrolase, a potential plant-defence suppressor that may inhibit the production of salicylic acid, which is important for plant defence response signaling. Much progress is still needed not only in the identification but in the mechanistic action of the genes and proteins identified.

4.1.2 In the plant

The role of photosynthesis in plant defence is a fundamental question awaiting further molecular and physiological elucidation. Different pathogens, based on life cycle (biotroph vs necrotroph), develop different pathogenesis mechanisms that impact differently on the plant's photosynthesis efficiency as well as on its gene and proteome profiling. *Xanthomonas axonopodis pv. citri*, the bacterial pathogen responsible for citrus canker encodes a plant-like natriuretic peptide (XacPNP) that is expressed specifically during the infection process and prevents deterioration of the physiological condition of the infected tissue to the benefit of the invaders (Nembaware et al., 2004, Gehring and Irving, 2003). The wild pathogen expressing the XacPNP peptide maintains the plant in a condition that prevents chlorosis and no significant drop of photosynthesis. In contrast, citrus leaves infected with a XacPNP deletion mutant (DeltaXacPNP) resulted significant reduction of photosynthesis efficiency, and proteomic assays revealed a major reduction in photosynthetic proteins such as Rubisco, Rubisco activase and ATP synthase as a compared with infection with wild type bacteria (Garavaglia et al., 2010). Similarly, *Pyrenophora tritici-repentis*, is an important foliar disease of wheat. The fungus produces the host-specific, chlorosis-inducing toxin Ptr ToxB. Kim et al (Kim et al., 2010) examined the effects of Ptr ToxB on sensitive wheat. Photosynthesis was significantly reduced within 12 h of toxin treatment, prior to the development of chlorosis at 48-72 h. Proteomics analysis by 2-DE revealed a total of 102 protein spots with significantly altered intensities 12-36 h after toxin treatment, of which 66 were more abundant and 36 were less abundant than in the buffer-treated control. In the last decade, an abundant literature has treated large dataset gene expression profiling of plant-pathogen interactions (Bilgin et al., 2010, Eichmann et al., 2006, Fofana et al., 2007, Lee et al., 2004, Zou et al., 2005) among many others. Of interest was the study by Fofana et al (Fofana et al., 2007) where difference in temporal gene expression profiling of the wheat leaf rust pathosystem was reported in compatible and incompatible defence pathways using cDNA microarray. Gene ontology assignment of differentially expressed genes showed alterations in gene expression for different molecular functions, cellular location and biological process for genes (Figure 3). The authors observed changes in the expression of genes involved in different biological processes such as photosynthesis, redox control, resistance and resistance-related genes (NBS-LRR, cyclophilin-like protein, MLo4-like gene, MRP1), components of the shikimate-phenylpropanoid pathway as well as genes involved in signal transduction (Myb-like transcription factors, calmodulin MAPKK, PI4PK), heat shock proteins, osmotic control genes and metabolisms (Fofana et al., 2007). Six hours after inoculation, a coordinated decrease in transcription of photosynthesis genes (photosystemmII phosphoprotein, ribulose-1,5 biphosphate carboxylase/oxygenase small unit, Type III LHCII CAB precursor protein, photosystem II type II chlorophyll A/B-binding protein, ribulose-1,5 biphosphate carboxylase activase) in the resistant but not susceptible interactions was observed in agreement with the general trends of photosynthesis inhibition. Biotic stress globally downregulates photosynthesis genes (Bilgin et al., 2010). By comparing transcriptomic data from microarray experiments after 22 different forms of biotic damage on eight different plant species, Bilgin et al (Bilgin et al., 2010) reported that transcript levels of photosynthesis light reaction, carbon reduction cycle and pigment synthesis genes decreased regardless of the type of biotic attack.

Genes coding for the synthesis of jasmonic acid and those involved in the responses to salicylic acid and ethylene were upregulated. The upregulation of JA and SA genes suggest that the downregulation of photosynthesis-related genes was part of a defence response. Analysis of gene clusters revealed that the transcript levels of 84% of the genes that carry a chloroplast targeting peptide sequence were decreased (Bilgin et al., 2010). The concept of computational network analysis (Liu et al., 2010) appears to be of good relevance as it could assist in identifying not only networks specific to the plant, to the pathogen but also genes that interact between the plant and the pathogen.

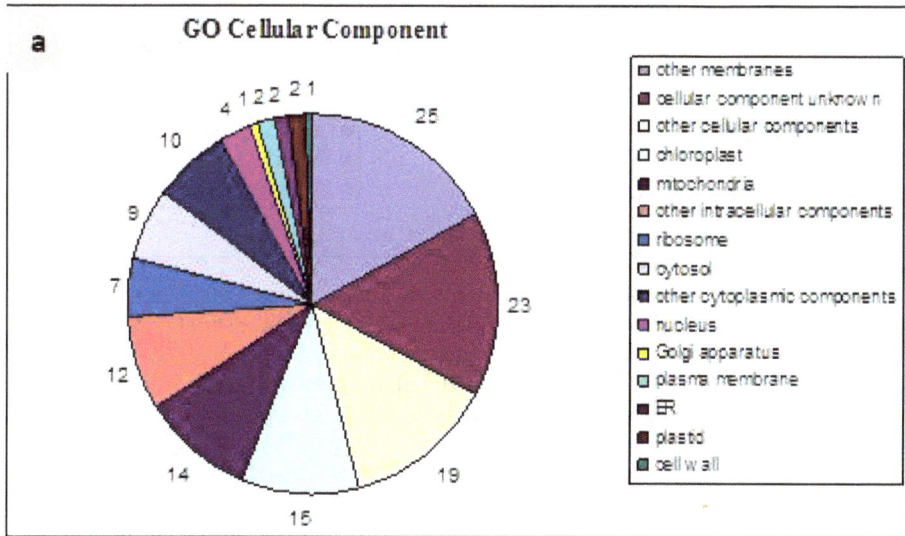

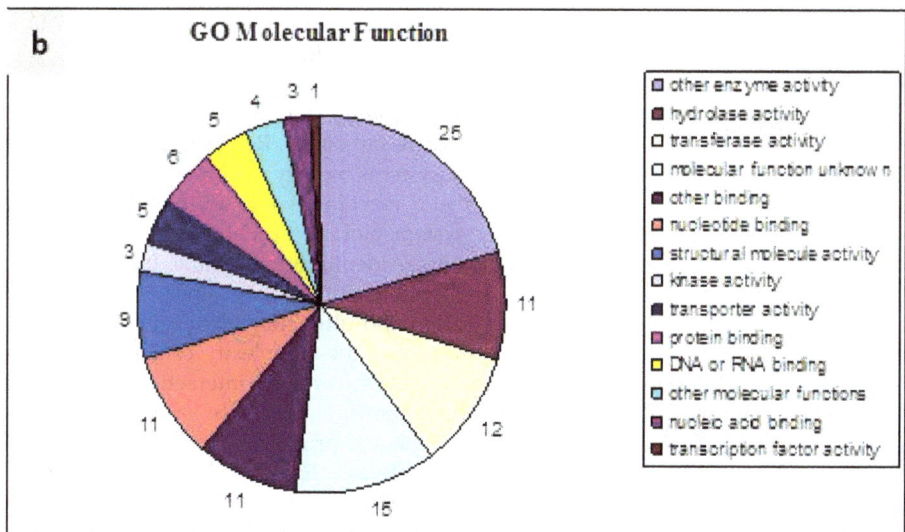

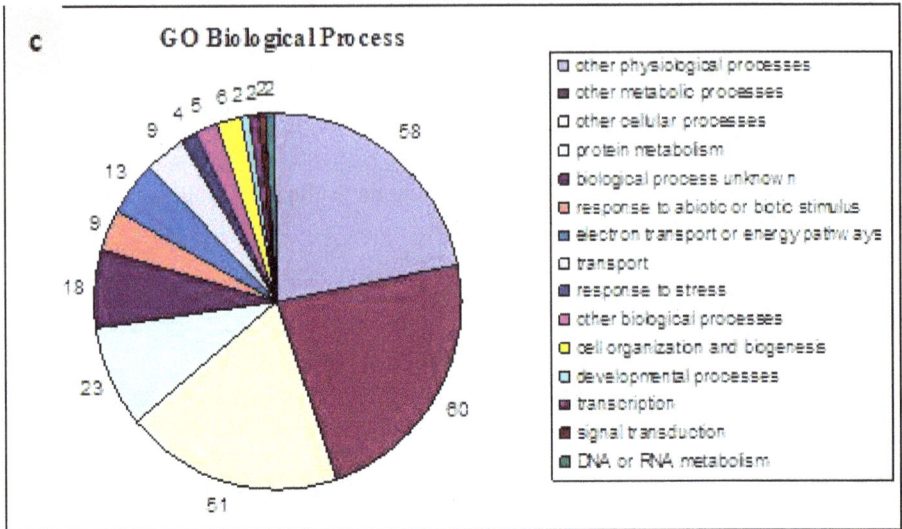

Fig. 3. Gene ontology assignment of differentially expressed genes (Fofana et al., 2007). A BLASTX search of the differentially expressed sequences against the set of predicted *Arabidopsis thaliana* proteins was used to assign gene ontology. The first hit with an *E* value less than or equal to 1×10^{-5} was used as a functional assignment and the TAIR GO annotation tool was used to bin the genes into the ontology groupings; a) cellular location b) molecular function and c) biological process.

4.2 What could be the future strategies?

Plant productivity depends on the plant's ability to produce higher biomass and seed, which relies on its photosynthetic capacity. However, as mentioned above, this inherent potential is constantly challenged and compromised by phytopathogens. The main question facing plant biologists remains our ability to improve plant productivity under increasing biotic pressure. One of the avenues could consist of emphasizing on gene networks discovery through both computational network discovery strategy (Liu et al., 2010) and gene and proteome analysis in living plants challenged with pathogens. Recently, Zhu et al (Zhu et al., 2010) proposed C_4 rice as an ideal arena for systems biology research. This group raised the possibility of engineering C_4 photosynthetic machinery into C_3 plant such as rice. However, the pivotal role to be played by system biology in identifying key regulatory elements controlling development of C_4 features, identifying essential biochemical and anatomical features required to achieve high photosynthetic efficiency, elucidating the genetic mechanisms underlining C_4 differentiation and ultimately identifying viable routes to engineer C_4 rice has been emphasized to decipher the complexity of such engineering (Zhu et al., 2010). A second level complexity comes from the interaction between two organisms as is the case in plant-pathogen interactions. It will be of great interest to put emphasis on the identification of a) more plant gene and protein network clusters and their interactomes, b) more pathogen gene and protein network clusters and their interactomes, c) more plant-pathogen gene and protein network clusters and the interacting genes and proteins linking both organisms, for a better understanding of key target points. This would allow the design of strategies to suicide specifically pathogen vital interactome (such as a key component of virulence interactome or life cycle) and to dismantle any genes linking plant–pathogens network clusters through which the invader diverts the photosynthetates for its own. A second avenue could consist of combining system biology approach and agronomic practices that can contribute to increased plant photosynthetic capacity. Recently, Zhang et al (ZhangXie et al., 2008) described a soil symbiotic bacteria that augments photosynthesis in *Arabidopsis* by decreasing glucose sensing and abscisic acid levels *in planta*. Would such symbiotic system be applicable in a field system? Would there be any such symbiont that could antagonistically interfere with plant pathogenetic soil born diseases and reduce their impact on plant productivity? Those are some of the questions, we believe, could be the focus for further investigations.

5. Conclusion

Photosynthesis is a process that converts solar energy to chemical energy in many different organisms, ranging from plants to bacteria. It provides all the food we eat and all the fossil fuel we use. Photosynthesis of terrestrial higher plants is however constantly challenged by abiotic and biotic stresses. In this review, we described briefly the general process of photosynthesis, its outcome and limiting factors; the complex plant-pathogen inter-relationships and their effects on photosynthesis; and the insights the genomics and proteomics era can shed into the elucidation of the many genes and protein clusters and networks that sustain the plant-pathogen interactions in general, and photosynthesis, in particular. Photosynthesis feeds the globe and pathogen threats are increasing. A system biology approach, using both computational gene network discovery and gene and proteome analysis in living plants challenged with pathogens, was proposed as one of the pivotal player in identifying key gene and protein network clusters and their interactomes in both the plant and pathogens towards the design of strategies to suicide specifically pathogen vital interactome and to dismantle any genes linking plant – pathogens network clusters. This approach could be complemented with agronomic practices contributing to increased plant photosynthetic capacity.

6. Acknowledgments

The authors warmly thank Dr. Kaushik Ghose (University of Prince Edwards Island) for his kind willingness to proof read this Manuscript. We also wish to acknowledge Dr. Cloutier and her lab (Cereal Research Centre, Agriculture and Agri-Food Canada, Winnipeg, Manitoba), the lab from which Dr. Fofana has performed his work on the leaf-rust pathosystem. This chapter is in recognition and gratitude to Professor Patrick du Jardin (Gembloux Agro-Bio Tech, Universite de Liege, Belgium) from whom I received most of my taste and flavour for molecular plant physiology.

7. References

Abrams, M. D., Kubiske, M. E. & Steiner, K. C. (1990) Drought adaptations and responses in five genotypes of *Fraxinus pennsylvanica* Marsh.: photosynthesis, water relations and leaf morphology. *Tree physiology*, 6, 305-15.

Ackerson, R. C. & Hebert, R. R. (1981) Osmoregulation in Cotton in Response to Water Stress : I. Alterations in photosynthesis, leaf conductance, translocation, and ultrastructure. *Plant physiology,* 67, 484-8.

Aguera, E., Ruano, D., Cabello, P. & de la Haba, P. (2006) Impact of atmospheric CO2 on growth, photosynthesis and nitrogen metabolism in cucumber (*Cucumis sativus* L.) plants. *Journal of plant physiology,* 163, 809-17.

Ainsworth, E. A. & Bush, D. R. (2011) Carbohydrate export from the leaf: a highly regulated process and target to enhance photosynthesis and productivity. *Plant physiology,* 155, 64-9.

Ainsworth, E. A. & Rogers, A. (2007) The response of photosynthesis and stomatal conductance to rising [CO2]: mechanisms and environmental interactions. *Plant, cell & environment,* 30, 258-70.

Albert, K. R., Mikkelsen, T. N. & Ro-Poulsen, H. (2008) Ambient UV-B radiation decreases photosynthesis in high arctic *Vaccinium uliginosum. Physiologia plantarum,* 133, 199-210.

Allakhverdiev, S. I., Kreslavski, V. D., Klimov, V. V., Los, D. A., Carpentier, R. & Mohanty, P. (2008) Heat stress: an overview of molecular responses in photosynthesis. *Photosynthesis research,* 98, 541-50.

Baker, N. R., Long, S. P. & Ort, D. R. (1988) Photosynthesis and temperature, with particular reference to effects on quantum yield. *Symposia of the Society for Experimental Biology,* 42, 347-75.

Baldridge, G. D., O'Neill, N. R. & Samac, D. A. (1998) Alfalfa (*Medicago sativa* L.) resistance to the root-lesion nematode, *Pratylenchus penetrans*: defense-response gene mRNA and isoflavonoid phytoalexin levels in roots. *Plant molecular biology,* 38, 999-1010.

Baldry, C. W., Bucke, C. & Walker, D. A. (1966) Temperature and photosynthesis. I. Some effects of temperature on carbon dioxide fixation by isolated chloroplasts. *Biochimica et biophysica acta,* 126, 207-13.

Barhoumi, Z., Djebali, W., Chaibi, W., Abdelly, C. & Smaoui, A. (2007) Salt impact on photosynthesis and leaf ultrastructure of *Aeluropus littoralis. Journal of plant research,* 120, 529-37.

Barrow, S. R. & Cockburn, W. (1982) Effects of light quantity and quality on the decarboxylation of malic Acid in crassulacean Acid metabolism photosynthesis. *Plant physiology,* 69, 568-71.

Bassham, J. A. (1977) Increasing crop production through more controlled photosynthesis. *Science (New York, N.Y.),* 197, 630-8.

Batista-Santos, P., Lidon, F. C., Fortunato, A., Leitao, A. E., Lopes, E., Partelli, F., Ribeiro, A. I. & Ramalho, J. C. (2011) The impact of cold on photosynthesis in genotypes of Coffea spp.-photosystem sensitivity, photoprotective mechanisms and gene expression. *Journal of plant physiology,* 168, 792-806.

Bauerle, W. L., Bowden, J. D. & Wang, G. G. (2007) The influence of temperature on within-canopy acclimation and variation in leaf photosynthesis: spatial acclimation to microclimate gradients among climatically divergent *Acer rubrum* L. genotypes. *Journal of experimental botany,* 58, 3285-98.

Berger, S., Sinha, A. K. & Roitsch, T. (2007) Plant physiology meets phytopathology: plant primary metabolism and plant-pathogen interactions. *Journal of experimental botany,* 58, 4019-26.

Berry, J. A. (1975) Adaptation of photosynthetic processes to stress. *Science,* 188, 644-650.

Bertamini, M. & Nedunchezhian, N. (2004) Photoinhibition and recovery of photosynthesis in leaves of *Vitis berlandieri* and *Vitis rupestris. Journal of plant physiology*, 161, 203-10.

Bhatt, R. K., Baig, M. J., Tiwari, H. S. & Roy, S. (2010) Growth, yield and photosynthesis of Panicum maximum and Stylosanthes hamata under elevated CO2. *Journal of environmental biology / Academy of Environmental Biology, India*, 31, 549-52.

Biemelt, S. & Sonnewald, U. (2006) Plant-microbe interactions to probe regulation of plant carbon metabolism. *Journal of plant physiology*, 163, 307-18.

Bilgin, D. D., Zavala, J. A., Zhu, J., Clough, S. J., Ort, D. R. & DeLucia, E. H. (2010) Biotic stress globally downregulates photosynthesis genes. *Plant, cell & environment*, 33, 1597-613.

Bilska, A. & Sowinski, P. (2010) Closure of plasmodesmata in maize (*Zea mays*) at low temperature: a new mechanism for inhibition of photosynthesis. *Annals of botany*, 106, 675-86.

Bischof, K., Hanelt, D. & Wiencke, C. (2000) Effects of ultraviolet radiation on photosynthesis and related enzyme reactions of marine macroalgae. *Planta*, 211, 555-62.

Bonfig, K. B., Schreiber, U., Gabler, A., Roitsch, T. & Berger, S. (2006) Infection with virulent and avirulent *P. syringae* strains differentially affects photosynthesis and sink metabolism in *Arabidopsis* leaves. *Planta*, 225, 1-12.

Chou, H. M., Bundock, N., Rolfe, S. A. & Scholes, J. D. (2000) Infection of *Arabidopsis thaliana* leaves with *Albugo candida* (white blister rust) causes a reprogramming of host metabolism. *Mol Plant Pathol* 1, 99–113.

Daayf, F., Schmitt, A. & Belanger, R. R. (1997) Evidence of Phytoalexins in Cucumber Leaves Infected with Powdery Mildew following Treatment with Leaf Extracts of *Reynoutria sachalinensis. Plant physiology*, 113, 719-727.

Dixon, R. A., Harrison, M. J. & Paiva, L. N. (1995) The isoflavonoid phytoalexin pathway: From enzymes to genes to transcription factors. *Physiol Plant*, 93, 385-392.

Dodds, P. N. & Rathjen, J. P. (2010) Plant immunity: towards an integrated view of plant-pathogen interactions. *Nature reviews. Genetics*, 11, 539-48.

Duggan, D. J., Bittner, M., Chen, Y., Meltzer, P. & Trent, J. M. (1999) Expression profiling using cDNA microarrays. *Nature genetics*, 21, 10-4.

Eichmann, R., Biemelt, S., Schafer, P., Scholz, U., Jansen, C., Felk, A., Schafer, W., Langen, G., Sonnewald, U., Kogel, K. H. & Huckelhoven, R. (2006) Macroarray expression analysis of barley susceptibility and nonhost resistance to *Blumeria graminis. Journal of plant physiology*, 163, 657-70.

El-Bebany, A. F., Rampitsch, C. & Daayf, F. (2010) Proteomic analysis of the phytopathogenic soilborne fungus *Verticillium dahliae* reveals differential protein expression in isolates that differ in aggressiveness. *Proteomics*, 10, 289-303.

Elmore, J. M. & Coaker, G. (2011a) Biochemical purification of native immune protein complexes. *Methods in molecular biology (Clifton, N.J.)*, 712, 31-44.

Elmore, J. M. & Coaker, G. (2011b) The Role of the Plasma Membrane H+-ATPase in Plant-Microbe Interactions. *Molecular plant*, 4, 416-27.

Elmore, J. M., Lin, Z. J. & Coaker, G. (2011) Plant NB-LRR signaling: upstreams and downstreams. *Current opinion in plant biology*.

Erickson, S. J. & Hawkins, C. E. (1980) Effects of halogenated organic compounds on photosynthesis in estuarine phytoplankton. *Bulletin of environmental contamination and toxicology*, 24, 910-5.

Essmann, J., Bones, P., Weis, E. & Scharte, J. (2008) Leaf carbohydrate metabolism during defense: Intracellular sucrose-cleaving enzymes do not compensate repression of cell wall invertase. *Plant signaling & behavior*, 3, 885-7.

Eulgem, T. (2005) Regulation of the Arabidopsis defense transcriptome. *Trends in plant science*, 10, 71-8.

Faris, J. D., Zhang, Z., Lu, H., Lu, S., Reddy, L., Cloutier, S., Fellers, J. P., Meinhardt, S. W., Rasmussen, J. B., Xu, S. S., Oliver, R. P., Simons, K. J. & Friesen, T. L. (2010) A unique wheat disease resistance-like gene governs effector-triggered susceptibility to necrotrophic pathogens. *Proceedings of the National Academy of Sciences of the United States of America*, 107, 13544-9.

Fawe, A., Abou-Zaid, M., Menzie, J. G. & Belanger, R. R. (1998) Silicon-mediated accumulation of flavonoid phytoalexins in cucumber. *Phytopathology*, 88, 396-401.

Fofana, B., Banks, T. W., McCallum, B., Strelkov, S. E. & Cloutier, S. (2007) Temporal gene expression profiling of the wheat leaf rust pathosystem using cDNA microarray reveals differences in compatible and incompatible defence pathways. *International journal of plant genomics*, 2007, 17542.

Fofana, B., Benhamou, N., McNally, D., Labbe, C., Seguin, A. & Belanger, R. (2005) Suppression of Induced Resistance in Cucumber Through Disruption of the Flavonoid Pathway. *Phytopathology*, 95.

Fofana, B., McNally, D., Labbe, C., Boulanger, R., Benhamou, N., Seguin, A. & Belanger, R. (2002) Milsana-Induced Resistance in Powdery Mildew-Infected Cucumber Plants Correlates with the Induction of Chalcone Synthase and Chalcone Isomerase. *Physiol Plant Mol Pathol*, 61, 121-132.

Fu, J., Liu, H., Li, Y., Yu, H., Li, X., Xiao, J. & Wang, S. (2011) Manipulating broad-spectrum disease resistance by suppressing pathogen-induced auxin accumulation in rice. *Plant physiology*, 155, 589-602.

Garavaglia, B. S., Thomas, L., Gottig, N., Zimaro, T., Garofalo, C. G., Gehring, C. & Ottado, J. (2010) Shedding light on the role of photosynthesis in pathogen colonization and host defense. *Communicative & integrative biology*, 3, 382-4.

Garg, H., Li, H., Sivasithamparam, K., Kuo, J. & Barbetti, M. J. (2010) The infection processes of *Sclerotinia sclerotiorum* in cotyledon tissue of a resistant and a susceptible genotype of *Brassica napus*. *Annals of botany*, 106, 897-908.

Gehring, C. A. & Irving, H. R. (2003) Natriuretic peptides--a class of heterologous molecules in plants. *The international journal of biochemistry & cell biology*, 35, 1318-22.

Gest, H. (1993) Photosynthetic and quasi-photosynthetic bacteria. *FEMS Microbiol Lett*, 112, 1-6.

Gest, H. (2002) History of the word photosynthesis and evolution of its definition. *Photosynthesis research*, 73, 7-10.

Gonzalez-Teuber, M., Pozo, M. J., Muck, A., Svatos, A., Adame-Alvarez, R. M. & Heil, M. (2010) Glucanases and chitinases as causal agents in the protection of Acacia extrafloral nectar from infestation by phytopathogens. *Plant physiology*, 152, 1705-15.

Guldener, U., Mannhaupt, G., Munsterkotter, M., Haase, D., Oesterheld, M., Stumpflen, V., Mewes, H. W. & Adam, G. (2006) FGDB: a comprehensive fungal genome resource on the plant pathogen *Fusarium graminearum*. *Nucleic acids research*, 34, D456-8.

Guldener, U., Seong, K. Y., Boddu, J., Cho, S., Trail, F., Xu, J. R., Adam, G., Mewes, H. W., Muehlbauer, G. J. & Kistler, H. C. (2006) Development of a *Fusarium graminearum*

Affymetrix GeneChip for profiling fungal gene expression in vitro and in planta. *Fungal genetics and biology : FG & B*, 43, 316-25.

Halitschke, R., Hamilton, J. G. & Kessler, A. (2011) Herbivore-specific elicitation of photosynthesis by mirid bug salivary secretions in the wild tobacco *Nicotiana attenuata*. *The New phytologist*.

Hammerschmidt, R. & Kuc, J. (1982) Lignification as a mechanism for induced systemic resistance in cucumber *Physiol Plant Pathol* 20 61-71.

Hammerschmidt, R. & Yang-Cashman, P. (1995) Induced Resistance in Cucurbits. *Induced Resistance to disease in plants* (ed R. H. a. J. Kuć), pp. 63-85. Kluwer Academic Publishers, The Netherlands.

Horst, R. J., Engelsdorf, T., Sonnewald, U. & Voll, L. M. (2008) Infection of maize leaves with *Ustilago maydis* prevents establishment of C4 photosynthesis. *Journal of plant physiology*, 165, 19-28.

Hu, G. & Rijkenberg, F. H. (1998a) Subcellular localization of beta-1,3-glucanase in *Puccinia recondita* f.sp. tritici-infected wheat leaves. *Planta*, 204, 324-34.

Hu, G. & Rijkenberg, F. H. J. (1998b) Scanning electron microscopy of early infection structure formation by *Puccinia recondita* f. *sp. tritici* on and in susceptible and resistant wheat lines. *Mycological Research*, 102, 391-399.

Ibanez, A. J., Scharte, J., Bones, P., Pirkl, A., Meldau, S., Baldwin, I. T., Hillenkamp, F., Weis, E. & Dreisewerd, K. (2010) Rapid metabolic profiling of *Nicotiana tabacum* defence responses against *Phytophthora nicotianae* using direct infrared laser desorption ionization mass spectrometry and principal component analysis. *Plant methods*, 6, 14.

Ishihara, A., Hashimoto, Y., Miyagawa, H. & Wakasa, K. (2008) Induction of serotonin accumulation by feeding of rice striped stem borer in rice leaves. *Plant signaling & behavior*, 3, 714-6.

Ishihara, A., Hashimoto, Y., Tanaka, C., Dubouzet, J. G., Nakao, T., Matsuda, F., Nishioka, T., Miyagawa, H. & Wakasa, K. (2008) The tryptophan pathway is involved in the defense responses of rice against pathogenic infection via serotonin production. *The Plant journal : for cell and molecular biology*, 54, 481-95.

Ishihara, A., Nakao, T., Mashimo, Y., Murai, M., Ichimaru, N., Tanaka, C., Nakajima, H., Wakasa, K. & Miyagawa, H. (2011) Probing the role of tryptophan-derived secondary metabolism in defense responses against *Bipolaris oryzae* infection in rice leaves by a suicide substrate of tryptophan decarboxylase. *Phytochemistry*, 72, 7-13.

Jagadeeswaran, G., Raina, S., Acharya, B. R., Maqbool, S. B., Mosher, S. L., Appel, H. M., Schultz, J. C., Klessig, D. F. & Raina, R. (2007) *Arabidopsis* GH3-LIKE DEFENSE GENE 1 is required for accumulation of salicylic acid, activation of defense responses and resistance to *Pseudomonas syringae*. *The Plant journal : for cell and molecular biology*, 51, 234-46.

Kim, Y. M., Bouras, N., Kav, N. N. & Strelkov, S. E. (2010) Inhibition of photosynthesis and modification of the wheat leaf proteome by Ptr ToxB: a host-specific toxin from the fungal pathogen *Pyrenophora tritici-repentis*. *Proteomics*, 10, 2911-26.

King, E. G. & Caylor, K. K. (2010) Herbivores and mutualistic ants interact to modify tree photosynthesis. *The New phytologist*, 187, 17-21.

Kocal, N., Sonnewald, U. & Sonnewald, S. (2008) Cell wall-bound invertase limits sucrose export and is involved in symptom development and inhibition of photosynthesis during compatible interaction between tomato and *Xanthomonas campestris* pv vesicatoria. *Plant physiology*, 148, 1523-36.

Korves, T. M. & Bergelson, J. (2003) A developmental response to pathogen infection in *Arabidopsis. Plant physiology*, 133, 339-47.

Kumar, A., Li, C. & Portis, A. R., Jr. (2009) *Arabidopsis thaliana* expressing a thermostable chimeric Rubisco activase exhibits enhanced growth and higher rates of photosynthesis at moderately high temperatures. *Photosynthesis research*, 100, 143-53.

Kurek, I., Chang, T. K., Bertain, S. M., Madrigal, A., Liu, L., Lassner, M. W. & Zhu, G. (2007) Enhanced Thermostability of *Arabidopsis Rubisco* activase improves photosynthesis and growth rates under moderate heat stress. *The Plant cell*, 19, 3230-41.

Lee, S., Kim, S. Y., Chung, E., Joung, Y. H., Pai, H. S., Hur, C. G. & Choi, D. (2004) EST and microarray analyses of pathogen-responsive genes in hot pepper (*Capsicum annuum* L.) non-host resistance against soybean pustule pathogen (*Xanthomonas axonopodis* pv. glycines). *Functional & integrative genomics*, 4, 196-205.

Liu, J., Elmore, J. M., Lin, Z. J. & Coaker, G. (2011) A receptor-like cytoplasmic kinase phosphorylates the host target RIN4, leading to the activation of a plant innate immune receptor. *Cell host & microbe*, 9, 137-46.

Liu, X., Tang, W. H., Zhao, X. M. & Chen, L. (2010) A network approach to predict pathogenic genes for *Fusarium graminearum. PloS one*, 5.

McNally, D. J., Wurms, K. V., Labbe, C. & Belanger, R. R. (2003b) Synthesis of C-glycosyl flavonoid phytoalexins as a site-specific response to fungal penetration in cucumber. *Physiol Plant Mol Pathol*, 63, 293-303.

McNally, D. J., Wurms, K. V., Labbe, C., Quideau, S. & Belanger, R. R. (2003) Complex C-glycosyl flavonoid phytoalexins from *Cucumis sativus. Journal of natural products*, 66, 1280-3.

Murchie, E. H. & Niyogi, K. K. (2011) Manipulation of photoprotection to improve plant photosynthesis. *Plant physiology*, 155, 86-92.

Nabity, P. D., Zavala, J. A. & DeLucia, E. H. (2009) Indirect suppression of photosynthesis on individual leaves by arthropod herbivory. *Annals of botany*, 103, 655-63.

Navarro, L., Dunoyer, P., Jay, F., Arnold, B., Dharmasiri, N., Estelle, M., Voinnet, O. & Jones, J. D. (2006) A plant miRNA contributes to antibacterial resistance by repressing auxin signaling. *Science (New York, N.Y.)*, 312, 436-9.

Nembaware, V., Seoighe, C., Sayed, M. & Gehring, C. (2004) A plant natriuretic peptide-like gene in the bacterial pathogen *Xanthomonas axonopodis* may induce hyper-hydration in the plant host: a hypothesis of molecular mimicry. *BMC evolutionary biology*, 4, 10.

Nunes-Nesi, A., Sulpice, R., Gibon, Y. & Fernie, A. R. (2008) The enigmatic contribution of mitochondrial function in photosynthesis. *Journal of experimental botany*, 59, 1675-84.

OED (1989) Oxford English Dictionary. *The Oxford English Dictionary* (ed E. W. John Simpson). Clarendon Press, Oxford.

Paul, M. J. & Foyer, C. H. (2001) Sink regulation of photosynthesis. *Journal of experimental botany*, 52, 1383-400.

Paul, M. J. & Pellny, T. K. (2003) Carbon metabolite feedback regulation of leaf photosynthesis and development. *Journal of experimental botany*, 54, 539-47.

Petit, A. N., Vaillant, N., Boulay, M., Clement, C. & Fontaine, F. (2006) Alteration of photosynthesis in grapevines affected by esca. *Phytopathology*, 96, 1060-6.

Rekah, Y., Shtienberg, D. & Katan, J. (1999) Spatial distribution and temporal development of *Fusarium* crown and root rot of tomato and pathogen dissemination in field soil *Phytopathology* 89, 831-839.

Rhodes, M. J. (1994) Physiological roles for secondary metabolites in plants: some progress, many outstanding problems. *Plant molecular biology*, 24, 1-20.

Ripley, B. S., Abraham, T. I. & Osborne, C. P. (2008) Consequences of C4 photosynthesis for the partitioning of growth: a test using C3 and C4 subspecies of *Alloteropsis semialata* under nitrogen-limitation. *Journal of experimental botany*, 59, 1705-14.

Ripley, B. S., Gilbert, M. E., Ibrahim, D. G. & Osborne, C. P. (2007) Drought constraints on C4 photosynthesis: stomatal and metabolic limitations in C3 and C4 subspecies of Alloteropsis semialata. *Journal of experimental botany*, 58, 1351-63.

Roberntz, P. & Stockfors, J. (1998) Effects of elevated CO_2 concentration and nutrition on net photosynthesis, stomatal conductance and needle respiration of field-grown Norway spruce trees. *Tree physiology*, 18, 233-241.

Romeis, T. (2001) Protein kinases in the plant defence response. *Current opinion in plant biology*, 4, 407-14.

Romeis, T., Ludwig, A. A., Martin, R. & Jones, J. D. (2001) Calcium-dependent protein kinases play an essential role in a plant defence response. *The EMBO journal*, 20, 5556-67.

Scharte, J., Schon, H., Tjaden, Z., Weis, E. & von Schaewen, A. (2009) Isoenzyme replacement of glucose-6-phosphate dehydrogenase in the cytosol improves stress tolerance in plants. *Proceedings of the National Academy of Sciences of the United States of America*, 106, 8061-6.

Schenk, P. M., Kazan, K., Wilson, I., Anderson, J. P., Richmond, T., Somerville, S. C. & Manners, J. M. (2000) Coordinated plant defense responses in *Arabidopsis* revealed by microarray analysis. *Proceedings of the National Academy of Sciences of the United States of America*, 97, 11655-60.

Seo, Y. S., Cho, J. I., Lee, S. K., Ryu, H. S., Han, M., Hahn, T. R., Sonnewald, U. & Jeon, J. S. (2007) Current insights into the primary carbon flux that occurs in plants undergoing a defense response *Plant Stress* 1, 42–49.

Song, X., Rampitsch, C., Soltani, B., Mauthe, W., Linning, R., Banks, T., McCallum, B. & Bakkeren, G. (2011) Proteome analysis of wheat leaf rust fungus, *Puccinia triticina*, infection structures enriched for haustoria. *Proteomics*, 11, 944-63.

Strelkov, S. E., Lamari, L. & Ballance, G. M. (1998) Induced chlorophyll degradation by a chlorosis toxin from *Pyrenophora tritici-repentis Can. J. Plant Pathol.*, 20, 428–435.

Taiz, L. & Zeiger, E. (2010) Plant Physiology. Sinauer Associates Inc, Sunderland, Massachussetts, USA.

Tang, J., Zielinski, R., Aldea, M. & DeLucia, E. (2009) Spatial association of photosynthesis and chemical defense in *Arabidopsis thaliana* following herbivory by *Trichoplusia ni*. *Physiologia plantarum*, 137, 115-24.

Tariq, V. N. & Jeffries, P. (1986) Ultrastructure of penetration of *Phaseolus spp.* by *Sclerotinia sclerotiorum*. *Canadian Journal of Botany*, 64, 2909-2015.

Tariq, V. N. & Jeffries, P. (1984) Appressorium formation by *Sclerotinia sclerotiorum*: scanning electron microscopy. . *Transactions of British Mycological Society* 82.

Trotta, A., Wrzaczek, M., Scharte, J., Tikkanen, M., Konert, G., Rahikainen, M., Holmstrom, M., Hiltunen, H. M., Rips, S., Sipari, N., Mulo, P., Weis, E., von Schaewen, A., Aro, E. M. & Kangasjarvi, S. (2011) Regulatory subunit B'{gamma} of protein phosphatase 2A prevents unnecessary defense reactions under low light in *Arabidopsis thaliana. Plant physiology*, 156, 1464-1480.

Vidaver, A. K. & Lambrecht, P. A. (2004) Bacteria as plant pathogens. *The plant health Instructor*, DOI: 10.1094/PHI-I-2004-0809-01.

Voegele, R. T., Wirsel, S., Moll, U., Lechner, M. & Mendgen, K. (2006) Cloning and characterization of a novel invertase from the obligate biotroph *Uromyces fabae* and analysis of expression patterns of host and pathogen invertases in the course of infection. *Mol Plant Microbe Interact* 19, 625–634.

Ward, E. R., Uknes, S. J., Williams, S. C., Dincher, S. S., Wiederhold, D. L., Alexander, D. C., Ahl-Goy, P., Metraux, J. P. & Ryals, J. A. (1991) Coordinate Gene Activity in Response to Agents That Induce Systemic Acquired Resistance. *The Plant cell*, 3, 1085-1094.

Wen, Y., Wang, W., Feng, J., Luo, M. C., Tsuda, K., Katagiri, F., Bauchan, G. & Xiao, S. (2011) Identification and utilization of a sow thistle powdery mildew as a poorly adapted pathogen to dissect post-invasion non-host resistance mechanisms in *Arabidopsis. Journal of experimental botany*, 62, 2117-29.

Wurms, K., Labbe, C., Benhamou, N. & Belanger, R. (1999) Effects of Milsana and Benzothiadiazole on the Ultrastructure of Powdery Mildew Haustoria on Cucumber. *Phytopathology*, 89, 728-736.

Zhang, C. & Turgeon, R. (2009) Downregulating the sucrose transporter VpSUT1 in *Verbascum phoeniceum* does not inhibit phloem loading. *Proceedings of the National Academy of Sciences of the United States of America*, 106, 18849-54.

Zhang, H., Xie, X., Kim, M. S., Kornyeyev, D. A., Holaday, S. & Pare, P. W. (2008) Soil bacteria augment *Arabidopsis* photosynthesis by decreasing glucose sensing and abscisic acid levels in planta. *The Plant journal : for cell and molecular biology*, 56, 264-73.

Zhang, X., Wollenweber, B., Jiang, D., Liu, F. & Zhao, J. (2008) Water deficits and heat shock effects on photosynthesis of a transgenic *Arabidopsis thaliana* constitutively expressing ABP9, a bZIP transcription factor. *Journal of experimental botany*, 59, 839-48.

Zhang, Z., Friesen, T. L., Xu, S. S., Shi, G., Liu, Z., Rasmussen, J. B. & Faris, J. D. (2011) Two putatively homoeologous wheat genes mediate recognition of SnTox3 to confer effector-triggered susceptibility to *Stagonospora nodorum. The Plant journal : for cell and molecular biology*, 65, 27-38.

Zhao, X. M., Zhang, X. W., Tang, W. H. & Chen, L. (2009) FPPI: *Fusarium graminearum* protein-protein interaction database. *Journal of proteome research*, 8, 4714-21.

Zhu, X. G., Shan, L., Wang, Y. & Quick, W. P. (2010) C4 rice - an ideal arena for systems biology research. *Journal of integrative plant biology*, 52, 762-70.

Zou, J., Rodriguez-Zas, S., Aldea, M., Li, M., Zhu, J., Gonzalez, D. O., Vodkin, L. O., DeLucia, E. & Clough, S. J. (2005) Expression profiling soybean response to *Pseudomonas syringae* reveals new defense-related genes and rapid HR-specific downregulation of photosynthesis. *Molecular plant-microbe interactions : MPMI*, 18, 1161-74.

9

Photosynthetic Adaptive Strategies in Evergreen and Semi-Deciduous Species of Mediterranean Maquis During Winter

Carmen Arena[1] and Luca Vitale[2]
[1]Department of Structural and Functional Biology,
University of Naples Federico II,
[2]Istituto per I Sistemi Agricoli e Forestali del Mediterraneo,
(ISAFoM – CNR)
Italy

1. Introduction

Mediterranean-type ecosystems are characterised by a particular temperature and rainfall regime that limits plant growth in both summer and winter seasons (Mitrakos, 1980; Larcher, 2000). Mediterranean plant community is very heterogeneous and include many evergreen and semi-deciduous species that present a complex mixture of elements, some deriving from *in situ* evolution, others having colonized the area from adjacent regions in different periods in the past (Blondel & Aronson 1999; Gratani & Varone, 2004). The result of this evolution is that the Mediterranean maquis species are well adapted to environmental stress conditions and successfully overcome them (Sànchez-Blanco et al., 2002; Varone & Gratani, 2007).

Structural and physiological adaptations consist in a mixture of characteristics that make these species very resistant to stresses. High leaf consistency, leaf tissue density, leaf thickness, and reduced leaf area are traits improving drought resistance by decreasing photochemical damages to the photosynthetic system (Abril & Hanano 1998; Castro-Díez et al. 1998; Gratani & Ghia 2002).

In this study we have focused our attention on photosynthetic adaptive strategies in Mediterranean evergreen and semi-deciduous species subjected to winter temperatures.

Winter depression of photosynthetic activity, occurring between December and February, is the consequence of low temperatures which are responsible for slowing down metabolic processes and cessation of growth (Rhizopoulou et al., 1989; Larcher, 2000).

Under these conditions, photosynthetic performance may decline and may be restored when the environmental conditions become favourable for growth in spring (Larcher, 2000; Oliveira & Peñuelas, 2004). The combination of low temperatures and high light, may induce a reduction in photochemical efficiency, increasing the sensitivity of photosystems to photoinhibition (Powles, 1984). Mediterranean plant communities comprise many evergreen and semi-deciduous species that cope with winter cold through different strategies that include biochemical, physiological, anatomical and cytological modifications (Huner et al., 1981; Boese & Huner 1990; Long et al., 1994; Oliveira & Peñuelas, 2000; Tattini et al., 2000).

The chilling-induced photosynthetic decline can be attributed both to a reduced activity of enzymes involved in the photosynthetic carbon reduction cycle (Sassenrath et al., 1990; Hutchinson et al., 2000), or to a photoinhibitory process. In fact, when chilling is protracted for a long time, the reduction of carbon assimilation can lead to an increase of excitation energy to reaction centres, that if not safely dissipated, induces damages at photosystems level compromising the whole photosynthetic apparatus (Baker, 1994; Tjus et al., 1998). However, in nature, the photosynthetic decrease as well as the reduction of photochemical activity at low temperatures, often represent a regulatory mechanism associated with photoprotective strategies that promote the dissipation of excess excitation energy avoiding irreversible damages to photosystems (Long et al., 1994; D'Ambrosio et al., 2006). Several mechanisms have evolved in plants in order to protect photosystems against photodamages; they include thermal dissipation, chloroplasts movements, chlorophyll concentration changes, increases in the capacity for scavenging the active oxygen species and the PSII ability to transfer electrons to acceptors different from CO_2 (Niyogi, 2000).

It has been reported that the resistance of Mediterranean maquis evergreen species to photoinhibition is associated mainly to the increase in scavenging capacity and thermal dissipation processess, as well as to the increment of carotenoids pool or reduction in chlorophyll content (Garcìa-Plazaola et al., 1999, 2000; Arena et al., 2008). On the other hand, the semi-deciduous species such as *Cistus* rely on pheno-morphological features such as short lifetime of leaves and leaf pubescence to protect leaves from the excess of light and, thus, reduce the investment in other physiological mechanisms (Werner et al., 1999; Oliveira & Peñuelas, 2001, 2002, 2004). Previous studies have demonstrated that the resistance to environmental constraints such as low temperature or high irradiance can depend on leaf age (Shirke, 2001; Bertamini & Nedunchezhian, 2003). Young and mature leaves may differ both in photosynthetic performance and some leaf functional traits such as the sclerophylly index LMA (leaf mass per area) and its opposite leaf specific area (SLA), leaf dry matter content (LDMC) and relative water content (RWC). These properties affect significantly the whole plant physiology. More specifically, LMA variations are linked to biomass allocation strategies (Wilson et al., 1999) and to photosynthetic acclimation under different conditions, RWC is a good indicator to evaluate the plant water status (Cornelissen et al., 2003; Teulat et al., 1997) and LDMC represent an index of resource use by plant (Garnier et al., 2001). LDMC is related to leaf lifespan and it is involved in the trade-off between the quick production of biomass and the efficient conservation of nutrients (Poorter & Garnier, 1999; Ryser & Urbas, 2000). Generally young leaves appears more vulnerable than mature leaves to stress, since have a reduced degree of xeromorphism (lower LMA). In this chapter has been examined the photosynthetic and photochemical behaviour of young and mature leaves of different species of the Mediterranean maquis, grown during the winter, in response to low temperatures. In particular our attention has been focused on the evergreen species *Laurus nobilis* L., *Phillyrea angustifolia* L. and *Quercus ilex* L. and on the semi-deciduous species *Cistus incanus* L. that are widespread in Southern Italy area. Our specific purposes were: 1) to focus on eco-physiological strategies adopted by the different species to optimize the carbon gain during winter and minimize the photoinhibitory damage risks; 2) to compare the behaviour of young and mature leaves under low winter temperature in order to elucidate if the photoprotective mechanisms may be influenced by the leaf age.

2. Material and methods

Two different experiments have been considered in this study; the first experiment has been carried out on evergreens *L. nobilis*, *P. angustifolia* and *Q. ilex* and analyzes the photosynthetic and the photochemical performance of young and mature leaves during the winter and of mature leaves during the winter and following spring. The second experiment is focused on the photochemical behaviour of young and mature leaves of the semi-deciduous species *C. incanus* during winter and of mature leaves during the winter and the following spring. It is well know that the *C. incanus* species produces two different typologies of leaves: winter leaves and summer leaves with dissimilar morpho-anatomical traits (Aronne & De Micco, 2001). In the present study only winter leaves have been examined. The experimental planning of the work is reported in Fig. 1.

2.1 The experimental planning schema

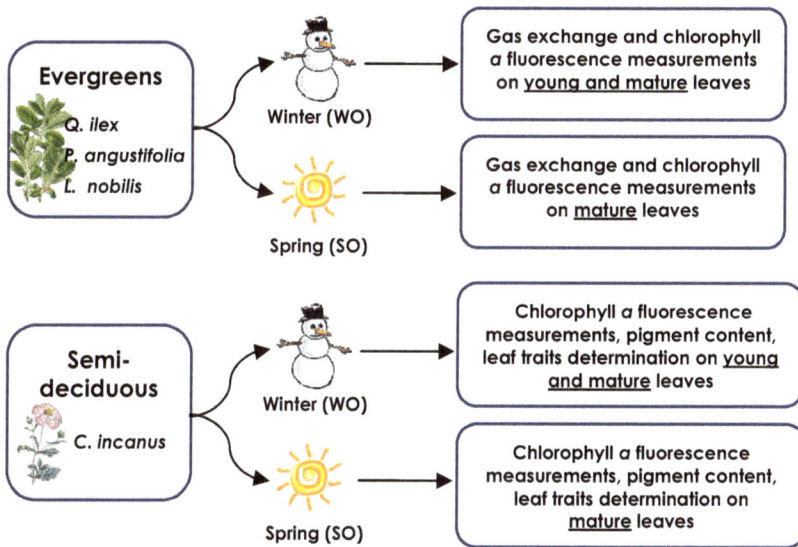

Fig. 1. The experimental planning of the work.

2.2 Plant material and growth conditions

First experiment. Two years old plants of *Q. ilex*, *P. angustifolia* and *L. nobilis* coming from the garden centre of Corpo Forestale dello Stato of Sabaudia (Latina, Italy) were transplanted in 15 L pots in January 2004 and placed outdoor in the Botanical Garden of Naples University for one year. Pots were large enough to avoid limitations in root growth and were filled with a mixture of peat and soil in the proportion 50:50. The temperature conditions at the experimental site during plant growth were typical of the Mediterranean region with cold winters and warm summers (Fig. 2A). Gas exchange and chlorophyll *a* fluorescence measurements were performed in winter (early March 2005) and in spring (during May 2005); for winter measurements, 8 mature leaves of one year old and 8 young leaves sprouted in late October of the previous year, were selected randomly for each species from 4 different plants. The photosynthetic behaviour of one-year old leaves in winter was compared with that of one-year old leaves in spring.

Second experiment. In November 2007, eight plants of *C. incanus*, of three years old, were collected in the field in the Castel Volturno Natural reserve (Naples, Italy). The climate of the reserve is typically Mediterranean, with dry summers and rainy autumns and winters. The main vegetation type is maquis often opening into garrigue formations dominated by evergreen sclerophylls and seasonally dimorphic species.

The collected plants were excavated *in situ* and quickly transplanted in 15 L pots filled with native soil, then were carried to the Department of Structural and Functional Biology of Naples University and placed outdoors in a open area of the Department. The temperature conditions experienced by plants during growth are shown in Fig. 2B. Outdoor temperatures at the experimental site, during the experimental period, ranged between minimum values of 2 °C and maximum values of 16 °C.

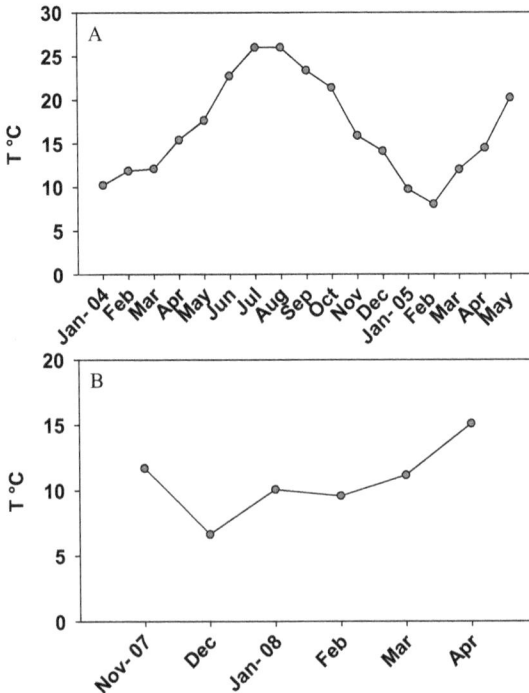

Fig. 2. Monthly mean air temperature (T °C) at the two experimental sites during the evergreens growth (A) and *C. incanus* growth (B). Data have been collected from Naples Largo San Marcellino weather station.

At the beginning of February, three healthy plants were selected for eco-physiological analyses; young leaves of about 15 days old and mature leaves of about 30 days old were chosen for photochemical measurements and photosynthetic pigments and leaf functional traits determinations. At the end of April, mature leaves were analysed again and compared to mature leaves in winter on basis of chlorophyll *a* fluorescence measurements, chlorophyll content and leaf functional traits determinations. All analyses were carried out on six leaves from 3 different specimens.

2.3 Gas exchange and chlorophyll a fluorescence measurements

In the first experiment, on the three evergreen species, gas exchange and chlorophyll *a* fluorescence measurements were performed simultaneously in winter (March 2008) and in spring (May 2005) by a portable gas exchange system (HCM-1000, Walz, Germany) in a climatized cuvette equipped with a fiber optic connected with a portable pulse amplitude modulated fluorometer (Mini-PAM, Walz, Germany). All measurements were performed at midday under clear-sky conditions. In winter both young and mature leaves were analysed, in spring only mature leaves were considered.

In the second experiment, on the semi-deciduous species *C. incanus* L. measurements of chlorophyll *a* fluorescence were performed by a portable pulse amplitude modulated fluorometer (Mini-PAM, Walz, Germany) equipped with a leaf-clip holder (Leaf-Clip Holder 2030-B, Walz, Germany), which allows the simultaneous recording of the incident photosynthetic photon flux density on the leaf and abaxial leaf temperature. Measurements were performed at midday, under natural light and temperature conditions, on young and mature leaves, during winter (February 2008), and on mature leaves during spring (April 2008). The air temperature (T_{air}) and the Photosynthetic Photon Flux Densities (PPFD) experienced by *C. incanus* leaves at midday, during the days of measurements, are reported in Table 1.

	Days of measurements	Young leaves	Mature leaves
T $_{air}$	2 Feb 2008	12 ± 0.12	11 ± 0.22
	29 Apr 2008	-	22 ± 0.05
PPFD	2 Feb 2008	693 ± 28	712 ± 29
	29 Apr 2008	-	1074 ± 28

Table 1. Air temperature (T_{air}, °C) and Photosynthetic photon flux density (PPFD, μmol photons m^{-2} s^{-1}) measured at midday on *C. incanus* plants at the experimental site in the days of measurements. Data reported are means ± SE (n=6).

For gas exchange measurements, each leaf has been kept in cuvette for 5-6 min. The acquisition of data was made when steady-state rate of net assimilation was achieved. A constant photosynthetic photon flux density (PPFD) of 1000 μmol photons m^{-2} s^{-1} was provided to the leaves by an external light source (1050-H, Walz, Germany) positioned on the cuvette plane. The PPFD of 1000 μmol photons m^{-2} s^{-1} was selected in order to obtain the values of light-saturated net photosynthetic rate for each species.

Net photosynthetic rate (A_N), stomatal conductance to water (g_{H2O}) and intercellular CO_2 concentration (C_i) were calculated by the software operating in HCM-1000 using the von

Caemmerer and Farquhar equations (1981). The ratio of intercellular to ambient CO_2 concentration, C_i/C_a, was used to calculate the apparent carboxylation efficiency.

As concerns chlorophyll a fluorescence measurements, in the early morning, on 30 min dark-adapted leaves, the background fluorescence signal, F_o, was induced by light of about 0.5 μmol photons m^{-2} s^{-1} at the frequency of 0.6 kHz. In order to determine the maximal fluorescence level in the dark-adapted state, F_m, a 1s saturating light pulse (10000 μmol photons m^{-2} s^{-1}) was applied by previously setting the frequency at 20 kHz; the maximum PSII photochemical efficiency (F_v/F_m) was calculated as:

$$[F_v/F_m=(F_m-F_o)/F_m]$$

The saturating pulse intensity was chosen in order to saturate the fluorescence yield but avoiding photoinhibition during the pulse.

At midday, the steady-state fluorescence signal (F_t) and the maximal fluorescence (F_m') under illumination were measured, setting the light measure at a frequency of 20 kHz. F_m' was determined by a 1s saturating light pulse (10000 μmol photons m^{-2} s^{-1}). The partitioning of absorbed light energy was calculated following the model of Kramer et al. (2004). The quantum yield of PSII linear electron transport (Φ_{PSII}) was estimated following Genty et al. (1989) as:

$$\Phi_{PSII} = (F_m' - F_t)/F_m'$$

The yields of regulated energy dissipation was calculated as:

$$\Phi_{NPQ} = 1 - \Phi_{PSII} - 1/(NPQ + 1 + q_L \times (F_m/F_o-1)$$

whereas the non-regulated energy dissipation in PSII was calculated as:

$$\Phi_{NO} = 1/(NPQ + 1 + q_L \times (F_m/F_o-1))$$

The coefficient of photochemical quenching, qL, was defined and calculated following Kramer et al. (2004) as:

$$(F_m' - F_t)/(F_m' - F_o') \times F_o'/F_t = q_p \times F_o'/F_t$$

The value of F_o' was estimated as: $F_o' = F_o/(F_v/F_m + F_o/F_m')$ (Oxborough & Baker, 1997). Non-photochemical quenching was expressed according to Bilger & Björkman (1990) as:

$$[NPQ = (F_m-F_m')/F_m']$$

The statistical analysis of the data was performed by one-way ANOVA followed by Student-Newman-Keuls test (*Sigma-Stat 3.1*) based on a significance level of $P < 0.05$. Data are means ± SE (at least n = 6).

2.4 Photosynthetic pigment content and functional leaf traits determination

After fluorescence measurements, leaves were detached from *C. incanus* plants and carried to the laboratory for the photosynthetic pigment content determination. Pigments were extracted with a mortar and pestle in ice-cold 100% acetone and quantified by a spectrophotometer according to Lichtenthaler (1987). A different group of leaves of comparable age to those used for fluorescence measurements and pigment determinations, was collected and utilized for the specific leaf area (SLA) and leaf dry matter content

(LDMC) measurements. Specific leaf area was calculed as the ratio of leaf area to leaf dry mass and expressed as cm² g⁻¹ dw (dry weight). For dry mass determination, leaves were dried at 70 °C for 48 h. Leaf dry matter content (LDMC) was measured as the oven-dry mass of a leaf divided by its water-saturated fresh mass and expressed as g g⁻¹ wslm (water saturated leaf mass). Leaf dry matter content is related to the average density of the leaf tissues (Cornelissen et al., 2003).

3. Results

3.1 Young and mature leaves of *L. nobilis* L., *P. angustifolia* L. and *Quercus ilex* L. during winter

During winter, mature leaves of all species showed an higher (P<0.001) net photosynthetic rate (A_N) compared to young leaves. In both young and mature leaves the highest (P<0.001) A_N values were measured in *Q. ilex* whereas the lowest (P<0.001) in *L. nobilis* (Fig. 3A, D).

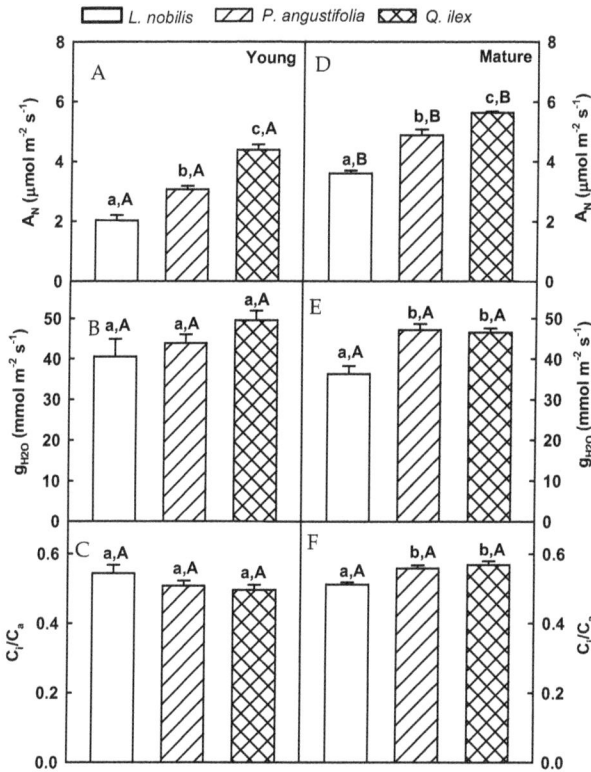

Fig. 3. Net photosynthetic rate (A_N), stomatal conductance to water (g_{H2O}) and ratio of intercellular to ambient CO_2 concentration (C_i/C_a) in young and mature leaves of *Laurus nobilis*, *Phillyrea angustifolia* and *Quercus ilex*, during winter. Different letters indicate statistical differences between young and mature leaves (small letters) and among species (capital letters). Values are means ± SD (n=8).

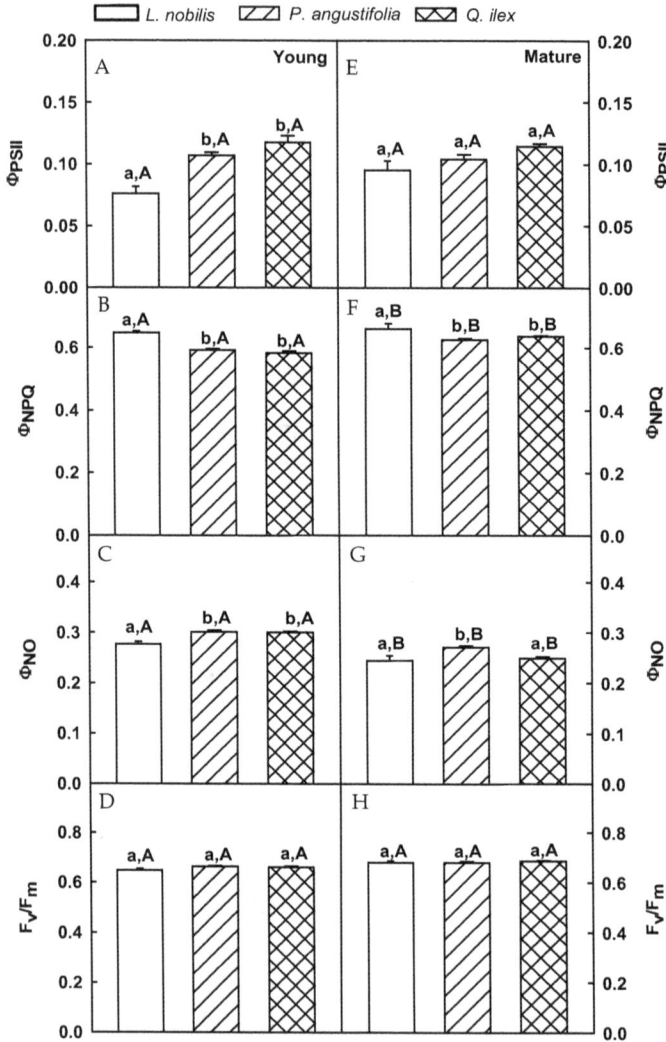

Fig. 4. Quantum yield of linear PSII electron transport (Φ_{PSII}), regulated energy dissipation (Φ_{NPQ}), non-regulated energy dissipation (Φ_{NO}) and maximum PSII photochemical efficiency (F_v/F_m) in young and mature leaves of *Laurus nobilis*, *Phillyrea angustifolia* and *Quercus ilex*, during winter. Different letters indicate statistical differences between young and mature leaves (small letters) and among species (capital letters). Values are means ± SD (n=8).

In young leaves, stomatal conductance to water (g_{H2O}) did not differ among the species; on the contrary, in mature leaves g_{H2O} was significantly lower (P<0.01) in *L. nobilis* than *P. angustifolia* and *Q. ilex*. No significant difference in g_{H2O} between young and mature leaves of the same species was measured (Fig. 3B, E).

The ratio of intercellular to ambient CO_2 concentration (C_i/C_a) was similar for young leaves of all species, conversely in mature leaves was lower ($P<0.001$) in *L. nobilis* compared to *P. angustifolia* and *Q. ilex*. No significant difference within young and mature leaves of the same species was observed in C_i/C_a ratio (Fig. 3 C, F).

The analysis of photochemistry showed that, among young leaves of different species, the quantum yield of PSII linear electron transport (Φ_{PSII}) was higher ($P<0.005$) in *P. angustifolia* and *Q. ilex* compared to *L. nobilis* (Fig. 4A) on the contrary *L. nobilis* showed the highest regulated energy dissipation, Φ_{NPQ}, ($P<0.05$) and the lowest ($P<0.005$) non-regulated energy dissipation, Φ_{NO}, compared to other species. No difference was detected in Φ_{NPQ} and Φ_{NO} between *P. angustifolia* and *Q. ilex* (Fig. 4B, C). All mature leaves exhibited no significant difference in Φ_{PSII} (Fig. 4E) but leaves of *L. nobilis* showed again the highest Φ_{NPQ} ($P<0.05$); the highest ($P<0.005$) Φ_{NO} was found in *P. angustifolia* (Fig. 4 F, G). No variation in maximum PSII photochemical efficiency (F_v/F_m) among different species and between young and mature leaves were found (Fig. 4 D, H). The comparison between young and mature leaves evidenced no difference in Φ_{PSII} and lower ($P<0.001$) and higher ($P<0.005$) values of Φ_{NPQ} and Φ_{NO}, respectively, in mature leaves.

3.2 Mature leaves of *L. nobilis* L., *P. angustifolia* L. and *Quercus ilex* L. during winter and spring

During winter, within different species, *Q. ilex* showed higher net photosynthetic rate (A_N) ($P<0.001$) and stomatal conductance to water (g_{H2O}) ($P<0.05$) as well as a lower ($P<0.005$) intercellular to ambient CO_2 concentration ratio (C_i/C_a) compared to *L. nobilis* and *P. angustifolia* (Fig. 5A, B, C). The lowest values of A_N and g_{H2O} was found in *L. nobilis*. No significant difference between *L. nobilis* and *P. angustifolia* in C_i/C_a ratio was found. During spring, among species, *Q. ilex* exhibited again the highest ($P<0.001$) net photosynthetic rate (A_N) and the lowest C_i/C_a ratio ($P<0.05$) compared to *L. nobilis* and *P. angustifolia* (Fig. 5 D, F), but similar values of g_{H2O} (Fig. 5E).

The comparison between winter and spring showed that, during spring, an increase in A_N ($P<0.001$) and g_{H2O} ($P<0.05$) were observed in all species compared to winter (Fig. 5D, E); on the other hand, no significant difference in C_i/C_a ratio was found (Fig. 5F).

During winter the photochemical performance varied among species (Fig. 6).

In particular, *L. nobilis* showed the lowest ($P<0.001$) quantum yield of PSII linear electron transport (F_{PSII}) and non-regulated energy dissipation (Φ_{NO}), as well as the highest ($P<0.01$) regulated energy dissipation (Φ_{NPQ}) (Fig. 6A, B, C). No difference in F_v/F_m values was observed among species (Fig. 6 D).

During spring, *Q. ilex* and *P. angustifolia* showed an higher ($P<0.001$) Φ_{PSII} than *L. nobilis* (Fig. 6E). The lowest ($P<0.01$) Φ_{NPQ} was detected in *Q. ilex*, whereas the highest ($P<0.01$) F_{NO} was found in *L. nobilis* (Fig. 6F, G). Similar values of maximum PSII photochemical efficiency, F_v/F_m, were observed among species (Fig. 6H).

The comparison between the two campaign of measurements has evidenced that in all species F_{PSII} and Φ_{NPQ} were respectively higher and lower ($P<0.001$) in spring than in winter (Fig. 6A, E, B, F). In spring compared to winter, Φ_{NO} increased ($P<0.01$) only in *L. nobilis*, whereas decreased ($P<0.05$) in *P. angustifolia* and remained unvaried in *Q. ilex* (Fig. 6C, G). The maximum PSII photochemical efficiency F_v/F_m was lower in winter as compared to spring ($P<0.005$) for all species (Fig. 6D, H).

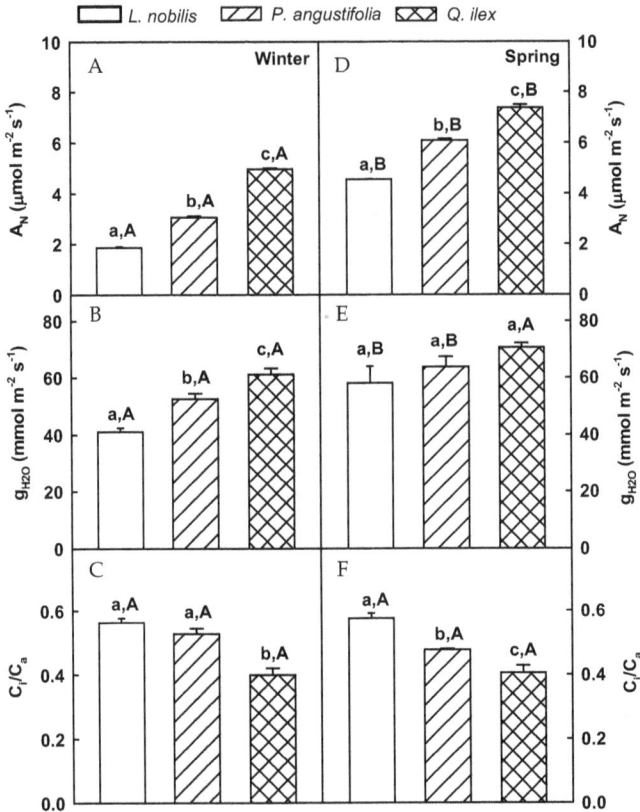

Fig. 5. Net photosynthetic rate (A_N), stomatal conductance to water (g_{H2O}) and ratio of intercellular to ambient CO_2 concentration (C_i/C_a) in mature leaves of *Laurus nobilis, Phillyrea angustifolia* and *Quercus ilex*, during winter and spring. Different letters indicate statistical differences among species (small letters) and between seasons (capital letters). Values are means ± SD (n=8).

3.3 The semi-deciduous species *Cistus incanus* L.

The comparison between young and mature leaves of the semi-deciduous species *C. incanus* evidenced that the quantum yield of PSII linear electron transport (Φ_{PSII}) was lower in mature as compared to young leaves (P<0.001) whereas the quantum yield of regulated energy dissipation (Φ_{NPQ}) showed an opposite tendency (P<0.05) (Fig. 7A, B). No significant difference in non regulated energy dissipation (Φ_{NO}) and maximum photochemical efficiency (F_v/F_m) was detected (P<0.05) between the two leaf typologies (Fig. 7C, D).

The photochemical behavior of mature *C. incanus* leaves was different during winter and the following spring. More specifically, in spring leaves showed higher values of Φ_{PSII} (P<0.001) and lower values of Φ_{NPQ} and Φ_{NO} (P<0.005) compared to winter (Fig. 7E, F, G), whereas no significant difference in F_v/F_m between the two seasons was observed (Fig. 7H).

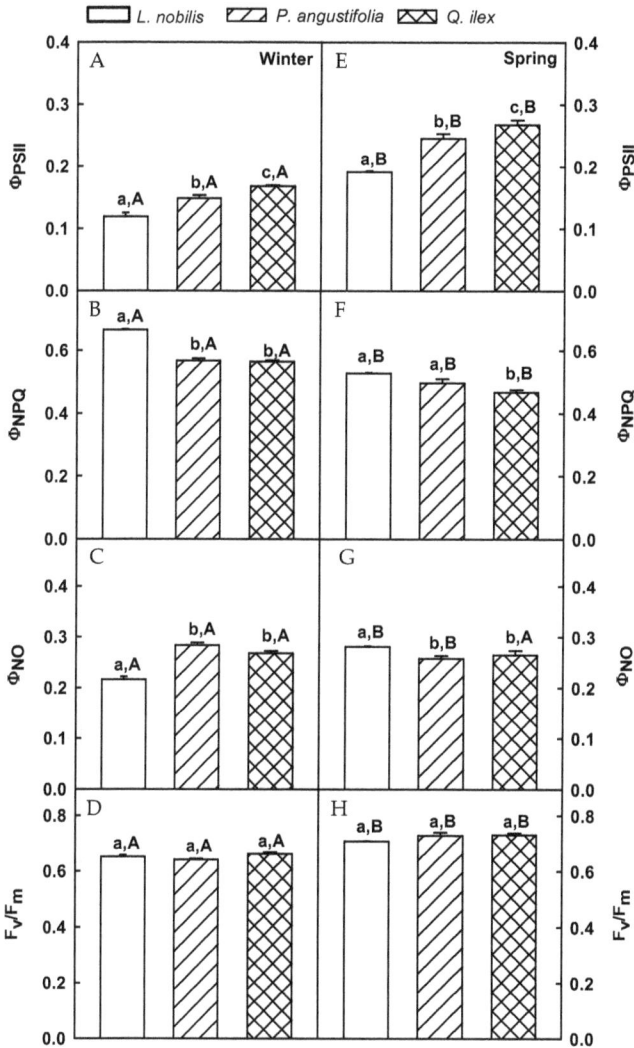

Fig. 6. Quantum yield of linear PSII electron transport (Φ_{PSII}), regulated energy dissipation (Φ_{NPQ}), non-regulated energy dissipation (Φ_{NO}) and maximum PSII photochemical efficiency (F_v/F_m) in mature leaves of *Laurus nobilis, Phillyrea angustifolia* and *Quercus ilex*, during winter and spring. Different letters indicate statistical differences among species (small letters) and between seasons (capital letters). Values are means ± SD (n=8).

The results relative to leaf functional traits and photosynthetic pigment content are reported in the table 2. The analysis of functional leaf traits has evidenced that, as compared to mature leaves, young leaves showed lower values (P<0.05) of leaf area (LA), but no difference in specific leaf area (SLA) and leaf dry matter content (LDMC). Functional leaf traits did not show any difference between mature leaves in both winter and spring campaigns. The total chlorophyll content, chl (a+b), as well as the total carotenoid content, car (x+c), were higher in mature than in (P<0.01) young leaves, that showed a lower (P<0.05) chl a/b ratio. No difference in total chlorophyll and carotenoid content, between winter and spring, in mature leaves was detected.

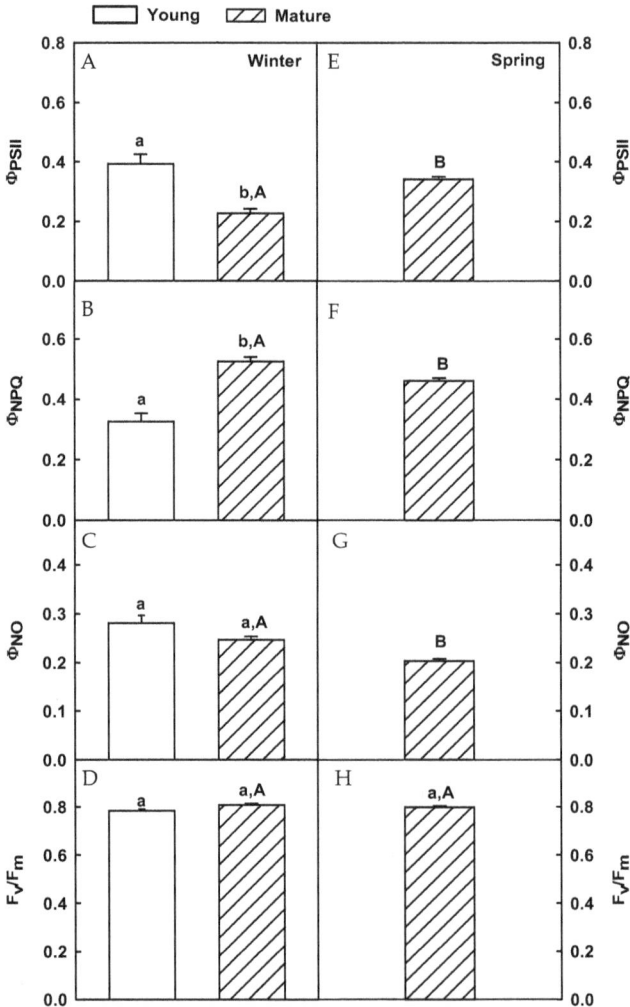

Fig. 7. Quantum yield of linear PSII electron transport (Φ_{PSII}), regulated energy dissipation (Φ_{NPQ}), non-regulated energy dissipation (Φ_{NO}) and maximum PSII photochemical efficiency (F_v/F_m) in C. *incanus* young and mature leaves during winter and in mature leaves during spring. Different letters indicate statistical differences between young and mature leaves (small letters) and between seasons (capital letters). Values are means ± SD (n=6).

| | Winter | | Spring |
	Young leaves	Mature leaves	Mature leaves
LA (cm²)	3.02±0.14 a	8.01±0.28 b	8.32±0.44 b
SLA (cm² g⁻¹ dw)	127.13±11.68 a	120.40±4.96 a	134.03±8.88 a
LDMC (g g⁻¹ wslm)	0.22±0.01 a	0.20±0.02 a	0.21±0.01 a
chl (a+b) (µg cm⁻²)	57.90±1.18 a	76.61±5.8 b	88,01±6 b
car (x+c) (µg cm⁻²)	11.09±0.29 a	14.22±1.05 b	16±2.32 b
Chl a/b	3.03±0.01 a	2.37±0.23 b	2.5±0.34 b

Table 2. Leaf Area (LA), Specific Leaf Area (SLA), Leaf Dry Matter Content (LDMC), total chlorophyll (chl a+b), total carotenoids (car x+c) and chlorophyll a/b ratio in C. *incanus* young and mature leaves during winter and in mature leaves during spring. Data reported are means ± SE (n=6). Different letters indicate statistically significant differences.

4. Discussion

4.1 Young and mature leaves of *Laurus nobilis* L., *Phillyrea angustifolia* L. and *Quercus ilex* L. in winter

In disagreement with data reported in literature for other species (Urban et al., 2008), young leaves of all species showed lower A_N values compared to mature ones, indicating a marked sensitivity to winter temperatures. It is likely to hypothesize that this could be attributable to a reduced capacity of the mesophyll to assimilate CO_2 because no difference in apparent carboxilation efficiency (C_i/C_a) between young and mature leaves was found. The significant differences between the two leaf populations, indicate the higher resistance of mature leaves photosynthetic machinery to low temperature. However, despite photosynthesis reduction, no variation in Φ_{PSII} between young and mature leaves was detected; thus the lower A_N values in young leaves may be due either to limitations in photosynthetic dark reactions or to additional dissipative processes, other than CO_2 assimilation, active in consuming the reductive power of the electron transport chain (*e.g.* photorespiration and/or Mehler reaction). The fluorescence analysis has evidenced that in young leaves the excess of absorbed light was dissipated more by photochemical processes than by thermal dissipation associated to xanthophylls cycle, as indicated by lower Φ_{NPQ} values compared to mature leaves. Although such photochemical processes are useful to protect the photosynthetic apparatus by photoinhibitory damage risks, it is well known that they can lead to an overproduction of reactive oxygen species (ROS). Even if ROS are continuously produced and removed during normal physiological events, when plants experience severe stress conditions, more O_2 molecules are expected to be used as alternative electron acceptors disturbing the ROS production-removal balance and promoting the accumulation of ROS (Osório et al., 2011). Our results indicate that, in young leaves, under winter temperature, a large part of absorbed energy was diverted to non-regulated energy conversion processes (increase in Φ_{NO}) than in mature leaves, a circumstance that favors the production of ROS.

On the contrary, in mature leaves, more absorbed light was dissipated by thermal dissipation processes associated to xanthophylls cycle (higher Φ_{NPQ}). This result is in contrast with data reported by other authors who found a reduction in thermal dissipation by xanthophylls cycle as the leaves expanded (Choinski & Eamus, 2003; Jiang et al., 2005). Our data suggest that leaf age influences the photoprotection mechanisms. More specifically, young and mature leaves regulate in a different way the dissipation of absorbed light energy in order to maintain high the photochemical efficiency. The absence of significant differences in F_v/F_m ratio between the two leaf population indicates that both thermal dissipation and the alternative electron sink and/or additional quenching mechanism(s) are suitable for photoprotection, assuming a similar weight in photoprotection.

Among species, the higher A_N rates in *Q. ilex* compared to *P. angustifolia* and *L. nobilis* in both young and mature leaves indicates *Q. ilex* as the species with more efficient photosynthetic process at low temperature (Ogaya & Peñuelas, 2003). This is likely due to the highest utilization of reductive power of electron transport chain in C fixation rather than in dissipative processes under low temperature. Our data demonstrate that under low temperatures, the strategies utilized to dissipate the excess of absorbed light vary among species. In particular in both young and mature leaves, *L. nobilis*, as compared to *P. angustifolia* and *Q. ilex*, diverts more excitation energy to regulated energy dissipation processes than to non-regulated energy dissipation processes (higher Φ_{NPQ}, lower Φ_{NO}). These different mechanisms seem equally important in maintaining an elevated maximum PSII photochemical efficiency, as confirmed by comparable F_v/F_m ratio in all species.

4.2 Mature leaves of *L. nobilis* L., *P. angustifolia* L. and *Quercus ilex* L. during winter and spring

Equinoctial periods, characterized by the absence of drought and cold stress, are the most favorable seasons for the photosynthetic activity of Mediterranean vegetation (Savè et al., 1999). Data presented in this section are consistent with literature, indeed in spring, compared to winter, high rates of gas exchanges and a better photochemical efficiency were measured for all species. The highest values of A_N and g_{H2O} measured during winter in *Q. ilex*, suggest for this species a better resistance to low temperature (Ogaya & Peñuelas, 2003), differently from *L. nobilis* that showed the lowest photosynthetic activity and stomatal conductance and the highest C_i/C_a ratio. This latter constitutes a proxy tool to evaluate the occurrence of non-stomatal limitations to photosynthesis. In *L. nobilis*, the similar C_i/C_a values found in winter compared to spring, despite the low photosynthetic activity, denote the presence of non-stomatal limitation to photosynthetic process likely due to a reduced activity of Rubisco (Sage & Sharkey, 1987), and/or of other carbon assimilation enzymes (Sassenrath et al., 1990) at low temperatures. The analysis of photosynthetic energy partitioning evidenced that in winter, when net CO_2 assimilation was limited by low temperatures, more absorbed energy was converted into regulated energy dissipation (higher Φ_{NPQ}) compared to spring. On the contrary, in spring when air temperature became favourable for photosynthesis, the absorbed energy was diverted mainly to net CO_2 assimilation (higher Φ_{PSII}) and only a little in non-regulated energy dissipation (low Φ_{NO}). The higher thermal dissipation and the low F_v/F_m values in winter compared to spring were likely the result of a photoprotective mechanisms by which plants cope with winter stress. This strategy is probably based on maintaining PSII primed for energy dissipation and engaged in diurnal energy dissipation throughout the night (Adams et al., 2001).

4.3 *Cistus incanus* L. young and mature leaves in winter

Under winter temperature, *C. incanus* young leaves exhibit a higher photochemical activity than mature leaves. The utilization of reductive power of electron transport in photochemistry reduces the need for the thermal dissipative process, in particular the fraction of the regulated thermal energy dissipation (low Φ_{NPQ} values). Mature leaves showed an opposite tendency. However in both leaf typologies no variation of non-regulated energy dissipation component (Φ_{NO}) was found. High values of Φ_{NPQ} are indicative of a high photoprotective capacity, whereas high values of Φ_{NO} may reflect the inability of a plant to protect itself against photodamage (Klughammer & Schreiber, 2008; Osório et al., 2011). In our opinion, as maximum PSII photochemical efficiency (F_v/F_m) and Φ_{NO} are similar in the two leaf populations, we suppose that the different strategies adopted by young and mature leaves are equally helpful in leaf photoprotection under winter temperatures.

The acclimation of plants in relation to the environmental conditions is expressed, among other factors, also by their leaf characteristics (Bussotti et al., 2008) and photosynthetic pigment adjustments.

Functional leaf traits analyses indicate that, even if specific leaf area (SLA) as well as the leaf dry matter content (LDMC) do not vary between young and mature C. *incanus* leaves, mature leaves present a greater leaf blade and have a higher total chlorophyll and carotenoid contents per unit leaf area. The adjustment of photosynthetic pigment composition in mature leaves could be interpreted as further strategy in order to enhance the light harvest and thus compensate for the reduction in allocation of absorbed light in photochemistry.

4.4 *Cistus incanus* L. mature leaves in winter and spring

The behaviour of C. *incanus* mature leaves differ in winter and spring. The analysis of photochemistry showed that temperatures of 11 °C does not injure the photosynthetic apparatus, but affects significantly its efficiency. Indeed, the low values of Φ_{PSII} evidenced a decline in photochemical activity that may lead to an increase of excitation pressure in photosystem II with important consequence for the plant cells in terms of decrease of intracellular ATP and NADP production. On the other hand, the fraction of the regulated energy dissipation (Φ_{NPQ}) higher in leaves during winter compared to spring, indicates that the regulated thermal dissipation for winter leaves was enhanced under low temperature to compensate for reduced photochemistry. Nevertheless during winter, leaves show also an higher non-regulated energy dissipation in PSII (Φ_{NO}), indicating the occurrence of a stress condition for photosynthetic apparatus (Osório et al., 2011). It is reasonable to hypothesize that leaves during winter cope with low temperature by means of flexible component of thermal energy dissipation and the alternative electron sink and/or additional quenching mechanism(s). These factors may contribute to the high stress resistance of C. *incanus* leaves and allow photosynthetic apparatus to maintain during winter a high maximal PSII photochemical efficiency (F_v/F_m).

The F_v/F_m values found in leaves during winter were close to those reported for winter leaves of other *Cistus* species as well as to those of unstressed plants of other Mediterranean species (Oliveira & Peñuelas, 2001, 2004). In spring, after the return to mild temperatures (*i.e.* 22 °C), an increase of (Φ_{PSII}) was observed.

These results suggest that during February the reduction in photochemistry found at temperatures of 11 °C and at PPFD of about 700 μmol photons m^{-2} s^{-1} (table 1) was due to a downregulation of PSII reaction centres, rather than to an impairment of photosynthetic apparatus. This strategy may represent a safety mechanism against the photoinhibitory damage risk as a consequence of combined effect of low temperature and moderately high irradiances on photosystems. In this view, the lack of significant differences in maximum PSII photochemical efficiency (F_v/F_m), as well as in total chlorophylls and carotenoids content between mature leaves in winter and spring supports this hypothesis, confirming that photochemical apparatus of *C. incanus* remained stable and effective at winter temperatures.

5. Conclusions

The results of the present study indicate that leaf age influences the photoprotection mechanisms. Under saturating irradiance and low winter temperature mature leaves of all evergreen species, by higher CO_2 assimilation rates and higher thermal energy dissipation linked to the flexible component, cope more efficiently with the excess of absorbed light and result to be less sensitive to photoinhibition. On the other hand young leaves utilize the reducing power mainly in processes other than photosynthesis and show higher values of non-regulated energy dissipation in PSII. However both different mechanisms are useful in maintain the maximum PSII photochemical efficiency at comparable values in young and mature leaves.

Among species both young and mature leaves of *Q. ilex* exhibited the highest photosynthetic performance indicating a better resistance to low temperatures.

The comparison between mature leaves in winter and spring shows higher values of net photosynthesis and photochemical efficiency in all evergreen species during spring and a lower contribute of flexible and sustained thermal dissipation in winter. At low temperature, the significant increase of thermal and photochemical processes other than photosynthesis allow mature leaves of evergreen species to maintain an elevated photochemical efficiency, despite the strong reduction of carbon assimilation. Among species, *Q. ilex* showed the best photosynthetic performance under winter, indicating a better acclimation capability of photosynthetic apparatus.

In *C. incanus* species, during winter, young leaves showed a higher photochemical efficiency than mature leaves. The increase in photochemistry leads to a reduction of thermal dissipative processes. On the other hand, the mature leaves exhibited an opposite tendency. However, both strategies are useful in leaf photoprotection under winter since maximum PSII photochemical efficiency is high and similar in the two leaf populations.

The comparison between mature leaves in winter and spring has evidenced a lower quantum yield of PSII linear electron transport and an increase of regulated thermal dissipation processes during winter. The recovery of photochemical activity in spring under mild temperature, indicates that the drop in photochemistry in winter was due to the balance between energy absorbed and dissipated at PSII level rather than to an impairment of photosynthetic apparatus. In this context, the higher thermal dissipation in winter compensate for the reduced photochemistry, allowing maximum PSII photochemical efficiency to remain unchanged compared to spring. This may be interpreted as a dynamic regulatory process protecting the photosynthetic apparatus from severe damage by excess light at low temperature.

6. Acknowledgments

The authors are grateful to Prof. Mazzarella of the Department of Geophysic and Vulcanology (University Federico II Naples) for providing meteorological data and to Corpo

Forestale dello Stato of Sabaudia (Latina, Italy) for supplying the plants used in the experiments.

7. References

Abril, M. & Hanano, R. (1998). Ecophysiological responses of three evergreen woody Mediterranean species to water stress. *Acta oecologica*, Vol. 19, pp. 377-387, ISSN 1146-609X

Adams, W.W.; Demmig-Adams, B., Rosentiel, T.N., & Ebbert, V. (2001). Dependence of photosynthesis and energy dissipation activity upon growth form and light environment during the winter. *Photosynthesis Research*, Vol. 67, pp. 51–62, ISSN (printed) 0166-8595

Arena, C.; Vitale, L. & Virzo De Santo A. (2008). Photosynthesis and photoprotective strategies in *Laurus nobilis* L. and *Quercus ilex* L. under summer drought and winter cold. *Plant Biosystems*, Vol. 142, pp. 472-479, ISSN 1126-3504

Aronne, G. & De Micco, V. (2001). Seasonal dimorphism in the Mediterranean *Cistus incanus* L. subsp. *incanus*. *Annals of Botany*, Vol. 87, pp. 789-794, ISSN 0305-7364

Baker, N.R. (1994). Chilling stress and photosynthesis. *Causes of Photooxidative Stress and Amelioration of Defences Systems in Plants* (eds C.H Foyer & P.M. Mullineaux), pp 127-154. CRC Press, Boca Raton, Florida.

Bertamini, M. & Nedunchezhian, N. (2003). Photoinhibition of photosynthesis in mature and young leaves of grapevine (*Vitis vinifera* L.). *Plant Science*, Vol. 164, No. 4, pp. 635-644, ISSN 0168-9452

Bilger, W. & Björkman, O. (1990). Role of the xanthophyll cycle in photoprotection elucidated by measurements of light-induced absorbance changes, fluorescence and photosynthesis in leaves of *Hedera canariensis*. *Photosynthesis Research*, Vol. 25, pp. 173-185, ISSN (printed) 0166-8595

Blondel, J. & Aronson, J. (1999). Biology and Wildlife of the Mediterranean Region. ISBN 0 19 850035 1, Oxford University Press, New York.

Boese, S.R. & Huner, N.P.A. (1990). Effect of growth temperature and temperature shift on spinach leaf morphology and photosynthesis. *Plant Physiology*, Vol. 94, pp. 1830-1836, ISSN (printed) 0032-0889

Bussotti, F. (2008). Functional leaf traits, plant communities and acclimation processes in relation to oxidative stress in trees: a critical overview. *Global Change Biology*, Vol. 14, pp. 2727–2739, ISSN (printed) 1354-1013

Caemmerer, S. von & Farquhar, G.D. (1981). Some relationship between the biochemistry of photosynthesis and the gas exchange of leaves. *Planta*, Vol. 153, pp. 376-387, ISSN (printed) 0032-0935

Castro-Díez, P.; Villar-Salvador, P., Pérez-Rontomé, C., Maestro-Martínez, M. & Montserrat-Martí, G. (1998) Leaf morphology, leaf chemical composition and stem xylem characteristics in two *Pistacia* (Anarcardiaceae) species along climatic gradient. *Flora*, Vol. 193, pp. 195-202, ISSN 0367-2530

Choinski, Jr. & Eamus, D. (2003) Changes in photosynthesis during leaf expansion in *Corymbia gummifera*. *Australian Journal of Botany*, Vol 51, pp. 111-118, ISSN (printed) 0067-1924

Cornelissen, J.H.C.; Lavorel, S., Garnier, E., Díaz, S., Buchmann, N., Gurvich, D.E., Reich, P.B., ter Steege, H., Morgan, H.D., van der Heijden, M.G.A., Pausas, J.G. & Poorter

H. (2003). Handbook of protocols for standardised and easy measurements of plant functional traits worldwide. *Australian Journal of Botany*, Vol. 51, pp. 335-380, ISSN (printed) 0067-1924

D'Ambrosio, N.; Arena, C. & Virzo De Santo, A. (2006). Temperature response of photosynthesis, excitation energy dissipation and alternative electron sinks to carbon assimilation in *Beta vulgaris* L. *Environmental and Experimental Botany*, Vol. 55, pp. 248-257, ISSN 0098-8472

Demming-Adams, B.; Adams, W. W., Barker, D. H., Logan, B. A., Bowling, D. R. & Verhoeven, A. S. (1996). Using chlorophyll fluorescence to assess the fraction of absorbed light allocated to thermal dissipation of excess excitation. *Physiologia Plantarum*, Vol. 98, pp. 253-264, ISSN (printed) 0031-9317

Garnier, E.; Shipley, B., Roumet, C. & Laurent, G. (2001). A standardized protocol for the determination of specific leaf area and leaf dry matter content. *Functional Ecology* Vol. 15, pp. 688-695. ISSN (printed) 0269-8463

Garcìa-Plazaola, J.L.; Artetxe, U. & Becerril, J.M. (1999). Diurnal changes in antioxidant and carotenoid composition in Mediterranean schlerophyll tree *Quercus ilex* (L) during winter. *Plant Science*, Vol. 143, pp. 125-133, ISSN 0168-9452

Garcìa-Plazaola, J.L.; Hernández, A. & Becerril, J.M. (2000). Photoprotective responses to winter stress in evergreen Mediterranean ecosystems. *Plant Biology*, Vol. 2, pp. 530-535, ISSN 1438-8677

Genty, B.; Briantais, J. M. & Baker, N.R. (1989). The relationship between the quantum yield of photosynthetic electron transport and quenching of chlorophyll fluorescence. *Biochimica and Biophysica Acta*, Vol. 990, pp. 87-92, ISSN 0006-3002

Gratani, L. & Ghia, E. (2002). Adaptive strategy at the leaf level of *Arbutus unedo* L. to cope with Mediterranean climate. *Flora*, Vol. 197, pp. 275-284, ISSN 0367-2530

Gratani, L. & Varone, L. (2004). Adaptive photosynthetic strategies of the Mediterranean maquis species according to their origin. *Photosynthetica*, Vol. 42, No.4, pp. 551-558, ISSN (printed) 0300-3604

Huner, N.P.A.; Palta, J.P., Li, P.H. & Carter, J.V. (1981). Anatomical changes in leaves of *Puma rye* in response to growth at cold-hardening temperatures. *Botanical Gazette* Vol. 142, pp. 55-62, ISSN 0006-8071

Hutchinson, R.S.; Groom, Q. & Ort, D.R. (2000). Differential effects of chilling-induced photooxidation on the redox regulation of photosynthetic enzymes. *Biochemistry*, Vol. 39, pp. 6679-6688, ISSN (printed) 0006-2960

Jiang, C.D.; Li, P.M., Gao, H.Y., Zou, Q., Jiang, G.M. & Li, L.H. (2005). Enhanced photiprotection at the early stage of leaf expansion in field-grown soybean plants. *Plant Science*, Vol. 168, pp. 911-919, ISSN (printed) 0168-9452

Kramer, D.M.; Johnson, G.; Kiirats, O. & Edwards, G.E. (2004). New fluorescence parameters for the determination of Q_A redox state and excitation energy fluxes. *Photosynthesis Research*, Vol. 79, pp. 209-218, ISSN (printed) 0166-8595

Klughammer, C. & Schreiber U. (2008). Complementary PS II quantum yields calculated from simple fluorescence parameters measured by PAM fluorometry and the Saturation Pulse method. *PAM Application Notes*, Vol. 1, pp. 27 -35

Larcher, W. (2000). Temperature stress and survival ability of Mediterranean sclerophyllous plants. *Plant Biosystems*, Vol. 134, pp. 279-295, ISSN: 1126-3504

Long, S.; Humphries, S. & Falkowski, P.G. (1994) Photoinhibition of photosynthesis in nature. *Annual Review of Plant Physiology and Plant molecular Biology*, Vol. 45, pp. 633-662, ISSN 1040-2519

Mitrakos, K.A. (1980). A theory for Mediterranean plant life. *Acta Oecologica*, Vol. 1, pp. 245-252, ISSN 1146-609X

Niyogi, K.K. (2000). Safety valves of photosynthesis. *Current Opinion in Plant Biology*, Vol. 3, pp. 445-460, ISSN 1369-5266

Ogaya, R. & Peñuelas, J. (2003). Comparative seasonal gas exchange and chlorophyll fluorescence of two dominant woody species in a Holm Oak forest. *Flora*, Vol. 198, pp. 132-141, ISSN (printed) 0367-2530

Oliveira, G. & Peñuelas, J. (2000). Comparative photochemical and phenomorphological responses to winter stress of an evergreen (*Quercus ilex* L.) and a semi-deciduous (*Cistus albidus* L.) Mediterranean woody species. *Acta Oecologica*, Vol. 21, pp. 97-107, ISSN 1146-609X

Oliveira, G. & Peñuelas, J. (2001). Allocation of absorbed light energy into photochemical and dissipation in a semi-deciduous and an evergreen Mediterranean woody species during winter. *Australian Journal of Plant Physiology*, Vol. 28, pp. 471-480, ISSN 0310-7841

Oliveira, G. & Peñuelas, J. (2002). Comparative protective strategies of *Cistus albidus* and *Quercus ilex* facing photoinhibitory winter conditions. *Environmental and Experimental Botany*, Vol. 47, pp. 281-289, ISSN 0098-8472

Oliveira, G. & Peñuelas, J. (2004). Effects of winter cold stress on photosynthesis and photochemical efficiency of PSII of the Mediterranean *Cistus albidus* L. and *Quercus ilex* L. *Plant Ecology*, Vol. 174, pp. 179-191, ISSN (printed) 1385-0237

Ort, D.R. & Baker, N.R. (2002). A photoprotective role for O_2 as an alternative electron sink in photosynthesis? *Current Opinion in Plant Biology*, Vol. 5, pp. 193-198, ISSN (printed) 1369-5266

Osório, M. L., Osório, J., Vieira, A.C., Gonçalves, S. & Romano, A. (20011). Influence of enhanced temperature on photosynthesis, photooxidative damage, and antioxidant strategies in Ceratonia siliqua L. seedlings subjected to water deficit and rewatering. *Photosynthetica*, Vol. 49 (1), pp. 3-12, ISSN (printed) 0300-3604

Oxborough, K. & Baker, N.R. (1997). Resolving chlorophyll a fluorescence images of photosynthetic efficiency into photochemical and non-photochemical components - calculation of q_p and F_v'/F_m' without measuring F_0'. *Photosynthesis Research*, Vol. 54, pp. 135-142, ISSN (printed) 0166-8595

Poorter, H. & Garnier, E. (1999). Ecological significance of inherent variation in relative growth rate and its components. *Handbook of functional plant ecology* (eds F.I. Pugnaire & F. Valladares), pp. 81-120, New York, Marcel Dekker.

Powles, S.B. (1984). Photoinhibition of photosynthesis induced by visible light. *Annual Review of Plant Physiology*, Vol. 35, 15-44, ISSN 0066-4294

Rhizopoulou, S.; Angelopulos, K. & Mitrakos, K. (1989). Seasonal variations of accumulated ions, soluble sugars and solute potential in the expressed sap from leaves of evergreen sclerophyll species. *Acta Oecologica/Oecologia Plantarum*, Vol. 10, pp. 311-319, ISSN 1146-609X

Ryser, P. & Urbas, P. (2000). Ecological significance of leaf life span among Central European grass species. *Oikos*, Vol. 91, pp. 41–50, ISSN (printed) 0030-1299

Sage, R.F. & Sharkey, T.D. (1987). The effect of temperature on the occurrence of O_2 and CO_2 insensitive photosynthesis in field grown plants. *Plant Physiology*, Vol. 84, pp. 658–664, ISSN (printed) 0032-0889

Sánchez-Blanco, M. J.; Rodríguez, M. J., Morales, M. A., Ortuño, M. F. & Torrecillas A. (2002). Comparative growth and water relations of *Cistus albidus* and *Cistus monspeliensis* plants during water deficit conditions and recovery. *Plant Science*, Vol. 162, pp. 107-113, ISSN 0168-9452

Sassenrath, G.F.; Ort, D.R. & Portis, A.R. Jr (1990). Impaired reductive activation of stromal bisphosphatases in tomato leaves following high light. *Archives of Biochemistry and Biophysics*, Vol. 282, pp. 302-30, ISSN 0003-9861

Savé, R.; Castell, C. & Terradas J. (1999). Gas exchange and water relations, In: *Ecology Mediterranean Evergreen Oak Forest. Ecological Studies*, F. Rodà, J. Retana, C.A. Gracia, J. Bellot (eds.), Vol. 137, pp. 135-147, Spring Verlag, Berlin & Heidelberg, ISSN 0070-8356

Shirke, P.A. (2001) Leaf photosynthesis, dark respiration an fluorescence as influenced by leaf age in an evergreen tree, *Prosopis Juliflora*. *Photosynthetica*, Vol. 311, No. 7, pp. 305-311, ISSN (printed) 0300-3604

Tattini, M.; Gravano, E., Pinelli, P., Mulinacci, N. & Romani A. (2000). Flavonoids accumulate in leaves and glandular trichomes of *Phillyrea latifolia* exposed to excess solar radiation. *New Phytologist*, Vol. 148, pp. 69-77, ISSN (printed) 0028- 646X

Teulat, B.; Monneveux, P., Wery, J., Borries, C., Sourys, I., Charrier, A. & This, D. (1997). Relationships between relative water content and growth parameters under water stress in barley: a QTL study. *New Phytologist*, Vol. 137, pp. 99–107, ISSN (printed) 0028- 646X

Tjus, S.E.; Moller, B.L. & Scheller, H.V. (1998). Photosystem I is an early target of photoinhibition in barley illuminated at chilling temperatures. *Plant Physiology*, Vol. 116, pp. 755-764, ISSN (printed) 0032-0889

Urban, O.; Sprtová, M., Kosvancová, M., Tomásková, I., Lichtenthaler, H.K. & Marek, M.V. (2008). Comparison of photosynthetic induction and transient limitations during the induction phase in young and mature leaves from three poplar clones. *Tree Physiology*, Vol. 28, pp. 1189-1197, ISSN (printed) 0829-318X

Varone, L. & Gratani, L. (2007). Physiological response of eight Mediterranean maquis species to low air temperatures during winter. *Photosynthetica*, Vol. 45, No.3, pp. 385-391, ISSN (printed) 0300-3604

Werner, C.; Correia, O. & Beyschlag, W. (1999). Two different strategies of Mediterranean macchia plants to avoid photoinhibitory damage by excessive radiation levels during summer drought. *Acta Oecologica*, Vol. 20, pp. 15-23, ISSN 1146-609X

Wilson, P.J.; Thompson, K. and Hodgson, J.G. (1999). Specific leaf area and leaf dry matter content as alternative predictors of plant strategies. *New Phytologist*, Vol. 143, pp. 155-162, ISSN (printed) 0028- 646X

10

The Core- and Pan-Genomes of Photosynthetic Prokaryotes

Jeffrey W. Touchman and Yih-Kuang Lu
Arizona State University
USA

1. Introduction

Genome sequencing projects are revealing new information about the distribution and evolution of photosynthesis and phototrophy, particularly in prokaryotes. Although coverage of the five phyla containing photosynthetic prokaryotes (Chlorobi, Chloroflexi, Cyanobacteria, Proteobacteria and Firmicutes) is limited and uneven, full genome sequences are now available for 82 strains from these phyla. In this chapter, we present data and comparisons that reflect recent advances in phototroph biology as a result of insights from genome sequencing. By performing a comprehensive analysis of the core-genome (the pool of genes shared by all phototrophic prokaryotes) and pan-genome (the global gene repertoire of all phototrophic prokaryotes: core genome + dispensable genome) along with available biological data for each organism, we address the following key questions: 1) what are the principal drivers behind the evolution and distribution of phototrophy and 2) how do environmental parameters correlate with genomic content to define niche partitioning and ecotype distributions in photic environments?

Over a decade has passed since the first phototrophic prokaryote, the cyanobacterium *Synechocystis* sp. PCC 6803, was completely sequenced (Kaneko et al., 1996). Since then, availability of an increasing diversity of newly sequenced species is accumulating in public databases at a sustained pace and there is little indication that this trend will level off in the near future (Raymond & Swingley, 2008). A deepening archive of complete genomes has enabled comparative genomic analyses, which has heavily influenced our views of genome evolution and uncovered the extent of gene sharing between organisms (Pallen & Wren, 2007). The analysis of pan-and core-genomes in particular allows us to link genome content to the relationship of organisms to one another and to their physical surroundings. For example, a low pan-genome diversity due to extensive overlap of metabolic function among groups of bacteria could reflect shared environmental habitats and resource utilization, while distinctive species that adapt to disparate environments would be expected to have a high pan-genome diversity. This approach was first developed by Tettelin et al. (2005) and Hogg et al. (2007) for tracking the number of unique genes among multiple strains of *Streptococcus agalactiae* and *Haemophilus influenzae*, respectively. Such analysis resulted in the determination of core-genes that encode functions related to the basic metabolism and phenotype of the species, and a pan-genome that consists of dispensable or unique genes that impart specific functionalities to individual strains.

Within prokaryotes, photosynthetic capability is present within five major groups, which include heliobacteria, green filamentous bacteria (*Chloroflexus sp.*), green sulfur bacteria (*Chlorobium sp.*), *Proteobacteria,* and *Cyanobacteria* (Blankenship, 1992; Gest & Favinger, 1983; Olson & Pierson, 1987; Vermaas, 1994). While only *Cyanobacteria,* which contain two distinct reaction centers linked to each other, are capable of oxygenic photosynthesis, other photosynthetic bacteria primarily carry out anoxygenic photosynthesis with a single reaction center. Traditionally, the phylogenetic relationship of these five distinct photosynthetic groups has been constructed by comparing sequences of the small subunit 16S rRNA gene (Ludwig & Klenk, 2001). But the use of the 16S rRNA gene is unable to resolve the relationships among these phototrophs with confidence, which is central to understanding their evolution. For example, phylogenetic trees based on a comparison of different combinations of 527 shared genes amongst all five photosynthetic prokaryote groups shows that no less than 15 different tree topologies can be constructed depending on the subset of genes used in the analysis, only one of which matches the traditional 16S rDNA tree (Raymond et al., 2002). In fact, comparing just those genes involved in photosynthesis supports no coherent relationship among the different photosynthetic bacteria either, indicating that such genes may have been subjects of lateral gene transfers (ibid).

Recent genome sequencing efforts have made whole genome data available for many more representatives of each of the five phyla of bacteria with photosynthetic members. To resolve the complicated relationship between bacterial phototrophy and evolutionary history, we describe an analysis of the 82 fully-sequenced photosynthetic prokaryotes to construct the pan- and core-genomes across all available strains. We present results showing various gene-based indicators of the relationship between genome and phenotype among these organisms. Not surprisingly, our findings describe new relationships between gene content and environmental habitat. These results add to a complete gene-based functional annotation of the phototrophic prokaryotes, and set the groundwork for continuing studies on genetic and evolutionary dynamics of this important photosynthetic community.

2. Whole-genome analysis of phototrophic prokaryotes

The list and summary details of 82 fully-sequenced photosynthetic species used in this study are shown in Table 1 (Liolios et al., 2006). Every species exhibits common characteristics with other relatives in the same phylum. For example, the *Chlorobia* and heliobacteria (*Firmicutes*) are strictly anaerobic while the *Chloroflexia* and *Proteobacteria* are facultatively anaerobic. The *Chloroflexia* are alkali-trophic thermophiles whereas other phylyl members are neutral pH mesophiles. Genome size is generally uniform among the *Chlorobia* and *Chloroflexia*, but varies widly among the *Cyanobacteria* and *Proteobacteria*. Furthermore, both *Chloroflexia* and *Proteobacteria* possess a pheophytin-quinone reaction center, while *Heliobacteria* and *Chlorobia* use an iron-sulfur reaction center. *Cyanobacteria* exclusively possesses two types of reaction centers. Both *Chlorobia* and *Cyanobacteria* are two phyla comprised entirely of photosynthetic representatives. Although most of the photosynthetic species are free-living organisms, *Nostoc sp.* PCC 7120, *Nostoc punctiforme* PCC 73102, and *Acaryochloris marina* MBIC11017 in the *Cyanobacteria* and *Bradyrhizobium* BTAi1, ORS278 and some *Methylobacterium* strains in the *Proteobacteria* form a mutual relationship with terrestrial plants and coral. The *Heliobacteria* (e.g., *Heliobacterium modesticaldum*) are the only photosynthetic members of the *Firmicutes*. The genome of *Heliobacillus mobilis*, the strain most studied biochemically, still remains proprietary and was not included in our analysis.

GENUS	ORGANISM NAME	SYMBOL	SIZE(Mb)	REACTION CENTER	OXYGEN REQUIREMENT	TEMPERATURE RANGE	PH	BIOTIC RELATIONSHIPS	MOTILITY	GENOME DATA
Heliobacterium	Heliobacterium modesticaldum Ice1	heliom	3.1	Fe-S	Obligate anaerobe	Thermophile	6 - 7	Free living	Motile	NC_010337
Chlorobaculum	Chlorobaculum parvum NCIB 8327	cpnci	2.3	Fe-S	Anaerobe	Mesophile	6.5 - 7.0	Free living	Nonmotile	NC_011027
Chlorobaculum	Chlorobaculum tepidum TLS	cttls	2.2	Fe-S	Obligate anaerobe	Thermophile	6.9 - 7.0	Free living	Nonmotile	NC_002932
Chlorobium	Chlorobium chlorochromatii CaD3	cccad	2.6	Fe-S	Anaerobe	Mesophile	7.0 - 7.3	Free living	Motile	NC_007514
Chlorobium	Chlorobium limicola DSM 245	cl245	2.8	Fe-S	Anaerobe	Thermophile	6.8 - 7.1	Free living	Nonmotile	NC_010803
Chlorobium	Chlorobium phaeovibrioides DSM 265	cp265	2	Fe-S	Facultative	Mesophile	6.8 - 7.0	Free living	Nonmotile	NC_009337
Chlorobium	Chlorobium phaeobacteroides DSM 266	cp266	3.1	Fe-S	Facultative	Mesophile	6.7 - 7.0	Free living	Nonmotile	NC_008639
Chlorobium	Chlorobium phaeobacteroides BS1	cpbs1	2.7	Fe-S	Facultative	Mesophile	6.7 - 7.0	Free living	Nonmotile	NC_010831
Chloroherpeton	Chloroherpeton thalassium ATCC 35110	ct35110	3.3	Fe-S	Facultative	Mesophile	6.8	Free living	Nonmotile	NC_011026
Pelodictyon	Pelodictyon luteolum DSM 273	cl273	2.4	Fe-S	Anaerobe	Mesophile	6.5 - 7.0	Free living	Nonmotile	NC_007512
Pelodictyon	Pelodictyon phaeoclathratiforme BU-1	ppbu1	3	Fe-S	Anaerobe	Mesophile	6.5 - 7.0	Free living	Nonmotile	NC_011060
Prosthecochloris	Prosthecochloris aestuarii DSM 271	pa271	2.5	Fe-S	Anaerobe	Mesophile	6.5 - 7.0	Free living	Nonmotile	NC_011059 NC_011061
Chloroflexus	Chloroflexus aggregans DSM 9485	ca9485	4.7	Pheo-Q	Facultative	Thermophile	8 - 9	Free living	Motile	NC_011831
Chloroflexus	Chloroflexus aurantiacus J-10-fl	caj10	5.3	Pheo-Q	Facultative	Thermophile	8 - 9	Free living	Motile	NC_010175
Chloroflexus	Chloroflexus sp. Y-400-fl	cy400	5.3	Pheo-Q	Facultative	Thermophile	8 - 9	Free living	Motile	NC_012032
Roseiflexus	Roseiflexus castenholzii DSM 13941	rc13941	5.7	Pheo-Q	Facultative	Thermophile	7.8 - 8.2	Free living	Motile	NC_009767
Roseiflexus	Roseiflexus sp. RS-1	rrs1	5.8	Pheo-Q	Facultative	Thermophile	7.8 - 8.3	Free living	Motile	NC_009523
Anabaena	Anabaena variabilis ATCC 29413	av29413	6.4	Fe-S, Pheo-Q	Aerobe	Mesophile	6 - 8	Free living	Motile	NC_007410 NC_007411 NC_007412 NC_007413 NC_014030
Nostoc	Nostoc sp. PCC 7120	n7120	6.4	Fe-S, Pheo-Q	Aerobe	Mesophile	6 - 8	Symbiotic	Motile	NC_003240 NC_003241 NC_003267 NC_003270 NC_003270 NC_003272 NC_003273 NC_003276
Nostoc	Nostoc punctiforme PCC 73102	n73102	8.2	Fe-S, Pheo-Q	Aerobe	Mesophile	6 - 8	Symbiotic	Motile	NC_010628 NC_010629 NC_010630 NC_010631 NC_010632 NC_010633
Prochlorococcus	Prochlorococcus marinus subsp marinus str. CCMP1375	pm1375	1.8	Fe-S, Pheo-Q	Aerobe	Mesophile	6 - 8	Free living	Nonmotile	NC_005042
Prochlorococcus	Prochlorococcus marinus subsp pastoris str. CCMP1986	pm1986	1.7	Fe-S, Pheo-Q	Aerobe	Mesophile	6 - 8	Free living	Nonmotile	NC_005072
Prochlorococcus	Prochlorococcus marinus str. MIT 9211	pm9211	1.7	Fe-S, Pheo-Q	Aerobe	Mesophile	6 - 8	Free living	Nonmotile	NC_009976
Prochlorococcus	Prochlorococcus marinus str. MIT 9215	pm9215	1.7	Fe-S, Pheo-Q	Aerobe	Mesophile	6 - 8	Free living	Nonmotile	NC_009840
Prochlorococcus	Prochlorococcus marinus str. MIT 9301	pm9301	1.6	Fe-S, Pheo-Q	Aerobe	Mesophile	6 - 8	Free living	Nonmotile	NC_009091
Prochlorococcus	Prochlorococcus marinus str. MIT 9303	pm9303	2.7	Fe-S, Pheo-Q	Aerobe	Mesophile	6 - 8	Free living	Nonmotile	NC_008820
Prochlorococcus	Prochlorococcus marinus str. MIT 9312	pm9312	1.7	Fe-S, Pheo-Q	Aerobe	Mesophile	6 - 8	Free living	Nonmotile	NC_007577
Prochlorococcus	Prochlorococcus marinus str. MIT 9313	pm9313	2.4	Fe-S, Pheo-Q	Aerobe	Mesophile	6 - 8	Free living	Nonmotile	NC_005071
Prochlorococcus	Prochlorococcus marinus str. MIT 9515	pm9515	1.7	Fe-S, Pheo-Q	Aerobe	Mesophile	6 - 8	Free living	Nonmotile	NC_008817
Prochlorococcus	Prochlorococcus marinus str. AS9601	pm9601	1.7	Fe-S, Pheo-Q	Aerobe	Mesophile	6 - 8	Free living	Nonmotile	NC_008816
Prochlorococcus	Prochlorococcus marinus subsp pastoris str. NATL1A	pmnatl1	1.9	Fe-S, Pheo-Q	Aerobe	Mesophile	6 - 8	Free living	Nonmotile	NC_008819
Prochlorococcus	Prochlorococcus marinus subsp pastoris str. NATL2A	pmnatl2	1.8	Fe-S, Pheo-Q	Aerobe	Mesophile	6 - 8	Free living	Nonmotile	NC_007335
Synechococcus	Synechococcus sp. PCC 7002	s7002	3.0	Fe-S, Pheo-Q	Facultative	Mesophile	6 - 8	Free living	Motile	NC_010475 NC_010476 NC_010477 NC_010478 NC_010479 NC_010480 NC_010474
Synechococcus	Synechococcus sp. CC9311	s9311	2.6	Fe-S, Pheo-Q	Aerobe	Mesophile	6 - 8	Free living	Motile	NC_008319
Synechococcus	Synechococcus sp. CC9605	s9605	2.5	Fe-S, Pheo-Q	Aerobe	Mesophile	6 - 8	Free living	Motile	NC_007516
Synechococcus	Synechococcus sp. CC9902	s9902	2.2	Fe-S, Pheo-Q	Aerobe	Mesophile	6 - 8	Free living	Motile	NC_007513
Synechococcus	Synechococcus sp. JA-2-3Ba	sja23b	3.0	Fe-S, Pheo-Q	Facultative	Thermophile	6 - 8	Free living	Motile	NC_007775
Synechococcus	Synechococcus sp. JA-3-3Ab	sja33a	2.9	Fe-S, Pheo-Q	Facultative	Thermophile	6 - 8	Free living	Motile	NC_007770
Synechococcus	Synechococcus sp. RCC307	srcc307	2.2	Fe-S, Pheo-Q	Facultative	Mesophile	6 - 8	Free living	Motile	NC_009482
Synechococcus	Synechococcus sp. WH 7803	swh7803	2.4	Fe-S, Pheo-Q	Aerobe	Mesophile	6 - 8	Free living	Motile	NC_009481
Synechococcus	Synechococcus sp. WH 8102	swh8102	2.4	Fe-S, Pheo-Q	Aerobe	Mesophile	6 - 8	Free living	Motile	NC_005070
Synechococcus	Synechococcus elongatus PCC 6301	se6301	2.7	Fe-S, Pheo-Q	Facultative	Mesophile	6 - 8	Free living	Motile	NC_006576
Synechococcus	Synechococcus elongatus PCC 7942	se7942	2.7	Fe-S, Pheo-Q	Facultative	Mesophile	6 - 8	Free living	Motile	NC_007604 NC_007595
Synechocystis	Synechocystis sp. PCC 6803	s6803	3.6	Fe-S, Pheo-Q	Facultative	Mesophile	6 - 8	Free living	Motile	NC_000911 NC_005231 NC_005232 NC_005229 NC_005230
Acaryochloris	Acaryochloris marina MBIC11017	ammbic	6.5	Fe-S, Pheo-Q	Aerobe	Mesophile	6 - 8	Symbiotic	Nonmotile	NC_009925 NC_009926 NC_009927 NC_009928 NC_009929 NC_009930 NC_009931 NC_009932 NC_009933 NC_009934
Cyanothece	Cyanothece sp. ATCC 51142	c51142	5.5	Fe-S, Pheo-Q	Facultative	Mesophile	6 - 8	Free living	Nonmotile	NC_010546 NC_010542 NC_010543 NC_010539 NC_010547 NC_010541
Cyanothece	Cyanothece sp. PCC 7424	c7424	5.9	Fe-S, Pheo-Q	Facultative	Mesophile	6 - 8	Free living	Nonmotile	NC_011729 NC_011737 NC_011730 NC_011732 NC_011733 NC_011734 NC_011738
Cyanothece	Cyanothece sp. PCC 7425	c7425	5.4	Fe-S, Pheo-Q	Facultative	Mesophile	6 - 8	Free living	Nonmotile	NC_011884 NC_011885 NC_011882 NC_011880
Cyanothece	Cyanothece sp. PCC 8801	c8801	4.7	Fe-S, Pheo-Q	Aerobe	Mesophile	6 - 8	Free living	Nonmotile	NC_011726 NC_011723 NC_011727 NC_011721
Cyanothece	Cyanothece sp. PCC 8802	c8802	4.7	Fe-S, Pheo-Q	Aerobe	Mesophile	6 - 8	Free living	Nonmotile	NC_013161 NC_013160 NC_013167 NC_013168 NC_013163
Gloeobacter	Gloeobacter violaceus PCC 7421	gv7421	4.7	Fe-S, Pheo-Q	Aerobe	Mesophile	6 - 8	Free living	Nonmotile	NC_005125
Microcystis	Microcystis aeruginosa NIES-843	ma843	5.8	Fe-S, Pheo-Q	Aerobe	Mesophile	6 - 8	Free living	Nonmotile	NC_010296
Thermosynechococcus	Thermosynechococcus elongatus BP-1	tebp1	2.6	Fe-S, Pheo-Q	Aerobe	Thermophile	6 - 8	Free living	Nonmotile	NC_004113
Trichodesmium	Trichodesmium erythraeum IMS101	teims	7.8	Fe-S, Pheo-Q	Aerobe	Mesophile	6 - 8	Free living	Motile	NC_008312
Methylocella	Methylocella silvestris BL2	msbl2	4.3	Pheo-Q	Aerobe	Psychrotolerant	5 - 6.5	Free living	Nonmotile	NC_011666
Bradyrhizobium	Bradyrhizobium sp. BTAi1	bbtai	8.3	Pheo-Q	Facultative	Mesophile	6 - 8	Symbiotic	Motile	NC_009485 NC_009475
Bradyrhizobium	Bradyrhizobium sp. ORS278	bor278	7.5	Pheo-Q	Facultative	Mesophile	6 - 8	Symbiotic	Motile	NC_009445
Rhodopseudomonas	Rhodopseudomonas palustris BisA53	rpa53	5.5	Pheo-Q	Facultative	Mesophile	6 - 8	Free living	Motile	NC_008435
Rhodopseudomonas	Rhodopseudomonas palustris BisB18	rpb18	5.5	Pheo-Q	Facultative	Mesophile	6 - 8	Free living	Motile	NC_007925
Rhodopseudomonas	Rhodopseudomonas palustris BisB5	rpb5	4.9	Pheo-Q	Facultative	Mesophile	6 - 8	Free living	Motile	NC_007958
Rhodopseudomonas	Rhodopseudomonas palustris CGA009	rpcga	5.5	Pheo-Q	Facultative	Mesophile	6 - 8	Free living	Motile	NC_005296 NC_005297
Rhodopseudomonas	Rhodopseudomonas palustris HaA2	rpha	5.3	Pheo-Q	Facultative	Mesophile	6 - 8	Free living	Motile	NC_007778
Rhodopseudomonas	Rhodopseudomonas palustris TIE-1	rptie	5.7	Pheo-Q	Obligate aerobe	Mesophile	6 - 8	Free living	Motile	NC_011004
Methylobacterium	Methylobacterium sp. 4-46	m446	7.7	Pheo-Q	Aerobe	Mesophile	6 - 8	Free living	Motile	NC_010511 NC_010374 NC_010373
Methylobacterium	Methylobacterium chloromethanicum CM4	mccm4	5.8	Pheo-Q	Aerobe	Mesophile	6 - 8	Free living	Motile	NC_011757 NC_011760 NC_011758
Methylobacterium	Methylobacterium extorquens AM1	meam1	6.9	Pheo-Q	Aerobe	Mesophile	6 - 8	Free living	Motile	NC_012808 NC_012807 NC_012809 NC_012810 NC_012811
Methylobacterium	Methylobacterium extorquens DM4	medm4	6.1	Pheo-Q	Aerobe	Mesophile	6 - 8	Free living	Motile	NC_012988 NC_012989 NC_012987
Methylobacterium	Methylobacterium extorquens PA1	mepa1	5.5	Pheo-Q	Facultative	Mesophile	6 - 8	Free living	Motile	NC_010172
Methylobacterium	Methylobacterium populi BJ001	mpbj	5.8	Pheo-Q	Aerobe	Mesophile	6 - 8	Free living	Motile	NC_010725 NC_010721 NC_010727
Methylobacterium	Methylobacterium radiotolerans JCM 2831	mrjcm	6.1	Pheo-Q	Aerobe	Mesophile	6 - 8	Free living	Motile	NC_010505 NC_010509 NC_010510 NC_010514 NC_010502 NC_010504 NC_010517 NC_010518 NC_010507
Dinoroseobacter	Dinoroseobacter shibae DFL 12	dsdfl	3.8	Pheo-Q	Facultative	Mesophile	6.5 - 9	Free living	Motile	NC_009952
Jannaschia	Jannaschia sp. CCS1	jccs1	4.3	Pheo-Q	Obligate aerobe	Mesophile	6 - 8	Free living	Nonmotile	NC_007802 NC_007801
Rhodobacter	Rhodobacter capsulatus SB1003	rc1003	3.7	Pheo-Q	Facultative	Mesophile	6 - 8	Free living	Motile	NC_014034 NC_014035
Roseobacter	Roseobacter denitrificans OCh 114	rd114	4.1	Pheo-Q	Aerobe	Mesophile	6 - 8	Free living	Motile	NC_008209 NC_008386 NC_008387 NC_008388 NC_008389
Rhodobacter	Rhodobacter sphaeroides KD131	rs131	4.5	Pheo-Q	Facultative	Mesophile	6 - 8	Free living	Motile	NC_011963 NC_011958 NC_011962 NC_011960
Rhodobacter	Rhodobacter sphaeroides ATCC 17025	rs17025	3.2	Pheo-Q	Facultative	Mesophile	6 - 8	Free living	Motile	NC_009428 NC_009430 NC_009431 NC_009432 NC_009433 NC_009429
Rhodobacter	Rhodobacter sphaeroides ATCC 17029	rs17029	4.4	Pheo-Q	Facultative	Mesophile	6 - 8	Free living	Motile	NC_009049 NC_009050 NC_009040
Rhodobacter	Rhodobacter sphaeroides 2.4.1	rs241	4.1	Pheo-Q	Facultative	Mesophile	6 - 8	Free living	Motile	NC_007493 NC_007494 NC_009007 NC_007488 NC_007489 NC_007490 NC_009008
Rhodospirillum	Rhodospirillum centenum SW	rcsw	4.4	Pheo-Q	Facultative	Mesophile/Thermo tolerar	6 - 8	Free living	Motile	NC_011420
Rhodospirillum	Rhodospirillum rubrum ATCC 11170	rr11170	4.4	Pheo-Q	Facultative	Mesophile	6 - 8	Free living	Motile	NC_007643 NC_007641
Allochromatium	Allochromatium vinosum DSM 180	av180	3.6	Pheo-Q	Anaerobe	Mesophile	6 - 8	Free living	Motile	NC_013851 NC_013852 NC_013862
Halorhodospira	Halorhodospira halophila SL1	hhsl1	2.7	Pheo-Q	Anaerobe	Mesophile	6 - 8	Free living	Motile	NC_008789

Table 1. Summary of 82 photosynthetic prokaryotes with whole-genome sequences

2.1 Clustering of ortholog groups of photosynthetic prokaryotes

All of the 312,254 protein sequences from 82 photosynthetic prokaryote genomes were collected and clustered with the Markov clustering algorithm Ortho-MCL (Chen et al., 2006). Ortho-MCL is a graph-clustering algorithm designed to identify homologous proteins based on sequence similarity and distinguish true orthologs from paralogous relationships without computationally intensive phylogenetic analysis. Upon clustering, 41,824 proteins (13.3%) were removed due to the absence of detectable sequence similarities (BLASTP; $E=10^{-5}$) and 272,686 (86.7%) were assigned to clusters. To assess clustering performance we modified a method described by Frech and Chen (2010) whereby both false-positive (the number of proteins that are found in two or more separate clusters) and false-negative (number of proteins that are found in wrong clusters) results were calculated using both the KEGG (Kanehisa & Goto, 2000) and COG (Natale et al., 2000) databases as a reference. An inflation index is then calculated that controls cluster granularity and gene family size while limiting error (Huerta-Cepas et al., 2008). The inflation parameter impacts the calculation of the number of shared orthologss in each phylum.

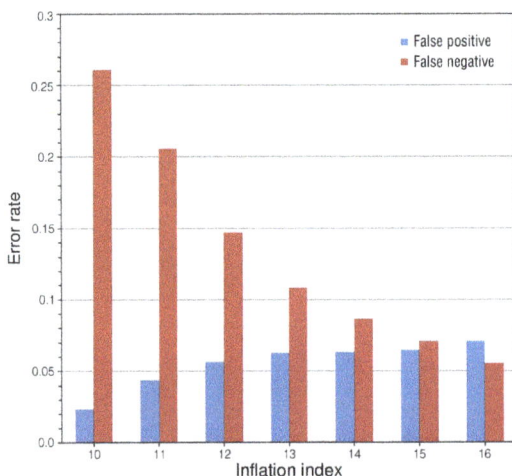

Fig. 1. Estimate of false positive and false negative error rate during ortholog clustering

As Figure 1 shows, an increasing false-positive rate is anti-parallel to decreasing false-negative rate in the inflation parameter. In order to obtain an adequate clustering result, we adjusted the Ortho-MCL program parameters such that reference ortholog clusters compared to both KEGG and COG are classified correctly to minimize erroneous clustering of orthologous groups (inflation index of 15). In our analysis, each predicted orthologous group was evaluated and corrected based on information from both KEGG and COG databases.

2.2 The assembly of core- and pan-genomes

The pan-genome of all 82 species contains 312,254 genes that form 23,362 ortholog clusters. Based on the clustering results, we observed that every photosynthetic prokaryote shares large portions of its genes with others. 204,074 genes that represent 74.8% of the entire data were found to co-exist in at least two organisms from any phyla ("multi-shares"; Figure 2).

The number of gene clusters specific to a particular phylum is much smaller. Both *Cyanobacteria* and *Proteobacteria* possess 32,316 (11.8%) and 30,717 (11.3%) phylyl-specific gene clusters, respectively, whereas both *Chlorobi* and *Chloroflexi* have 2,290 (0.8%) and 3,123 (1.1%) gene clusters, respectively. Additionally, 16,665 genes of all species (6.1%) are in common (that is, are contained in the phototrophic prokaryote core-genome). On the surface, this result suggests a remarkable degree of overlap in the gene composition across all five major phyla of photosynthetic prokaryotes.

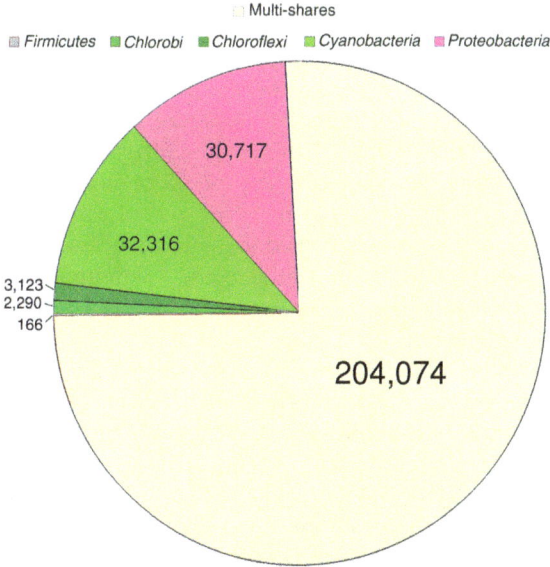

Fig. 2. Distribution of clustered genes within the pan-genome of the five photosynthetic bacterial phyla. Numbers indicate the number of genes specific to each group.

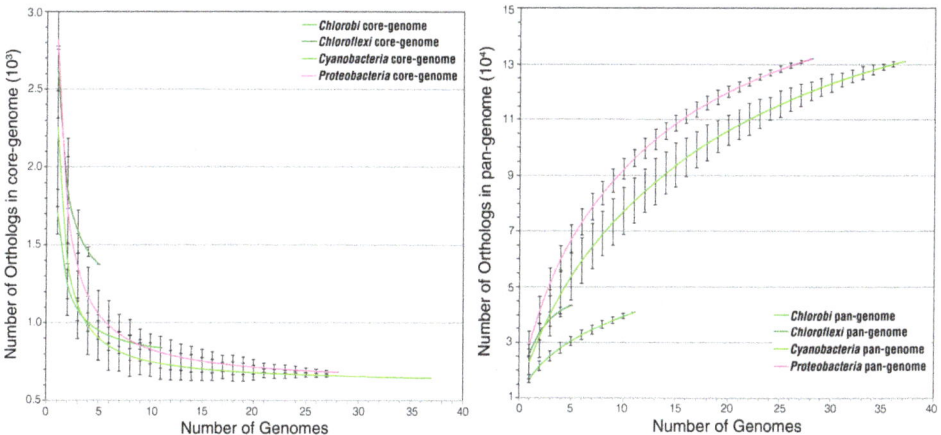

Fig. 3. Plot of the contraction of the core-genome (left) or expansion of the pan-genome (right) as the number of photosynthetic prokaryote genomes analyzed is increased.

To estimate the change of the core-genome size within a particular phyla upon sequential addition of each new genome sequence, a plot was extrapolated by fitting a power law function to the data (Figure 3). As more genomes are compared, there is an asymptotic decline in the number of core orthologs in every phyla, similar to observations for *Streptococcus* (Lefebure & Stanhope, 2007) and *Prochlorococcus* genomes (Kettler et al., 2007). The pan-genome, in contrast, was determined by the plot of the numbers of new orthologs, which fit a decaying exponential curve (Figure 3). The gene accumulation curve for each phyla is clearly far from saturated.

2.3 The core- and pan-genomes of phototrophic prokaryotes
The results of the clustering analysis to determine the core- and pan-genome sizes for each phylum and all phyla together is shown in Figure 4. The size of the phylum core-genomes are: 819 genes in the *Chlorobi*, 1,392 in the *Chloroflexi*, 619 in the *Cyanobacteria*, and 644 in the *Proteobacteria*. The core-genome of all 82 phototrophs considered together consists of 268 genes shared by all organisms. This overall core-genome encompasses a large number of housekeeping genes involved in genetic processes and metabolism and a small number of genes involved in cellular and environmental processes. The housekeeping genes involved in genetic processes include DNA polymerase, ligase, and helicase for DNA replication; RNA polymerase, ribosomal proteins, and tRNA synthetases for translation; and chaperones and signal peptidase for post-translational processes. The housekeeping genes involved in metabolism are mainly involved in the biosynthesis of amino acids, nucleotides, and coenzymes, and a few key enzymes such as transketolase, phosphoglycerate mutase, phosphoglycerate kinase of the glycolysis, acetyl-CoA carboxylase of the tricarbxylic acid (TCA) cycle, H+-transporting ATPase, acyl carrier protein, and UDP-N-acetylmuramate-L-alanine ligase for the biosynthesis of bacterial cell wall are preserved. Moreover, we identified the chlorophyll-synthesizing enzymes that include porphobilinogen synthase, oxygen-independent coproporphyrinogen III oxidase, magnesium chelatase, chlorophyll synthase, magnesium-protoporphyrin O-methyltransferase, and light-independent protochlorophyllide reductase. For both cellular and environmental processes, glycosyltransferase for cell membrane biogenesis, phosphate transport system proteins, signal recognition SRP54, and sec-independent protein TatC for membrane transport were identified, suggesting that transferring phosphate and translocating membrane proteins are universal in photosynthetic organisms. The large proportion of housekeeping genes responsible for nearly all major genetic functions and the biosynthesis for both amino acids and nucleotides is understandable since these genes are essential for basic life functions. We observed a paucity of genes involved in both cellular and environmental processes in the overall core-genome. This observation supports the view in which essential life functions are unchanging while nonessential or environment-specific functions are found in a flexible genome (Kettler et al., 2007).

Core-genomes were also calculated in a pairwise fashion between photosynthetic phyla to gauge the number of shared orthologs in a given pair of phylyl pan-genomes (Figure 5). Each circle in the figure is proportional to the size (number in the circle) of the shared orthologs. Although these results are heavily influenced by the size of the dataset for an individual phyla, it provides a provisional measure of shared genes between phyla.

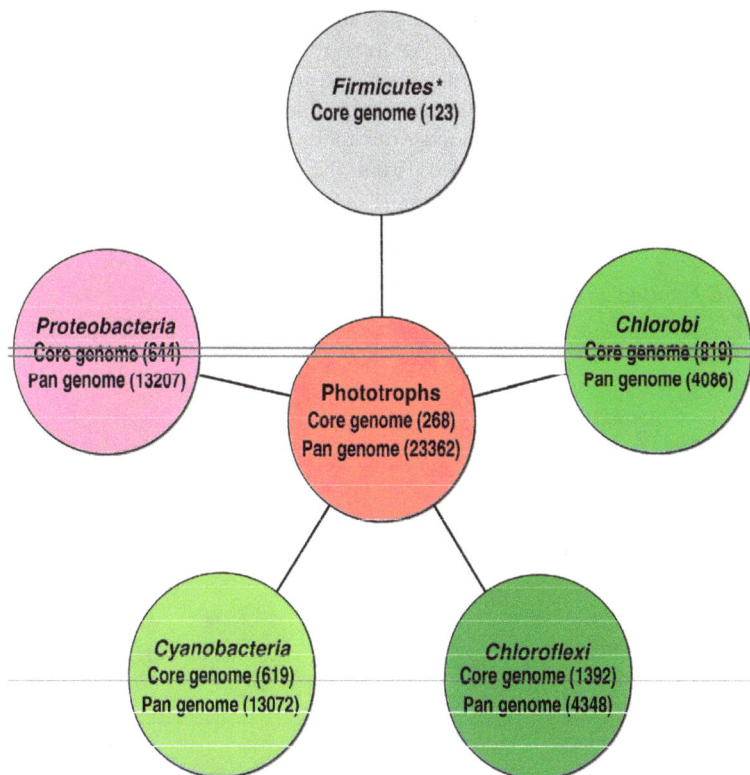

Fig. 4. Both core- and pan-genomes present in all five photosynthetic phyla. Each colored circle represents a phylum: *Firmicutes* (grey), *Chlorobi* (green), *Chloroflexi* (dark-green), *Proteobacteria* (purple), *Cyanobacteria* (light-green), and all phyla together (red). Numbers represent the ortholog clusters contained within the core genome or pan genome of each phyla. * The Firmicutes, with only a single sequenced genome, lack a pan genome.

2.3.1 *Heliobacteria (Firmictues)*

Given that there was only a single fully-sequenced genome from the phototrophic *Heliobacteria (Firmicutes)* available for study, the core genome of *Heliobacteria* was provisionally constructed by excluding those genes that are homologous to any known genes from the other sequence-available *Firmicutes*. It is worth noting that although there are four heliobacteria genera containing a total of ten species that have been formally described: *Heliobacterium*, *Heliobacillus*, *Heliophilum*, and *Heliorestis*, the phototrophic *Heliobacterium modesticaldum* is the only sequenced bacteria representing them (Sattley et al., 2008). When *H. modesticaldum* was compared with the available bacterial genomes of the *Firmicutes*, we identified 123 ortholog clusters tentatively assigned to the core-genome of this organism. Genes encoding proteins involved in major genetic, cellular, and environmental processes and metabolism are very limited. This may be partly due to their mutualistic relationship with plants. Other major ortholog groups in this core-genome are involved in sporulation. The previous examination of several other *Heliobacteria* species for sporulating genes has indicated that sporulation gene presence may be universal within the heliobacteria (Kimble-Long & Madigan, 2001). It should be noted that a set of genes involved in bacteriochlorophyll (Bchl) g biosynthesis, not found in other phototrophs, were frequently reported in other heliobacteria species (reviewed in Asao & Madigan, 2010). These enzymes were not clustered into the core-genome due to their absence in the non-phototrophic *Firmicutes*. Additional genome sequences from the heliobacteria group will aid in our understanding of their specific core-genome.

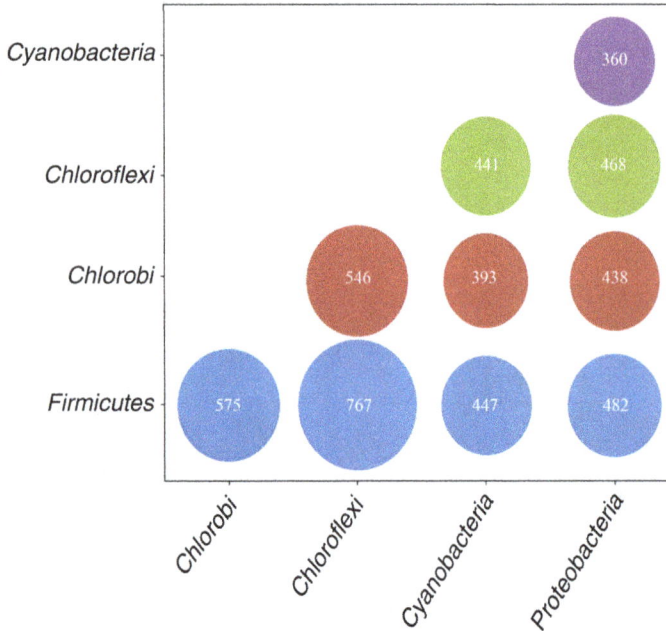

Fig. 5. Pairwise comparison of shared ortholog groups between phyla

2.3.2 *Chlorobi*
The *Chlorobia* core-genome contains 819 genes representing 30-40% of the total genes in a given *Chlorobia* genome. As a phyla, they are very similar with respect to gene content compared to the other phototrophic prokaryotes. In addition to the components of the core-genome for all species, the *Chlorobia* core orthologs are composed of major metabolism genes such as the electron transport chain that supports photosynthesis and sulfur oxidation, the reductive TCA cycle supporting carbon fixation and transport, and others for the biosynthesis of amino acids, lipids, and coenzymes. The core-genome also contains the type I reaction center unique to the *Chlorobi*. Our findings are similar to other recent reports (Davenport et al., 2010). In addition to those identified orthologs for central metabolism, we also identified genes involved in the biosynthesis of Bchl, carotenoids, and the photosynthetic "chlorosome" apparatus. Most pigment-synthesizing enzymes operate downstream of the metabolic pathways for final products like Bchl c, d, chlorobactene, and γ-carotene, which are located on the chlorosome to harvest light. A few metal and inorganic compound transporters for iron, nickel, and molybdate as well as the major facilitator

superfamily (MFS) transporter, were identified. Since the *Chlorobi* species are capable of fixing nitrogen, preserving these transport systems is necessary to support this process.

A total of 1,774 ortholog clusters are assigned to the pan-genome of *Chlorobia*. Most of these are associated with phylogenetically close species and have functions such as secretion, extracellular constituents, and cell wall biogenesis. These are conspicuous features of the genuses *Chlorobium* and *Prosthecochloris*. Although the *Chlorobia* have been well-characterized biochemically and microbiologically (Frigaard & Bryant, 2004), our finding that the *Chlorobia* possess a relatively uniform core-genome complemented by a relatively limited set of accessory genes enhancing cellular activities provides further insight into their anoxygenic phototrophic lifestyle. The core-and pan-genome pattern suggests a largely vertical inheritance that has preserved the core-genome needed for major cellular activities-- a result of living in environmentally stable niches.

2.3.3 *Chloroflexi*

The *Chloroflexi* core-genome contains 1,392 ortholog clusters, the largest size among the five phototrophic groups. It reflects roughly 35% of the genes of a *Chloroflexus* specie's genome. The functional composition of this core-genome is somewhat similar to that of the *Chlorobi* core-genome, since many core genes involved in both genetic and cellular processes were cross-identified. However, the core set also contains type II reaction centers, NADH dehydrogenase, and cytochrome *c* oxidase, similar to the *Proteobacteria* but different from the *Chlorobi*. Moreover, many transporters for metal ions, inorganic and organic compounds, as well as two-component histidine kinases for signal transduction were identified in the *Chloroflexi* but not seen in the *Chlorobi*. The conservation of functionally diverse transporters with signal-transduction histidine kinases may be related to a more dynamic life-style. Generally, *Chloroflexus* is a photoheterotroph and usually found in the lower layers of microbial mats with cyanobacteria growing above it that provide organic byproducts. The *Chloroflexi* core-genome possesses numerous heat shock proteins, chaperones, and signal peptidases involved in protein folding and translocating processes that likely serve to reinforce protein structures in the thermophilic *Chloroflexus* species. For genes involved in major metabolic pathways, the core-genes appear to be largely conserved across all photosynthetic phyla. Although the *Chloroflexi* are found to be distinctive from the *Chlorobi*, they do have some common characteristics such as the absence of intra-cytoplasmic membrane structures and chlorosomes on their plasma membranes. They also use the same Bchl *a* and *c* biosythesis pathways.

The *Chloroflexi* pan-genome contains 4,348 genes, and in contrast to the *Chlorobi* pan-genome it is comprised of more putative genes for extracelluar constituents, inter-cellular communication, and other physiological and biochemical activities. For example, genes involved in the 3-hydroxypropionate pathway for carbon fixation were found. But generally, the *Chloroflexi* core-genome equips most of the major functional genes for a wide range of metabolisms such as synthesis of organic compounds, energy production, transport, genetic processing, etc. Such coverage throughout most cellular activities makes the core-genome of the *Chloroflexi* similar in character to that of *Chlorobia*.

2.3.4 *Cyanobacteria*

Representing the largest sampled phylogenetic clade of the phototrophic prokaryotes, the *Cyanobacteria* have 37 completely sequenced genomes available for analysis, resulting in the smallest core- (619 genes) and largest pan-genome (13,072 genes) of all five phyla. The proportion of genes designated "core" with respect to any cyanobacterial genome varies from less than 10% (in the the non-*Prochlorococcus/Synechococcus* genomes) to nearly 38% for *Prochlorococcus* and *Synechococcus* strains. The core orthologs are responsible for several major reactions such as the Calvin cycle, glycolysis, the incomplete TCA cycle, and pathways to synthesize amino acids and cofactors. Two types of photosystems (PS I and PS II) and the participating electron transport chain for oxygenic photosynthesis are also included.

The large pan-genome of cyanobacteria appears to support diverse abilities and processes. There are many genes found in the pan-genome that carry out metabolic activities unrelated to photosynthesis. For example, *M. aeruginosa* NIES-843 produces a diverse range of toxins with the non-ribosomal peptide synthetases (Kaneko et al., 2007). These enzymes produce neurotoxins and hepatotoxins that cause a variety of human illnesses, and are responsible for deaths in native and domestic animals. *T. erythraeum* IMS101 can perform nitrogen fixation in the presence of oxygen (Sandh et al., 2011). *Nostocaceae* species generally have an unbranched filamentous cell type, develop heterocysts, and possess multiple plasmids. In contrast, *Prochlorococcus* and *Synechococcus* species have a small round shape with no plasmids. Finally, a large number of genes identified in members of the *Nostocaceae* and *A. marina* MBIC11017 have unknown functions. Judging by the life style of these cyanobacterial species, which have a mutualistic relationship with terrestrial plants (Baker et al., 2003) and coral (Marquardt et al., 1997), it is possible that these genes are involved in supporting inter-communication and mutualism with their host.

2.3.5 Proteobacteria

The *Proteobacteria* contain 644 core gene clusters and 13,207 non-redundant genes in the pan genome. The percentage of core genes in any of the *Proteobacteria* genomes varies from 10% to 25%, similar to the results obtained for the *Cyanobacteria*. This is because the *Proteobacteria* is the second major photosynthetic group with 28 completely sequenced genomes from phylogenetically distinct clades. The *Proteobacteria* core-genome preserves most of the key enzymes essential to major cellular activities, similar to other core-genomes. The type II reaction center and light-harvesting proteins are in the core genome, the former of which was also identified in the *Chloroflexi*. Nevertheless, the additional orthologs coding for the bacterial flagella, chemotaxis, and respiratory electron transfer chain proteins unique to the *Proteobacteria* were also identified. Both flagella and chemotaxis help cells move either toward nutrients or away from unfavourable living conditions and both anaerobic and aerobic respiration supports chemo-heterotrophic growth when phototrophic growth is not possible. Thus, integrating both cell mobility and respiration to the *Proteobacteria* core genome suggests an ecological advantage of adaptation to a broader range of living environments than other phototrophic phyla.

In contrast to the core-genome, the characteristics of the pan-genome are widely diverse-- from variant types of nitrogen assimilation, carbon assimilation, and hydrogen metabolism to inter-cellular communications and nodulation. Such vast variety in the functional repertoire associated with the pan-genome can give the *proteobacteria*, such as *Rhodobacter*, *Rhodopseudomonas*, and *Rhodospillium* genera, a broad range of growth conditions for anaerobic phototrophy and aerobic chemoheterotrophy in the absence of light (Larimer et al., 2004; Lu et al., 2010; Mackenzie et al., 2007). Over 50 genes associated with nodulation were identified in the diazotrophic *Bradyrhizobiaceae* and *Rhizobiaceae*. Most photosynthetic bacteria in these two orders are capable of forming mutualistic symbiosis with terrestrial plants by fixing nitrogen inside special structures called legumes. Genes for hydrogen production or metabolizing C_1-compounds such as methane were identified in *Rhodopseudomonas*, some *Rhodobacteraceae*, and *Methylobacterium*. These traits have garned much-warranted attention for their potential ability to reduce CH_4 (greenhouse gas) emission (Eller & Frenzel, 2001; Lidstrom & Chistoserdova, 2002). Based on the construction of both core- and pan-genomes, photosynthetic members in the *Proteobacteria* exhibit the greatest gene diversity amongst all phyla studied. This diversity reflects their ability to grow chemoheterophically as well as phototrophically, which makes them better at living in a broader range of environments than the *Cyanobacteria*.

Taken together, we have identified core-genes responsible for phylum-specific reactions. We have also observed a wide variety of accessory functions supporting smaller groups of bacteria. The core-genome assembled from a group of closely related bacteria represents a backbone of essential components regulating the adaptability to specific niches. Our results indicate that the gene content of each phylum-specific core is distinctive and can exemplify the very different evolutionary histories of the major photosynthetic groups, where the accessory components comprising the pan-genomes provide fitness advantages in distinct habitats.

2.4 Phylogeny of photosynthetic prokaryotes using the pan-genome

Construction of both core- and pan-genomes of all photosynthetic bacteria provides a novel opportunity to determine the phylogenetic relationship among these prokaryotes. Several methods have been used to evaluate the phylogenies of different bacterial groups such as single-gene phylogenies (e.g., 16S rDNA), concatenated sequences of photosynthesis-related proteins (Rokas et al., 2003), and signature sequences of house-keeping proteins (Gupta, 2003). The sequences compared in these methods are necessarily present in all analyzed species. Here, we present a phylogeny that is formulated using the clustered pan-genome that does not rely on a universally shared collection of genes. Hierarchical clustering with resampling 100 times was performed based on a relative Manhattan distance calculated on the presence/absence of an ortholog between a given pair of genomes (Snipen & Ussery, 2010). It in essence generates a tree based on shared gene content. Figure 6 shows the resulting tree. There is broad agreement with this tree and traditional single-gene phylogenies. But surprisingly, the topology of the tree shows that both *A. vinosum* DSM 180 and *H. halophila* SL1, both belonging to the γ-*Proteobacteria* class, are situated outside of the *Proteobacteria* clade and positioned between the *Chlorobi* and *Firmicutes*. We investigated in detail the gene content of these two organisms and found that *A. vinosum* DSM 180 has lost most of the *Proteobacteria*-specific orthologs, while *H. halophila* SL1 contains more shared orthologs with *A. vinosum* DSM 180 than between the other purple bacteria species. Another unusual topology is found in the *Proteobacteria* clade where both *Rhodobacter* and *Rhodospirillum* families are closer to the *Rhodopesudomonas* genus, which belongs phylogenetically to the *Rhizobia* within the *Brydorhizobia* and *Methylobacteria* families.

We further utilized the pan-genome to reveal a three-dimensional relationship between individual species and the major photosynthetic lineages. By performing a multidimensional scaling analysis of the ortholog distribution across all 82 species, we found that related species were clustered in groups reflecting their phyla (Figure 7). While the *Heliobacteria*, *Chlorobi*, and *Chloroflexi* species occupied a central space, the *Proteobacteria* and *Cyanobacteria* were greatly separated and located on opposing poles. The relative position of the *Proteobacteria* and *Cyanobacteria* groups apart from each other indicate that their ortholog profiles have diverged substantially. Additionally, the distribution of both *Cyanobacteria* and *Proteobacteria* species is also consistent with their phylogenetic positions in Figure 6. However, the γ-*Proteobacterial* species, *A. vinosum* DSM 180 and *H. halophila* SL1 were exceptionally close to both *Chlorobi* and *Chloroflexi*, a result similar to the two-dimensional pan-genome-based phylogenetic tree. Clustering organisms by determining the occurrence of the specific patterns of orthologs shared by a group of species reveals an overall pattern consistent with both 16s rDNA- and pan-genome-based phylogenies. Yet, the observation of shared orthologs in one or a group of species can highlight functional divergence or convergence in groups that can be quantified by gene analysis but missed by single-gene-based phylogenies.

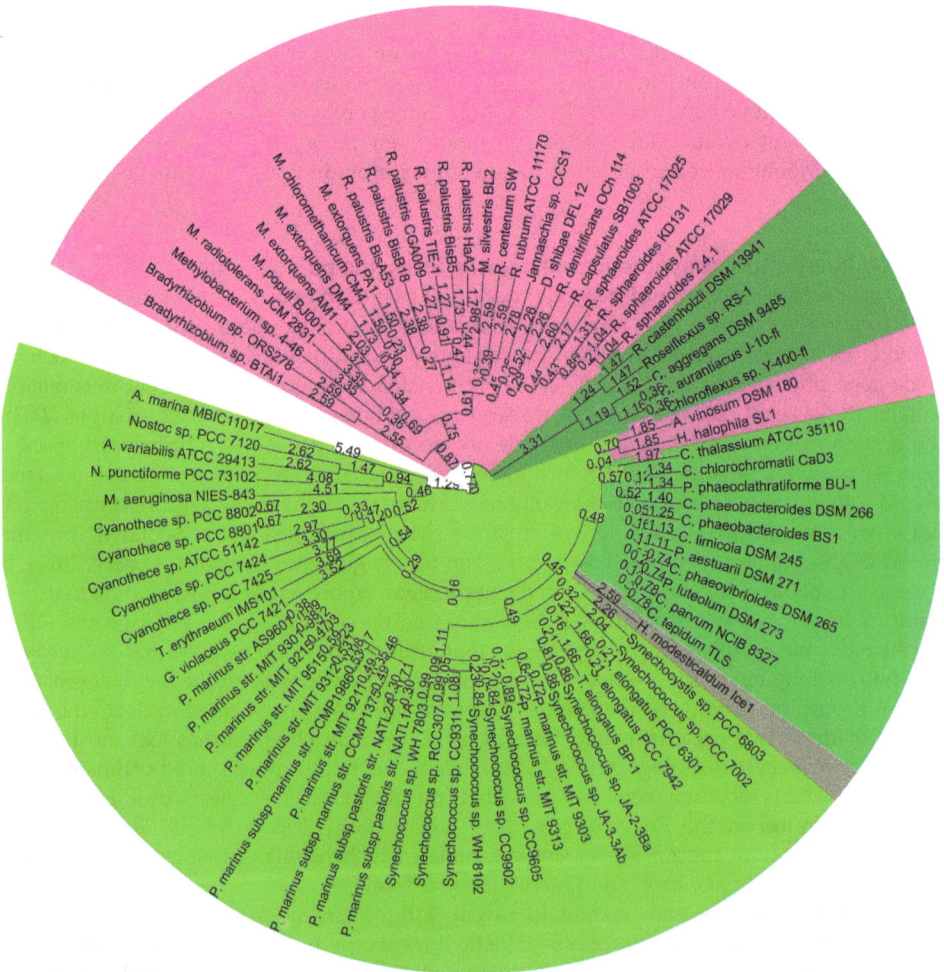

Fig. 6. Phylogenetic relationship of 82 photosynthetic prokaryotes reconstructed with the pan-genome

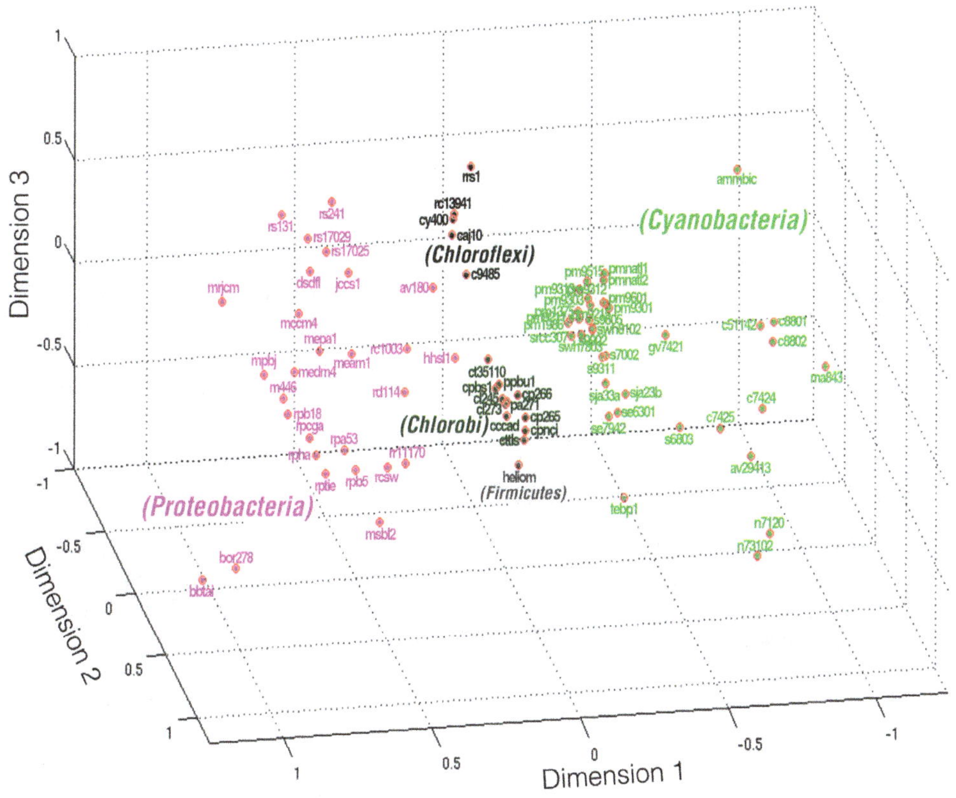

Fig. 7. Multidimensional scaling analysis showing the organization of 82 photosynthetic prokaryotes in a pan-genomic distribution.

3. Acknowledgements

This work was supported by the U.S. National Science Foundation Phototrophic Prokaryotes Sequencing Project, grant number 0950550.

4. References

Asao, M., & Madigan, M.T. (2010). Taxonomy, phylogeny, and ecology of the heliobacteria. *Photosynthesis Research*, Vol.104, No.2-3, (2010), pp. 103-111, ISSN 1573-5079; 0166-8595

Baker, J.A., Entsch, B., & McKay, D.B. (2003). The cyanobiont in an Azolla fern is neither Anabaena nor Nostoc. *FEMS microbiology letters*, Vol.229, No.1, (2003), pp. 43-47, ISSN 0378-1097; 0378-1097

Blankenship, R.E. (1992). Origin and early evolution of photosynthesis. *Photosynthesis Research*, Vol.33, No.2, (1992), pp. 91-111, ISSN 0166-8595

Chen, F., Mackey, A.J., Stoeckert, C.J., Jr., & Roos, D.S. (2006). OrthoMCL-DB: querying a comprehensive multi-species collection of ortholog groups. *Nucleic Acids Res*, Vol.34, No.Database issue, (Jan 1 2006), pp. D363-368, ISSN 1362-4962; 0305-1048

Davenport, C., Ussery, D.W., & Tummler, B. (2010). Comparative genomics of green sulfur bacteria. *Photosynthesis Research*, Vol.104, No.2-3, (2010), pp. 137-152, ISSN 1573-5079; 0166-8595

Eller, G., & Frenzel, P. (2001). Changes in activity and community structure of methane-oxidizing bacteria over the growth period of rice. *Applied and Environmental Microbiology*, Vol.67, No.6, (2001), pp. 2395-2403, ISSN 0099-2240

Frech, C., & Chen, N. (2010). Genome-wide comparative gene family classification. *PLoS One*, Vol.5, No.10, (2010), pp. e13409, ISSN 1932-6203

Frigaard, N.U., & Bryant, D.A. (2004). Seeing green bacteria in a new light: genomics-enabled studies of the photosynthetic apparatus in green sulfur bacteria and filamentous anoxygenic phototrophic bacteria. *Archives of Microbiology*, Vol.182, No.4, (2004), pp. 265-276, ISSN 0302-8933; 0302-8933

Gest, H., & Favinger, J.L. (1983). Heliobacterium chlorum, an anoxygenic brownish-green photosynthetic bacterium containing a "new" form of bacteriochlorophyll. *Archives of Microbiology*, Vol.136, No.1, (1983), pp. 11-16, ISSN 0302-8933

Gupta, R.S. (2003). Evolutionary relationships among photosynthetic bacteria. *Photosynth Res*, Vol.76, No.1-3, (2003), pp. 173-183, ISSN 0166-8595

Hogg, J.S., Hu, F.Z., Janto, B., Boissy, R., Hayes, J., Keefe, R., Post, J.C., & Ehrlich, G.D. (2007). Characterization and modeling of the Haemophilus influenzae core and supragenomes based on the complete genomic sequences of Rd and 12 clinical nontypeable strains. *Genome Biol*, Vol.8, No.6, (2007), pp. R103, ISSN 1465-6914; 1465-6906

Huerta-Cepas, J., Bueno, A., Dopazo, J., & Gabaldon, T. (2008). PhylomeDB: a database for genome-wide collections of gene phylogenies. *Nucleic acids research*, Vol.36, No.Database issue, (2008), pp. D491-496, ISSN 1362-4962; 0305-1048

Kanehisa, M., & Goto, S. (2000). KEGG: kyoto encyclopedia of genes and genomes. *Nucleic acids research*, Vol.28, No.1, (2000), pp. 27-30, ISSN 0305-1048; 0305-1048

Kaneko, T., Nakajima, N., Okamoto, S., Suzuki, I., Tanabe, Y., Tamaoki, M., Nakamura, Y., Kasai, F., Watanabe, A., Kawashima, K., et al. (2007). Complete genomic structure of the bloom-forming toxic cyanobacterium Microcystis aeruginosa NIES-843. *DNA research,* Vol.14, No.6, (2007), pp. 247-256, ISSN 1756-1663; 1340-2838

Kaneko, T., Sato, S., Kotani, H., Tanaka, A., Asamizu, E., Nakamura, Y., Miyajima, N., Hirosawa, M., Sugiura, M., Sasamoto, S., et al. (1996). Sequence analysis of the genome of the unicellular cyanobacterium Synechocystis sp. strain PCC6803. II. Sequence determination of the entire genome and assignment of potential protein-coding regions. *DNA Research,* Vol.3, No.3, (Jun 30 1996), pp. 109-136, ISSN 1340-2838

Kettler, G.C., Martiny, A.C., Huang, K., Zucker, J., Coleman, M.L., Rodrigue, S., Chen, F., Lapidus, A., Ferriera, S., Johnson, J., et al. (2007). Patterns and implications of gene gain and loss in the evolution of Prochlorococcus. *PLoS genetics,* Vol.3, No.12, (2007), pp. e231, ISSN 1553-7404; 1553-7390

Kimble-Long, L.K., & Madigan, M.T. (2001). Molecular evidence that the capacity for endosporulation is universal among phototrophic heliobacteria. *FEMS microbiology letters,* Vol.199, No.2, (2001), pp. 191-195, ISSN 0378-1097; 0378-1097

Larimer, F.W., Chain, P., Hauser, L., Lamerdin, J., Malfatti, S., Do, L., Land, M.L., Pelletier, D.A., Beatty, J.T., Lang, A.S., et al. (2004). Complete genome sequence of the metabolically versatile photosynthetic bacterium Rhodopseudomonas palustris. *Nature biotechnology,* Vol.22, No.1, (2004), pp. 55-61, ISSN 1087-0156; 1087-0156

Lefebure, T., & Stanhope, M.J. (2007). Evolution of the core and pan-genome of Streptococcus: positive selection, recombination, and genome composition. *Genome biology,* Vol.8, No.5, (2007), pp. R71, ISSN 1465-6914; 1465-6906

Lidstrom, M.E., & Chistoserdova, L. (2002). Plants in the pink: cytokinin production by methylobacterium. *Journal of Bacteriology,* Vol.184, No.7, (2002), pp. 1818, ISSN 0021-9193; 0021-9193

Liolios, K., Tavernarakis, N., Hugenholtz, P., & Kyrpides, N.C. (2006). The Genomes On Line Database (GOLD) v.2: a monitor of genome projects worldwide. *Nucleic Acids Res,* Vol.34, No.Database issue, (Jan 1 2006), pp. D332-334, ISSN 1362-4962; 0305-1048

Lu, Y.K., Marden, J., Han, M., Swingley, W.D., Mastrian, S.D., Chowdhury, S.R., Hao, J., Helmy, T., Kim, S., Kurdoglu, A.A., et al. (2010). Metabolic flexibility revealed in the genome of the cyst-forming alpha-1 proteobacterium Rhodospirillum centenum. *BMC Genomics,* Vol.11, No.1, (May 25 2010), pp. 325, ISSN 1471-2164; 1471-2164

Ludwig, W., & Klenk, H.P. (2001). Overview: A Phylogenetic Backbone and Taxonomic Framework for Prokaryotic Systematics. In: *The Archaea and the Deeply Branching and Phototrophic Bacteria,* D.R. Booke, & R.W. Casteholz, eds., Springer-Verlag, pp. 49-65. ISBN 0387987711, 9780387987712, Berlin

Mackenzie, C., Eraso, J.M., Choudhary, M., Roh, J.H., Zeng, X., Bruscella, P., Puskas, A., & Kaplan, S. (2007). Postgenomic adventures with Rhodobacter sphaeroides. *Annual Review of Microbiology,* Vol.61, (2007), pp. 283-307, ISSN 0066-4227; 0066-4227

Marquardt, J., Senger, H., Miyashita, H., Miyachi, S., & Morschel, E. (1997). Isolation and characterization of biliprotein aggregates from Acaryochloris marina, a Prochloron-like prokaryote containing mainly chlorophyll d. *FEBS letters*, Vol.410, No.2-3, (1997), pp. 428-432, ISSN 0014-5793; 0014-5793

Natale, D.A., Galperin, M.Y., Tatusov, R.L., & Koonin, E.V. (2000). Using the COG database to improve gene recognition in complete genomes. *Genetica*, Vol.108, No.1, (2000), pp. 9-17, ISSN 0016-6707; 0016-6707

Olson, J.M., & Pierson, B.K. (1987). Evolution of Reaction Centers in Photosynthetic Prokaryotes. In International Review of Cytology, K.W.J. G.H. Bourne, & M. Friedlander, eds. (Academic Press), pp. 209-248.

Pallen, M.J., & Wren, B.W. (2007). Bacterial pathogenomics. *Nature*, Vol.449, No.7164, (Oct 18 2007), pp. 835-842, ISSN 1476-4687; 0028-0836

Raymond, J., & Swingley, W.D. (2008). Phototroph genomics ten years on. *Photosynth Res*, Vol.97, No.1, (Jul 2008), pp. 5-19, ISSN 0166-8595

Raymond, J., Zhaxybayeva, O., Gogarten, J.P., Gerdes, S.Y., & Blankenship, R.E. (2002). Whole-genome analysis of photosynthetic prokaryotes. *Science*, Vol.298, No.5598, (Nov 22 2002), pp. 1616-1620, ISSN 1095-9203

Rokas, A., Williams, B.L., King, N., & Carroll, S.B. (2003). Genome-scale approaches to resolving incongruence in molecular phylogenies. *Nature*, Vol.425, No.6960, (2003), pp. 798-804, ISSN 1476-4687; 0028-0836

Sandh, G., Ran, L., Xu, L., Sundqvist, G., Bulone, V., & Bergman, B. (2011). Comparative proteomic profiles of the marine cyanobacterium Trichodesmium erythraeum IMS101 under different nitrogen regimes. *Proteomics*, Vol.11, No.3, (2011), pp. 406-419, ISSN 1615-9861; 1615-9853

Sattley, W.M., Madigan, M.T., Swingley, W.D., Cheung, P.C., Clocksin, K.M., Conrad, A.L., Dejesa, L.C., Honchak, B.M., Jung, D.O., Karbach, L.E., et al. (2008). The genome of Heliobacterium modesticaldum, a phototrophic representative of the Firmicutes containing the simplest photosynthetic apparatus. *J Bacteriol*, Vol.190, No.13, (Jul 2008), pp. 4687-4696, ISSN 1098-5530

Snipen, L., & Ussery, D.W. (2010). Standard operating procedure for computing pangenome trees. *Standards in genomic sciences*, Vol.2, No.1, (2010), pp. 135-141, ISSN 1944-3277; 1944-3277

Tettelin, H., Masignani, V., Cieslewicz, M.J., Donati, C., Medini, D., Ward, N.L., Angiuoli, S.V., Crabtree, J., Jones, A.L., Durkin, A.S., et al. (2005). Genome analysis of multiple pathogenic isolates of Streptococcus agalactiae: implications for the microbial "pan-genome". *Proceedings of the National Academy of Sciences of the United States of America*, Vol.102, No.39, (2005), pp. 13950-13955, ISSN 0027-8424; 0027-8424

Vermaas, W. (1994). Evolution of heliobacteria: Implications for photosynthetic reaction center complexes. *Photosynthesis Research*, Vol.41, No.1, (1994), pp. 285-294, ISSN 0166-8595

11

The Stringent Response in Phototrophs

Shinji Masuda
Center for Biological Resources & Informatics, Tokyo Institute of Technology
Japan

1. Introduction

Organisms must respond to environmental changes if they are to survive. As a result, species have evolved numerous intracellular and intercellular regulatory systems that often reflect an organism's environment. In bacteria, one of the most important regulatory systems is the stringent response (Cashel et al., 1996). Signaling *via* this response is mediated by guanosine 5'-triphosphate 3'-diphosphate (pppGpp) and guanosine 5'-diphosphate 3'-diphosphate (ppGpp), which function as second messengers. The stringent response was first discovered over 40 years ago in *Escherichia coli*. When *E. coli* cells are grown under nutrient-rich conditions but then transferred to a nutrient-limited environment, intracellular levels of pppGpp and ppGpp ((p)ppGpp) rapidly increase (Cashel et al., 1996). (p)ppGpp controls many vital cellular processes, including transcription and translation. For example, (p)ppGpp directly binds RNA polymerase and alters its promoter-binding affinity (Chatterji et al., 1998; Toulokhonov et al., 2001; Artsimovitch et al., 2004). When nutrient availability changes, therefore, the stringent response simultaneously adjusts the level of transcription for many genes. In *E. coli*, synthesis and degradation of (p)ppGpp are catalyzed by two enzymes RelA and SpoT (Cashel et al., 1996).

Deficiencies in iron, phosphate, nitrogen, or carbon each represent environmental stresses that trigger (p)ppGpp accumulation (Cashel et al., 1996). For photosynthetic bacteria, sunlight is also an important "nutrient". Characterization of a SpoT homolog in the purple photosynthetic bacterium, *Rhodobacter capsulatus*, showed that the stringent response also regulates photosynthesis (Masuda & Bauer, 2004). Genes that encode (p)ppGpp synthases and hydrolases are highly conserved in plants (van der Biezen et al., 2000; Kasai et al., 2002; Yamada et al., 2003; Givens et al., 2004; Tozawa et al., 2007; Masuda et al., 2008a; Kim et al., 2009) and are called RSHs (RelA/SpoT homologs). All known plant RSHs are targeted to chloroplasts, suggesting that they may control chloroplast function. Here we summarize our current understanding of the stringent response in phototrophs. For details concerning the mechanisms of the stringent response itself, several recent reviews are available (Magnusson et al., 2005; Braeken et al., 2006; Jain et al., 2006; Ochi, 2007; Potrykus & Cashel, 2008; Srivatsan & Wang, 2008).

2. (p)ppGpp synthases and hydrolases in bacteria

In *E. coli*, the level of (p)ppGpp is controlled by two enzymes, RelA and SpoT (Cashel et al., 1996). Both enzymes synthesize (p)ppGpp by transferring the pyrophosphate of ATP to the ribose of GTP or GDP at the 3' hydroxyl position. (p)ppGpp is hydrolyzed by SpoT, but not RelA, as RelA has only (p)ppGpp synthase activity. Biochemical and crystallographic studies have revealed two distinct domains in SpoT that mediate either synthesis or hydrolysis of (p)ppGpp (Fig. 1) (Cashel et al., 1996). As expected, the (p)ppGpp hydrolysis domain (HD) is not conserved in RelA. The (p)ppGpp HD is found in a superfamily of metal-dependent phosphohydrolases (Aravind & Koonin, 1998; Hogg et al., 2004). In fact, SpoT-like proteins require a divalent cation such a Mn^{2+} for their hydrolase activity (Wendrich et al., 2000).

A large number of bacterial genomes have been sequenced. These data have revealed that SpoT-like proteins, which contain both (p)ppGpp synthase and (p)ppGpp hydrolase domains, are generally conserved among bacterial species (Mittenhuber, 2001a). In contrast, RelA-like proteins, which contain only a (p)ppGpp synthase domain, have only been found in β- and γ- *Proteobacteria*. Phylogenetic analyses suggest that RelA branched off from a SpoT-like protein following the divergence of α- and β- *Proteobacteria* (Mittenhuber, 2001a).

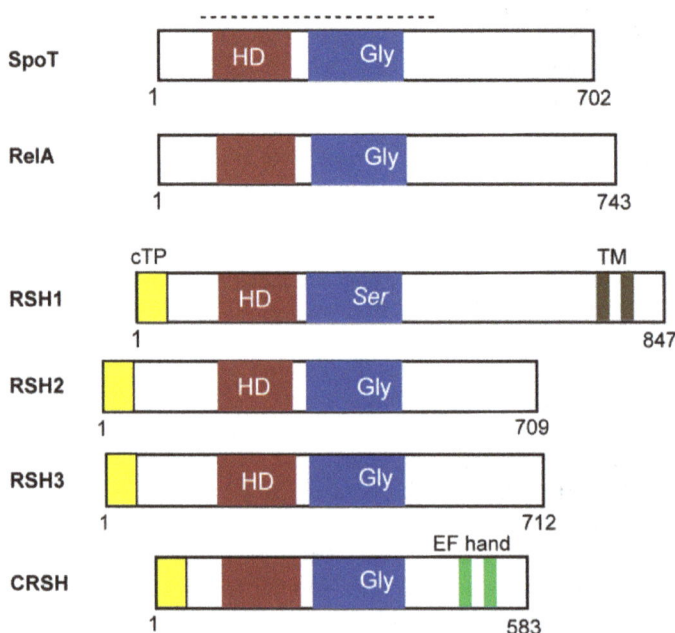

Fig. 1. Schematic depiction of the domain structure of SpoT and RelA from *E. coli* and of Arabidopsis RSHs. The region used to construct the phylogenetic tree (Fig. 3) is indicated by a dashed line. TM: putative transmembrane region; cTP: chloroplast transit peptide; EF-hand: Ca^{2+}-binding domain; HD: HD domain responsible for (p)ppGpp degradation. RelA and SpoT do not conserve several critical amino acids in the HD domain. RSH1 does not conserve the critical Gly residue (changed to Ser) that is necessary for (p)ppGpp synthase activity of RelA. For more detals, see text.

A small (p)ppGpp synthase protein, which lacks an HD domain, has also been found in some *Firmicutes* bacteria (Lemos et al., 2007; Nanakiya et al., 2008). Biochemical analyses indicate that these small enzymes have (p)ppGpp synthase activity.

3. (p)ppGpp synthases and hydrolases in photosynthetic bacteria

To date, six bacterial phyla have been shown to contain species capable of chlorophyll-based photosynthesis: *Cyanobacteria*, *Proteobacteria* (purple bacteria), *Chlorobi* (green sulfur bacteria), *Chloroflexi* (anoxygenic filamentous bacteria), *Firmicutes* (heliobacteria), and *Acidobacteria* (Bryant & Frigaard, 2006; Bryant et al., 2007). All of these photosynthetic bacteria produce SpoT and/or RelA proteins, but a careful analysis of these enzymes has only been done for two species: *Rhodobacter capsulatus* (a purple bacterium) (Masuda & Bauer, 2004), and *Anabaena* sp. PCC7120 (a cyanobacterium) (Ning et al., 2011). Genomic analysis has indicated that both of these bacterial species encode a single SpoT-like protein, but not a RelA-like protein. Based on complementation tests using *E. coli relA* and *relA/spoT* mutants, the SpoT-like enzyme of each of these bacteria likely have (p)ppGpp synthase activity. In wild-type strains of both bacteria, cellular levels of (p)ppGpp increase upon amino-acid starvation (caused by addition of serine hydroxamate), suggesting that these SpoT-like proteins induce the stringent response upon amino-acid starvation. In addition, these *spoT*-like genes are essential for cell viability, as loss-of-function mutations are lethal (Masuda & Bauer, 2004; Ning et al., 2011).

R. capsulatus is one of the most extensively studied photosynthetic purple bacteria (Bauer, 2004). It exhibits remarkable bio-energetic versatility and is capable of aerobic respiratory growth, anaerobic respiratory growth (using dimethyl sulfoxide as an electron donor), and photosynthetic growth. As such, this bacterium can adjust its mode of growth in response to environmental conditions (e.g., oxygen concentration, light intensity). In fact, sophisticated regulatory systems that respond to changes in redox and light have been identified (Bauer et al., 2003). Masuda et al. (2004) asked whether the stringent response affects growth-mode control in *R. capsulatus* by functionally characterizing *spoT* of this organism. They found that the lethality associated with a *spoT* mutation could be rescued by loss of *hvrA*, a gene that encodes a nucleoid protein (Masuda & Bauer, 2004). HvrA was originally identified as a *trans*-acting factor that represses transcription of photosynthesis genes under intense light conditions (Buggy et al., 1994). In intense light, *R. capsulatus* down-regulates components of the photosynthetic apparatus (e.g., the light-harvesting complexes, and the photosynthetic reaction center) to avoid photo-damage. *hvrA* mutants, however, cannot reduce photopigment synthesis under intense light. HvrA is a typical bacterial nucleoid protein that resembles *E. coli* H-NS and StpA (Bertin et al., 1999; Masuda & Bauer, 2004). Nucleoid proteins bind curved DNA with low sequence specificity and affect a large number of genes involved in multiple physiological processes (McLeod & Johnson, 2001; Dorman & Deighan, 2003). In fact, *R. capsulatus* HvrA transcriptionally regulates (positively and negatively) genes involved in photosynthesis, nitrogen fixation, and electron transfer, for example (Buggy et al., 1994; Kern et al., 1998; Swem & Bauer, 2002). Although it is not entirely clear why loss of *hvrA* rescues the lethality associated with *spoT*-like loss-of-function, this result suggests a functional link between the (p)ppGpp-dependent stringent response and the nucleoid protein HvrA. A similar phenomenon exists in *E. coli* (Johansson et al., 2000); strains lacking the nucleoid proteins H-NS and StpA have a slow-growth phenotype, which can be partially suppressed by mutations in *spoT* and *relA*. These results suggest that genes regulated by (p)ppGpp are also regulated by the nucleoid structure. Notably, *hvrA* expression itself is controlled by RegA and RegB, a redox-sensitive, two-component system in *R. capsulatus* (Du et al., 1999). Specifically, *hvrA* transcription is repressed or activated under aerobic or anaerobic conditions, respectively. Taken together, these results suggest that HvrA and SpoT-like proteins are functionally linked in R. *capsulatus* to efficiently utilize energy sources (e.g., oxygen, light, amino acids, nitrogen, and carbon) in response to changing environmental conditions (Fig. 2). In support of this hypothesis, *hvrA* and *spoT*-like double mutants produce significantly lower levels of photopigments (Masuda & Bauer, 2004). Importantly, exogenously added carbon compensates for pigmentation loss in these double mutants (Masuda & Bauer, 2004), suggesting that the stringent response (induced by the SpoT-like protein) promotes photopigment synthesis in R. *capsulatus* specifically during starvation.

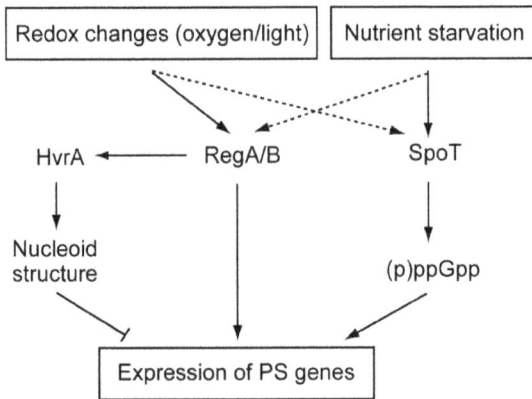

Fig. 2. A model for the coordinated regulation of photosynthesis (PS) gene expression in response to redox, light, and nutrient conditions in the purple bacterium R. *capsulatus*. Putative signaling pathways are indicated by dashed arrows. Solid arrows represent signaling pathways supported by experimental evidence.

Anabaena sp. PCC7120 is a filamentous cyanobacterium that is frequently used as a model organism to characterize nitrogen fixation in heterocysts, which are specialized cells that fix nitrogen. It is well established that nitrogen depletion induces heterocyst formation. In *Anabaena* sp. PCC7120, however, the stringent response is not likely involved in heterocyst formation, as neither SpoT-like proteins nor (p)ppGpp levels increase upon nitrogen depletion (Ning et al., 2011). In these experiments, however, bulk levels of (p)ppGpp were measured, so it is possible that (p)ppGpp levels rise specifically in heterocyst-forming cells. In fact, (p)ppGpp accumulation was previously observed upon nitrogen depletion in other cyanobacteria (Akinyanju & Amith, 1979; Friga et al., 1981). If this is true, then the activity of the SpoT-like protein would be controlled post-translationally (as in E. *coli*) because nitrogen depletion does not affect *spoT*-like expression. To understand the stringent response in cyanobacteria, therefore, genetic analysis of the *spoT*-like gene is necessary.

4. (p)ppGpp synthases and hydrolases in plants and algae

van der Biezen et al. (2000) identified RelA/SpoT-like proteins in the model plant *Arabidopsis thaliana* and designated them as RSHs. Since then, genes encoding RSHs have been found in many plant and algal species (Kasai et al., 2002; Yamada et al., 2003; Givens et al., 2004; Tozawa et al., 2007; Masuda et al., 2008a; Kim et al., 2009). Figure 3 shows a phylogenetic tree that is based on amino-acid sequences of (p)ppGpp synthase and (p)ppGpp hydrolase domains of SpoT-like and RSH proteins. In this tree, bacterial SpoT-like proteins and plant RSHs are clearly separated, although SpoT-like proteins of the bacterial phyla *Deinococcus-Thermus* form a branch with plant RSH families (RSH1). It has been suggested that plant RSHs were introduced into a proto-plant cell by endosymbiosis of an ancestral cyanobacterium (Givens et al., 2004). Our phylogenetic analysis, however, does not clearly support this hypothesis, although cyanobacterial SpoT-like proteins are relatively similar to plant RSHs. Our results may agree with a previous phylogenetic analysis, which suggested that plant RSHs were introduced into a proto-plant cell by lateral gene transfer from a pathogenic bacterium (van der Biezen, 2000). Additional experiments are necessary to clarify the origin of plant RSHs.

Arabidopsis has four RSHs, RSH1, RSH2, RSH3, and CRSH (Ca²⁺-activated RSH) (van der Biezen et al., 2000; Masuda et al., 2008a). Primary structures of these RSHs are shown in Fig. 1. Each of these four proteins has a putative chloroplast transit peptide at the N-terminus, suggesting that each functions in plastids. CRSH has two Ca^{2+}-binding domains (EF-hand motifs) at the C-terminus. Sequences similar to the (p)ppGpp synthase and hydrolase domains of *E. coli* SpoT are found in the central region of RSHs. However, the conserved Gly residue, which is necessary for RelA (p)ppGpp synthase activity, is not conserved in RSH1 (changed to Ser). In addition, the HD domain that mediates (p)ppGpp hydrolase activity in SpoT is not conserved in CRSH (Masuda et al., 2008a). Both the Gly residue and the HD domain are conserved in RSH2 and RSH3. These results suggest that RSH1 and CRSH may have only (p)ppGpp hydrolase and synthase activity, respectively, whereas RSH2 and RSH3 may have both activities.

Protein domain structures (Fig. 1) and the phylogenetic tree (Fig. 3) clearly show that Arabidopsis RSHs can be classified into three distinct families: RSH1, RSH2/3, and CRSH. Mining existing databases, all three types of RSHs can be found in another dicotyledon plant (*Nicotiana tabacum*), a monocotyledon plant (*Oryza sativa*), and a moss (*Physcomitrella patens*) (Fig. 3). RSH amino acid sequence alignments reveal that the RSH1, RSH2/3, and CRSH families each have a conserved linear arrangement of domains (Fig. 1). In addition, all RSH1-family members lack the conserved Gly residue that is critical for (p)ppGpp synthesis, all CRSH-family members lack the HD domain required for (p)ppGpp hydrolysis but have two EF-hand motifs instead, and all RSH2/3-family members contain both the Gly residue and the HD domain. These results suggest that the functional roles for these three RSH families may have been conserved between plant species. The green algae *Chlamydomonas reinhardtii* has a single *RSH* gene that does not cluster within any other plant RSH (Fig. 3), suggesting that the three plant RSH families (RSH1, RSH2/3 and CRSH) diverged after the separation of algae and mosses but before the separation of mosses and seed plants. Perhaps these three families were established when plant species adapted to terrestrial growth. Importantly, the *Chlamydomonas* RSH is not similar to bacterial SpoT-like proteins (Fig. 3), suggesting that RSH genes were not introduced into plant cells by endosymbiosis.

Recently, a small family of proteins that contain a domain resembling the SpoT HD domain was identified in metazoa, including *Drosophila melanogaster, Caenorhabditis elegans* and human (Sun et al., 2010). The protein was named Mesh1 (metazoan SpoT homolog 1). Biochemical analyses indicated that *Drosophila* Mesh1 has (p)ppGpp hydrolase activity. Arabidopsis also has a Mesh1 homolog (Sun et al., 2010), found from a cDNA clone submitted to GenBank as part of the Arabidopsis full-length cDNA cloning project (accession no. BAF00616). However, I found that the nucleotide sequence of Arabidopsis *Mesh1* is identical, at least in part, to *RSH2*. Compared to full-length *RSH2*, *Mesh1* lacks ~200 bp from the 5' end, and it contains the first and third intron sequences. These results indicate that *Mesh1*, if it represents a functional transcript, is likely a splice variant of *RSH2*. Given that no *Mesh1* homologs have been identified in other plant species, and no data concerning Mesh1 function have been reported, the putative (p)ppGpp hydrolase, Mesh1, will not be discussed here.

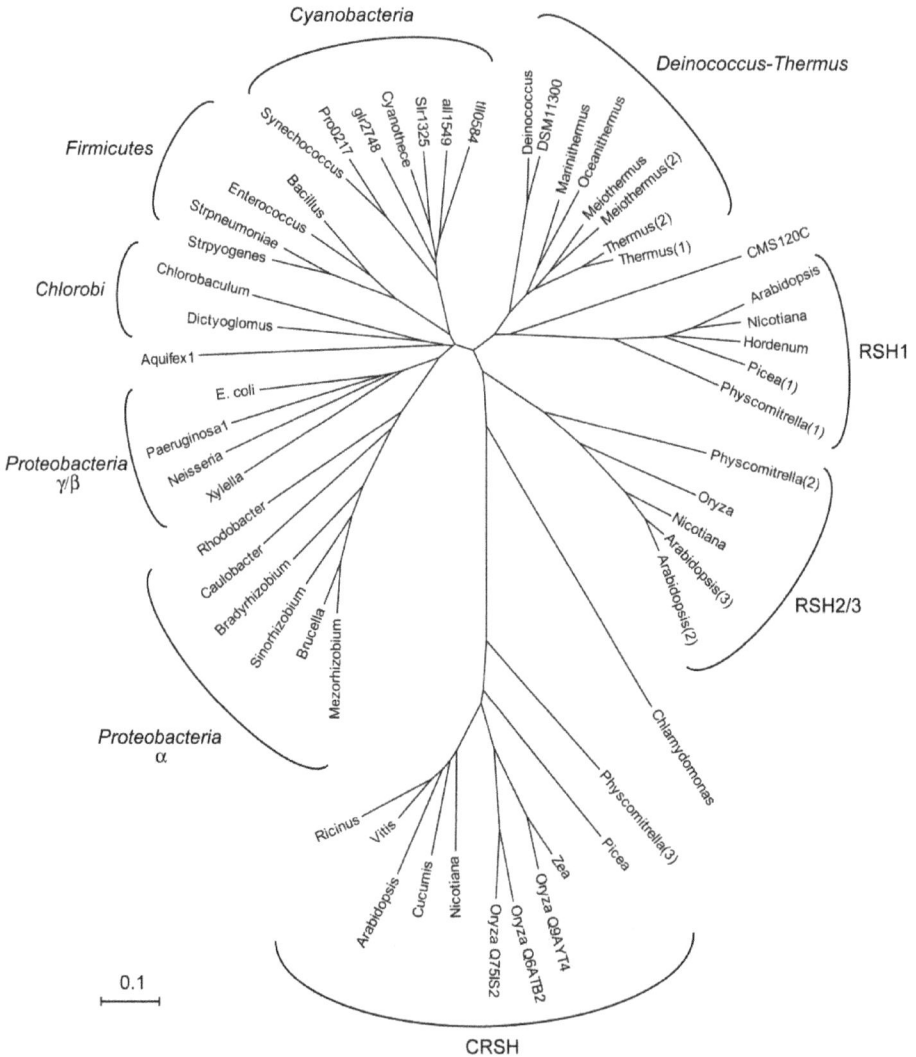

Fig. 3. A phylogenetic tree based on (p)ppGpp synthase and (p)ppGpp hydrolase domains of bacterial SpoT-like proteins and plant RSHs. The region used to construct the phylogenetic tree corresponds to amino acid residues 25–325 of *E. coli* SpoT (indicated by the dashed line in Fig. 1). All gaps in the sequence alignment were omitted, and the tree was constructed using the neighbor-joining method.

Recently, a novel (p)ppGpp degradation system was found in the bacterium *Thermus thermophilus* (Ooga et al., 2009). Specifically, a Nudix (nucleoside diphosphates linked to some moiety X) pyrophosphatase degrades ppGpp (both *in vivo* and *in vitro*) to maintain proper levels of ppGpp. Over 20 genes that encode Nudix pyrophosphatases have been identified in plants (Ogawa et al., 2005; Gunawardana et al., 2009). These proteins have pyrophosphatase activity and degrade a variety of substrates that include (d)NTPs, nucleotide sugars, NADH, and NADPH. Given that some Nudix pyrophosphatases localize to chloroplasts (Ogawa et al., 2008), these proteins could be involved in (p)ppGpp degradation by cooperating with the RSH1 and RSH2/3 families of enzymes. If this is the case, Nudix proteins could degrade (p)ppGpp to ppGp or pGpp or pGp, which are strong inhibitors of purine biosynthesis (Pao et al., 1980; Pao and Dyess, 1981). This suggests that the degradation products of (p)ppGpp may participate in the stringent response in plastids.

5. Physiological functions of RSHs in plants

Complementation analyses using *relA* and *relA/spoT* mutants in *E. coli* indicate that plant RSH2/3 and CRSH family members can functionally replace these bacterial enzymes. This suggests that plant RSHs have (p)ppGpp synthase activity (Givens et al., 2004; Tozawa et al., 2007; Masuda et al., 2008a; Mizusawa et al., 2008). This result was confirmed biochemically, clearly demonstrating that plant RSHs synthesize (p)ppGpp *in vitro* (Givens et al., 2004; Tozawa et al., 2007; Masuda et al., 2008a). These *in vitro* studies also showed that the (p)ppGpp synthase activity of CRSH is activated by Ca^{2+} (Tozawa et al., 2007; Masuda et al., 2008a). As mentioned previously, RSH1 may not have (p)ppGpp synthase activity, as it lacks the critical Gly residue found in bacterial RelA proteins. Mizusawa et al. (2008) showed that expression of Arabidopsis RSH1 does not rescue the *E. coli relA* and *relA/spoT* mutants, supporting this hypothesis. van der Biezen et al. (2000), however, reported the opposite result, concluding that RSH1 can complement the *relA* mutation. In addition, the (p)ppGpp hydrolase activities of RSH1, RSH2/3, and CRSH have not been confirmed. Clearly, these two issues require further investigation.

Protein import experiments indicate that the *Chlamydomonas* RSH is targeted to isolated chloroplasts, suggesting that the protein primarily localizes to plastids (Kasai et al., 2002). In addition, when Arabidopsis RSH1 and CRSH are fused to green fluorescent protein, they also localize to chloroplasts (Masuda et al., 2008a; Mizusawa et al., 2008). Finally, the localization of *N. tabacum* RSH2 has been studied by western blotting, which indicated that the enzyme is highly enriched in the chloroplast fraction (Givens et al., 2004). These results suggest that proteins from each RSH family in plants localize to plastids. For each plant RSH, however, the exact localization within chloroplasts seems to differ. *N. tabacum* RSH2 is in the insoluble fraction of chloroplasts (Givens et al., 2004), whereas Arabidopsis CRSH is in the soluble fraction (Masuda et al., 2008a). These localization differences may reflect different modes of enzymatic regulation. It has been suggested that membrane-associated ribosomes interact with *N. tabacum* RSH2 and regulate its activity, as is the case with bacterial RelA proteins (Givens et al., 2004). There is evidence to suggest that CRSH activity depends on Ca^{2+} levels in the chloroplast stroma (Tozawa et al., 2007; Masuda et al., 2008a). Finally, RSH1 contains two putative transmembrane helices in its C-terminal region (Fig. 1) (van der Biezen et al., 2000), suggesting that it localizes to the thylakoid and/or envelope membrane of plastids. This, however, has not been demonstrated experimentally.

The expression patterns of Arabidopsis *RSH* genes also suggest that there are functional differences between RSHs (Mizusawa et al., 2008). Microarray data indicate that *RSH* genes have diurnal rhythms of expression. Peak levels of expression are reached at noon, evening, and midnight for Arabidopsis *RSH2/3*, *RSH1*, and *CRSH*, respectively (Mizusawa et al., 2008). Given that RSH2/3 enzymes have both synthase and hydrolase activities, they likely maintain plastidial (p)ppGpp levels during daylight hours. At evening, RSH1 may then degrade (p)ppGpp, as it seems to have only (p)ppGpp hydrolase activity. At night, (p)ppGpp levels are likely kept low because keeping plants in the dark reduces cellular (p)ppGpp levels (Takahashi et al., 2004). These data suggest that (p)ppGpp is required to control light-dependent plastidial activities, such as photosynthesis. At night, CRSH likely maintains (p)ppGpp levels based on Ca^{2+} levels. Plastidial Ca^{2+} concentrations change in response to environmental stimuli, such as light conditions (Johnson et al., 1995; Sai & Johnson, 2002), allowing CRSH to translate these environmental changes into (p)ppGpp synthesis.

Expression of *RSH2* genes in Arabidopsis, rice and *N. tabacum* is elevated in response to cold and/or exogenous application of the plant hormone jasmonic acid (or its precursor 12-oxo-phytodienoic acid) (Xiong et al., 2001; Givens et al., 2004; Lee et al., 2005; Mizusawa et al., 2008). (p)ppGpp levels similarly increase in response to jasmonic acid (Takahashi et al., 2004). These observations suggest that the RSH2/3 family synthesizes (p)ppGpp in response to abiotic stresses. The plastidial stringent response may play a role in these types of plant defense responses, although the mechanisms remain unclear.

Histochemical analyses indicate that Arabidopsis *RSH2* and *RSH3* are expressed in all green tissue (Mizusawa et al., 2008), suggesting that the RSH2/3 family controls light-dependent plastidial function. Interestingly, all *RSH* genes are expressed at high levels in flower tissues, such as the pistil and stamen, suggesting an important role for (p)ppGpp in flower development (Mizusawa et al., 2008). Genetic knockdown of *CRSH* disrupts flower morphology, as pistil and stamen development are not coordinated. This defect results in infertility (Masuda et al., 2008a), indicating that RSH-dependent (p)ppGpp production is required for plant reproduction. As discussed below, (p)ppGpp may control the biosynthesis of amino acids, fatty acids, and nucleotides in plastids, as has been observed in bacteria. Because these compounds serve as precursors for plant hormones (e.g., jasmonates, cytokinin, and auxin), the plastidial stringent response may control host plant development by regulating hormone biosynthesis.

6. Which plastidial functions are regulated by (p)ppGpp?

In bacteria, (p)ppGpp regulates the transcription of a large number of genes *via* two distinct mechanisms. First, (p)ppGpp allosterically controls RNA polymerase activity through direct association with the β- or β'- subunit of the polymerase (Chatterji et al., 1998; Toulokhonov et al., 2001; Artsimovitch et al., 2004). Second, RelA- and SpoT-dependent (p)ppGpp synthesis uses ATP and GTP (or GDP) as substrates. This significantly reduces the amount of NTPs available for RNA synthesis, thereby indirectly decreasing RNA polymerase activity (Krasny & Gourse, 2004). In chloroplasts, two RNA polymerases transcribe the plastid genome (Shiina et al., 2005; Liere et al., 2011). One is the bacterial type of RNA polymerase called plastid-encoded plastid RNA polymerase (PEP). The other is the T7-phase type of RNA polymerase called nuclear-encoded plastid RNA polymerase (NEP). As with *E. coli* RNA polymerase, (p)ppGpp directly binds to PEP (Sato et al., 2009) and can inhibit PEP- mediated transcription when exogenously added *in vitro* (Sato et al., 2009). This suggests that (p)ppGpp directly controls PEP activity. The second "indirect" control of plastidial RNA polymerases needs to be studied in more detail. If this mechanism is real, both PEP and NEP activities should be affected. It has recently been reported that some plant hormones, including jasmonates and auxin, affect transcription of chloroplast genes (Zubo et al., 2011). Because *RSH2* expression is induced by jasmonates (Givens et al., 2004; Mizusawa et al., 2008), it is possible that RSH2-dependent (p)ppGpp synthesis mediates the effects of plant hormones on plastidial gene expression.

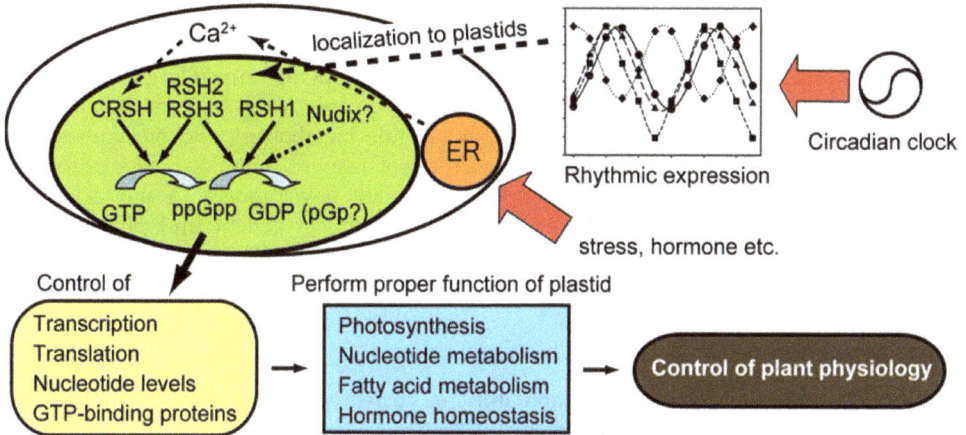

Fig. 4. A model for the stringent response in higher plants.

In addition to regulating transcription, (p)ppGpp also controls translation in bacteria (Milon et al., 2006). The translation initiation factor, IF2, binds and hydrolyzes GTP to initiate translation. (p)ppGpp binds to the GTP-binding pocket of IF2, thereby inhibiting translation initiation (Milon et al., 2006). Given that a bacterial IF2 homolog is found in chloroplasts (Miura et al., 2007), (p)ppGpp may also control translation of plastid genes; the chloroplast genome encodes genes involved in photosynthesis, electron transfer, and fatty-acid biosynthesis, for example. These various plastidial functions, therefore, should be regulated by the (p)ppGpp-dependent stringent response.

Bacteria produce several GTP-binding proteins, some of which are conserved in plants and function in chloroplasts (Mittenhuber, 2001b; Masuda et al., 2008b). Because (p)ppGpp interacts with the GTP-binding pocket of IF2, chloroplast GTP-binding proteins may also be regulated by (p)ppGpp. Several enzymes involved in nucleotide biosynthesis are regulated by (p)ppGpp in an allosteric manner (Gallant et al., 1971; Hou et al., 1999). It is thought that nucleotide biosynthesis can take place in plastids because one of the enzymes that catalyzes phosphoribosyl diphosphate synthesis (the first step in purine and pyrimidine biosynthesis) localizes to plastids (Krath & Hove-Jensen, 1999). As a result, nucleotide biosynthesis in plastids may be directly regulated by (p)ppGpp. Furthermore, consumption of GTP (GDP) and ATP during (p)ppGpp synthesis may also indirectly influence nucleotide metabolism in plastids.

7. Concluding remarks

It has been almost one-half century since (p)ppGpp was first discovered in *E. coli*. Since then, the physiological roles for (p)ppGpp in controlling bacterial cell metabolism have been well documented. The role of the stringent response in photosynthetic bacteria, however, remains unclear, even though photosynthesis is one of the most important anabolic reactions in biology. Future studies are needed if we are to understand how the stringent response controls different types of photosynthesis.

Many plant and algal species produce (p)ppGpp synthases and hydrolases called RSHs. Studies in Arabidopsis indicate that RSHs can be classified into three distinct families, RSH1, RSH2/3 and CRSH, all of which function in plastids. *RSH* gene expression profiles and the domain structures of RSHs suggest that RSH families are functionally diverse. Furthermore, these functional differences are likely necessary to properly regulate plastidial (p)ppGpp levels. Although the specific target proteins of (p)ppGpp remain largely unknown in plastids, the RSH-dependent stringent response regulates many aspects of plastidial function, including transcription, translation, nucleotide metabolism, and biosynthesis of amino acids and fatty acids. As a result, the stringent response may also regulate plant hormone biosynthesis, which is required for host plant development. A model of the stringent response in plastids is shown in Fig. 4. Additional genetic and physiological experiments are needed if we are to understand the precise roles of the (p)ppGpp-mediated stringent response in higher plants.

8. Acknowledgment

The work cited from the author's laboratory is currently funded by the Ministry of Education, Culture, Science and Technology of Japan.

9. References

Akinyanju, J. & Smith, R. J. (1979). Accumulation of ppGpp and pppGpp during nitrogen deprivation of the cyanophyte *Anabaena cylindrica*. *FEBS Lett.* 107: 173-176.

Aravind, L. & Koonin, E. V. (1998). The HD domain defines a new superfamily of metal-dependent phosphohydrolases. *Trends Biochem. Sci.* 23: 469-472.

Artsimovitch, I.; Patlan, V.; Sekine, S.; Vassylyeva, M. N.; Hosaka, T.; Ochi, K.; Yokoyama, S. & Vassylyev, D. G. (2004). Structural basis for transcription regulation by alarmone ppGpp. *Cell* 117: 299-310.

Bauer, C. E. (2004). Regulation of photosystem synthesis in *Rhodobacter capsulatus*. *Photosynth. Res.* 80: 353-360.

Bauer, C.; Elsen, S.; Swem, L. R.; Swem, D. L. & Masuda, S. (2003). Redox and light regulation of gene expression in photosynthetic prokaryotes. *Phil. Trans. R. Soc. Lond. B* 358: 147-154.

Bertin, P.; Benhabiles, N.; Krin, E.; Laurent-Winter, C.; Tendeng, C.; Turlin, E.; Thomas, A.; Danchin, A. & Brasseur, R. (1999). The structural and functional organization of H-NS-like proteins is evolutionarily conserved in gram-negative bacteria. *Mol. Microbiol.* 31: 319-329.

Braeken, K.; Moris, M.; Daniels, R.; Vanderleyden, J. & Michiels, J. (2006). New horizons for (p)ppGpp in bacterial and plant physiology. *Trends Microbiol.* 14: 45-54.

Bryant, D.A. & Frigaard, N.U. (2006). Prokaryotic photosynthesis and phototrophy illuminated. *Trends Microbiol.* 14: 488-496.

Bryant, D.A.; Costas, A.M.; Maresca, J.A.; Chew, A.G.; Klatt, C.G.; Bateson, M.M.; Tallon, L.J.; Hostetler, J.; Nelson, W.C.; Heidelberg, J.F. & Ward, D.M. (2007). Candidatus *Chloracidobacterium thermophilum*: an aerobic phototrophic *Acidobacterium*. *Science* 317: 523-526.

Buggy, J. J.; Sganga, M. W. & Bauer, C. E. (1994). Characterization of a light-responding trans-activator responsible for differentially controlling reaction center and light-harvesting-I gene expression in *Rhodobacter capsulatus*. *J. Bacteriol.* 176: 6936-6943.

Chatterji, D.; Fujita, N. & Ishihama, A. (1998). The mediator for stringent control, ppGpp, binds to the beta-subunit of *Escherichia coli* RNA polymerase. *Genes Cells* 3: 279-287.

Cashel, M.; Gentry, D. R.; Hernandez, V. J. & Vinella, D. (1996). The stringent response, In: *Escherichia coli and Salmonella: Cellular and Molecular Biology*, 2nd ed., Neidhardt, F. C.; Curtiss, III R.; Ingraham, J. L.; Lin, E. C. C.; Low, K. B.; Magasanik, B.; Reznikoff, W. S.; Riley, M.; Schaechter, M. & Umbarger, H. E., pp 1458-1496, AMS Press, Washington D.C.

Dorman, C. J. & Deighan, P. (2003). Regulation of gene expression by histone-like proteins in bacteria. *Curr. Opin. Genet. Dev.* 13: 179-184.

Du, S.; Kouadio, J. L. & Bauer, C. E. (1999). Regulated expression of a highly conserved regulatory gene cluster is necessary for controlling photosynthesis gene expression in response to anaerobiosis in *Rhodobacter capsulatus*. *J. Bacteriol.* 181: 4334-4341.

Friga, G. M.; Borbely, G. & Farkas, G. L. (1981). Accumulation of guanosine tetraphosphate (ppGpp) under nitrogen starvation in *Anacystis nidulans*, a cyanobacterium. *Arch. Microbiol.* 129: 341-343.

Gallant, J.; Irr, J. & Cashel, M. (1971). The mechanism of amino acid control of guanylate and adenylate biosynthesis. *J. Biol. Chem.* 246: 5812-5816.

Givens, R. M.; Lin, M.H.; Taylor, D.J.; Mechold, U.; Berry, J.O. & Hernandez, V.J. (2004). Inducible expression, enzymatic activity, and origin of higher plant homologues of bacterial RelA/SpoT stress proteins in *Nicotiana tabacum*. *J. Biol. Chem.* 279: 495-504.

Gunawardana, D.; Likic, V. A. & Gayler, K. R. (2009). A comprehensive bioinformatics analysis of the Nudix superfamily in *Arabidopsis thaliana*. *Comp. Funct. Genomics.* 820381.

Hogg, T.; Mechold, U.; Malke, H.; Cashel, M. & Hilgenfeld, R. (2004). Conformational antagonism between opposing active sites in a bifunctional RelA/SpoT homolog modulates (p)ppGpp metabolism during the stringent response. *Cell* 117: 57-68.

Hou, Z.; Cashel, M.; Fromm, H. J. & Honzatko, R. B. (1999). Effectors of the stringent response target the active site of *Escherichia coli* adenylosuccinate synthetase. *J. Biol. Chem.* 274: 17505-17510.

Jain, V.; Kumar, M. & Chatterji, D. (2006). ppGpp: Stringent response and survival. *J. Microbiol.* 44: 1-10.

Johansson, J.; Balsalobre, C.; Wang, S. Y.; Urbonaviciene, J.; Jin, D. J.; Sonden, B. & Uhlin, B. E. (2000). Nucleoid proteins stimulate stringently controlled bacterial promoters: a link between the cAMP-CRP and the (p)ppGpp regulons in *Escherichia coli*. *Cell* 102: 475-485.

Johnson, C. H.; Knight, M. R.; Kondo, T.; Masson, P.; Sedbrook, J.; Haley, A. & Trewavas, A. (1995). Circadian oscillations of cytosolic and chloroplastic free calcium in plants. *Science* 29: 1863-1865

Kasai, K.; Usami, S.; Yamada, T.; Endo, Y.; Ochi, K. & Tozawa, Y. (2002). A RelA-SpoT homolog (Cr-RSH) identified in *Chlamydomonas reinhardtii* generates stringent factor *in vivo* and localizes to chloroplasts *in vitro*. *Nucleic. Acids Res.* 30: 4985-4992.

Kern, M.; Kamp, P. B.; Paschen, A.; Masepohl, B. & Klipp, W. (1998). Evidence for a regulatory link of nitrogen fixation and photosynthesis in *Rhodobacter capsulatus* via HvrA. *J. Bacteriol.* 180: 1965-1969.

Kim, T-H.; Ok, S. H.; Kim, D.; Suh, S-C.; Byun, M. O. & Shin, J. S. (2009). Molecular characterization of a biotic and abiotic stress resistance-related gene RelA/SpoT homologue (*PepRSH*) from pepper. *Plant Sci.* 176: 635-642.

Krasny, L. & Gourse, R. L. (2004). An alternative strategy for bacterial ribosome synthesis: *Bacillus subtilis* rRNA transcription regulation. *EMBO J.* 23: 4473-83.

Krath, B. N. & Hove-Jensen, B. (1999). Organellar and cytosolic localization of four phosphoribosyl diphosphate synthase lsozymes in spinach. *Plant Physiol.* 119: 497-505.

Lee, B-h.; Henderson, D. A. & Zhu, J-K. (2005). The Arabidopsis cold-responsive transcriptome and its regulation by ICE1. *Plant Cell* 17: 3155-3175.

Lemos, J. A.; Lin, V. K.; Nascimento, M. M.; Abranches, J. & Burne, R. A. (2007). Three gene products govern (p)ppGpp production by *Streptococcus mutans*. *Mol. Microbiol.* 65: 1568-1581.

Liere, K.; Weihe, A. & Borner, T. (2011). The transcription machineries of plant mitochondria and chloroplasts: Composition, function, and regulation. *J. Plant Physiol.* 168: 1345-1360.

Magnusson, L. U.; Farewell, A. & Nystrom, T. (2005). ppGpp: a global regulator in *Escherichia coli*. *Trends Microbiol.* 13: 236-242.

Masuda, S. & Bauer, C. E. (2004). Null mutation of HvrA compensates for loss of an essential *relA/spoT*-like gene in *Rhodobacter capsulatus*. *J. Bacteriol.* 186: 235-239.

Masuda, S.; Mizusawa, K.; Narisawa, T.; Tozawa, Y.; Ohta, H. & Takamiya, K. (2008a). The bacterial stringent response, conserved in chloroplasts, controls plant fertilization. *Plant Cell Physiol.* 49: 135-141

Masuda, S.; Tozawa, Y. & Ohta, H. (2008b). Possible target of "magic spots" in plant signaling. *Plant Signaling Behav.* 3: 1021-1023.

Milon, P.; Tischenko, E.; Tomsic, J.; Caserta, E.; Folkers, G.; Teana, A. L.; Rodnina, M. V.; Pon, C. L.; Boelens, R. & Gualerzi, C. O. (2006). The nucleotide-binding site of bacterial translation initiation factor 2 (IF2) as a metabolic sensor. *Proc. Natl. Acad. Sci. USA* 103: 13962-13967.

Miura, E.; Kato, Y.; Matsushima, R.; Albrecht, V.; Laalami, S. & Sakamoto, W. (2007). The balance between protein synthesis and degradation in chloroplasts determines leaf variegation in Arabidopsis yellow variegated mutants. *Plant Cell* 19: 1313-1328.

Mizusawa, K.; Masuda, S. & Ohta, H. (2008). Expression profiling of four RelA/SpoT-like proteins, homologues of bacterial stringent factors, in Arabidopsis. *Planta* 228: 553-562.

McLeod, S. M. & Johnson, R. C. (2001). Control of transcription by nucleoid proteins. *Curr. Opin. Microbiol.* 4: 152-159.

Mittenhuber, G. (2001a). Comparative genomics and evolution of genes encoding bacterial (p)ppGpp synthetases/hydrolases (the Rel, RelA and SpoT proteins). *J. Mol. Microbiol. Biotechnol.* 3: 585-600.

Mittenhuber, G. (2001b). Comparative genomics of prokaryotic GTP-binding proteins (the Era, Obg, EngA, ThdF (TrmE), YchF and YihA families) and their relationship to eukaryotic GTP-binding proteins (the DRG, ARF, RAB, RAN, RAS and RHO families). *J. Mol. Microbiol. Biotechnol.* 3: 21-35.

Nanamiya, H.; Kasai, K.; Nozawa, A.; Yun, C. S.; Narisawa, T.; Murakami, K.; Natori, Y.; Kawamura, F. & Tozawa, Y. (2008). Identification and functional analysis of novel (p)ppGpp synthetase genes in Bacillus subtilis. *Mol. Microbiol.* 67: 291-304.

Ning, D.; Qian, Y.; Miao, X. & Wen, C. (2011). Role of the all1549 (ana-rsh) gene, a relA/spot homolog, of the cyanobacterium *Anabaena* sp. PCC7120. *Curr. Microbiol.* 62: 1767-1773.

Ochi, K. (2007). From microbial differentiation to ribosome engineering. *Biosci. Biotechnol. Biochem.* 71: 1371-1386.

Ogawa, T.; Ueda, Y.; Yoshimura, K. & Shigeoka, S. (2005). Comprehensive analysis of cytosolic Nudix hydrolases in Arabidopsis thaliana. *J. Biol. Chem.* 280: 25277-25283.

Ogawa, T.; Yoshimura, K.; Miyake, H.; Ishikawa, K.; Ito, D.; Tanabe, N. & Shigeoka, S. (2008). Molecular characterization of organelle-type Nudix hydrolases in Arabidopsis. *Plant Physiol.* 148: 1412-1424.

Ooga, T.; Ohashi, Y.; Kuramitsu, S.; Koyama, Y.; Tomita, M.; Soga, T. & Masui, R. (2009). Degradation of ppGpp by nudix pyrophosphatase modulates the transition of growth phase in the bacterium *Thermus thermophiles. J. Biol. Chem.* 284: 15549-15556.

Pao, C. C.; Dennis, P. P. & Gallant, J. A. (1980). Regulation of ribosomal and transfer RNA synthesis by Guanosine 5'-diphosphate-3'-monophosphate. *J. Biol. Chem.* 255: 1830-1833.

Pao, C. C. & Dyess, B. T, (1981). Effect of unusual guanosine nucleotides on the activities of some *Escherichia coli* cellular enzymes. *Biochim. Biophys. Acta* 677: 358-362.

Potrykus, K. & Cashel, M. (2008). (p)ppGpp: still magical? *Annu. Rev. Microbiol.* 62: 35-51.

Sai, J. & Johnson, C. H. (2002). Dark-stimulated calcium iron fluxes in the chloroplast stroma and cytosol. *Plant Cell* 14: 1279-1291.

Sato, M.; Takahashi, K.; Ochiai, Y.; Hosaka, T.; Ochi, K. & Nabeta, K. (2009). Bacterial alarmone, guanosine 5'-diphosphate 3'-diphosphate (ppGpp), predominantly binds the □' subunit of plastid-encoded plastid RNA polymerase in chloroplasts. *ChemBioChem* 10: 1227-1233.

Shiina, T.; Tsunoyama, Y.; Nakahira, Y. & Khan, M. S. (2005). Plastid RNA polymerases, promoters, and transcription regulators in higher plants. *Int. Rev. Cytol.* 244: 1-68.

Srivatsan, A. & Wang, J. D (2008). Control of bacterial transcription, translation and replication by (p)ppGpp. *Curr. Opin. Microbiol.* 11: 100-105.

Sun, D.; Lee, G.; Lee, J. H.; Kim, H. Y.; Rhee, H. W.; Park, S. Y.; Kim, K. J.; Kim, Y.; Kim, B. Y.; Hong, J. I.; Park, C.; Choy, H. E.; Kim, J. H.; Jeon, Y. H. & Chung, J. (2010). A metazoan ortholog of SpoT hydrolyzes ppGpp and functions in starvation responses. *Nat. Struct. Mol. Biol.* 17: 1188-1194.

Swem D. L. & Bauer, C. E. (2002). Coordination of ubiquinol oxidase and cytochrome cbb(3) oxidase expression by multiple regulators in *Rhodobacter capsulatus. J. Bacteriol.* 184: 2815-2820.

Takahashi, K.; Kasai, K. & Ochi, K. (2004). Identification of the bacterial alarmone guanosine 5'-diphosphate 3'-diphosphate (ppGpp) in plants. *Proc. Natl. Acad. Sci. USA* 101: 4320-4324.

Toulokhonov, II; Shulgina, I. & Hernandez, V. J. (2001). Binding of the transcription effector ppGpp to *Escherichia coli* RNA polymerase is allosteric, modular, and occurs near the N terminus of the beta'-subunit, *J. Biol. Chem.* 276: 1220-1225

Tozawa, Y.; Nozawa, A.; Kanno, T.; Narisawa, T.; Masuda, S.; Kasai, K. & Nanamiya, H. (2007). Calcium-activated (p)ppGpp synthetase in chloroplasts of land plants. *J. Biol. Chem.* 282: 35536-35545.

van der Biezen, E. A.; Sun, J.; Coleman, M. J.; Bibb, M. J. & Jones, J. D. (2000). *Arabidopsis* RelA/SpoT homologs implicate (p)ppGpp in plant signaling. *Proc. Natl. Acad. Sci. USA* 97: 3747-3752.

Wendrich, T. M.; Beckering, C. L. & Marahiel, M. A. (2000). Characterization of the *relA/spoT* gene from Bacillus stearothermophilus. *FEMS Microbiol. Lett.* 190: 195-201.

Xiong, L.; Lee M-W.; Qi, M. & Yang, Y. (2001) Identification of defense-related rice genes by suppression subtractive hybridization and differential screening. *Mol. Plant Micro. Interct.* 14: 685-692.

Yamada, A.; Tsutsumi, K.; Tanimoto, S. & Ozeki, Y. (2003). Plant RelA/SpoT homolog confers salt tolerance in *Escherichia coli* and *Saccharomyces cerevisiae*. *Plant Cell Physiol.* 44: 3-9.

Zubo, Y. O.; Yamburenko, M. V.; Kusnetsov, V. V. & Borner, T. (2011). Methyl jasmonate, gibberellic acid, and auxin affect transcription and transcript accumulation of chloroplast genes in barley. *J. Plant Physiol.* 168: 1335-1344.

12

Transglutaminase is Involved in the Remodeling of Tobacco Thylakoids

Nikolaos E. Ioannidis[1], Josep Maria Torné[2],
Kiriakos Kotzabasis[1] and Mireya Santos[2]
[1]Department of Biology, University of Crete, Heraklion, Crete,
[2]Departament de Genètica Molecular, Centre for Research in Agricultural Genomics,
CRAG-CSIC-IRTA-UAB, Barcelona,
[1]Greece
[2]Spain

1. Introduction

Photosynthesis light reactions are among the more fast, complex and important processes in the ecosystem. They take place in specific membranes the so-called thylakoids and they produce O_2, energy (ATP) and reducing equivalents (NADPH). In this chapter we will discuss recent findings that shed light in important aspects of thylakoid architecture and functional organization. Key role for the remodeling of thylakoids plays a plastidal transglutaminase that was recently cloned from maize. Transglutaminases (TGases, EC 2.3.2.13) are intra- and extra-cellular enzymes that catalyze post-translational modification of proteins by establishing ε-(γ-glutamyl) links and covalent conjugation of polyamines. Transglutaminase (TGase) activity is present in chloroplasts of higher plants being PSII antenna proteins the enzyme's natural substrates. Although the functionality of this plastidial enzyme is not clear, a role in antenna regulation has been hypothesized. The isolation, for the first time in plants, of two related complementary maize DNA clones, *tgz15* and *tgz21*, encoding active maize (*Zea mays L*) chloroplastic TGase (chlTGZ) has contributed to deep on the role of this enzyme in plants (Torné et al. 2002; Villalobos et al 2004). In addition, the main polyamines, putrescine (Put), spermidine (Spd) and spermine (Spm) are normally produced and oxidized in chloroplasts. Thus, all types of post-translational modifications (i.e. mono-Put, mono-Spd, mono-Spm, bis-Put, bis-Spd and bis-Spm) are in theory probable for the target proteins. These modifications may alter charge and/or conformation of the target protein as well as their linking with other proteins (Kotzabasis et al. 1993; Del Duca et al. 1994; Della Mea et al. 2004).

A strong tool for a deeper study of gene functionality is the effect of its over-expression in an heterologous plant system. Here we will discuss in detail the information about the recent chlTGZ over-expression in tobacco (*Nicotiana tabacum* var. Petit Havana) chloroplasts (Ioannidis et al. 2009) and its characterization. After chloroplast transformation, transglutaminase activity in TGZ-over-expressers was up-regulated 4-fold with respect to the wild-type plants, which in turn rised its thylakoid-associated polyamine content about 90%. A major increase in the granum size (i.e. increase in the number of stacked layers) accompanied by a concomitant decrease of stroma thylakoids in the TGase over-expressers was observed. Functional comparison between wild type tobacco and chlTGZ over-expressers was according to these observations, and illustrated in terms of fast fluorescence induction kinetics, non-photochemical quenching of the singlet excited state of chlorophyll a and antenna heterogeneity of PSII. Both *in vivo* probing and extensive electron microscopy studies indicated thylakoid remodeling. PSII antenna heterogeneity in vivo changes in the over-expressers to a great extent, with an increase of the centers located in grana-appressed regions (PSIIα) at the expense of centers located mainly in stroma thylakoids (PSIIβ). Finally, late stages of plant development present alterations in the photosynthetic apparatus, chloroplast ultrastructure, and, particularly, oxidative and antioxidative metabolism pathways are induced (Ortigosa et al 2010). At the same time, the over-expressed TGZ protein, accumulated progressively in chloroplast inclusion bodies (Villar-Piqué et al. 2010). These results are discussed in line with chlTGZ involvement in chloroplast functionality.

2. Thylakoids and photosynthesis

The chloroplasts of higher plants are bounded by two envelope membranes that surround an aqueous matrix, the stroma, and the internal photosynthetic membranes, the thylakoids (Staehelin & van der Staay 1996). Chloroplasts have an apparently periodic ultrastructure: cylindrical grana stacks of about 10–20 layers with a diameter of 300–600 nm, interconnected by lamellae of several hundred nm in length (Mustardy & Garab 2003). Although our understanding regarding architecture of thylakoids is advanced, many issues such as self-assembly and structural flexibility, still remain to be explored (Mustardy & Garab 2003).

The two photosystems are spatially separated in thylakoids in vivo : photosystem II and its main chlorophyll a/b light-harvesting complex, (LHCII), are found predominantly in the stacked membranes; this region is largely deficient in photosystem I (PSI), LHCI and ATPase, which are enriched in the stroma membranes (Andersson & Andersson 1980). Separation of the two pigment systems is probably important in preventing unregulated excitation energy flow between the two photosystems (Andersson & Andersson 1988). Without this, PSI, which is much faster than PSII, would disturb the balance of the energy distribution between the two photosystems (Trissl & Wilhelm 1993). Also PSII exhibits a heterogeneity in terms of antenna size with centers of large chlorophyll antenna size termed PSIIα (occur in grana) and of smaller antenna termed PSIIβ (occur in stroma lamellae) (Melis & Homann 1976; Melis 1989; Kirschhoff et al 2007; Kaftan et al. 1999).

The abundance of LHCII in the granum suggests that these antenna complexes also play a structural role. Indeed, LHCII has been shown to stabilize the granum ultrastructure, and to participate in the cation-mediated stacking of the membranes (Staehelin & van der Staay 1996; Kirchhoff et al. 2007; Arnzten 1978; Duniec et al 1981; Barber 1982). These light-harvesting complexes have also been shown to be involved, via electrostatic and osmotic forces, in the lateral organization of the membranes (Garab et al. 1991). Previous studies showed that the strength of stacking is affected by the phosphorylation of LHCII and of several other phosphoproteins (Allen et al. 1981). LHCII is largely responsible for the organization of the plant photosynthetic system by maintaining the tight appression of thylakoid membranes in chloroplast grana (Allen & Forsberg 2001). An important role for this effect plays the stromal surface of the LHCII trimer which is mainly flat and negatively charged as demonstrated by recent structural studies in higher plants (Standfuss et al. 2005). This complex collects excitation energy and transfers it to the reaction centres of PSII and PSI (van Amerongen & Dekker 2003). Also, LHCII prevents damage to the photosynthetic system by several different mechanisms when there is too much light. Potentially harmful chlorophyll (Chl) triplets are quenched by carotenoids in the complex while a special mechanism, referred to as nonphotochemical quenching (NPQ), has evolved in plants to dissipate excess energy as heat (Pascal et al. 2005).

The interplay between grana and stroma lamellae regions is of exceptional importance because it defines the available space for photosystems and the other supercomplexes of the photosynthetic apparatus such as ATPase. It is well established that "sun" and "shade" plants show distinct differences in the organization of their thylakoid system (Staehelin & van der Staay 1996). In turn, this affect the efficiency with which light is harvested and utilized.

With respect to thylakoid membrane biogenesis, Wang et al. 2004 showed that the *Thf1* gene product played a crucial role in a dynamic process of vesicle-mediated thylakoid membrane biogenesis in *Arabidopsis*. Recently, Chi et al. 2008 have reported that a rice thioredoxin *m* isoform (*Ostrxm*) seems to be required for chloroplast biogenesis and differentiation. However, the factors that determine grana formation are not yet fully understood.

3. Transglutaminases and polyamines

Transglutaminases (TGases) are intracellular and extra cellular enzymes that catalyse post-translational modification of proteins by establishing ε-(γ-glutamyl) links and covalent conjugation of polyamines (Lorand & Graham 2003). However, the role of TGases in chloroplast is not fully understood yet. Maize (*Zea mays L.*) TGase was immunodetected in meristematic calli and their isolated chloroplasts, as a unique 58 kDa band. The activity was shown to be light sensitive, affected by hormone deprivation and with a light/dark rhythm (Bernet 1997; Bernet et al 1999). Subcellular localization studies showed that, in adult plants, the enzyme was specifically localized in the chloroplast grana-appressed thylakoids and close to LHCII and its abundance depended on the degree of grana development (Villalobos et al 2001; Villalobos 2007; Santos et al. 2007). An important step for the elucidation of the plastidal TGase role in plants was the isolation for the first time in plants of two related complementary maize DNA clones, *tgz15* and *tgz21,* encoding active maize TGase (Torné et al. 2002; Villalobos et al. 2004). Interestingly, their expression is dependent on the duration of light exposure, indicating a role for adaptation in different light environmental conditions including natural habitats (Pintó-Marijuan et al 2007; Carvajal et al. 2007-2011). Proteomic studies indicates that plastidial maize TGase is a peripheral thylakoid protein forming part of a specific PSII protein complex which includes LHCII, ATPase and PsbS proteins, its expression pattern changing according to chloroplast developmental stage and light regime (Campos et al. 2010). Tacking into account all the described results, it has been hypothesized that TGases are implicated in the photosynthetic process (Villalobos et al 2004; Pintó-Marijuan et al 2007; Serafini-Fracassini & Del Duca, 2008).

A rather overlooked post-translational modification of LHCII that might be important for stacking of thylakoids is its polyaminylation. Polyamines (PAs) are low molecular weight aliphatic amines that are almost fully protonated under normal pH values and thus possess a net charge of up to +4. The main polyamines putrescine (Put), spermidine (Spd) and spermine (Spm) are normally found in the LHCII of higher plants (Kotzabasis et al. 1993a). Plastidal Transglutaminases might attach covalently polyamines of all thylakoid proteins specifically in LHCII, CP29, CP26 and CP24 (Del Duca et al 1994). Recently, it was demonstrated that a plastidial TGase activity in maize polyaminylates purified LHCII catalyzing the production of mono and bis glutamyl PAs in a light dependent way (Della Mea et al 2004). As commented elsewhere, authors indicated that the additional positive charges inserted on proteins by the protein-bound PAs might induce conformational changes by conjugation of the two terminal amino-groups of PAs to one or two glutamine residues of LHCII and they discussed if light sensitivity is due to the enzyme or to the substrate. In the work of Carvajal et al. (2011), when purified plastidial maize TGase (TGZ) was added to maize thylakoid protein extracts,TGase activity was significantly higher (in a light dependent manner) than that of the same extract without TGZ addition, indicating that thylakoid proteins are the specific substrate of TGZ. However, in the same work it is demonstrated that, if a non-plant protein was used as TGZ substrate, TGase activity was not light-dependent. These last results indicate that light dependence of plastidial TGase activity is probably related to its specific substrate (thylakoid proteins) and not to the enzyme itself.

First evidence for a role of plastidial TGase in the thylakoids 3D architecture comes from tobacco chloroplasts over expressing maize TGase (TGZ) (Ioannidis et al 2009). In that work, we hypothesized that TGase is implicated in the ratio regulation of grana to stroma thylakoids (Villalobos et al. 2004). A combination of genetic engineering and *in vivo* probing approach was used to test this hypothesis. Here we discuss in detail the information about the effect of maize *tgz* gene over-expression and its characterization in tobacco chloroplasts via plastid transformation, where the transgene is integrated in the plastid genome by homologous recombination (Maliga 2004; Fernández-San Millán et al 2007, 2008).

4. Maize transglutaminase over-expressed in tobacco chloroplasts

4.1 Vector construction, chloroplast transformation and plant regeneration

To introduce the *tgz13* gene into tobacco Wt chloroplasts, the *tgz* gene was PCR amplified, fused to the promoter and 5'untranslated region of the *psbA* gene and finally introduced into the multiple cloning site of the pAF vector, rendering the final vector, pAF-*tgz13* (Fig. 1A). The pAF vector was specifically constructed for tobacco plastid transformation and includes the *trnI* and *trnA* border sequences, homologous to the inverted repeat regions of the tobacco plastid genome (Fernández-San Millán et al 2008). The regulatory sequences of the *psbA* gene were chosen due to the high levels of heterologous gene expression they confer in transplastomic plants (Fernández-San Millán et al 2003; Molina et al 2004). After that, leaves of tobacco Wt plants were bombarded with gold microprojectils coated with plasmid DNA containing the *tgz* gene and plants were regenerated in the selective spectinomycin medium. Southern blot analysis performed on shoots developed after the second round of selection with spectinomycin revealed some plants that were homoplasmic for the *tgz* gene (Fig. 1) (Ioannidis et al. 2009). These experiments were carried out in J. Veramendi laboratory (Public Univ. Navarra. Spain).

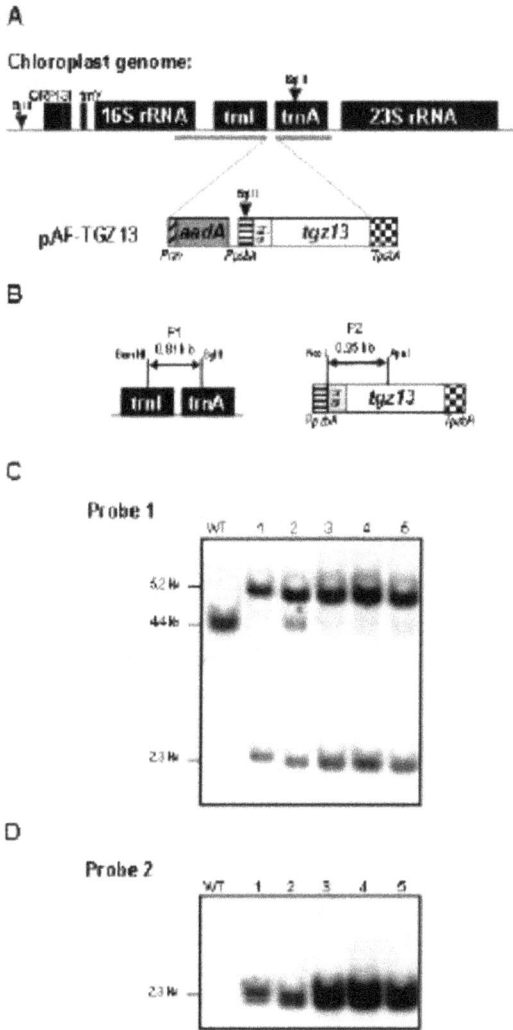

Fig. 1. Schematic representation of tobacco plastid genome transformation using the maize transglutaminase *tgz* gene. A, map of the wild-type and *tgz*-transformed genomes. Regions for homologous recombination are underlined in the native chloroplast genome; B, the 0.81 kb fragment (P1) of the targeting region for homologous recombination and the 0.95 kb *tgz* sequence (P2) were used as probes for Southern blot analysis; C, D, Southern blot analysis of five independent transgenic lines is shown. Blots were probed with P1 (C) and P2 (D). ORF131, trnV, 16S rRNA, trnI, trnA, 23S rRNA: original sequences of the chloroplast genome; aadA: aminoglycoside 3′- adenylyltransferase; Prrn: 16S rRNA promoter; PpsbA: psbA promoter; TpsbA: terminator region of the psbA gene; WT: wild-type plant. Phenotype of typical leaves used for this study from plants of TGZ-transplastomic tobacco (PG) and wild-type tobacco (WT). From Ioannidis et al. 2009.

4.2 Transglutaminase activity and thylakoid associated polyamines

The TGase activity in tobacco leaves over-expressing maize *TGZ* was nearly four times higher than that of the Wt plants (Table 1). This result was corroborated by the presence of TGZ protein in the over-expressers detected by western blot and analyzed by mass spectrometry (data not presented). By a sensitive HPLC method we have estimated the amount of associated polyamines in thylakoids (Kotzabasis et al 1993b). Plants over-expressing TGZ showed a total increase of 90% in the titer of thylakoid associated polyamines (Put, Spd and Spm) on a Chl basis (Fig. 2). Bound Put was increased about 3 times and the higher polyamines about 60% in comparison to the Wt.

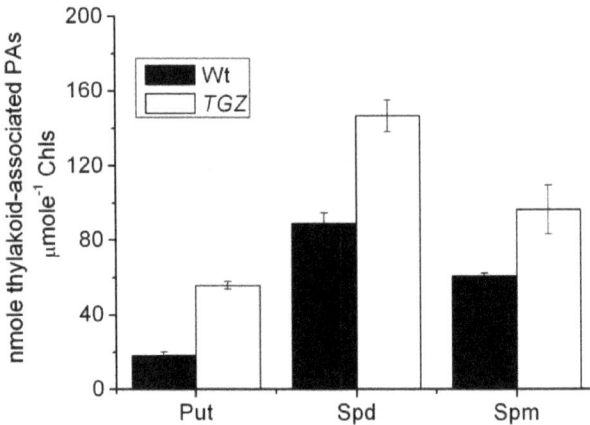

Fig. 2. Thylakoid associated polyamines of *Nicotiana tabacum* Wt and tobacco over-expressing *tgz* from maize. Data are presented on a chlorophyll basis because protein titer was substantially higher in transformed tobacco due to the over-expression of *tgz*. Vertical bars denote standard deviation (n=3). From Ioannidis et al. 2009.

4.3 Thylakoid ultrastructure and pigment content

Transmission electronic microscopy revealed important differences between Wt and TGZ over-expressing chloroplasts. Wt chloroplasts exhibit a normal thylakoid network architecture (Fig. 3, A and C), exhibiting grana and stroma lamellae in a normal proportion. Over_expression of *tgz* resulted in a severe depletion of chloroplast stroma lamellae and, interestingly, a grana dominance (Fig. 3, B and D), the granum size (number of stacked layers) being increased up to nearly 1000 nm, the double that of the Wt granum size (Table 1 and Fig. 3D). Furthermore, a reduction in the total Chl content was evident from 1.86 (mg·g^{-1}FW) in Wt to 0.6 (mg·g^{-1}FW) in over-*tgz* with a parallel decrease of the Chla/Chlb ratio (Table 1). The total carotenoid titer was also reduced (Table 1). In fact, as commented in the next paragraph, at later stages of development TGZ-plants are severely chlorotic (Ortigosa et al. 2010).

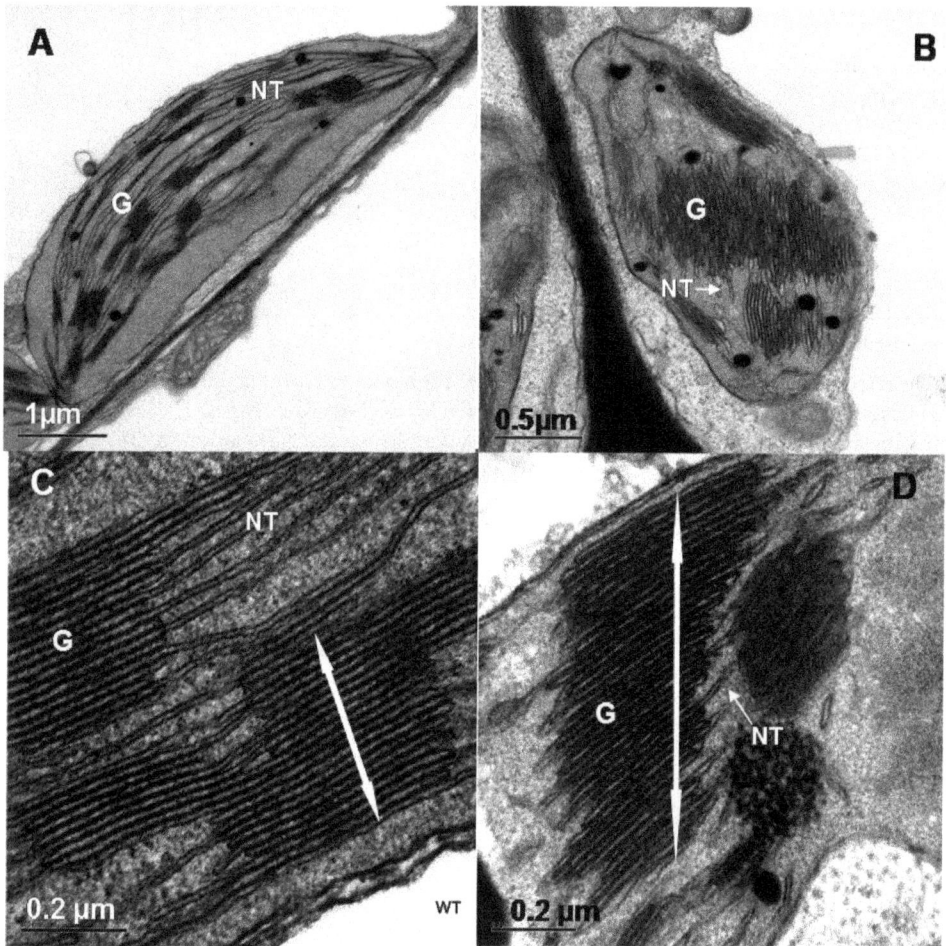

Fig. 3. Ultrastructure of chloroplasts from tobacco Wt and over-expressing TGZ.
Transformed tobacco (B, D) shows an increased stacking of thylakoids and a reduced stroma
thylakoid network. Large grana appear with a size many hundreds nm bigger than Wt
plants (A, C). G= grana; NT= non-appressed thylakoids; p= plastoglobuli; pbl= prolamellar
body lattice. Arrow in C= 0.35 μm approx.; arrow in D= 0.9 μm approx. From Ioannidis et al.
2009.

4.4 Fluorescence induction kinetics

Over-expression of maize *tgz* in tobacco has a small effect (about 13% decrease) in the
structure and functionality of PSII, as judged by the F_V/F_M values. Maximum quantum
efficiency of PSII in the transformants is about 0.7, whereas Wt tobacco exhibits optimal
values of about 0.81 (Table 1).

	Wt tobacco	over TGZ	Oldest over TGZ
F_V/F_M	0.812 (0.031)	0.702 (0.070)	0,122 (0.02)
qE	0.16 (0.02)	1.008 (0.08)	-ND
$t_{1/2DCMU}$ (ms)	166 (9)	110 (12)	-ND
Chl a (mg·g^{-1}FW)	1.38 (0.11)	0.42 (0.08)	0.1 (0.01)
Chl b (mg·g^{-1}FW)	0.48 (0.04)	0.18 (0.05)	0.06 (0.01).
Total Chls (mg·g^{-1}FW)	1.86 (0.15)	0.60 (0.13)	0.16 (0.02)
Chla/Chlb	2.87 (0.06)	2.33 (0.17)	1.73 (0.25)
Carotenoids (mg·g^{-1}FW)	0.29 (0.02)	0.13 (0.02)	0.03
Maximum granum size (nm)	400 (*)	1000 (*)	----
Transglutaminase activity pmol Put ·mg protein·h^{-1}	758.9 (89.2)	3067.3 (661)	1636.6 (172.4)

*measured from 50 chloroplasts

Table 1. Comparison of fluorescence parameters, pigment content, maximum granum size and transglutaminase activity in *Nicotiana tabacum* Wt and overexpressing *tgz* (over TGZ and oldest TGZ) leaves. Numbers in parenthesis denote standard deviation (n=3). Transformed tobacco values (right column) were statistically different in comparison to the Wt values (left column) at p<0.05. From Ioannidis et al. 2009.

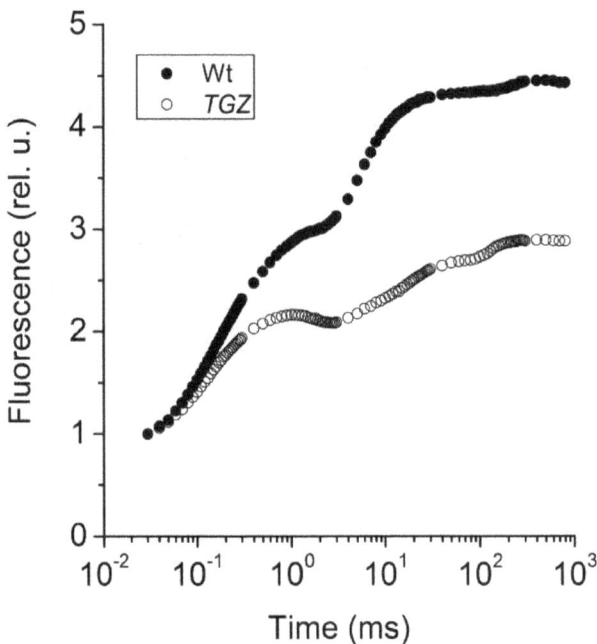

Fig. 4. Fluorescence induction curves of *Nicotiana tabacum* Wt and tobacco over-expressing *tgz* from maize (TGZ). Samples were dark adapted for 20 min and were illuminated with 3000 μmol [photons] m^{-2}s^{-1}. The time axis is semi-logarithmic for clarity and data are normalized to Fo.

More pronounced differences appear at later stages of development. A detailed transient kinetics of fluorescence induction shows that there is a major difference both in the shape and in the amplitude between Wt and over-TGZ tobacco plants (Fig. 4). The maximal difference in F_V during fluorescence induction is at 10 ms (about 70% higher values for Wt in comparison to the transformed) and there are also large differences in the F_M values (about 50% higher for the wild type). The effective PSII antenna size increased also in the over-expressers as indicated by the shortest closure time of their reaction centers (see Table 1 parameter $t_{1/2DCMU}$) in comparison to that of the Wt. The value of the energy-dependent component of the non-photochemical quenching (qE) in the case of the transformed tobacco is about 6 times higher than that of the Wt (Table 1).

4.5 Oxidative stress symptoms and leaf aging
The results obtained with later stages of leaf development revealed that photochemistry impairment and oxidative stress increased with transplastomic leaf age. These alterations included decrease in pigment levels, changes in the photosynthetic apparatus, in the

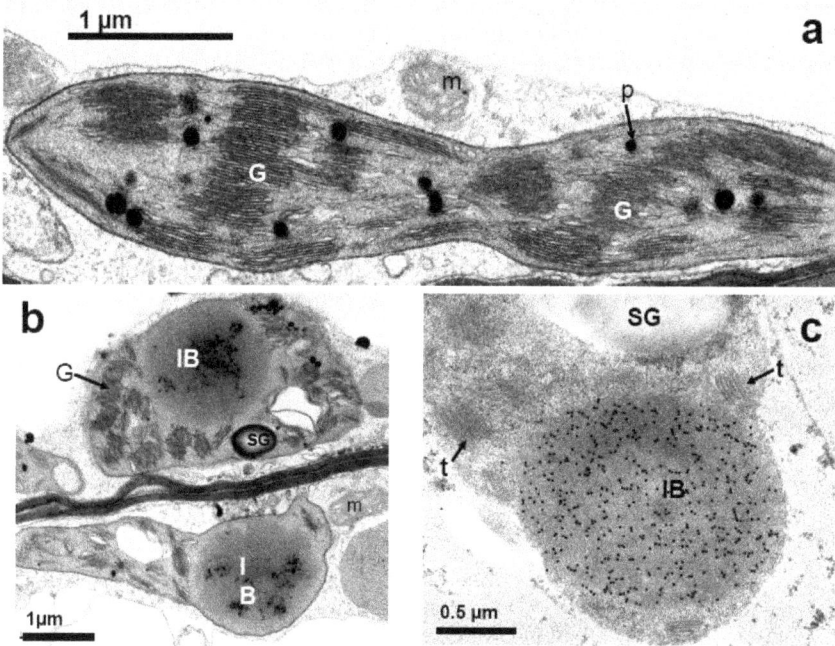

Fig. 5. Ultrastructure of tobacco chloroplasts over-expressing TGZ and TGZ immunolocalization. **a** A dividing chloroplast, showing increased grana appression and a reduced stroma thylakoid network. **b** Two oldest-leaf-chloroplasts containing large inclusion bodies. In the upper chloroplast an over-appressed granum is still visible. In the lower chloroplast the thylakoid network is disorganized. IBs larger than 1 μm are usually present. **c** Subcellular immunolocalization of TGZ-protein in the IB, using an anti-TGZ4 antibody. IB, inclusion body; G, grana; m, mitochondria; p, plastoglobuli; SG, starch grains; t, thylakoids. From Ortigosa et al. Planta 2010

chloroplast ultrastructure, and, particularly, the activation of oxidative and antioxidative metabolism pathways (see Tables 2 and 3). At the same time, the over-expressed TGZ protein accumulated progressively in chloroplast inclusion bodies. These traits were accompanied by thylakoid scattering, membrane degradation and reduction of thylakoid interconnections (Fig. 5). Consequently, the electron transport between photosystems decrease dramatically in the old leaves. In spite of these alterations, transplastomic plants can be maintained and reproduced *in vitro* (Ortigosa et al. 2010). These experiments were carried out in J. A. Hernandez laboratory (CEBAS-CSIC, Murcia, Spain).

5. Current and future developments

The over-expression of a heterologous gene could be a valuable tool for the understanding of the corresponding protein functionality. As mentioned above, the over-expression of TGZ resulted in a 4-fold increase of plastidial TGase activity, causing a significant increase in grana size and about 90% increase in thylakoid-associated polyamines (Ioannidis et al 2009). Interestingly, transformed plants exhibit increased ability to induce NPQ, a small decrease in maximal quantum yield of PSII and about 6 times higher qE, in comparison to the Wt. These results are in line with recent studies showing that elevation of Spd and Spm titers could lead to an increase in NPQ in tobacco (Ioannidis et al. 2007). Also, the effect of TGZ over PSII antenna is showed in the decrease of Chl a/Chl b ratio, which is an indicator for changes in the stoichiometry of the photosystems (in particular their LHCs) (Table 1) and the effective PSII antenna size increase. These results are in line with accumulating data showing that a plastidial TGase activity specifically polyaminylate PSII antenna proteins such as LHCII, CP29, CP26 and CP24 (Del Duca et al. 1994; Della Mea et al 2004). Chl b is found in LHCII and, consequently, a decrease in the Chl a/Chl b ratio suggests an increase in the abundance of LHCs of PSII relative to PSI similar to that suggested for a hyperstacking mutant of *Arabidopsis* (Häussler et al. 2009). Also noteworthy is the fact that OJIP transients indicate an increase in the connectivity of PSII centers in the transformed tobacco. *In vitro* investigations such us microscopy of thylakoids at high resolution hopefully will shed light on this matter in the near future.

	H_2O_2 nmol g^{-1} FW	TBARS nmol g^{-1} FW
Wt		
Upper leaves	2.57c	0.108bc
Middle leaves	2.30 c	0.331b
Transformed Plants		
PG leaves	4.37a	1.11a
Y leaves	3.42 b	0.97a

Table 2. Effect of TGZ over-expression on H_2O_2 content (nmol g^{-1} FW) and lipid peroxidation (TBARS) (nmol g^{-1} FW) in tobacco plant leaves. Different letters indicate significant differences according to Tukey's test ($P \leq 0.05$). From Ortigosa et al. 2010.

Regarding the apparent PSII antenna increase, two possible explanations can be hypothesized: either both PSIIα and PSIIβ increase their antenna size or the portion of the large antenna centers (PSIIα) is increased. By using a non destructive method (Andersson

& Melis 1983), we have *in vivo* estimated the poise between PSIIα and PSIIβ centers. The results indicate that PSIIα centers are accumulating in over-TGZ plants and, simultaneously, the number of PSIIβ centers is declining. PSIIα centers are of large antenna size and are considered to occur in grana regions (Melis 1989; Kirchhoff et al. 2007). PSIIβ centers possess a smaller antenna size and are considered to occur in stroma lamellae (Melis 1989; Kirchhoff et al. 2007). As the phenotype of the transformed plants is getting more intense, the portion of PSIIα increases at the expense of PSIIβ centers approaching 100% (Fig. 6). Furthermore, the remarkable increase in PSIIα/PSIIβ ratio indicates diminishing of stroma thylakoids. In order to crosscheck this hypothesis we studied the ultra-thin structure of the chloroplast. Transmission electron microscopy revealed that *tgz* over-expression resulted in an increase of grana stacking and a decrease of stroma lamellae. Remarkably, the size of the granum (number of stacked layers) in TGZ- chloroplasts is up to 1000 nm, whereas in Wt chloroplasts it was not larger than 400 nm. On the ground that in higher plants granum diameter is up to 600 nm (Mustardy and Garab 2003) the over-expression of *tgz* caused a significant and relative uncommon increase in granum size.

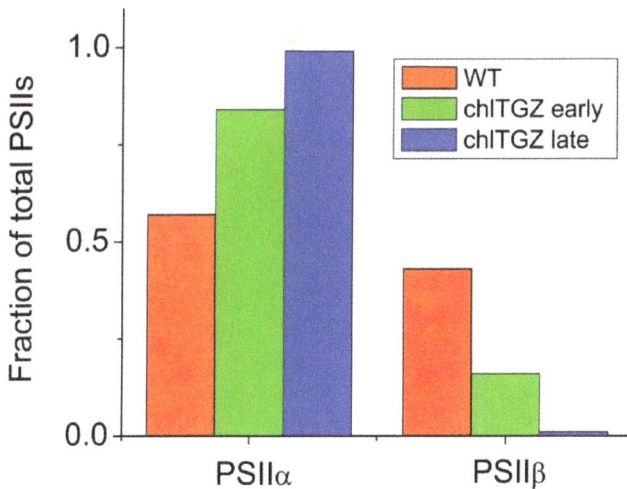

Fig. 6. The comparison of PSII antenna heterogeneity of *Nicotiana tabacum* Wt and tobacco overexpressing *tgz* from maize, as estimated by the fraction of PSIIα (white) and PSIIβ (black). Relative amount of PSIIα and PSIIβ for early and later stages of development of TGZ-transplastomic tobacco. From Ioannidis et al. 2009.

On the other hand, the reduction in the amount of stroma thylakoids leads to a number of problems regarding the functionality of the photosynthetic apparatus. Stroma lamellae are among others the major site of ATPase and a chloroplast with severely reduced stroma lamellae would not accommodate as many ATPases as Wt. Given that ATPases allow lumen protons to escape in stroma less "proton channels" means higher ΔpH between stroma and lumen during illumination (Kramer et al. 2003). Consistent with this view, the light induced energization of the thylakoid was higher (i.e. higher qE for *tgz*). At this point it should be noted that the high NPQ of the over-expressers is not fully understood at the moment. Perhaps the increased stacking (Goss et al. 2007) or the increased antenna of PSII (Pascal et al. 2005) are also contributing factors to the high NPQ values since total carotenoids in transplastomic plants are less than in Wt. First evidence i.e. the elevated qE (Table 1) and the F_M' value (end of light phase) that is close to F_0 value (Fig 4B open triangles Ioannidis et al. 2009 for transformed tobacco, indicate that the lumen-pH induced dissipative conformation of antenna and/or PSII reaction center is more efficiently formed in TGZ than in Wt. A possible interpretation -which still needs experimental verification- on the ground that LHCII, CP29, CP26 and CP24 are normal substrates of the plastidal TGase (Del Duca et al. 1994; Della Mea et al. 2004) and putative sites of qE (Pascal et al. 2005; Kovacs et al 2006 and refs therein) is that they are changing their conformation upon polyaminylation which in turn promotes dissipation. However, further research is on going for the elucidation of this phenomenon. In addition, structural and biochemical changes that appeared only sparsely in early phases (Fig 3D) and progressively appear more frequently in the latest phases of plant development are indicative of oxidative stress (Austin et al 2006; Ortigosa et al. 2010), due to impairment of photochemistry, as indicated also by the decreased F_V/F_M (Table 1).

On the contrary the underlying causes of increased stacking are better understood. Polyaminylation of proteins result in significant change in the charge of the target protein (Della Mea et al 2004). It is well established that negative charges of chlorophyll binding proteins must be neutralized by positive cations in order adjacent membranes to stack and in turn grana formation to occur (Standfuss et al. 2005). This kind of charge neutralization is feasible with monovalent or divalent inorganic cations (Kirchhoff et al. 2007; Barber 1982) or with organic cations such as polyamines (Ioannidis et al. 2007). Noteworthy, fluorescence transients of tobacco thylakoids indicate that the higher polyamines are much more efficient in stacking than Mg^{+2} (Ioannidis et al. 2007). Although the later works quantified the coulombic effects of non-covalently bound polyamines it seems that bound polyamines can also cause stacking (Ioannidis et al. 2009). The self-assembly of the thylakoids into grana was suggested to occur upon *in vitro* cation addition, and migration of minor LHCIIs from PSIIβ to PSIIα (Kirchhoff et al. 2007). Our *in vivo* results showing a PSIIβ reduction and a thylakoid-stacking increase in the *tgz*-transformants are in line with this view, Recent electron microscope tomography results and proposed models for the three-dimensional organisation of thylakoids are also in agreement with our results (Shimoni et al 2005). In addition, lower chlorophyll content and lower Chl a/Chl b ratio was also the case for a mutant of Arabidopsis (*adg1-1/tpt-1*) that exhibit increased stacking (Häusler et al. 2009). This phenomenon is also present in our tgz-transformants that presented less chlorophyll content per leaf basis than the Wt and, at later stages of development, this phenomenon is more intense (Ortigosa et al. 2010).

Enzymatic activity	Wt		Transformed plants	
	Upper leaves	Middle leaves	PG leaves	Y leaves
APX nmol min⁻¹ mg⁻¹ prot	1053b	961b	1557a	1479a
MDHAR nmol min⁻¹ mg⁻¹ prot	57.98c	60.18c	75.48b	93.62a
DHAR nmol min⁻¹ mg⁻¹ prot	15.11b	4.74c	26.67a	20.22b
GR nmol min⁻¹ mg⁻¹ prot	54.37a	57.73a	62.83a	57.25a
CATALASE μmol min⁻¹ mg⁻¹ prot	172.1a	178.2a	88.3b	74.9c
POX nmol min⁻¹ mg⁻¹ prot	118.6c	164.2c	423.0b	822.4a
NADH-POX nmol min⁻¹ mg⁻¹ prot	13.64c	28.02c	211.6b	330.3a
GST nmol min⁻¹ mg⁻¹ prot	6.17a	6.13a	5.47ab	4.25b
GPX nmol min⁻¹ mg⁻¹ prot	nd	nd	nd	nd
G6PDH nmol min⁻¹ mg⁻¹ prot	10.55b	6.12c	16.15a	16.74a
SOD U mg-1 prot	31.0b	39.7b	57.6a	54.9a

Table 3. Effect of chlTGZ over-expression on antioxidative enzyme activities in tobacco plant leaves. Different letters indicate statistical significance according to Tukey's test ($P \leq 0.05$); nd, not detectable. From Ortigosa et al. 2010.

5.1 Implications of the work

Thylakoid architecture is a major factor which affects functionality and efficiency of the photosynthetic apparatus. Light conditions in terms of quality and intensity define thylakoid architecture, but the details of the molecular mechanism which is responsible for this regulation is largely unknown (Anderson 1999; Mullineaux 2005). We provide evidence that the remodeling of the grana could be feasible through over-expression of a single enzyme. Therefore, we suggest that *tgz* has an important functional role in the formation of the grana stacks. Moreover TGZ over-expression, due to the enormous and stable granum size, may provide a powerful tool for the study and understanding of grana function that has long been debated (Mullineaux 2005).

i.   Insight into the role of thylakoid bound polyamines

Polyamines are ubiquitous molecules with an ill defined mode of action. Although thousands of papers appeared the last decades concerning their effects, their role remains obscure. The interest is still high because polyamines are essential for cell growth and important for plant tolerance to stress. The fact that, in plants, free, bound and phenolic-conjugated polyamine forms are present, make their role more puzzling. This work significantly improve our understanding by sheding light mainly on the role of bound polyamines that will facilitate to understand the implication of the other polyamine forms.

ii.   Transglutaminases in thylakoids and photosynthetic implications

Transglutaminase activity depends on Ca^{2+}, GTP and light (Villalobos et al. 2004; Del Duca & Fracassini 2008), which are key factors for chloroplast energetics. Transglutaminase activity was shown to be light sensitive, affected by hormone deprivation and with a light/dark rhythm (Bernet 1997; Bernet et al. 1999). Subcellular localization studies showed that, the enzyme was specifically localized in the chloroplast grana-appressed thylakoids and close to LHCII (Villalobos et al. 2001; Villalobos 2007; Santos et al. 2007). Finally, proteomic studies indicates that maize chloroplastic TGase is a peripheral thylakoid protein forming part of a specific PSII protein complex which includes LHCII, ATPase and PsbS proteins (Campos et al. 2010). With the presented results, we give important *in vivo* and *in vitro* data that reinforce the idea that the role of TGase in thylakoids is the modification of LHCII antenna proteins by polyaminilation, giving new properties to the complex, in particular under low light or stress conditions.

Why the photosynthetic apparatus has enzymes with TGase action near the reaction centers of PSII? A plausible hypothesis is that biological glues such as transglutaminases have a "polymerizing" and/or a "stabilizing" role. More particularly, crosslink of LHCIIs could increase the absorption cross section of PSII which in turn will increase photon harvesting by PSII. The latter could account for the significant increase of PSIIα centers in TGZ over-expressers. On the other hand, attachment of polyamines increase the positive charge of the protein as well as the connections intra and inter molecularly, stabilizing more firmly loosely aggregated complexes. This stabilization may be of importance during stress conditions conferring tolerance to the photosynthetic apparatus (Lütz et al 2005; Navakoudis et al 2007; Demetriou et al 2007; Sfichi et al 2008). In consequence, a possible role for polyamines on LHCII could be the activation of the dissipative antenna conformation (Ioannidis et al 2011). Furthermore, although thylakoid localization of ADC (arginine decarboxylase, Put producer) was long ago reported (Borrell et al 1995) and its importance for stress tolerance acknowledged (Galston 2001), only recently it becomes apparent that Put, and higher polyamines derived from Put, could modulate the photosynthesis protonic circuit, which is central for plant life and stress tolerance (Ioannidis et al 2011 submitted). An enzyme as TGase, that modulates the poise between free and bound polyamine forms in the following equilibrium may have a key role for the fine tuning of these processes.

$$\text{Pas} + \text{LHCII} \xrightleftharpoons{\text{TGase}} \text{Pas-LHCII} \qquad (1)$$

5.2 Future experiments

This chapter summarized recent results showing that over-expression of TGZ in tobacco, dramatically alter the organization of the thylakoid network. TGZ acted as a grana making

enzyme and increased granum size more than 100%. PSIIα centers increased, and ,
concomitantly, stroma thylakoids were depleted. At the same time, thylakoid associated
polyamines increased 90%.

On the grounds that TGases have LHCbs as a natural substrate it is plausible that
polyamines increase in thylakoids were due to LHCII modification. In future works, we will
test whether LHCII has a different profile of bound polyamines due to TGZ over-expression.
If this is the case (if more polyamines are LHCII-attached), then, PSIIα centers increase could
be the direct outcome of LHCII polyaminylation. First results show a 80% Spd and Spm
increase in isolated LHCII antenna proteins from tobacco TGZ over-expressers (Ioannidis et
al. in preparation). TGases may affect, not only the thylakoid structure, but also the
architecture of the thylakoid network. This enzyme could alter the function of
photosynthetic complexes and affect photosynthesis in multiple ways. Given that LHCII has
a key role in light harvesting, photoprotective qE and state transitions, a highly
polyaminylated LHCII *in vivo* should be tested for every one of these processes. First results
show that antenna down regulation is much more sensitive under these conditions. Future
experiments should also reveal the exact residue(s) of polyaminylation and increase further
our understanding regarding the structure and plasticity of the thylakoid network. Last but
not least, TGases may cross link the complexes of PSII outer antenna with the core. Newly
engineered plants will help to elucidate these issues.

6. Conclusion

Overexpression of chlTGZ in tobacco increased the activity of plastidal transglutaminase,
the thylakoid associated polyamines, the fraction of PSIIα centers and thylakoid stacking.
We suggest that chlTGZ has an important role in the remodeling of the thylakoid
network.

7. Acknowledgements

NEI thanks Greek Fellowship Foundation for funding (UOC). Authors thanks all the groups
that contributed to a part of the revised results: J. Veramendi (Publ. Univ. Navarra), J.A.
Hernandez (CEBAS-CSIC, Murcia), I. Fleck (Fac. Biology, Univ. Barcelona), A. V. Coelho
(ITQB, Univ. Lisboa). This study was supported by the Spanish projects MEC BFU2006-
15115-01/BMC, BFU 2009-08575, CGL2005-03998/BOS and BIO2005-00155. Also, CERBA
(Generalitat de Catalunya) supported partially this work.

8. References

Allen JF, Bennett J, Steinback KE & Arntzen CJ, (1981) Chloroplast protein phosphorylation
    couples plastoquinone redox state to distribution of excitation energy between
    photosystems, *Nature* 291: 25–29.
Allen JF & Forsberg J, (2001) Molecular recognition in thylakoid structure and function.
    Trends Plant Sci 6: 317–326.
Anderson J. M (1999) Insights into the consequences of grana stacking of thylakoid
    membranes in vascular plants: a personal perspective. *Aust J. Plant Physiol* 26: 625-
    639

Andersson B. & Anderson JM, (1980) Lateral heterogeneity in the distribution of chlorophyll–protein complexes of the thylakoid membranes of spinach chloroplasts, *Biochim Biophys Acta* 593: 427–440.

Anderson JM, & Andersson B, (1988) The dynamic photosynthetic membrane and regulation of solar energy conversion. *Trends Biochem Sci* 13: 351–355.

Anderson JM & Melis A, (1983) Localization of different photosystems in separate regions of chloroplast membranes. *Proc Natl Acad Sci U S A* 80: 745–749.

Arntzen CJ, (1978) Dynamic structural features of chloroplast lamellae, *Curr Top Bioenerg* 8: 111–160.

Austin II JR, Frost E, Vidi P.-A, Kessler F & Staehelin L.A., (2006) Plastoglobules Are Lipoprotein Subcompartments of the Chloroplast That Are Permanently Coupled to Thylakoid Membranes and Contain Biosynthetic Enzymes, *The Plant Cell*, 18: 1693–1703

Barber J., (1982) Influence of surface charges on thylakoid structure and function. *Annu. Rev. Plant Physiol* 33: 261–295.

Bernet (1997) Studies on putrescine metabolism and related enzymes during the differentiation of Zea mays meristematic callus, *PhD Thesis*, University of Barcelona, Barcelona, Spain

Bernet E., Claparols I., Dondini L., Santos M., Serafini-Fracassini D. & Torné JM, (1999) Changes in polyamine content, arginine and ornithine decarboxylases and transglutaminase activities during light/dark phases in maize calluses and their chloroplasts, *Plant Physiol Biochem* 37: 899-909.

Borrell, A., Culianez-Macia F.A., Altabella T., Besford R.T., Flores D. & Tiburcio A.F. (1995) Arginine Decarboxylase Is Localized in Chloroplasts. *Plant Physiol*. 109: 771-776.

Campos A, Carvajal-Vallejos P.K., Villalobos E., Franco C.F., Almeida A.M., Coelho A.V., Torné J.M. & Santos M., (2010) Characterization of Zea mays L. plastidial transglutaminase: interactions with thylakoid membrane proteins. *Pl. Biol.* 12: 708-716

Carvajal-Vallejos, P. K., Campos, A., Fuentes-Prior, P., Villalobos, E., Almeida, A. M., Barbera, E., Torne, J. M. & Santos, M. (2007) Purification and in vitro refolding of maize chloroplast transglutaminase over-expressed in Escherichia coli. *Biotechnology Letters.* 29: 1255-1262.

Carvajal P, Gibert J., Campos N., Lopera O., Barberá E., Torne J. & Santos M. (2011). Activity of maize transglutaminase over-expressed in Escherichia coli inclusion bodies: an alternative to protein refolding. *Biotech.Progress* 27(1): 232-240.

Chi YH, Moon JCh, Park JH, Kim HS, Zulfugarov IS, Fanata WI., Jang HH, Lee JR, Kim ST, Chung YY, Lim ChO, Kim JY, Yun DJ, Lee Ch, Lee KO, & Lee SY, (2008) Abnormal chloroplast development and growth inhibition in rice Thiredoxin m Knock-Down plants, *Plant Physiol* 148: 808-817.

Del Duca S., Tidu V., Bassi R., Esposito C., & Serafini-Fracassini D, (1994) Identification of chlorophyll-a/b proteins as substrates of transglutaminase activity in isolated chloroplasts of Helianthus tuberosus, *Planta* 193: 283-289.

Della Mea M, Di Sandro A, Dondini L, Del Duca S, Vantini F., Bergamini C., Bassi R, & Serafini-Fracassini D, (2004) A Zea mays 39-kDa thylakoid transglutaminase catalyses the modification by polyamines of light-harvesting complex II in a light-dependent way, *Planta* 219: 754-764.

Demetriou G., C. Neonaki, E. Navakoudis & K. Kotzabasis (2007). Salt stress impact on the molecular structure und function of the photosynthetic apparatus – The protective role of polyamines. *Biochim. Biophys. Acta* 1767: 272-280.

Duniec JT, Israelachvili JN, Ninham BW, Pashley RM, & Thorne SW, (1981) An ion-exchange model for thylakoid stacking in chloroplasts. *FEBS Lett* 129: 193-196.

Fernandez-San Millan A, Farran A, Molina I, Mingo-Castel AM, & Veramendi J, (2007) Expression of recombinant proteins lacking methionine as N-terminal amino acid in plastids: human serum albumin as a case study, *J Biotechnol* 127: 593-604.

Fernandez-San Millan A, Mingo-Castel AM, Miller M, & Daniell H, (2003) A chloroplast transgenic approach to hyper-express and purify human serum albumin, protein highly susceptible to proteolytic degradation, *Plant Biotechnol J* 1: 71-79.

Fernandez-San Millan A, Ortigosa SM, Hervas-Stubbs S, Corral-Martínez P, eguí-Simarro JM, J. Gaétan J, P. Coursaget P, & J. Veramendi J, (2008) Human papillomavirus L1 protein expressed in tobacco chloroplasts self-assembles into virus-like particles that are highly immunogenic, *Plant Biotech J* 6: 427-441.

Galston AW (2001) Plant biology — retrospect and prospect. *Curr Sci* 80: 150-152.

Garab G, Kieleczawa J, Sutherland JC, Bustamante C, & Hind G, (1991) Organization of pigment–protein complexes into macrodomains in the thylakoid membranes of wild-type and chlorophyll b-less mutant of barley as revealed by circular dichroism, *Photochem Photobiol* 54: 273–281.

Goss R, Oroszi S, & Wilhelm C, (2007) The importance of grana stacking for xanthophyll cycle-dependent NPQ in the thylakoid membranes of higher plants, *Physiol Plant* 131: 496-507

Häusler RE, Geimer S, Henning Kunz H, Schmitz J, Dörmann P, Bell K, Hetfeld S, Guballa A, & Flügge UI, (2009) Chlororespiration and Grana Hyperstacking: How an Arabidopsis Double Mutant Can Survive Despite Defects in Starch Biosynthesis and Daily Carbon Export from Chloroplasts, *Plant Physiology*, 149: 515–533

Ioannidis NE, Cruz JA, Kotzabasis K, & Kramer DM (2011) Evidence that putrescine modulates the higher plant photosynthetic proton circuit (submitted to EMBOJ-2011-78540)

Ioannidis NE, & Kotzabasis K, (2007) Effects of polyamines on the functionality of photosynthetic membrane in vivo and in vitro, *Biochim Biophys Acta* 1767: 1372-1382.

Ioannidis NE, Ortigosa SM, Veramendi J, Pintó-Marijuan M, Fleck I, Carvajal P, Kotzabasis K, Santos M, & Torné JM (2009) Remodeling of tobacco thylakoids by over-expression of maize plastidial transglutaminase. *Biochim Biophys Acta* 1787: 1215-1222.

Ioannidis N.E., L. Sfichi-Duke L, & K. Kotzabasis K (2011) Polyamines stimulate non-photochemical quenching of chlorophyll a fluorescence in Scenedesmus obliquus. *Photosynth. Res.* 107 : 169-175.

Kirchhoff H, Winfried H, Haferkamp S, Schoot T, Borinski M, Kubitscheck U, & Rögner M (2007) Structural and functional self-organization of Photosystem II in grana thylakoids, *Biochim Biophys Acta* 1767: 1180–1188.

Kaftan D, Meszaros T, Whitmarsh J, & Nebdal L, (1999) Characterization of Photosystem II activity and heterogenicity during the cell cycle of the green alga Scenedesmus quadricauda, *Plant Physiol* 120 (1999) 433-441.

Kotzabasis K, Fotinou C, Roubelakis-Angelakis KA, & Ghanotakis D (1993) Polyamines in the photosynthetic apparatus, *Photosynth Res* 38: 83–88.

Kotzabasis K., M.D. Christakis-Hampsas & K.A. Roubelakis-Angelakis (1993). A narrow bore HPLC method for the identification and quantitation of free, conjugated and bound polyamines. *Analytical Biochemistry* 214: 484-489.

Kramer DM, Cruz JA, & Kanazawa A, (2003) Balancing the central roles of the thylakoid proton gradient. *Trends Plant Sci* 8: 27–32.

Kovacs L, Damkjaer J, Kereiche S, Ilioaia C, Ruban AV, Boekema EJ, Jansson S, & Horton P, (2006) Lack of the light-harvesting complex CP24 affects the structure and function of the grana membranes of higher plant chloroplasts, *Plant Cell* 18: 3106–3120.

Lorand L & Graham RM, (2003) Transglutaminases: crosslinking enzymes with pleiotropic functions, *Nature Rev Mol Cell Biol* 4: 140-156.

Lütz C., Navakoudis, H.Seidlitz K, & Kotzabasis K (2005). Simulated solar irradiation with enhanced UV-B adjust plastid- and thylakoid-associated polyamine changes for UV-B protection. *Biochim. Biophys. Acta* 1710: 24-33.

Maliga P. (2004) Plastid transformation in higher plants, *Annu Rev Plant Biol* 55: 289-313.

Melis A, Spectroscopic methods in photosynthesis: photosystem stoichiometry and chlorophyll antenna size, *Phil Trans R Soc Lond* 323: 397-409.

Melis A, & Homann PH (1976) Heterogeneity of the photochemical centers in system II of chloroplasts, *Photochem Photobiol.* 23: 343–350.

Molina A, Hervás-Stubbs S, Daniell H, Mingo-Castel AM, & Veramendi J, (2004) High-yield expression of a viral peptide animal vaccine in transgenic tobacco chloroplasts, *Plant Biotech J* 2: 141-153

Mullineaux C.W. (2005) Function and evolution of grana *Trends Plant Sci* 10: 521-525

Mustardy L, & Garab G, (2003) Granum revisited. A three-dimensional model - where things fall into place, *Trends Plant Sci* 8: 117-125.

Navakoudis E, Vrentzou K, & Kotzabasis K (2007) A polyamine- and LHCII protease activity-based mechanism regulates the plasticity and adaptation status of the photosynthetic apparatus. *Biochim Biophys Acta* 1767: 261–271

Ortigosa SM, Díaz-Vivancos P, Clemente Moreno MJ, Pintó-Marijuan M, Fleck I, Veramendi J, Santos M, Hernandez JA & Torné JM. (2010) Oxidative stress induced in tobacco leaves by chloroplast over-expression of maize plastidial Transglutaminases. *Planta.* 232:593-605

Pascal AA, Liu Z, Broess K, van Oort B, van Amerongen H, Wang C, Horton P, Robert B, Chang W, Ruban A, (2005) Molecular basis of photoprotection and control of photosynthetic light-harvesting, *Nature* 436: 134–137.

Pintó-Marijuan , de Agazio M, Zacchini M, Santos MA, Torné JM, & Fleck I, (2007) Response of transglutaminase activity and bound putrescine to changes in light intensity under natural and controlled conditions in Quercus ilex leaves, *Physiol Plant* 131: 159-169.

Santos M, Villalobos E, Carvajal-Vallejos P, Barberá E, Campos A, Torné JM, (2007) in: Modern Research and Educational Topics in Microscopy Mendez-Villas A. & Diaz J. (Eds.) Immunolocalization of maize transglutaminase and its substrates in plant cells and in Escherichia coli transformed cells. *Modern Research and Educational Topics in Microscopy* (2007) pp. 212-223.

Serafini-Fracassini D & Del Duca S (2008) Transglutaminases: Widespread Cross-linking Enzymes in *Plants Ann Bot:* 102 (2): 145-152.

Sfichi L., Ioannidis, & Kotzabasis (2008). Fast and reversible response of thylakoid-associated polyamines during and after UV-B stress – a comparative study of the wild type and a mutant lacking chlorophyll b of unicellular green alga Scenedesmus obliquus. *Planta* 228: 341-353

Shimoni E, Rav-Hon O, Ohad I, Brumfeld V, & Reich Z, (2005) Three-Dimensional Organization of Higher-Plant Chloroplast Thylakoid Membranes Revealed by Electron Tomography, *The Plant Cell*, 17: 2580–2586

Staehelin LA, & van der Staay GWM, in: DR Ort, CF Yocum, (1996) (Eds.) The Light Reactions, Structure, composition, functional organization and dynamic properties of thylakoid membranes. In Oxygenic Photosynthesis: Kluwer Academic Publishers, Dordrecht, 1996, pp. 11–30.

Standfuss, J., Terwisscha van Scheltinga, A. C., Lamborghini, M., & Kuhlbrandt, W. (2005) Mechanisms of photoprotection and nonphotochemical quenching in pea light-harvesting complex at 2.5Å resolution, *EMBO J* 24: 919-928.

Torné JM, Santos M, Talavera D, & Villalobos E, Maize nucleotide sequence coding for a protein with transglutaminase activity and use thereof. (2002) Patent WO03102128 A1.

Trissl HW, & Wilhelm C, (1993) Why do thylakoid membranes from higher plants form grana stacks?, *Trends Biochem Sci* 18: 415–419.

van Amerongen H, & Dekker JP (2003), Light-harvesting in photosystem II. In Green BR, Parson WW eds, Light-Harvesting Antennas in Photosynthesis, Kluwer Academic Publishers, Dordrecht, pp 219–251

Villalobos E, Torné JM, Rigau J, Ollés I, Claparols I, & Santos M, (2001) Immunogold localization of a transglutaminase related to grana development in different maize cell types, *Protoplasma* 216: 155-163.

Villalobos E (2007) Study of maize transglutaminases, PhD Thesis, Univ. Barcelona, Spain.

Villalobos E, Santos M, Talavera D, Rodriguez-Falcón M,. Torné JM (2004) Molecular cloning and characterization of a maize transglutaminase complementary DNA. *Gene* 336 : 93-104

Villar-Piqué A., Sabaté R, Lopera O, Gibert J, Torné J.M., Santos M & Ventura S. (2010) Amyloid-like protein inclusion bodies in tobacco transgenic plants. *PLoS ONE* 5 (10): e13625.

Wang Q, Sullivan RW, Kight A, Henry HJ, Huang J, & Jones AM, (2004) Deletion of the chloroplast-localized Thylakoid Formation1 gene product in Arabidopsis leads to deficient thylakoid formation and variegated leaves, *Plant Physiol* 136: 3594-3604.

13

Morphological and Physiological Adjustments in Juvenile Tropical Trees Under Contrasting Sunlight Irradiance

Geraldo Rogério Faustini Cuzzuol and Camilla Rozindo Dias Milanez
Universidade Federal do Espírito Santo
Brasil

1. Introduction

Luminosity is considered one of the most relevant environmental factors in plant growth, and it is closely associated to forest succession. It controls from morphogenetic processes of germination to morphological and physiological patterns of plant growth in different classes of forest succession.

There are several proposals to define succession classes of tree based on the placement of the species in the forest. Overall, two extreme successional groups are distinguished: a) species of the early-succession category (pioneers), which germinate, survive, and grow only in glades; and b) species of the final or late-succession category (climax), which require shady environments in the understory to grow. However, a large number of species that occupy intermediate status between these two succession classes has already been acknowledged.

In the 1980s, early species began to be categorized into sun plants; and late species into shade plants. In the beginning of this century, the concept of sun-requiring species and shade-requiring species has been adopted for plants that need high irradiance and intense shading, respectively, in order to develop. Few species suit these limitations. Studies show that most species are able to tolerate intermediate irradiance conditions. They are, therefore, categorized as facultative sun and facultative shade species.

Other terms that have often been used are shade-tolerant plants and sun-tolerant plants. Shade-tolerant plants correspond to shade-requiring plants, or simply shade plants. Sun-tolerant plants are the sun-requiring plants, also called sun plants.

Since the term *tolerance* suggests better performance under optimum environmental conditions, but able to acclimatize under conditions that are less favorable to growth, in this chapter we adopt the term *sun plants* (pioneer, sun-requiring or sun-tolerant plants), and *shade plants* (non-pioneer, shade-requiring or shade tolerant). The species in-between these two categories we call: facultative sun plants (early intermediate plants), which develop under full sunlight but tolerate moderate irradiance; and facultative shade plants (late intermediate plants), which prefer intense shading but are able to grow under moderate shading.

The initiative of categorizing the species into different status of succession is based on quantitative and qualitative criteria of the luminous spectrum occurring in natural environments, which varies significantly from the edge to the interior forest. However, this categorization does not always correspond to the results obtained under controlled irradiance conditions. Some species considered ombrophytes of tropical forest have shown high phenotypic plasticity at the juvenile stage. They have been able to survive and grow under full sunlight. Nevertheless, the best performance occurred under moderate shading — typical characteristic of facultative shade species. Another aspect to be taken into account is phenology, because there could be different responses from juvenile to adult stage.

With so many environmental and ontogenetic variables influencing the morphological and physiological responses, it is difficult to find a scale for growth, biochemical, and physiological patterns which is able to characterize sun and shade requiring and facultative species.

Due to their high sensitivity to luminosity, the shade species have received special attention. Studies carried out with shade plants have shown that this kind of plant has lower plasticity under contrasting irradiance, which, in some cases, can compromise its growth and survival under full sunlight. When exposed to high sun irradiance, shade species suffer immediate and irreversible damages such as chlorosis, burns, and necrosis (Figure 1), followed by leaf abscission. If they are not capable of adapting to the new environment, they can collapse because of photoinhibition.

Morphological and physiological responses to variations in light intensity are well documented regarding leaves of arboreal vegetation in temperate areas. Studies on tropical tree have increasingly focused on medium term responses, leaving a gap concerning short term responses to light stress. Especially, regarding shade tolerant, semidecidual species. Based on the few tropical shade tolerant species in this study, we understand that the damages appear in the first seven days of exposure to direct solar radiation. In this period, there are photoinhibition and photo-oxidation followed by partial or complete abscission of leaves. Even so, they are able to sprout new leaves with new morphological and physiological characteristics without compromising survival.

This chapter aims at presenting up-to-date and unpublished results about the morphological, biochemical, and physiological adjustment of tropical shade arboreal vegetation after exposure to full sunlight. These data may encourage revisions to the status in forest succession of tropical species, because the descriptions of their ecological preferences concerning luminosity are quite contradictory.

2. Morphology and growth measurement

2.1 Growth

Species of the same succession group and even ecotypes of the same species have different reactions to irradiance alterations. In general, shade species of temperate climates do not survive or have low survival rates when exposed to full sunlight.

Regardless of their status in the forest succession, tropical forest plants under limiting irradiance have low root: shoot ratio (R:S); and higher leaf area ratio (LAR), leaf mass ratio (LMR), and specific leaf area (SLA). These responses provide higher photosynthetic activity in relation to breathing, allowing these species to be established inside the forest, where luminosity represents only between 2 to 8% of sunlight irradiance in the canopy.

Aiming at relating the succession status of 15 semideciduous tropical trees (Table 1) to growth measurements (Figure 2, 3 e 4), Souza & Válio (2003) verified that early-succession species (pioneer or sun plant) kept higher relative growth rate (RGR), even in the shade.

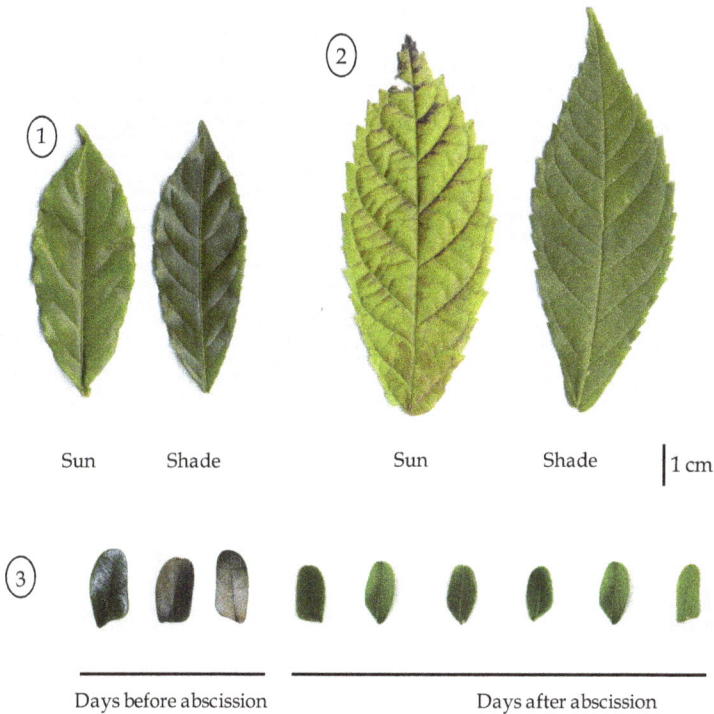

Fig. 1. Morphological features of leaves of Brazilian tropical species grown under full sunlight and shading (20% of photosynthetically active radiation). ① *Cariniana estrellensis* (Lecythidaceae), sun-tolerant; ② *Paratecoma peroba* (Bignoniaceae), shade-tolerant; ③ *Caesalpinia echinata* (Fabaceae) moderately shade-tolerant. Notice the little difference in coloration of *C. estrellensis* leaves under the two luminosity conditions. The leaves of *P. peroba* under full sunlight, however, presented chlorosis and burns at the veins. Notice the burn at the *C. echinata* pinnules before abscission. In the new pinnules sprouted after abscission, the reduced leaf area and lighter color can be noticed. Photographs provided by Paradyzo (2011) and Mengarda (2011).

For tropical forest species in the early succession stage or sun plants, RGR ranged between 40 and 60 mg.g^{-1}.day^{-1} under full sunlight, whereas for late or shade plants RGR was around 20 mg.g^{-1}.day^{-1} under high irradiance (Figure 2). Nevertheless, the species *Caesalpinia echinata,* considered moderately shade-tolerant (facultative shade or early intermediate plant) tropical tree, showed higher RGR under full sunlight than under shading. This shows that RGR does not always follow the value decreasing from early species (sun plants) to late plants (shade or climax plants).

As far as net assimilation rate (NAR) is concerned, the early species showed values under full sunlight in which NAR ranged from 0.3 to 0.6mg.cm^2.day^{-1} (Figure 2). When under shading, the early and late species almost did not present differences regarding NAR, which was under 0.2 mg.cm^2.day^{-1}. In some cases, NAR can reach very low values (around 0.01 mg.cm^2.day^{-1}), as seen in *C. echinata,* a species moderately shade-tolerant, under high shade.

This inconsistent pattern in terms of growth and forest succession status can be attributed to ontogenesis. Therefore, one should be careful not to extrapolate results obtained in the juvenile stage to adult stage. There is also the climatic factor in which the experiments were carried out. Overall, the experiments with tropical tree plants have been carried out in areas that differ in terms of irradiance intensity, precipitation, humidity, altitude, and average temperature. Another aspect that hinders comparisons in the analysis of results regards the lack of standard growth measurements, especially for growth rates expressed using different units of measurement.

Species	Successional status
Solanum granuloso-leprosum (Solanaceae)	E
Trema micrantha (Ulmaceae)	E
Cecropia pachystachya (Cecropiaceae)	E
Bauhinia forficata (Caesalpiniaceae)	E
Senna macranthera (Caesalpiniaceae)	E
Schizolobium parahyba (Caesalpiniaceae)	E
Piptadenia gonoacantha (Mimosaceae)	E
Chorisia speciosa (Bombacaceae)	I
Pseudobombax grandiflorum (Bombacaceae)	I
Ficus guaranitica (Moraceae)	L
Esenbeckia leiocarpa (Rutaceae)	L
Pachystroma longifolium (Euphorbiaceae)	L
Myroxylon peruiferum (Fabaceae)	L
Hymenaea courbaril (Caesalpiniaceae)	L

Table 1. Species studied, classification according to the successional status (E = early-uccessional; I = intermediate, L = late-successional). Souza & Válio (2003).

For tropical trees, low RGR values for early species under low irradiance have been associated to reduction in photosynthetic activity, as indicated by low NAR (Figure 2). However, RGR is not always related to NAR (physiological component of RGR). In some cases, RGR can be related to LAR (morphological component of RGR). These relations between RGR, NAR, and LAR depend much on the intensity of solar radiation. Not taking succession status into account, the 15 species analyzed by Souza & Válio (2003) showed that RGR of plants under full sunlight and under natural shading is related to NAR, but not to LAR (Figure 4). This shows higher influence of the physiological component on growth rate. However, under artificial shading, the RGR was correlated to LAR, but not to NAR (Figure 4). In this case, the morphological component, particularly leaf area, had greater effect on growth rate. These differences in the correlation between RGR and NAR or LAR concerning artificial and natural shading have also been attributed to the luminous spectrum quality. Under natural shading, the red light: distant red light ratio is low, suggesting the involvement of phytochrome in the increase in SLA, component of LAR, and in LMR. These results indicate that leaf thickness and allocation of biomass to the leaves are the most pronounced morphological alterations, regardless of the species' forest succession status.

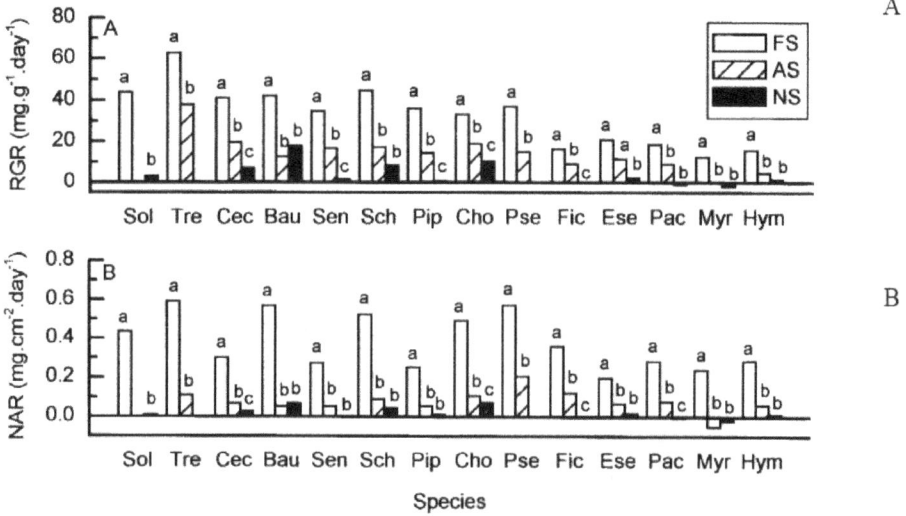

Fig. 2. A. Relative growth rate (RGR) and B. net assimilation rate (NAR) of the studied tree species under full sun (FS), artificial shade (AS) and natural shade (NS) treatments. Measurements for 0-100 days time interval. Sol = *Solanum*, Tre = *Trema*, Cec = *Cecropia*, Bau = *Bauhinia*, Sen = *Senna*, Sch = *Schizolobium*, Pip = *Piptadenia*, Cho = *Chorisia*, Pse = *Pseudobombax*, Fic = *Ficus*, Ese = *Esenbeckia*, Pac = *Pachystroma*, Myr = *Myroxylon*, Hym = *Hymenaea*. Values followed by the same letter are not significantly different. Souza & Válio (2003).

No difference in the R:S ratio has been noticed among early and late species, both under full sunlight and artificial shading (Figure 3). In general, R:S ratio ranged between 0.25 and 0.5. LMR showed higher plasticity for early species whose value ranged between 0.3 and 0.7 g.g^{-1}, especially under effect of shading (Figure 3). These results can be confirmed by the higher SLA values of early species under artificial shading (6 dm^2.g^{-1}). Under full sunlight, almost no difference has been found in terms of SLA of early and late plants. Early species under shading tended to present increased LAR values; around 3.4 dm^2.g^{-1}. Under full sunlight, the early and late species did not show significant LAR differences.

Although there are data maintaining that late species or shade plants show better performance than pioneer or sun plants under low luminous intensity, it does not always happen. Some species that are considered sun plants can show low RGR; typical of shade plants. The opposite can also happen, as seen in *C. echinata*, a moderately shade-tolerant species that showed higher RGR under full sunlight than in the shade.

Overall, the results have shown that morphological variations of tropical arboreal plants have higher influence on RGR when sun-tolerant species are under effect of shading.

Tropical tree shade species are able to develop in long periods of shading, keeping low growth rate, which favors the formation of a seedling bank. Due to their tolerance to higher irradiance, these plants show to be able to develop under increased luminosity, when glades are formed.

Therefore, the task of establishing a relation between growth measurement and successional status of tropical arboreal plants is complex. There is a paucity of more consistent data that allow defining sun and shade plants, as well as characterizing facultative sun and shade plants (intermediate plants in forest succession).

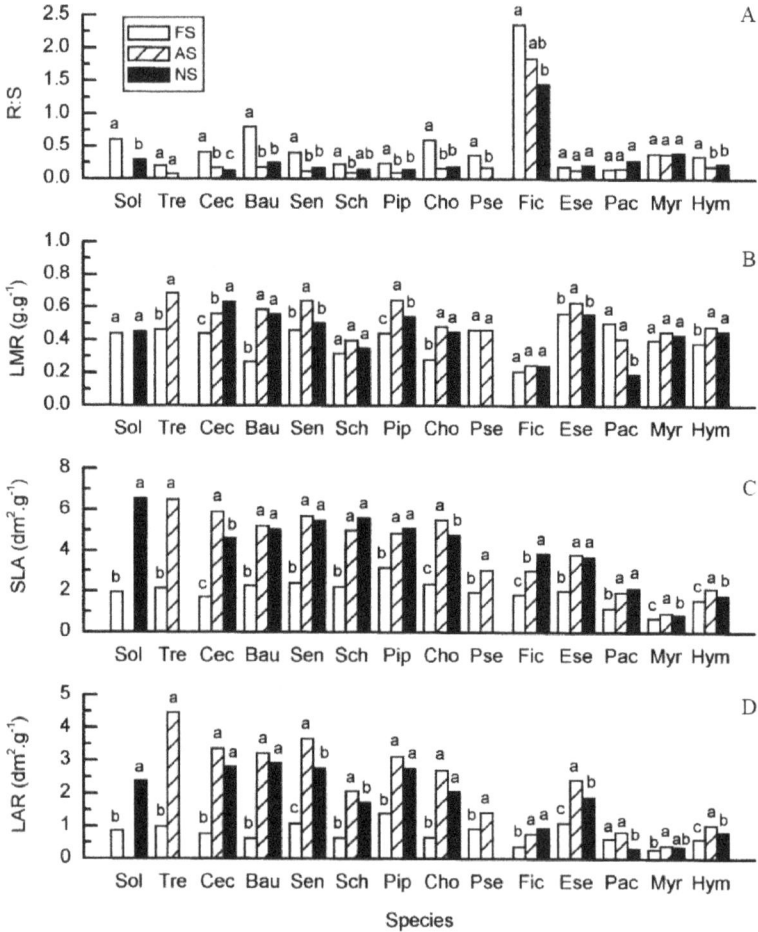

Fig. 3. A. Root:shoot ratio (R:S); B. leaf mass ratio (LMR); C. specific leaf area (SLA); and D. leaf area ratio (LAR) of the studied tree species under full sun (FS), artificial shade (AS), and natural shade (NS) treatments. Measurements after 100 days. Sol = *Solanum*, Tre = *Trema*, Cec = *Cecropia*, Bau = *Bauhinia*, Sen = *Senna*, Sch = *Schizolobium*, Pip = *Piptadenia*, Cho = *Chorisia*, Pse = *Pseudobombax*, Fic = *Ficus*, Ese = *Esenbeckia*, Pac = *Pachystroma*, Myr = *Myroxylon*, Hym = *Hymenaea*. Values followed by the same letter are not significantly different. Souza & Válio (2003).

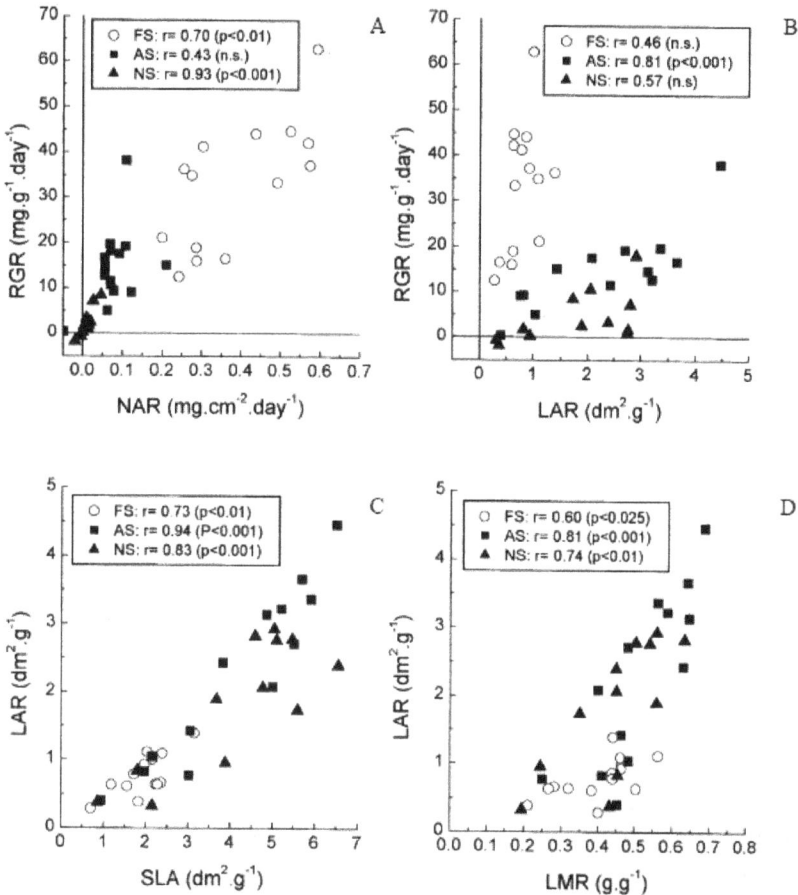

Fig. 4. Correlation between RGR and NAR (A); RGR and LAR (B); LAR and SLA (C); and LAR and LMR (D). Pooled data of all species under each one of the treatments (FS = full sun; AS = artificial shade, NS = natural shade). RGR = relative growth rate, NAR = net assimilation rate, LAR = leaf area ratio, LMR = leaf mass ratio. Souza & Válio (2003).

2.2 Leaf morphology

The ability a plant has to overcome the alarming stage is a result of physiological adjustments combined with morphological adaptations. This interaction has been considered the most relevant factor to acclimatization and survival of shade plants, when exposed to high irradiance. The morphological adjustment can start in existing plants. However, they are most pronounced in young leaves sprouted after high irradiance exposure.

In shade species, the damage caused by intense irradiance takes place already on the first days, resulting in leaf abscission. In *C. echinata*, total leaf abscission took place on the first seven days. However, in other tropical species this effect can come later, as observed in *Minquartia guianensis*, a shade species of the Amazon forest, in which 30% of its leaves collapsed before the end of the second week under full sunlight.

Differently from what had been speculated about understory species, the results have shown different degrees of sensitivity after these plants were exposed to high irradiance. Some understory species of humid tropical forest such as *Ouratea lucens* showed moderate photoinhibition, preserving most of their leaves. For *Hybanthus prunifolius*, however, there was severe photoinhibition and almost total loss of leaves.

The phenotypic plasticity of tropical arboreal plants to luminosity involves characteristics that are related to higher efficiency in capture or dissipation of light through the leaves. This essentially depends on the adjustments of morphological and anatomical components.

Among the most significant anatomical adjustments observed in shad species under high sun irradiance, we can highlight the thickening of cuticle, palisade parenchyma, and increase in stomatic density, and trichomes. For *C. echinata*, the new leaves sprouted after abscission showed thickening of palisade parenchyma (Table 2), which suggests an efficient morphological strategy to reduce photo-oxidative damage. In general, the highest stomatal density is associated to reduction in the stomatal opening area and, consequently, resistance to water loss through transpiration. Cuticle and adaxial epidermal cell thickening is also one of the adjustments often seen in tropical shade species, when exposed to full sunlight. These adaptations minimize leave surface heating by promoting of light reflection.

It is important to high light that the intensity of these responses may vary significantly among the leaves before and after abscission. In *C. echinata*, exposure to full sunlight induced limb thinning because of thickness reduction in adaxial epidermis and palisade parenchyma during the first seven days of exposure preceding leaf abscission (Table 2). However, the new leaves sprouted after abscission showed increased thickness in palisade and spongy parenchyma, which were the main contributors to limb thickening (Table 2). In this aspect, the palisade parenchyma increased 142% under full sunlight, whereas the spongy parenchyma increased 58.3% and the adaxial epidermis 12.5% compared to plants under shading. The higher elongation of chlorophyllian tissue in the new leaves reflected the higher water content; 50% higher compared to plants under shading. These data suggest that *C. echinata* is a species that uses water effectively under full sunlight.

The reduction in SLA after solar radiation exposure is common among tropical arboreal plants. This response was observed, for example, in *M. guianensis* and *C. echinata* (Table 2). Reduction in SLA means smaller solar radiation interception area, contributing to water loss reduction and improvement of photosynthetic performance, growth, and survival of the plants under full sunlight.

Variables	7 days		60 days	
	Shade	Sun	Shade	Sun
SLA (mg.cm^2)	250±19	170±16	210±21	150±14
Limb (mm)	120±14	135±18	137±11	195±18
Palisade parenchyma (mm)	32±06	44±08	40±08	80±10
Lacunary parenchyma (mm)	60±10	60±13	67±14	88±09
H$_2$O (mg.cm^2)	18±02	19±03	18±02	27±04

Table 2. SLA values, limb thickness, palisade parenchyma, lacunary parenchyma, and water content in leaves of *Caesalpinia echinata* after 7 and 60 days of transfer of plants from shade to full sun. ± represents standard error of the mean (n=6). Data provided by Mengarda (2010).

Besides the anatomic alterations, variations in secondary metabolite content may take place in plants under intense solar radiation. Phenolic and flavonoid compounds tend to accumulate in the epidermis and mesophyll of tropical tree shade plants under higher solar intensity. Leaves of *C. echinata* under shading have shown accumulation of phenols only in the epidermis, whereas under full sunlight, they also accumulated these compounds in mesophyll cells (Figure 5). Phenol accumulation indicates that the existence of an efficient antioxidative defense system working on the sequestration of several reactive oxygen species (ROS) and O_2 singlet in chroloplasts of plants under intense solar radiation. The stress caused by excessive solar radiation also induces biosynthesis of polyphenols, among them, flavonoids. Probably, using ROS as molecular signals. Also, an increase in flavonoid concentration in leaves of arboreal plants lessens the penetration of UV wavelength.

Shade   Sun

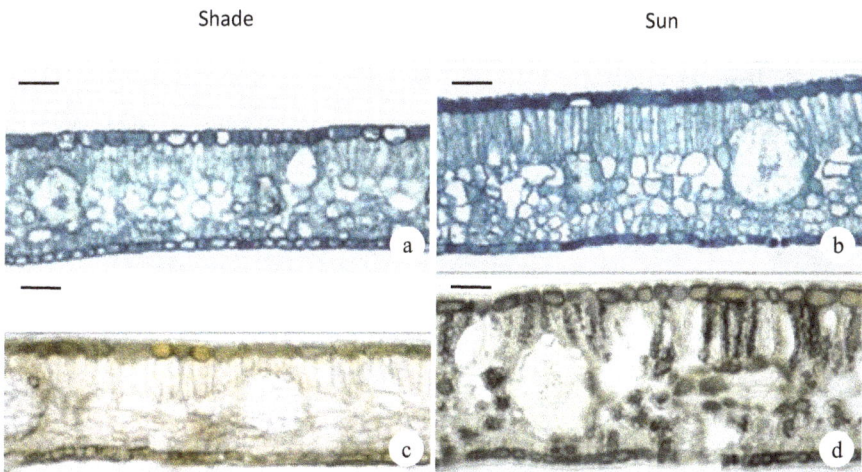

Fig. 5. Cross section of *Caesalpinia echinata* pinnules in the shade (a and c) and under full sunlight (b and d). The mesophyll of leaves under full sunlight showed chrolophyll parenchyma and adaxial epidermis thickening. Notice the higher accumulation of phenolic compounds in the limb of plants under full sunlight (d). Bar = 50µm. Data provided by Mengarda (2011).

3. Photosynthesis

The acclimatization strategy to high irradiance varies among species, and even among ecotypes of the same species. The physiological adjustments of shade plants exposed to high irradiance involve decrease in total chrolophyll concentration (Chl_{tot}) or increase in ratio between violaxanthin cycle pigments and Chl_{tot}. Violaxanthin and carotenoids reduce photoinhibition risks, oxidative damage, and increase dissipation of excessive energy through non-photochemical processes.

In the stage of light stress signalization of tropical shade tree species, the photoinhibition signals can be seen already in the first 24 hours of exposure to full sunlight. In *C. echinata*, a photosynthetic carbon assimilation (A), maximum quantum yield of photosystem II (F_v/F_M), water-use efficiency (WUE), stomatic conductance (g) e transpiration (E) decreased in the first three hours (Figure 6, 7 e 8) until they reached the lowest values in 48 hours, in a 192 hour period. During this period, it was not possible to identify the restitution stage that precedes the resistance stage.

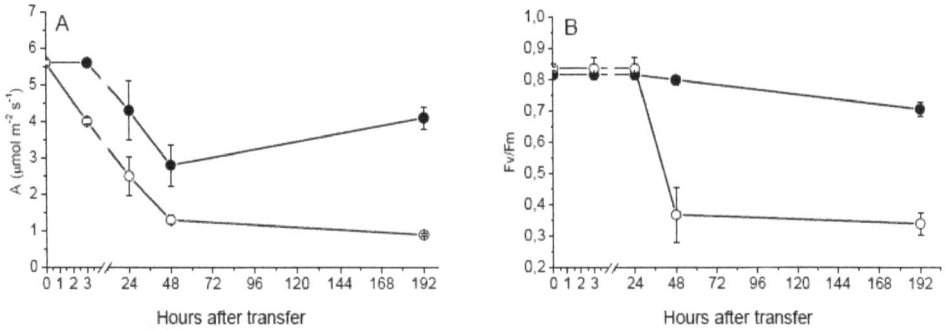

Fig. 6. Photosynthetic carbon assimilation (A) and maximum quantum yield of photosystem II (B) of *C. echinata* plants subjected to constant artificial shade of 50% (●) and transferred from shade to full sunlight (○) at 0, 3, 24, 48 and 192 h. after the start of the experiment. Vertical bars indicate standard error. Mengarda et al. (2009).

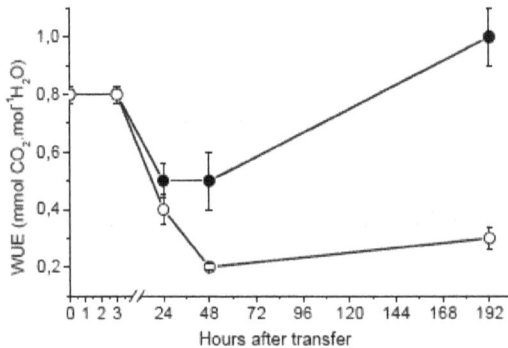

Fig. 7. Water-use efficiency of *C. echinata* plants subjected to constant artificial shade of 50% (●) and transferred from shade to full sunlight (○). Vertical bars indicate stardad error. Mengarda et al. (2009).

The factors that limit photosynthesis vary according to irradiance intensity. Plants developing in shaded environments invest more in light-capturing complexes, whereas plants developing in the sun invest in Calvin cycle and electron transport proteins. Thus, irradiance variations cause alterations in A because of the differences in maximum velocity of Rubisco carboxylation ($V_{c-max.}$) and in the maximum rate of ribulose biphosphate regeneration. The results obtained from tropical shade species show limited capacity to increase A in environments under high irradiance, due to inability to increase V_{c-max}.

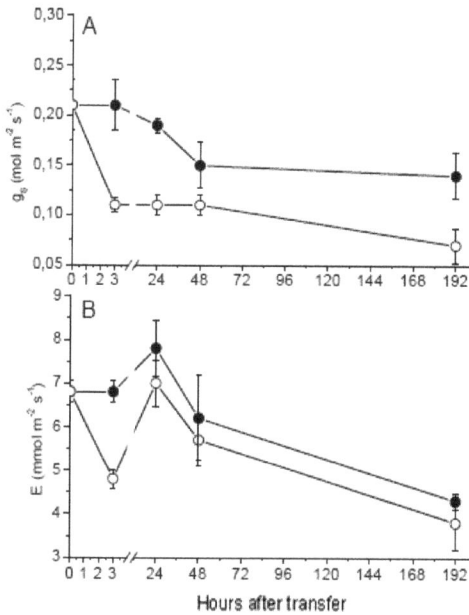

Fig. 8. Stomatic conductance (A) and transpiration (B) of *C. echinata* plants subjected to constant artificial shade of 50% (●) and transferred from shade to full sunlight (○). Vertical bars indicate stardad error. Mengarda et al. (2009).

The chrolophyll fluorescence parameters have been widely used to analyze responses from shade plants exposed to intense irradiance. Especially to detect alarming and resistance stages. Overall, tropical shade tree plants show higher density of absorbed photons per reaction center of PSII (ABS/RC) when exposed to high irradiance, as seen in *C. echinata* and *Aniba roseodora*. During the first three days of exposure to full sunlight, *C. echinata* plants (Figure 9) showed an increase in energy dissipated in the form of heat or fluorescence (DI_0/RC), in absorbed energy (ABS/RC), and in energy captured and converted into redox energy for electron transport (ET_0/TR_0). These responses can indicate inactivation of the reaction center (RC). On the second day of transfer of *C. echinata* plants from the shaded environment to full sunlight, there was a significant increase in the probability that the electron captured by the reaction center of PSII remain in the transport chain beyond QA- (ET_0/TR_0). Nevertheless, the significant reduction in the density of active reaction centers of PSII (RC/ABS) on the first seven days may have influenced the reduction in effective quantum efficiency of radiant energy conversion (F_V/F_0), and in the performance index (PI_{ABS}). The inactivation of 58 to 78% of RC in *C. echinata* (Figure 9), along with reduction in maximum quantum yield of photosystem II (F_V/F_M), indicate an increase in energy dissipation in the form of heat and fluorescence for some tropical tree plants exposed to intense irradiance. Although parameters such as RC/ABS, F_V/F_0 and PI_{ABS} have shown an increase after 20 days, the effective recovery of the photosynthetic apparatus was not observed during the 180 days of full sunlight exposure for *C. echinata* plants (Figure 9).

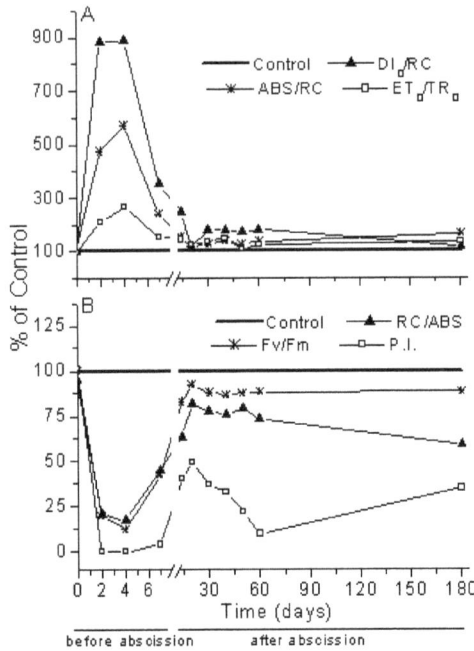

Fig. 9. Chrolophyll fluorescence emission *a* of *C. echinata* plants transferred from shade to full sunlight. Energy dissipation per reaction center (DI_0/RC), energy absorption per reaction center (ABS/RC), and probability that an excitron captured by the RC of FSII move an electron in the transport chain to beyond Q_A^- (ET_0/TR_0); (B) Ratio between number of active reaction centers of FSII and the amount of light absorbed by the antenna system (RC/ABS), maximum quantum yield of PSII (F_V/F_M), and performance index based on absorption (PI_{ABS}). Values expressed in percentage in relation to control (n=5). Data provided by Mengarda (2011).

The joint analysis of F_V/F_M and PI_{ABS} can be related to the ability plants have to transform luminous energy into metabolic reactions of the photosynthesis biochemical processes. Thus, taking into account the reduced F_V/F_M values and the significant drop in PI_{ABS} observed in plants under full sunlight, it can be concluded that the exposure of young *C. echinata* plants to full sunlight caused photoinhibition and reduction in primary photochemical efficiency, which can compromise photosynthesis.

In the first acclimatization weeks under high irradiance, the tropical tree plant *Minguartia guianensis* showed an increase in initial fluorescence (F_0), and reduction in maximum fluorescence (F_M) and F_V/F_M. Recovery only took place after four months of exposure to full sunlight, when F_V/F_M values reached 93% of the values observed for plants kept in the shade. For the tree plant *A. rosaeodora* of the Brazilian Amazon forest, reduction in F_V/F_M took place at 2 days of exposure to high irradiance, and recovery took place only at 60 days, when the values approached those observed for shaded plants.

For other tropical forest arboreal plants, the physiological adjustment process can slower. The transfer of Amazonian *M. guianensis* plants from a shaded environment to high irradiance caused a sharp decrease in F_V/F_M during the first three days, followed by gradual recovery until maximum values were reached on the 120th day, reaching the control group.

The stress condition is caused by excessive input of energy in the system, which hinders the use of this energy in the photosynthetic process. Under ideal conditions, the plants show low free energy in the system and optimum thermodynamic status, without compromising the photochemical stage in PSII. However, the photoinhibition observed in *C. echinata* plants subjected to direct solar radiation can represent a photoprotection mechanism, working on the balance between effective non-photochemical dissipation of excessive energy and the photosynthesis itself. This does nor decharacterizes a stress condition, but allows the adjustment and survival of the plant.

Chrlophyll fluorescence *a* can be considered a potential tool to differ species of particular successional status, which is useful to select species to be used for recovering degraded areas. In general, sun-requiring or facultative sun species (pioneer and early intermediate) show higher photochemical ability and increasing tendency to dissipate excessive energy as luminous intensity grows. A study carried out on three tropical tree leguminous plants (*C. echinata, C. ferrea and Machaerium obovatum*) observed higher electron transport rates and higher F_V/F_M with higher intensity of light saturation for *C. ferrea*, which suggests that this species is best adapted to environments with high irradiance, whereas *C. echinata* and *M. obovatum* do not show these responses. Therefore, *C. echinata* was characterized as late intermediate species, because it does not present photosynthetic acclimatization that is effective to high irradiance conditions. However, partial photosynthetic acclimatization and sufficient photoprotection mechanisms have been observed, allowing young plants to survive and grow under full sunlight. Thus, planting these species in reforestation has been suggested, adopting the intercropping system, in which plants receive moderate shading.

4. Soluble carbohydrates

One of the immediate metabolic signs in response to abrupt increase in irradiance is in the variation of soluble carbohydrates. These compounds as recognized as important molecular signals in the plant-environment relation and modeling agents of physiological and morphological processes.

Higher concentrations of glucose (Glu), fructose (Fru) and sucrose (Suc) were related to tolerance to water deficit during the winter for some tropical forest leguminous and tree plants. Polyols such as mannitol (Man) were also associated to water stress in *Fraxinus excelsior*. For *Olea europaea*, a Mediterranean tree, the accumulation of Man was associated to this species' tolerance to saline stress (Figure 10).

The accumulation of *monosaccharides* and Suc in leaf tissues of tree plants under water deficiency, increased salinity and intense irradiance has often been associated to osmoregulation. The relevant few studies published showed reduction in osmotic potential in leaf cells of Mediterranean tree plants such as *Ilex aquifolium* and *O. europaeae* under high irradiance. For *C. echinata*, a tropical forest specie, the leaves of plants under full sunlight showed higher water content (Table 2). This result was associated to higher content of total soluble carbohydrates (TCS), which possibly led to the reduction in water potential.

More recently, special attention has been paid to raffinose (Raf), which despite being found in low levels in leaves, has a significant role in the osmotic adjustment in herbaceous plants under stressing conditions, as observed in *Arabidopsis thaliana*. Raf is a trisaccharide that, along with stachyose and verbascose, is part of the oligosaccharides family of raffinose (RFO), which are synthesized from sucrose. Nevertheless, few are the publication on the involvement of Raf in tree plants' tolerance to high irradiance. Studies on *C. echinata* showed higher concentration of Raf in leaves after 60 days of exposure to full sunlight, suggesting an antioxidant action. Besides being osmoregulators, the monosaccharides, disaccharides such as Suc and RFOs have also been related to antioxidant actions in plants that are tolerant to water deficit and high irradiance. Therefore, investigating oxidative stress and its relation to carbohydrates could provide data to better understand the mechanisms of tolerance to light stress in tropical tree plants.

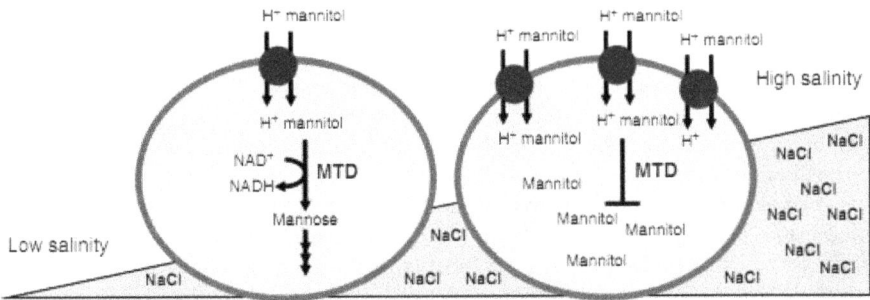

Fig. 10. Regulation of mannitol transport and metabolism as a mechanism providing salt tolerance in *O. europaea*. Conde et al. 2007.

Carbohydrates can also modify the morphological components of leaves. There are indications that the increase in palisade parenchyma thickness through elongation and periclinal cell divisions in tree plants under high irradiance is related to Suc concentration in leaves. *Chenopodium album* L. leaves, an annual herbaceous plant, showed increase in the number of layers of palisade parenchyma cells when exposed to high irradiance. This response is associated to concentration of Suc. In *C. echinata*, limb thickening because of palisade and lacunary parenchyma elongation has also been associated to higher concentration of leaf sucrose of plants under full sunlight.

There are indications that carbohydrates are also able to model plant growth, modifying biomass allocation patterns, growth rates, and R:S ratio. Under full sunlight, tree plants tend to invest more in root growth to the detriment of leaf area, resulting in RGR and NAR reduction.

In general, physiological and morphological responses associated to carbohydrates were obtained after exposing the plants to high irradiance for several days, leaving a gap concerning understanding the adjustment mechanisms at early stages of light stress. Studies on *C. echinata* have been carried out in order to categorize the alarming and resistance stages to luminous stress in a 180-day period. In this regard, carbohydrate contents and their relation to morphological and physiological adjustment have been analyzed at short periods of time (0, 2, 4, 7, 15, 30, 40, 50, 60, 120 and 180 days). The results show that the alarming stage was characterized by increased Glu and Fru contents with maximum peak at seven days. The resistance stage started with new leaves sprouting (at 15 days, Table 3), which stood out because of their higher concentrations of Suc and Raf. However, gradual reduction in Suc contents was observed by the 180th day (10 mg.g^{-1} DW), equal to the control group (shaded plants).

Carbohydrates	7 days		60 days	
	Shade	Sun	Shade	Sun
Glu	1.2±0.68	6.0±0.27	1.0±0.65	1.0±0.25
Fru	1.1±0.41	7.0±0.15	0.7±0.09	0.5±0.05
Suc	10.3±2.53	17.0±2.16	7.0±1.25	14.0±1.36
Raf	0.26±0.07	0.26±0.08	0.04±0.00	0.7±0.06
TSC	11.2±1.58	30.0±3.86	8.0±2.95	14.0±3.25

Table 3. Leaf concentration of soluble carbohydrates in *C. echinata* plants after 7 and 60 days of transfer from shade to full sunlight. Glucose (Glu); Fructose (Fru); Sucrose (Suc); Raffinose (Raf) and total soluble carbohydrates (TSC). ± represents standard error of the mean (n=6). Data provided by Mengarda (2010).

5. Conclusion

Some tropical tree plants considered shade plants or late intermediate in forest succession are able to survive under high irradiance. Even if at the alarming stage of light stress they show partial or total abscission of leaves, survival rate is high. This is due to these plants' ability to sprout new morphologically and physiologically adjusted leaves in environments with higher irradiance. Some tropical species considered shade plants show high plasticity to contrasting sunlight, at least in their early stages of growth. In part, physiological variables contribute more to plasticity index at the alarming stage of luminous stress. At the resistance stage, the new leaves sprouted show morphological alterations, such as SLA reduction and limb thickening. The most frequent morphological adjustment such as elongation and/or increase in the number of layers of palisade parenchyma, and reduction in SLA seem to be under the control of Suc. Photo-oxidative damage such as reduction in the *Chla:b* and *Chl*$_{total}$:carotenoids ratio, maximum photochemical yield of PSII, and PI_{ABS}. did not compromise the survival of shade plants when they were exposed to high solar radiation. The results point to the need of increasing the number of tropical species under controlled conditions, and assess their physiological and morphological mechanisms in field conditions at different stages of development; from seedling to adult stage. These data can support the proposal of a new forest succession classification for some tropical tree plants.

6. Acknowledgment

Thanks are owed to *Fundação Biodiversitas, Fundação de Apoio Científico e Tecnológico da Prefeitura Municipal de Vitória* (FACITEC-Process 38/2007) and *Fundação de Apoio à Pesquisa do Espírito Santo* (FAPES-Process 39044823/2007) for the financial support.

7. References

Aranda, I.; L. Gil & J. Pardos. (2001). Effects of thinning in a *Pinus sylvestris* L. stand on foliar water relations of *Fagus sylvatia* L. seedlings planted within the pinewood. *Trees*, Vol.15, (September, No.6), pp. 358-364, ISSN 1432-2285

Cao, K.F. 2000. Leaf anatomy and clorophyll contento f 12 woody species in contrasting light conditions in a Bornean heath forest. *Canadian Journal of Botany*, Vol.78, No.10, pp. 1245-1253

Conde, C., Silva, P., Agasse, A., Lemoine, R., Delrot, S., Tavares, R. & Gerós, H. (2007). Utilization and transporto f mannitol in *Olea europaea* and implications for salt stress tolerance. *Plant and Cell Physiology*, Vol.48, (January, No.1) No.1, pp. 42-53, ISSN 1471-9053

Coueé, I., Sulmon, C., Gouesbet, G. & Amrani, A.E. (2006). Involvement of soluble sugars in reactive oxygen species balance and responses to oxidative stress in plants. *Journal of Experimental Botany*, Vol.57, (Februray), pp. 449-459, ISSN 1460-2431

Demmig-Adams, B. & Adams, W.W. 2006. Photoprotection in an ecological context: the remarkable complexity of thermal energy dissipation. *New Physiologist*, Vol.172, (October), pp. 11-21, ISSN 1469-8137

Franco, A.C., Matsuba, S. & Orthen, B. (2007). Photoinhibition, carotenoid composition and the co-regulation of photochemical and non-phochemical quenching in neotropical savanna trees. *Tree Physiology*, Vol.27, (May), pp. 717-725, ISSN1758-4469

Geßler, A., Duarte, H.M., Franco, A.C., Lüttge, U., Mattos, E.A., Nahm, M., Rodrigues, P.J.E.P., Scarano, F.R. & Rennenberg, H. (2005). Ecophysiology of selected tree species in different plant communities at the periphery of the Atlantic Forest of SE – Brazil III. Three legume in a semi-deciduous dry forest. *Trees*, Vol.19, (September), pp. 523-530, ISSN 1432-2285

Gibson, S.I. 2004. Sugar and phytohormone response pathways: navigating a signalling network. *Journal of Experimental Botany*, Vol.55, No.395, (January) pp. 253-264, ISSN 1460-2431

Gleason, S.M. & Ares, A. 2004. Photosynthesis, carbohydrates storage and survival of a native and an introduced tree species in relation to light and defoliation. *Tree Physiology*, Vol.24, (October), pp. 1087-1097, ISSN 1758-4469

Gonçalves, J.F.C. & Santos-Jr, U.M. 2005. Utilization of the chlorophyll a fluorescence technique as a tool for selecting tolerant species to environments of high irradiance. *Brazilian Journal of Plant Physiology*, Vol.17, (Jul-Sept), pp. 307-313, ISSN1677-0420

Gonçalves, J.F.C.; Barreto, D.C.S.; Santos Jr, U.M.; Fernandes, A.V.; Sampaio, P.T.B. and Buckeridge, M.S. (2005). Growth, photosynthesis and stress indicators in young rosewood plants (*Aniba rosaeodora* Ducke) under different light intensities. *Brazilian Journal of Plant Physiology*, Vol.17, (Jul-Sept), pp. 325-334, ISSN1677-0420

Gonçalves, J.F.C.; Santos Jr., U.M.; Nina Jr., A.R. & Chevreuil, L.R. (2007). Energetic flux and performance índex in copaíba (*Copaifera multijuga* Hayne) and mahogany (*Swietenia macrophylla* King) seedlings grown under two irradiance environments. *Brazilian Journal of Plant Physiology*, Vol.19, (Jul-Sept), pp. 171-184, ISSN1677-0420

Guidi, L., Degl'Innocenti, E., Remorini, D., Massai, R. and Tattini, M. 2008. Interactions of water stress and solar irradiance on the physiology and biochemistry of *Ligustrum vulgare*. *Tree Physiology*, Vol.28, No.6, (June), pp. 873-883, ISSN 1758-4469

Hanba, Y.T., Kogami H. and Terashima I. 2002. The effect of growth irradiance on leaf anatomy and photosynthesis in *Acer* species differing in light demand. *Plant Cell and Environment*, Vol.25, No.8, (August), pp. 1021–1030, ISSN 1365-3040

Kitajima, K. & Hogan, K.P. (2003). Increases of chlorophyll *a/b* ratios during acclimation of tropical woody seedlings to nitrogen limitation and high light. *Plant Cell and Environment*, Vol.26, No.6, (June), pp. 857-865, ISSN 1365-3040

Kitajima, K. & Poorter, L. (2010). Tissue-level leaf toughness, but not lamina thickness, predicts sapling leaf lifespan and shade tolerance of tropical tree species. *New Phytologist*, Vol.186, No.1 (October), pp. 708-721, ISSN 1469-8137

Kitao, M.; Lei T. T.; Koike T.; Tobita H. & Maruyama Y. (2000). Susceptibility to photoinhibition of three deciduous broadleaf tree species with different successional traits raised under various light regimes. Plant *Cell and Environment*, Vol.23, No.1 (January, pp. 81–89, ISSN 1365-3040

Kuptz, D.; Grams, T.E.E. & Günter, S. (2009). Light acclimation of four native tree species in felling gaps within a tropical mountain rainforest. *Trees*, Vol.24, No1. (February), pp. 117-127, ISSN 1432-2285

Laisk, A.; Eichelmann, H.; Oja, V.; Rasulov, B.; Padu, E.; Bichele, I.; Pettai, H. & Kull, O. 2005. Adjustment of leaf photosynthesis to shade in a natural canopy: rate parameters. *Plant Cell and Environment*, Vol.28, No.3 (March), pp. 375-388, ISSN 1365-3040.

Marenco, R.A.; Gonçalves, J.F. & Vieira, G. (2001). Leaf gas exchange and carbohydrates in tropical trees differing in successional status in two light environments in central Amazonia. *Tree Physiology*, Vol.21, No.18 (December), pp. 1311-1318, ISNN 1758-4469

Marenco, R.A. & Vieira, G. (2005). Specific leaf area and photosynthetic parameters of tree species in the forest understorey as a function of the microsite light environment in Central Amazonia. *Journal of Tropical Forest Science*, Vol.17, No.2, pp. 265-278

Matsuki, S.; Ogawa, K.; Tanaka, A. & Hara, T. (2003). Morphological and photosynthetic responses of *Quercus crispula* seedlings to high-light conditions. *Tree Physiology*, Vol.23, No.11 (August), pp. 769-775, ISNN 1758-4469.

Melgar, J.C.; Guidi, L.; Remorini, D.; Agati, G.; Degl'Innocenti, E.; Castelli, S.; Baratto, M.C.; Faraloni, C. & Tattini, M. (2009). Antioxidant defences and oxidative damage in salt-treated olive plants under contrasting sunlight irradiance. *Tree Physiology*, Vol.29, No.9 (September), pp. 1187-1198, ISSN 1758-4469

Mendes, M.M.; Gazarini, L.C. & Rodrigues, M.L. (2001). Acclimation of *Myrtus communis* to contrasting Mediterranean light environments – effects on structure and chemical composition of foliage and plant water relations. *Environmental and Experimental Botany*, Vol.45, pp. 165-178

Mengarda, L.H.G.; Souza, R.L.F.; Campostrinin, E.; Reis, F.O.; Vendrame, W. & Cuzzuol, G.R.F. (2009). Ligth as an indicator of ecological succession in brazilwood (*Caesalpinia echinata* Lam.) *Brazilian Journal of Plant Physiology*, Vol.21, No.1, pp. 55-63, ISSN 1677-0420

Morais, R.R., Gonçalves, J.F.C., Santos Jr, U.M., Dünisch, O. and Santos, A.L.W. (2007). Chloroplastid pigment contents and chlorophyll *a* fluorescence in Amazon tropical three species. *Árvore*, Vol.31, pp. 959-966

Niinemets, Ü.; Kull, O. & Tenhunen, J.D. (1998). An analysis of light effects on foliar morphology, physiology and light interception in temperate deciduous woody species of contrasting shade tolerance. *Tree Physiology*, Vol.18, No.10 (October), pp. 681-696, ISNN 1758-4469

Niinemets, Ü.; A. Descatti; M. Rodeghiero & T. Tosens. (2005). Leaf internal difusion conductance limits photosynthesis more strongly in older leaves of Mediterranea

      evergreen broad-leaved species. *Plant Cell and Environment*, Vol.28, No.12
      (December), pp. 1552-1567, ISSN 1365-3040

Nishizawa, A.; Y. Yabuta & S. Shigeoka. (2008). Galactinol and raffinose constitute a novel
      function to protect plants from oxidative damage. *Plant Physiology*, Vol.147, No.3
      (May), pp. 1251–126, ISSN

Oguchi, R.; Hirosaka, K. & Hisose, T. (2003). Does the photosynthetic light-acclimation need
      change in leaf anatomy? *Plant Cell and Environment*, Vol.26, No.4 (April), pp. 505-
      512, ISSN 1365-3040

Oguchi, R.; Hirosaka, K. & Hisose, T. (2005). Leaf anatomy as a constraint for photosynthetic
      acclimation: differential responses in leaf anatomy to increasing growth irradiance
      among three deciduos trees. *Plant Cell and Environment*, Vol.28, No.7 (July), pp. 916-
      927, ISSN 1365-3040

Parida, A.K.; Das, A.B. & Mittra, B. (2004). Effects of salt on growth, ion accumulation,
      photosynthesis and leaf anatomy of the mangrove, *Bruguiera parviflora*. *Trees*,
      Vol.18, No.2 (March), pp. 167-174, ISSN1432-2285

Peters, S.; Mundree, J.A.; Thomson, J.M.; Farrant & F. Keller. (2007). Protection mechanisms
      in the resurrection plant *Xerophyta viscosa* (Baker): both sucrose and raffinose family
      oligosaccharides (RFOs) accumulate in leaves in response to water deficit. *Journal of*
      Experimental Botany, Vol.58, No.8, (June), pp. 1947–1956, ISSN 1460-2431

Reyes-Díaz, M.; Ivanov, A.G.; Huner, N.P.A.; Alberdi, M.; Corcuera, L.J. & Bravo, L.A.
      (2009). Thermal energy dissipation and its components in two developmental
      stages of shade-tolerants species, *Nothofagus nitida*, and a shade-intolerant species,
      Nothofagus dombeyi. *Tree Physiology*, Vol.29, No.5 (May), pp. 651-662, ISSN 1758-4469

Souza, R.P. & Válio, I. F.M. (2003). Seedling growth of fifteen Brazilian tropical tree species
      differing in sucessional status. *Revista Brasileira de Botânica*, Vol.26, No.1 (March),
      pp. 35-47, ISNN 0100-8404.

Terashima, I.; Hanba, Y.T.; Tazoe, Y.; Vyas, P. & Yano, S. (2006). Irradiance and phenotype:
      comparative eco-development of sun and shade leaves in relation to photosynthetic
      CO_2 diffusion. *Journal of Experimental Botany*, Vol.57, No.2, (January), pp. 343–354,
      ISSN 1460-2431

Urban, O.; Kosvancová, M.; Marek, M.V. & Lichtenthaler, H.K. (2007). Induction of
      photosynthesis and importance of limitations during the induction phase in sun
      and shade leaves of five ecologically contrasting tree species from the temperate
      zone. *Tree Physiolgy*, Vol.27, No.8 (August), pp. 1207-1215, ISNN 1758-4469

Valladares, F.; Arrieta, S.; Aranda, I.; Lorenzo, D.; Sánchez-Gómez, D.; Tena, D.; Suárez, F. &
      Pardos, J.A. (2005). Shade tolerance, photoinhibition sensitivity and phenotypic
      plasticity of *Ilex aquifolium* in continental Mediterranean sites. *Tree Physiology*,
      Vol.25, No.8 (August), pp. 1041-1052, ISSN 1758-4469

Wyka, T.; Robakowski, P. & Zytkowiak, R. (2007). Acclimation of leaves to contrasting
      irradiance in juvenile trees differing in shade tolerance. *Tree Physiology*, Vol.27, No.9
      (September), pp. 1293-1306, ISSN 1758-4469

Yano, S. & I. Terashima. 2004. Developmental process of sun and shade leaves in
      Chenopidium album L. *Plant, Cell and Environment*, Vol.27, No.6 (June), pp. 7181-793,
      ISSN 1365-3040

14

Primary Production in the Ocean

Daniel Conrad Ogilvie Thornton

Department of Oceanography, Texas A & M University, College Station, Texas,
USA

1. Introduction

Primary productivity is the process by which inorganic forms of carbon are synthesized by living organisms into simple organic compounds. Most carbon on Earth is in inorganic oxidized forms such as carbon dioxide (CO_2), bicarbonate (HCO_3^-), and carbonate (CO_3^{2-}). Inorganic carbon must be chemically reduced to form the organic molecules which are the building blocks of life and the mechanism by which energy is stored in living organisms. The reduction of inorganic carbon requires an investment of energy and this can come from light or from energy stored in some reduced inorganic compounds. Autotrophs are organisms capable of fixing inorganic carbon. Photoautotrophs use light energy to fix carbon, whereas chemoautotrophs use the energy released through the oxidation of reduced inorganic substrates to fix carbon into organic compounds.

Both photosynthesis and chemosynthesis contribute to the primary production of the oceans, however oxygenic photosynthesis is by far the dominant process in terms of the amount of carbon fixed and energy stored in organic compounds. Photosynthesis occurs in all parts of the ocean where there is sufficient light, whereas chemosynthesis is limited to locations where there are sufficient concentrations of reduced chemical substrates. Although the vast majority of the ocean's volume is too dark to support photosynthesis, organic carbon and energy is transferred to the dark waters via processes such as particle sinking and the vertical migrations of organisms. Almost all ecosystems in the ocean are fueled by organic carbon and energy which was initially fixed by oxygenic photosynthesis. Anoxygenic photosynthesis does occur in the ocean, however it is confined to anaerobic environments in which there is sufficient light or associated with aerobic anoxygenic photosynthesis (Kolber et al., 2000), the global significance of which is yet to be determined. Consequently, in this overview of primary production in the ocean I will focus on oxygenic photosynthesis.

Blankenship (2002) defined photosynthesis as: 'a process in which light energy is captured and stored by an organism, and the stored energy is used to drive cellular processes.' Oxygenic photosynthesis may be expressed as an oxidation-reduction reaction in the form:

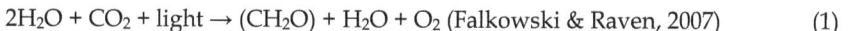

$$2H_2O + CO_2 + light \rightarrow (CH_2O) + H_2O + O_2 \text{ (Falkowski \& Raven, 2007)} \quad (1)$$

In this reaction carbohydrate is formed from carbon dioxide and water with light providing the energy for the reduction of carbon dioxide. Equation 1 is an empirical summary of the overall reaction, which in reality occurs in a number of steps. The light energy for the reaction is primarily absorbed by the green pigment chlorophyll.

2. Which organisms are important primary producers in the ocean?

In terms of number of species, phylogenetic diversity and contribution to total global primary production, the unicellular phytoplankton dominate primary production in the ocean (Falkowski et al., 2004). Almost all oxygenic photosynthetic primary producers in the ocean are either cyanobacteria (Cyanophyta) or eukaryotic algae. The eukaryotic algae are a diverse polyphyletic group, including both unicellular and multicellular organisms.

2.1 Multicellular primary producers

Most multicellular primary producers grow attached to substrates, therefore they are usually restricted to the coastal margins of the ocean in shallow waters where there are both attachment sites and sufficient light for photosynthesis. Important primary producers include seagrasses that form beds that are rooted in sediments in shallow water in tropical and temperate latitudes. Seagrasses (e.g *Zostera*) are flowering plants, unlike the macroalgae, which are not flowering plants and are phylogenetically diverse. Kelps, such as *Macrocystis* and *Laminaria* are locally significant macroalgae in shallow temperate and subpolar waters where there are suitable hard substrates for attachment. Geider et al. (2001) estimated that the net annual primary production by saltmarshes, estuaries and macrophytes was 1.2 Pg C, which is a relatively small proportion of total annual marine production (see section 6).

Some macroalgae are found in the open ocean; *Sargassum* is planktonic and forms rafts at the sea surface in tropical waters (Barnes & Hughes, 1988), mainly in the Gulf of Mexico and Sargasso Sea. The biomass of rafts rival phytoplankton biomass in the mixed layer in the Gulf of Mexico (on an areal basis) with a total standing stock of 2 - 11 million metric tons (Lapointe, 1995; Gower & King, 2008).

2.2 Phytoplankton

There are several groups of eukaryotic phytoplankton which make a significant contribution to global primary production. The most significant of these groups are the diatoms, dinoflagelletes and prymnesiophytes. Diatoms and dinoflagellates are usually found in the microphytoplankton (20 – 200 µm), whereas the prymnesiophytes are nanophytoplankton (2 – 20 µm). Photosynthetic bacteria contribute to the picophytoplankton and are < 2 µm in diameter. Oxygenic photosynthetic bacteria in the oceans belong to the division Cyanophyta, which contains the cyanobacteria (cyanophyceae) and the prochlorophytes (prochlorophyceae).

2.2.1 Photosynthetic bacteria

Arguably, the most important discovery of 20th century biological oceanography was the major role that prokaryotes have in nutrient cycling and production in the water column. As detection and enumeration methods improved it became apparent that photosynthetic bacteria are ubiquitous and make a significant contribution to biomass and primary productivity. The prochlorophytes possess the photosynthetic pigment divinyl chlorophyll *a*, but not chlorophyll *a* which is found in all other Cyanophyta and eukaryotic algae in the ocean. The prochlorophyte *Prochlorococcus marinus* is an abundant and significant primary producer in the open ocean (Chisolm et al., 1988; Karl, 2002) and can be found in concentrations in excess of 10^5 cells ml^{-1} (Chisolm et al., 1988). *Prochlorococcus* has been shown to contribute 9 % of gross primary production in the eastern equatorial Pacific, 39 %

in the western equatorial Pacific and up to 82 % in the subtropical north Pacific (Liu et al., 1997). *Prochlorococcus* is probably the most abundant photosynthetic organism on Earth. Important cyanophyceae include the coccoid *Synechococcus*, which makes a significant contribution to biomass and photosynthesis of the open ocean. For example, Morán et al. (2004) found that picophytoplankton dominated primary production in the North Atlantic subtropical gyre in 2001. *Synechococus* spp. contributed 3 and 10 % of the picophytoplankton biomass, respectively, in the subtropical and tropical domains. However, although *Synechococcus* spp. was significant, *Prochlorococcus* spp. dominated, contributing 69 % of biomass to the subtropical and 52 % to the tropical domain.

2.2.2 Diatoms

Diatoms (Heterokontophyta, Bacillariophyceae) are characterized by a cell wall composed of silica. Estimates of extant diatom species vary between 10,000 (Falkowski & Raven, 2007) and 100,000 (Falciatore & Bowler, 2002). They are found in a wide range of freshwater and marine environments, both in the water column and attached to surfaces. Diatoms make a significant contribution to global primary production on both a local and global scale. It is estimated that diatoms account for 40 to 45% of net oceanic productivity (approximately 20 Pg C yr^{-1}; 1 Pg = 10^{15} g) or almost a quarter of the carbon fixed annually on Earth by photosynthesis (Mann 1999; Falciatore & Bowler 2002; Sarthou et al. 2005), though in my opinion, this is probably an overestimate. Phytoplankton populations in relatively cool, well mixed waters are often dominated by diatoms in terms of productivity and biomass. In addition, diatoms often dominate the microphytobenthos (Thornton et al., 2002), which are populations of microalgae inhabiting the surface layers of sediments in shallow, coastal waters where there is sufficient light reaching the seabed to support photosynthesis. Diatoms may form monospecific blooms of rapidly growing populations; for example, *Skeletonema costatum* frequently forms blooms in coastal waters (Gallagher, 1980; Han et al., 1992; Thornton et al., 1999). Diatom blooms often terminate with aggregate formation, which in addition to the fecal pellets produced by grazers, can lead to the rapid flux of carbon and other nutrients from surface waters to deeper water and the seafloor (Thornton, 2002) (see section 7.1).

2.2.3 Prymnesiophytes

Prymnesiophytes (Prymnesiophyta) are motile, unicellular phytoplankton with two flagella. Most genera also have a filamentous appendage located between the flagella called a haptonema, the function of which is unknown. The cell surface of most prymnesiophytes is covered in elliptical organic scales, which are calcified in many genera. These scales of calcium carbonate are called coccoliths and the prymnesiophytes which possess them are coccolithophores. Coccolithophores are common in warm tropical waters characterized by a low partial pressure of carbon dioxide and saturated or supersaturated with calcium carbonate (Lee, 2008). The importance of the coccolithophores as primary producers during Earth's history is exemplified by the thick chalk deposits found in many parts of the world, such as the white cliffs along the coast of southern England. These deposits were formed from coccoliths that sank to the bottom of warm, shallow seas during the Cretaceous geological period. Moreover, calcium carbonate is the largest reservoir of carbon on Earth. Blooms of coccolithophores such as *Emiliania huxleyi* may be extensive and have been observed on satellite images as milky patches covering large areas of ocean (Balch et al.

1991). *Phaeocystis* is an important primary producer in coastal waters. This genus does not have coccoliths, it is characterized by a colonial life stage in which the cells are embedded in a hollow sphere of gelatinous polysaccharide. Colonies may be large enough to be seen by the naked eye. *Phaeocystis pouchetii* forms extensive blooms in the North Sea (Bätje & Michaelis, 1986) and *Phaeocystis antarctica* is an important primary producer in the Ross Sea (DiTullo et al., 2000).

2.2.4 Dinoflagellates

Dinoflagelletes (Dinophyta) are a largely planktonic division of motile unicellular microalgae that have two flagella. They can be found in both freshwater and marine environments. Generally, dinoflagellates are a more significant component of the phytoplankton in warmer waters. Some photosynthetic dinoflagellates form symbiotic relationships with other organisms, such as the zooxanthellae found in the tissues of tropical corals. Other dinoflagellates do not contribute to primary production as they are non-photosynthetic heterotrophs which are predatory, parasitic or saprophytic (Lee, 2008). Dinoflagellates often dominate surface stratified waters; in temperate zones there may be a succession from diatoms to dinoflagellates as the relatively nutrient rich, well mixed water column of spring stabilizes to form a stratified water column with relatively warm, nutrient poor surface waters. Dinoflagellates have a patchy distribution and may bloom to form 'red tides.' Some dinoflagellates are toxic and form harmful algal blooms; *Karenia brevis* blooms in the Gulf of Mexico and on the Atlantic coast of the USA, resulting in fish kills and human health problems (Magaña et al., 2003).

3. Measuring phytoplankton biomass

The best measure of biomass would be to determine the amount of organic carbon in the phytoplankton cells. However, such a measure is almost impossible in a natural seawater sample due to the presence of other organisms, detritus and dissolved organic matter. Consequently, photosynthetic pigments (usually chlorophyll *a*) are used as a proxy for the biomass of phytoplankton. There are a number of techniques for measuring the concentration of chlorophyll *a* and other photosynthetic pigments in water samples. These methods provide information that is relevant to that particular time and location, but these 'snapshots' have limited use at the regional or ocean basin scale. Over the last 30 years our understanding of the spatial and temporal distribution of phytoplankton biomass has been revolutionized by the measurement of ocean color from satellites orbiting the Earth. These instruments provide measurements over a short period of time of a large area, which is not possible from platforms such as ships.

3.1 Pigment analysis

Chlorophyll fluorescence has been used as tool to determine the distribution of phytoplankton biomass in the ocean since the development of flow-through flourometers (Lorenzen, 1966; Platt, 1972). Most oceanographic research vessels are fitted with flow-through chlorophyll fluorometers that provide continuous chlorophyll fluorescence data. However, there is not a truly linear relationship between *in vivo* flourescence and phytoplankton chlorophyll concentrations (Falkowski & Raven, 2007). The lack of linearity between *in vivo* fluorescence and chlorophyll is related to the fate of light energy absorbed by chlorophyll; light energy is either lost through fluorescence, heat dissipation or used in photochemistry (Maxwell & Johnson, 2000) and the balance between these processes changes depending on the physiological status of the phytoplankton, including rate of photosynthesis and prior light history (Kromkamp & Forster, 2003).

A more accurate estimate of photosynthetic biomass than *in vivo* chlorophyll fluorescence may be obtained if the photosynthetic pigments are extracted from the organism. For water samples containing phytoplankton, a known volume is filtered onto a glass fiber filter and the photosynthetic pigments are extracted using a known volume of organic solvent such as acetone or methanol. The concentration of chlorophyll *a* is then measured in the extract by spectrophotometry or fluorescence (Parsons et al., 1984; Jeffrey et al., 1997). While still widely used, these relatively simple techniques have a number of drawbacks. Firstly, chlorophyll degradation products may absorb light at the same wavelengths as chlorophyll, leading to an overestimation of chlorophyll concentration (Wiltshire, 2009). Secondly, the emission spectra of chlorophyll *a* and *b* overlap, which will result in inaccurate measurement of chlorophyll *a* in water containing chlorophyll *b* containing organisms (Wiltshire, 2009). For the accurate measurement of chlorophyll *a*, other photosynthetic pigments, and their derivatives, high performance liquid chromatography (HPLC) methods should be used (see Wiltshire (2009) for description). HPLC enables the pigments to be separated by chromatography and therefore relatively pure pigments pass through a fluorescence or spectrophotometric detector. In addition to chlorophyll *a*, algae contain multiple pigments. These accessory pigments are often diagnostic for major taxonomic groups (see Table 10.1; Wiltshire, 2009). CHEMTAX (Mackay et al., 1996) is a method by which the total amount of chlorophyll *a* can be allocated to the major taxonomic groups of algae based on the concentrations and ratios of accessory pigments. Thus, HPLC can be used to estimate phytoplankton biomass in terms of chlorophyll *a* and potentially determine the dominant groups of phytoplankton in the sample.

3.2 Ocean color

The Coastal Zone Color Scanner (1978-1986) was the first satellite mission that measured chlorophyll *a* concentrations using top of the atmosphere radiances (McClain, 2009). The success of this mission led to a number of missions to measure ocean color at either global or regional spatial scales by Japanese, European and United States space agencies. As a result of these missions, we now have over 30 years of ocean color measurements. The color of the ocean is affected by particulates and dissolved substances in the water and the absorption of light by water itself. Water is transparent at blue and green wavelengths, but strongly absorbs light at longer wavelengths (McLain, 2009). Chlorophyll *a* has a primary absorption peak near 440 nm and chromophoric dissolved organic matter (CDOM) absorbs in the UV (McLain, 2009). Thus, there is a shift from blue to brown water as pigment and particulate concentrations increase (McLain, 2009).

Measurements of ocean color enabled oceanographers to infer the spatial and temporal distribution of phytoplankton on ocean basin and global scales for the first time. The Sea-viewing Wide Field-of-view Sensor (SeaWIFS) and Moderate Resolution Imaging Spectroradiometers (MODIS) are currently active and have been collecting data since 1997 and 2002 (Aqua MODIS), respectively. MODIS collects data from 36 spectral bands from the entire Earth's surface very 1 to 2 days (http://modis.gsfc.nasa.gov). SeaWIFS also produces complete global coverage every two days (Miller, 2004). Goals for accuracy of satellite products are ± 5% for water-leaving radiances and ± 35 % for open-ocean chlorophyll *a* (McLain, 2009). In addition to rapid regional or global estimates of chlorophyll *a*, algorithms have been developed to estimate net primary productivity (NPP) based on ocean color (Behrenfield & Falkowski, 1997. See section 6 of this chapter).

4. Measuring marine photosynthesis

Photosynthesis results in primary productivity. The terms production and productivity are often used interchangeably and there no generally accepted definition of primary production (Underwood & Kromkamp, 1999). Falkowski & Raven (2007) define primary productivity as a time dependent process which is a rate with dimensions of mass per unit time; whereas primary production is defined as a quantity with dimensions of mass. In contrast, Underwood & Kromkamp (1999) define primary production as a *rate* of assimilation of inorganic carbon into organic matter by autotrophs. For the purposes of this review, I will use the definitions of Falkowski & Raven (2007).

There are a number of methods that are regularly used to measure rates of primary productivity. Techniques are based around gas exchange, the use of isotope tracers, or chlorophyll fluorescence. Primary productivity is usually expressed as the production of oxygen or the assimilation of inorganic carbon into organic carbon over time (equation 1). Carbon assimilation is a more useful measure as it can be directly converted into biomass and used to calculate growth. Common units for primary productivity in marine environment are mg C m^{-2} day^{-1} or g C m^{-2} $year^{-1}$. Primary productivity is often normalized to biomass, as it is useful to know how much biomass is responsible for the observed rates of productivity.

Different techniques will produce slightly different rates of productivity (Bender et al., 1987) as a result of the biases associated with each method. No single technique provides a 'true' measurement of primary productivity. Consequently, researchers should select their methodology based on what factors they want to relate their measurements to, time available to make the measurements, and which assumptions and sources of error are tolerable to answer their particular research questions. Most methods for measuring primary productivity in the ocean require that a sample of water is enclosed in a container, this in itself effects the primary producers. Phytoplankton may be killed on contact with the container or there may be an exchange of solutes between the walls of the container and the sample (Fogg & Thake, 1987). When working in oligotrophic waters contamination of the samples onboard ship is a serious problem. Williams & Robertson (1989) found that the rubber tubing associated with a Niskin sampling bottle severely inhibited primary productivity in samples taken in the oligotrophic Indian Ocean. Moreover, large areas of the ocean are iron limited (Boyd et al., 2000) and it is a challenge to prevent iron contamination in these areas given that oceanographers generally work from ships fabricated from steel.

4.1 Gas exchange methods

Changes in oxygen concentration over time in water samples can be used to calculate rates of photosynthesis and therefore primary productivity. This method involves enclosing water samples and incubating them in the dark and light either onboard ship or *in situ*. Bottles incubated in the dark are used to measure dark respiration rates. To ensure that the rates of photosynthesis are representative, the bottles should be incubated at *in situ* temperature and under ambient light. One way of doing this is to deploy the bottles on a line *in situ*; bottles deployed at different depths will be exposed to the ambient light and temperature at that depth. Changes in oxygen concentration can be monitored with oxygen electrodes or by taking water samples that are fixed and oxygen concentration is subsequently measured by Winkler titration (Parsons et al., 1984). In laboratory studies using pure cultures of phytoplankton oxygen electrode chambers have been used extensively in photosynthesis research (e.g. Colman & Rotatore, 1995; Johnston & Raven, 1996). These systems comprise of a small, optically clear, chamber (usually a few ml) which has a Clark-type oxygen electrode set in the base (see Allen and Holmes (1986) for a full description). Carbon assimilation rates based on oxygen production often assume a ratio of moles of O_2 produced for every mole of CO_2 assimilated, called the photosynthetic quotient, which usually deviates from the 1:1 ratio indicated by equation 1.

In sediments, profiles and changes in oxygen concentration over time may be made using oxygen microelectrodes (Revsbech & Jørgensen, 1983). The microphytobenthos is usually limited to the surface 2 or 3 mm of sediment, therefore high resolution measurements are required; photosynthesis is measured to a resolution of 100 μm and the sensing tips of the microelectrode have diameters of only 2 – 10 μm (Revsbech et al., 1989). While oxygen microelectrodes just measure oxygen concentration, it is possible to measure gross photosynthesis rates using the light-dark shift method (Revsbech & Jørgensen, 1983; Glud et al., 1992; Lassen et al., 1998; Hancke & Glud, 2004). Moreover, oxygen concentration profiles can be used to calculate respiration and net photosynthesis rates according to Kühl et al. (1996) and Hancke & Glud (2004) based on Fick's first law of diffusion. Estimates of benthic primary productivity are also made using oxygen exchange across the sediment-water interface using benthic chambers or sediment cores (Thornton et al., 2002).

Optodes have recently been used to measure changes in oxygen concentrations associated with photosynthesis. Optodes work by using fluorescence quenching by oxygen of a luminophore. The intensity of fluorescence is inversely proportional to the O_2 partial pressure at the luminophore (Glud et al., 1999). For example, Glud et al. (1999) used the luminophore ruthenium (III)-Tris-4,7-diphenyl-1,10-phena-throline, which absorbs blue light (450 nm), with the intensity of the emitted red light (650 nm) decreasing with increasing O_2 partial pressure. Unlike Clark-type oxygen electrodes, optodes do not consume oxygen. Two designs of optodes are used in photosynthesis measurements: optodes that are used in a similar way to oxygen microelectrodes (Miller & Dunton 2007), and planer optodes that produce a two-dimensional image of oxygen concentrations (Glud et al., 1999, 2001). Miller & Dunton (2007) used a micro-optode to measure photosynthesis-irradiance curves for the kelp *Laminaria hyperborea*. Planar optodes have been used to produce images of oxygen concentrations across the sediment-water interface in sediments colonized by photosynthetic biofilms (Glud et al., 1999, 2001). As planar optodes produce a two dimensional image, multiple oxygen profiles can be extracted from a single measurement (Glud et al., 2001). Moreover, the light-dark shift method can be used to measure gross photosynthesis rates (Glud et al., 1999).

4.2 Isotopes as tracers of aquatic photosynthesis

Carbon exists in three isotopes in nature. The most common isotope is ^{12}C, which makes up 98.9% of the natural carbon on Earth. Carbon also exists in another stable form as ^{13}C (1.1 %) and an insignificant amount of the radioactive isotope ^{14}C (< 0.0001 %) (Falkowski & Raven, 2007). The relatively low abundance of ^{14}C and ^{13}C means that these isotopes can potentially be used to measure photosynthesis rates and follow the passage of carbon through photosynthetic organisms when added as tracers. Uptake and assimilation of inorganic carbon into acid-stable organic carbon (Falkowski & Raven, 2007) is the most commonly employed method for measuring photosynthesis using the radioactive tracer ^{14}C (Steeman-Nielson, 1952). The rationale for the ^{14}C method is that the incorporation of radioactively labeled carbon is quantitatively proportional to the rate of incorporation of non-labeled inorganic carbon. Over relatively short incubations the results are a good approximation of gross photosynthesis and an approximation of net photosynthesis over longer time periods (Falkowski & Raven, 2007). This technique (described in Parsons et al., 1984) has been the primary method for measuring the primary productivity of phytoplankton for over fifty years. The method has the advantage of being relatively simple and sensitive. Although widely used, the technique is not without drawbacks and ambiguities. For example, there is an isotopic discrimination between ^{14}C and the natural isotope ^{12}C; less ^{14}C is fixed as it is heavier than ^{12}C, and a discrimination factor of 5 % is usually incorporated into the calculation of inorganic carbon fixation rates (Falkowski & Raven, 2007). Furthermore, the organic carbon, including the ^{14}C which has been fixed during the incubation, is usually separated from the sample by filtration. This can lead to a loss of ^{14}C labeled organic carbon due to rupture of cells on contact with the filter (Sharp, 1977) or exudation of photosynthetic products. There is also a continuing debate as to whether primary productivity measured with the ^{14}C method represents gross or net rates, or something in between the two (Underwood & Kromkamp, 1999).

The advantage of using ^{13}C as a tracer for photosynthesis is that it is not radioactive. This means that it is logistically simpler to use if one has access to an isotope ratio mass spectrometer. Moreover, unlike ^{14}C, ^{13}C can be added as tracer to natural ecosystems and used to trace the assimilation of carbon and transfer to higher trophic levels. Miller & Dunton (2007) used ^{13}C to measure the photosynthesis of the macroalga *Laminaria hyperborea*. Middelburg et al. (2000) and Bellinger et al. (2009) used ^{13}C as a tracer to trace carbon flow through intertidal benthic biofilms dominated by diatoms and cyanobacteria. The tracer was added to the sediment at low tide and followed through the ecosystem over a period of hours to days. Middelburg et al. (2000) showed that carbon fixed through photosynthesis was transferred to bacteria and nematodes within hours. Bellinger et al. (2009) examined the incorporation of the tracer into important biomolecules, including exopolymers (EPS) and phospholipid fatty acids (PLFAs).

Photosynthesis rates have also been measured with the stable isotope ^{18}O by adding labeled water as a tracer and measuring the production of ^{18}O labeled oxygen with a mass spectrometer (Bender et al., 1987; Suggett et al., 2003). The method produces a relatively precise measurement of gross photosynthesis (Falkowski & Raven, 2007). However, this technique has not been used extensively.

Oxygen exists in nature in the form of three isotopes; ^{16}O (99.76 % of the oxygen on Earth), ^{18}O (0.20 %), and ^{17}O (0.04 %) (Falkowski & Raven, 2007). Luz & Barken (2000) developed the triple isotope method using natural abundances of oxygen isotopes to estimate the production of photosynthetic oxygen using the isotopic composition of dissolved oxygen in seawater. The method was based on the ^{17}O anomaly ($^{17}\Delta$), which is calculated from $^{17}O/^{16}O$ and $^{18}O/^{16}O$ (Luz & Barkin, 2000, 2009). This innovative technique does not require water to be enclosed in bottles and therefore avoids bottle effects. The method is used to determine gross photosynthesis rates, enabling integrated productivity to be estimated on a time scale of weeks (Luz & Barkin, 2000). Luz & Barkin (2009) showed that combining $^{17}\Delta$ with O_2/Ar ratios enables gross and net oxygen production to be estimated.

4.3 Chlorophyll fluorescence

Chlorophyll a fluorescence can be used for more than estimating phytoplankton biomass (see 3.1) and there has been a wealth of research over the last 20 years on the application of variable chlorophyll a fluorescence to the measurement of photosynthesis and the physiological status of photosynthetic organisms. Energy absorbed by chlorophyll a may be used in photochemistry and stored in photosynthetic products, dissipated as heat, or lost as fluorescence. Chlorophyll a fluorescence is largely derived from the chlorophyll a associated with photosystem II (PSII); changes in the quantum yield of fluorescence directly relate to O_2 evolving capability as PSII is the oxygen evolving complex within the photosynthetic apparatus (Suggett, 2011). There are two main types of fluorometers that are used to measure variable chlorophyll a fluorescence; Pulse Amplitude Modulation (PAM) fluorometers (Schreiber et al., 1986) and Fast Repetition Rate (FRR) fluorometers (Kolber et al., 1998). These instruments use a modulated light source that allows measurements to be made in the presence of background light or under field light conditions (Maxwell & Johnson, 2000). The PAM approach is not sensitive enough to use in open ocean conditions (Suggett et al., 2003), although it is increasingly being used to measure photosynthetic parameters associated with the microphytobenthos (Underwood, 2002; Perkins et al., 2002; 2011; Serôdio, 2004), macrophytes (Enríquez & Borowitzka, 2011), and has been used with cultures of phytoplankton (Suggett et al., 2003; Thornton, 2009). FRR has been used in the open ocean (Babin et al., 1996; Suggett et al., 2001). The difference between PAM and FRR is beyond the scope of this review; for an overview see Huot & Babin (2011).

Modulated chlorophyll a fluorometers cannot be used to measure photosynthesis directly. One of the primary measurements made with modulated chlorophyll a fluorometers is the quantum yield of PSII photochemistry (Φ_{PSII}). Genty et al. (1989) demonstrated that Φ_{PSII} correlated with CO_2 assimilation in maize and barley, raising the possibility that variable chlorophyll a fluorescence could be used to estimate photosynthesis rates. Φ_{PSII} multiplied by the rate of light absorption by PSII is used to calculate electron transfer rate (ETR_{PSII}) through PSII (Enríquez & Borowitzka, 2011; Suggett et al., 2011; White & Critchley, 1999). ETR_{PSII} has been used as a proxy for photosynthesis. However, there are several reasons why the relationship between Φ_{PSII} (and therefore ETR_{PSII}) and CO_2 assimilation or O_2 production may not be constant (see Suggett, 2011). This effect may be compounded in the algae due to their taxonomic and resultant physiological diversity (Suggett, 2011). Suggett et al., (2009) used an FRR fluorometer to measure ETR_{PSII} and examined the relationship between ETR_{PSII} and photosynthesis measured by either gross O_2 production or $^{14}CO_2$ fixation. Measurements were made using six species of eukaryotic phytoplankton, representing a diversity of taxonomic groups. ETR_{PSII} was linearly related to the rate of gross O_2 production in all species; however, the slope of the relationship was significantly different for different species. ETR_{PSII} was also linearly related to $^{14}CO_2$ fixation; however, both the slope and intercept of the relationship was different for different species. These results highlight some of the challenges involved in using ETR_{PSII} to estimate photosynthesis rates, especially in natural populations of phytoplankton which are likely to be diverse both in terms of taxonomic composition and physiological status.

There are several advantages to variable fluorescence techniques; the techniques are not intrusive and do not harm the organisms, measurements can be made at high spatial and temporal resolution (Suggett et al., 2003), measurements do not require any wet chemistry, and the water sample does not have to be enclosed in a bottle (Kolber et al., 1998). Consequently, variable fluorescence instruments are suited to ocean observing programs; Yoshikawa & Furuya (2004) used a fluorometer moored *in situ* to monitor photosynthesis in coastal waters. Some of the disadvantages to variable fluorescence stem from the fact that this is a relatively young and rapidly evolving field. In the 1970s and 1980s technology was the limiting factor to the development of the field. Since the 1990s there has been a rapid evolution of the technology leading to a large number of commercially available instruments. However, an understanding of the physiology and development of theory associated with variable chlorophyll *a* fluorescence has arguably lagged behind instrument development in recent years. For new users, the large number of fluorescence parameters and their definitions can be confusing (Cosgrove & Borowitza, 2011). This is compounded by the fact that there is no standardized terminology and many fluorescence parameters have several synonyms in the literature. Attempts have been made to standardize terminology (Kromkamp & Forster, 2003).

Photoacoustics has also been used to study phytoplankton photosynthesis (Grinblat & Dubinsky, 2011). This is not a fluorescence technique, however it is based on the same principle that only a small and variable fraction of the energy absorbed by photosynthetic pigment is stored in photosynthetic products. While the preceding discussion has focused on fluorescence, only a few percent of the absorbed light energy is actually lost as fluorescence. The major loss of energy is through heat, which may account for over 60 % of the energy absorbed (Grinblat & Dubinsky, 2011). The photoacoustic method is based on the conversion of light energy to heat energy that results in a rise in temperature and an increase in pressure (photothermal effect) (Grinblat & Dubinsky, 2011). In practice, a suspension of phytoplankton is exposed to a laser pulse, some of the energy from the laser pulse is stored in the photochemical products of photosynthesis and the remainder is dissipated as heat, resulting in an acoustic wave which is measured by a detector (Grinblat & Dubinsky, 2011). This technique has not been used extensively; for further details see Grinblat & Dubinsky (2011).

5. Temporal and spatial variation in oceanic primary production

The mean chlorophyll *a* concentration in the global ocean is 0.32 mg m^{-3} (Falkowski & Raven, 2007). However, this is not evenly distributed throughout the ocean. Primary production at any one location will vary in space and time in response to factors limiting or stimulating photosynthesis and phytoplankton growth. Photosynthesis and growth in the sea are limited by nutrients, light or temperature. In the dynamic environment of a water column resources are patchy both in time and space. Consequently, phytoplankton may receive nutrients and light in pulses rather than a continuous supply. Generally, it is the interplay between nutrient and light availability that affects phytoplankton photosynthesis and primary production.

The traditional paradigm of biological oceanography was that bioavailable nitrogen is the nutrient limiting primary production in the ocean (Ryther & Dunstan, 1971; Howarth, 1988). This is an over simplification and is increasingly being challenged. For example, the Mediterranean Sea appears to be phosphorus limited (Thingstad & Rassoulzadegan, 1995) and there is evidence of phosphorus limitation in coastal ecosystems (Sundareshwar et al., 2003). In over 20 % of the ocean there are excess nutrients (nitrate, phosphate, silicate) and light, but the biomass of phytoplankton is relatively low (Martin et al., 1994). These areas are known as the high-nitrate, low-chlorophyll (HNLC) areas. They are located in the equatorial Pacific, the subarctic Pacific, and the Southern Ocean (Falkowski & Raven, 2007). Iron is an essential component of the nitrogenase enzyme, consequently iron limitation limits nitrogen fixation by cyanobacteria over large areas of the ocean (Falkowski et al., 1998) (see 7.3). Iron is supplied to the open ocean via wind blown dust from arid areas of the continents (Duce & Tindale, 1991). The upwelling of deep waters containing nitrate and phosphate produced from the remineralization of organic matter is important in maintaining high primary productivity in many areas of the ocean, such as along the western margins of Africa and South America. Conversely, thermal stratification and downwelling will limits primary production in the subtropical gyres as the sunlit surface waters are largely isolated from nutrient rich waters below the thermocline.

Light (solar radiation) provides the energy that drives photosynthesis. Light is variable on a number of spatial and temporal scales. Low latitudes receive more solar radiation than high latitudes and have less variation in solar radiation over the course of one year. At high latitudes there are pronounced seasons and variations in day length. Imposed on these month to month or season to season variations in solar radiation are short term fluctuations. The angle of the sun above the horizon affects how much light is reflected off the surface of the ocean. At midday, when the sun is at an angle of 90° to the sea surface, 2 % of the incoming solar radiation is reflected; this value increases to 40 % during the evening and early morning when the sun is at an angle of 5 ° (Trujillo & Thurman, 2005). Reflection off cloud cover also significantly reduces the input of solar radiation into the ocean. Conversely, net primary productivity may be inhibited by too much light, which can lead to photoinhibition or conditions conducive to photorespiration (Fig. 1).

The depth of the euphotic zone (surface layer in which there is enough light to support photosynthesis) is often less than 10 m and rarely greater than 100 m (Fogg & Thake, 1987), whereas the mean depth of the ocean is approximately 3,700 m. Therefore, photosynthesis and primary production is limited to a thin layer at the ocean surface and whether phytoplankton cells are mixed into the dark waters below will effect primary productivity. As the mixed layer of a water column increases the average photon flux density (i.e. light) to which the cells are exposed will decrease as the circulating cell will spend longer in darkness. Therefore the total gross productivity of the phytoplankton population will decrease (Fig. 1). However, the respiration rate of the population will be relatively constant, whatever the depth of mixing. This results in a *critical depth* in a mixed water column; if the cells are mixed below the critical depth then there will be no net productivity as the respiratory loss of carbon will exceed photosynthetic carbon gain. Net photosynthesis and the resulting net primary productivity will only occur in mixed water where the mixing is less than the critical depth (Sverdrup, 1953; Kirk, 1983).

Grazing also effects production; in a heavily grazed population of phytoplankton individual cells may show high rates of productivity, but there may be a low biomass of primary producers as a large proportion of the primary production is transferred to other trophic levels. In recent years there has been a realization that phytoplankton are subject to lytic viral infection (Suttle et al., 1990; Nagasaki et al., 2004), which will have an effect similar as grazing by reducing the biomass of primary producers in the water column.

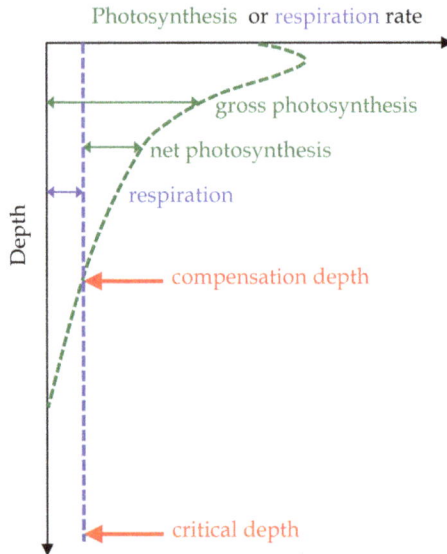

Fig. 1. Schematic of photosynthesis and respiration rates with depth in the ocean. The green line shows gross photosynthesis rate, which declines from a maximum just below the surface to zero in response to the availability of light. Phytoplankton respiration rate is constant with depth and is shown in blue; net photosynthesis is gross photosynthesis minus respiration. The compensation depth occurs where the net photosynthesis rate is zero as the respiration rate is equal to the gross photosynthesis rate. The critical depth occurs deeper in the water column and it is where depth integrated gross photosynthesis equals depth integrated respiration.

6. Global estimates of primary productivity

Estimates of net primary productivity in the oceans are on the order of 40 -55 Pg C yr-1 (1 Pg = 1 gigaton = 10^{15} g), which is approaching half of global annual net primary productivity. Falkowski et al. (1998), citing data from a number of papers, estimated that marine phytoplankton fix approximately 45 Pg C yr-1. Falkowski et al. (1998) estimated that 16 of the 45 Pg C yr-1 are exported from the surface to the ocean interior (see 7.2).

A key goal of satellite observations of ocean color has been to convert ocean color data to net primary productivity (NPP) and a variety of NPP algorithms exist (McCLain, 2009). Longhurst et al. (1995) estimated global net primary productivity of the oceans to be 45 – 50 Pg C yr-1 and Field et al. (1998) estimated a value of 48.5 Pg C yr-1. Since this initial work, many different algorithms have been developed to estimate NPP from ocean color data. Carr et al. (2006) compared 24 algorithms and found that the mean NPP estimate for the ocean was 51 Pg C yr-1, with the range of the estimates spanning 32 Pg C yr-1. Most of these algorithms have an empirical physiological parameter to account for phytoplankton physiological status (McClain, 2009), which is difficult to determine from space based observations (Behrenfeld et al., 2005).

Behrenfeld et al., (2005) developed a carbon-based model that does not require a physiological parameter. The logic of their approach was that laboratory experiments have shown that phytoplankton respond to changes in light, nutrients and temperature by adjusting cellular pigment content to meet requirements for photosynthesis. This physiological response can be seen in changes in the carbon: chlorophyll a ratio of the phytoplankton and therefore a remote sensing measure of the carbon: chlorophyll a ratio may provide an index of the physiological status of the phytoplankton. Westberry et al., (2008) recently refined this model and estimated a global ocean NPP of approximately 52 Pg C yr[-1].

Geider et al. (2001), based on the data of Longhurst et al. (1995) and Field et al. (1998), estimated NPP for different ocean environments, with values of: 1.2 Pg C yr[-1] for salt marshes, macrophytes and estuaries; 0.7 Pg C yr[-1] for coral reefs; 10.7 Pg C yr[-1] for the coastal domain; 6.4 Pg C yr[-1] for the polar regions; 29.3 Pg C yr[-1] for the remaining ocean. The average NPP of the land (excluding areas permanently covered in ice) is 426 g C m[-2] yr[-1], compared to 140 g C m[-2] yr[-1] for the ocean (Field et al., 1998). NPP per unit area of the land is three times greater than that of the ocean; however, the ocean contributes almost half of net global primary production as it covers approximately three quarters (70.8 %) of the surface area of Earth.

7. Oceanic primary production and the biogeochemical cycling of elements

Biogeochemical cycles occur as a result of the interplay between physical, geological, chemical and biological processes, which affect the distribution of elements between living and non-living reservoirs on Earth. Living organisms both react to changes in the environment and affect environmental change. This concept has led to 'Earth system science' in which the biosphere, chemosphere and geosphere are not regarded in isolation, but as an integrated whole. Phytoplankton play a crucial role in biogeochemical processes on the Earth. Phytoplankton and other marine primary producers are composed of elements in addition to carbon, oxygen, and hydrogen (equation 1). A 'typical' phytoplankton cell is composed of 25 to 50 % protein, 5 to 50 % polysaccharide, 5 to 20 % lipids, 3 to 20 % pigments, and 20 % nucleic acids (Emerson & Hedges, 2008). Therefore, the elements nitrogen, phosphorus and sulphur are required in relatively large quantities to build a functioning phytoplankton cell. Other elements are required in trace amounts, such as many metals. Consequently, phytoplankton photosynthesis and subsequent phytoplankton growth directly affect the cycling of biologically important elements on Earth. For example, cyanobacteria have been extant on Earth for at least 2.7 billion years (Knoll, 2003) and have been the dominant oxygenic photosynthetic organisms on Earth for most of that time. Photosynthesis by cyanobacteria led to an accumulation of oxygen in the atmosphere, replacing an anoxic atmosphere with something approaching the contemporary atmosphere by 2.2 billion years ago (Falkowski et al., 1998; Knoll, 2003).

7.1 Ecological stoichiometry of phytoplankton

Stoichiometric variation in the elemental composition of photosynthetic organisms profoundly affects biological productivity, food web dynamics and the biogeochemical cycling of elements across all spatial and temporal scales (Sterner & Elser, 2002). One of the earliest ecological stoichiometry paradigms was the Redfield ratio. Redfield showed that the ratio of C:N:P in organic material collected with a plankton net (i.e. mainly phytoplankton and zooplankton) was conservative and 106:16:1 (Redfield, 1958). Redfield proposed that the ratio of nitrate: phosphate in the ocean was determined by the requirements of phytoplankton, which release nitrogen and phosphorus into the environment as they are remineralized (Arrigo, 2005).

In the last few years there has been a resurgence of interest in measuring the elemental stoichiometry of marine phytoplankton (Quigg et al., 2003; Ho et al., 2003; Leonardos & Geider 2004; Bertilsson et al., 2003; Klausmeier et al., 2004), as determining the degree to which C:N:P stoichiometry can deviate from the Redfield ratio of 106:16:1 is seen as critical in the understanding of the role of phytoplankton in biogeochemistry (Falkowski, 2000; Geider & La Roche, 2002). For example, as we have seen, the Redfield ration of C:N:P of 106:16:1 represents a mean value for marine phytoplankton; there are actually significant differences in C:N, C:P and N:P ratios in different phyla of phytoplankton (Quigg et al., 2003). Klausmeier et al. (2004) used a modeling approach to show that N:P ratios depend on ecological conditions and that optimum N:P ratios vary between 8.2 and 45.0. Stoichiometric variation may occur between taxa or even within one species depending on growth conditions. A source of variation comes from the fact that different components of the cell's structure have different stoichiometric ratios (Arrigo, 2005). Cell machinery for acquiring resources (light and nutrients), such as chlorophyll and protein, are rich in N and contain low P. On the other hand, growth machinery such as ribosomal RNA are rich in both N and P. Consequently, changes in the relative proportions of key cell components will affect the stoichiometry of an individual phytoplankton (Arrigo, 2005). Different growth strategies will result in different cell stoichiometries; exponentially growing bloom-forming phytoplankton have an optimal N:P ratio of 8 as they are optimized for growth. Whereas the N:P ratio of phytoplankton in a resource scarce environmental is optimal at 36 to 45 as they contain more machinery for resource acquisition (Klausmeier et al., 2004; Arrigo, 2005).

7.2 Primary production and carbon cycling

As we have seen, marine photosynthesis supports approximately half of global primary production and therefore marine primary producers have a profound effect on the global carbon cycle. While there are some non-photosynthetic primary producers in the ocean, their contribution to carbon fixation in the ocean is negligible. Therefore, ultimately all the organic matter was originally fixed by oxygenic photosynthetic organisms. The pool of organic carbon associated with living organisms in the ocean is relatively small. Falkowski and Raven (2007) estimate that the amount of carbon in phytoplankton in the global ocean is between 0.25 and 0.65 Pg. To support an annual primary production of 50 Pg C yr^{-1}, the biomass of phytoplankton must turn over between 60 and 150 times per year, or every 2 to 6 days. This is very different from terrestrial primary productivity, which is dominated by multicellular woody organisms and has a mean turnover time of 12 to 16 years (Falkowski & Raven, 2007). The pool of non-living organic carbon in the ocean is much larger than the carbon associated with living organisms, with an estimate of 1000 Pg C (Falkowski et al. 2000). Most of the non-living carbon in the ocean is in the form of dissolved organic carbon (DOC) rather than particulate organic carbon (POC). Hansell et al. (2009) estimated that the oceans contain 662 Pg C as dissolved organic carbon (DOC). To put this into context, the atmosphere contained 612 Pg C in 1850 and 784 Pg C in 2000 (Emerson & Hedges 2008). However, the largest pool of carbon in the ocean is dissolved inorganic carbon, which contains 38,000 Pg C (Emerson & Hedges, 2008).

Given that the annual rate of marine primary productivity is around 50 Pg C year^{-1}, it can be seen that primary productivity in the ocean has the potential to affect the atmospheric content of carbon dioxide. For example, if there is a significant change in either marine photosynthesis or respiration, or the balance between the two processes, then the mixing ratio of carbon dioxide in the atmosphere will change. Consequently, determining the fate of primary production in the ocean is important in understanding the global carbon cycle and global climate change. An important question is how much organic carbon is exported from surface waters and into the deep ocean. Most of the carbon fixed through photosynthesis is remineralized back to carbon dioxide in surface waters. However, some organic carbon sinks into deeper water and a very small fraction of organic carbon will make it all the way to the seafloor. Even if the organic matter is remineralized at depth, it may take thousands of years for it to be returned to the atmosphere as CO_2 and carbon buried in the seafloor may be potentially sequestered for millions of years. The carbon pump refers to processes that affect a vertical gradient in dissolved inorganic carbon concentration in the ocean (Emerson & Hedges, 2008). These processes collectively remove carbon from the surface ocean and atmosphere and into the deep ocean. The carbon pump has two components, the solubility pump and biological carbon pump. The biological carbon pump is the biologically mediated processes that transport carbon from the surface ocean into the deep ocean, such as the sinking of phytoplankton aggregates (Thornton, 2002), fecal pellets, dead organisms, and DOC.

7.3 Primary production and nitrogen cycling

Cyanobacteria have an important role in the nitrogen cycle as many species are nitrogen fixing and consequently sustain primary productivity in nitrogen-limited areas of the ocean (Capone, 2000). The availability of nitrogen generally limits phytoplankton photosynthesis and growth in the ocean (Ryther & Dunstan, 1971; Howarth, 1988; Tyrell 1999). Although 78 % of the atmosphere is in the form of N_2, this can only be used by organisms possessing nitrogense enzymes to reduce N_2 to biologically available NH_4^+; this nitrogen fixing ability is limited to certain groups of prokaryotes, including many cyanophyceae. In the vast oligotrophic tropical and subtropical areas of ocean the colonial, filamentous cyanobacteria of the genus *Trichodesmium* are important primary producers as they can overcome nitrogen limitation by fixation (Capone et al., 1997). *Trichodesmium* may form extensive surface blooms (it is buoyant due to gas vesicles) in stable, clear, low nutrient water columns. It has a cosmopolitan distribution and is probably a significant contributor to global primary productivity (Capone et al., 1997). The remineralization of carbon associated with the cyanobacteria will release bioavailable nitrogen into the water column which may support the growth of other groups of primary producers. See Zehr & Paerl (2008) for a recent review of nitrogen fixation in the ocean.

The discussion above implies that cyanobacteria are constantly fixing nitrogen from the atmosphere and enriching the ocean with bioavailable nitrogen, which would accumulate in the ocean over time. This is not the case as there are processes that remove bioavailable nitrogen from ecosystems. A description of the marine nitrogen cycle is beyond the scope of this review (see Capone 2000; Thamdrup & Dalsgaard, 2008). However, nitrogen exists in a number of oxidation states in the ocean with the transformation between oxidation states largely conducted by prokaryotes. There are two major microbial processes that remove bioavailable nitrogen from the ocean and return it to the sink of the atmosphere: denitrification and anaerobic ammonium oxidation (anammox) (Thamdrup & Dalsgaard, 2008).

7.4 Primary production and phosphorus cycling

Concentrations of phosphorus in the ocean are controlled by geological processes; phosphorus is added to the ocean from the weathering of rocks and removed by burial in marine sediments. Unlike nitrogen, phosphorus has no atmospheric source. On geological time scales phosphorus is the *ultimate limiting nutrient* for primary production in the ocean (Tyrell, 1999). However, on shorter timescales bioavailable nitrogen is the *proximate limiting nutrient* in the sense of Liebig's law (Tyrell, 1999). The addition of just nitrogen (i.e. nitrate or ammonium) to surface waters from most of the ocean would result in an enhancement of phytoplankton production. Moreover, there is a slight excess of phosphate in the ocean compared to nitrate in surface waters as nitrate is depleted before phosphate (Tyrell, 1999).

7.5 Primary production and iron cycling

There was a vigorous debate as to what limited primary production in the HNLC areas of the global ocean (see section 5). John Martin (1992) proposed the hypothesis that iron availability limits the growth of phytoplankton in the HNLC ocean. The implications of this hypothesis are that the uptake of other nutrients is limited by iron availability and that the amount of carbon dioxide in the atmosphere varies as a function of iron transport to the surface of the ocean (Millero, 2006). This has implications for the global carbon cycle as it implies that iron supply to iron limited phytoplankton plays a role in determining how much carbon is sequestered in the deep ocean and underlying sediments. The iron hypothesis was tested in the laboratory and during ship board experiments. While these experiments showed iron limitation, the results do not necessary reflect the *in situ* response to iron fertilization as small samples enclosed in bottles are subject to bottle effects and do not capture the complexity of natural systems. Consequently, a number of experiments have been conducted in which large areas of ocean have been fertilized with iron (Martin et al., 1994; Coale et al., 1996; 2004; Boyd et al., 2000; Tsuda et al., 2003). These experiments have shown that HNLC are iron limited. There was a significant increase in phytoplankton biomass over the course of a few days in the iron fertilized patches resulting in a significant draw down of carbon dioxide due to photosynthetic carbon fixation. Moreover, the alleviation of iron limitation affected the efficiency of photosynthesis as there was an increase in Φ_{PSII} (Kolber et al., 1994; Coale et al., 2004). The addition of iron to HNLC areas has been proposed as a geoengineering approach to sequester atmospheric carbon dioxide in the deep ocean and thereby partially offset anthropogenic inputs of carbon dioxide from fossil fuel burning and land use change. This is controversial, as although the addition of iron to HNLC areas stimulates primary production and the drawdown of carbon dioxide, there is relatively little evidence that a significant proportion of the resulting organic carbon sinks into the deep ocean and is sequestered (Buesseler et al., 2004; Boyd et al., 2004).

8. Conclusion

Photosynthesis and primary production in the ocean is dominated by a phylogenetically diverse group of microalgae that make up the phytoplankton. A number of techniques are used to measure phytoplankton photosynthesis. These techniques are based on gas exchange, the use of radioactive or stable isotopes as tracers, and variable chlorophyll *a* fluorescence. The techniques are not equivalent, and different techniques will produce different rates of photosynthesis dependent on the biases and assumptions intrinsic in the methodology. Phytoplankton photosynthesis is a globally significant process. Using satellite data of ocean color, NPP in the global ocean is estimated to be around 50 Pg C year-1, which is similar to terrestrial NPP. Primary production in the ocean affects the cycling of biologically significant elements, such as carbon, nitrogen, and phosphorus. A key question is how phytoplankton production affects the biogeochemical cycling of elements (e.g. carbon, nitrogen, phosphorus, and iron) and climate change.

9. Acknowledgment

This material is based upon work supported by the National Science Foundation under Grant No. OCE 0726369 to Daniel C. O. Thornton. Any opinions, findings, and conclusions or recommendations expressed in this material are those of the author and do not necessarily reflect the views of the National Science Foundation.

10. References

Allen, J. F. & Holmes, N. G. (1986) Electron Transport and Redox Titration, In: *Photosynthesis Energy Transduction*, M. F. Hopkins & N R Baker (Eds.), pp.103-141, Oxford University Press, ISBN 978-0947946517, Oxford, United Kingdom

Arrigo, K. R. (2005) Marine Microorganisms and Global Nutrient Cycles. *Nature,* vol.437, No.7057, pp. 349-355, ISSN 0028-0836

Babin, M.; Morin, A.; Claustre, H.; Bricaud, A.; Kolber, Z. & Falkowski P. G. (1996) Nitrogen- and Irradiance-Dependent Variations of the Maximum Quantum Yield of Carbon Fixation in Eutrophic, Mesotrophic and Oligotrophic Marine Systems. *Deep-Sea Research I*, vol.43, No.8, pp.1241-1272, ISSN 0967-0637

Balch, W. M.; Holligan, P. M.; Ackleson, S. G. & Voss K. J. (1991) Biological and Optical Properties of Mesoscale Coccolithophore Blooms in the Gulf of Maine. *Limnology and Oceanography,* vol.36, No.4, pp. 629-643, ISSN 0024-3590

Barnes, R. S. K. & Hughes, R. N. (1987) *An Introduction to Marine Ecology* (second edition), Blackwell Scientific Publications, ISBN 0-632-02047-4, Oxford.

Bätje, M. & Michaelis, H. (1986) *Phaeocystis pouchetti* Blooms in the East Frisian Coastal Waters (German Bight, North Sea), *Marine Biology*, vol. 93, No.1, pp. 21-27, ISSN 0025-3162

Behrenfeld,M. J. & Falkowski, P. G. (1997) Photosynthetic Rates Derived from Satellite-Based Chlorophyll Concentration, *Limnology and Oceanography,* vol. 42, No.1, pp.1-20, ISSN 0024-3590

Behrenfeld, M. J.; Boss, E.; Siegel, D. A. & Shea, D. M. (2005) Carbon-Based Ocean Productivity and Phytoplankton Physiology from Space. *Global Biogeochemical Cycles*, vol.19, No.1, doi:10.1029/2004GB002299, ISSN 0886-6236

Bellinger, B. J.; Underwood, G. J. C.; Ziegler, S. E. & Gretz, M. R. (2009) Significance of Diatom-Derived Polymers in Carbon Flow Dynamics within Estuarine Biofilms Determined through Isotopic Enrichment, *Aquatic Microbiology Ecology*, vol.55, No.2, pp.169-187, ISSN 0948-3055

Bender, M.; Grande, K.; Johnson, K.; Marra, J.; Williams, P. J. LeB.; Sieburth, J.; Pilson, M.; Langdon, C.; Hitchcock, G.; Orchardo, J.; Hunt, C.; Donaghey, P. & Heinemann, K. (1987) A Comparison of Four Methods for Measuring Planktonic Community Production. *Limnology and Oceanography* vol.37, No.5, pp.1085-1098, ISSN 0024-3590

Bertilsson, S.; Berglund, O.; Karl, D. M. & Chisholm, S. W. (2003) Elemental Composition of Marine *Prochlorococcus* and *Synechococcus*: Implications for Ecological Stoichiometry of the Sea. *Limnology and Oceanography* vol.48, No.5, pp.1721-1731, ISSN 0024-3590

Blankenship, R. E. (2002) *Molecular Mechanisms of Photosynthesis*, Blackwell Science Ltd, ISBN 978-0-632-04321-7, Oxford, United Kingdom

Boyd, P. W.; Watson, A. J.; Law, C. S.; Abraham, E. R.; Trull, T.; Murdoch, R.; Bakker, D. C. E.; Bowie, A. R.; Buesseler, K. O.; Chang, H.; Charette, M.; Croot, P.; Downing, K.; Frew, R.; Gall, M.; Hadfield, M.; Hall, J.; Harvey, M.; Jameson, G.; LaRoche, J.; Liddicot, M.; Ling, R.; Maldonado, M.T.; McKay, R.M.; Nodder, S.; Pickmere, S.; Pridmore, R.; Rintoul, S.; Safi, K.; Sutton, P.; Strzepek, R.; Tanneberger, K.; Turner, S.; Waite, A & Zeldis, J. (2000) A Mesoscale Phytoplankton Bloom in the Polar Southern Ocean Stimulated by Iron Fertilization. *Nature*, vol.407, No.6805, pp.695-702, ISSN 0028-0836

Boyd, P. W.; Law, C. S.; Wong, C. S.; Nojiri, Y.; Tsuda, A; Levasseur, M.; Takeda, S.; Rivkin, R.; Harrison, P. J.; Strzepek, R.; Gower, J.; McKay, R. M.; Abraham, E.; Arychuk, M.; Barwell-Clarke, J.; Crawford, W.; Crawford, D.; Hale, M.; Harada, K.; Johnson, K.; Kiyosawa, H.; Kudo, I.; Marchetti, A.; Miller, W.; Needoba, J.; Nishioka, J.; Ogawa, H.; Page, J.; Robert, M.; Saito, H.; Sastri, A.; Sherry, N.; Soutar, T.; Sutherland, N.; Taira, Y.; Whitney, F.; Wong, S. K. E. & Yoshimura, T. (2004) The Decline and Fate of an Iron-Induced Subarctic Phytoplankton Bloom. *Nature*, vol.428, No.6982, pp.549-553, ISSN 0028-0836

Buesseler, K. O.; Andrews, J. E.; Pike, S. M. & Charette, M. A. (2004) The Effects of Iron Fertilization on Carbon Sequestration in the Southern Ocean. *Science*, Vol.304, No.5669, pp.414-417, ISSN 0036-8075

Capone, D. G. (2000) The Marine Microbial Nitrogen Cycle, In: *Microbial Ecology of the Oceans*, D. L. Kirchman (Ed.), pp. 455-493. Wiley-Liss, ISBN 0-471-29992-8, New York, United States

Capone, D. G., Zehr, J. P., Paerl, H. W., Bergman, B. & Carpenter, E. J. (1997). *Trichodesmium*, a Globally Significant Marine Cyanobacterium. *Science*, vol.276, No.5316, pp.1221-1229, ISSN 0036-8075

Carr, M. E.; Friedrichs, M. A. M.; Schmeltz, M.; Aita, M. N.; Antoine, D.; Arrigo, K. R.; Asanuma, I.; Aumont, O.; Barber, R.; Behrenfeld, M.; Bidigare, R.; Buitenhuis, E. T.; Campbell, J.; Ciotti, A.; Dierssen, H.; Dowell, M.; Dunne, J.; Esaias, W.; Gentili, B.; Gregg, W.; Groom, S.; Hoepffner, N.; Ishizaka, J.; Kameda, T.; Le Quere, C.; Lohrenz, S.; Marra, J.; Melin, F.; Moore, K.; Morel, A.; Reddy, T. E.; Ryan, J.; Scardi, M.; Smyth, T.; Turpie, K.; Tilstone, G.; Waters, K. & Yamanaka, Y. (2006) A Comparison of Global Estimates of Marine Primary Production from Ocean Color. *Deep-Sea Research Part II-Topical Studies in Oceanography*, vol. 53, No.5-7, pp.741-770, ISSN 0967-0645

Chisholm, S. W.; Olson, R. J.; Zettler, E. R.; Goericke, R.; Waterbury, J. B. & Welschmeyer, N. A. (1988) A Novel Free-Living Prochlorophyte Abundant in the Oceanic Euphotic Zone. *Nature*, vol.334, No.6180, pp.340-343, ISSN 0028-0836

Coale, K. H.; Johnson, K. S.; Fitzwater, S. E.; Gordon, R. M.; Tanner, S.; Chavez, F. P.; Ferioli, L,; Sakamoto, C.; Rogers, P.; Millero, F.; Steinberg, P.; Nightingale, P.; Cooper, D.; Cochlan, W. P.; Landry, M. R.; Constantinou, J.; Rollwagen, G.; Trasvina, A.& Kudela, R. (1996) A Massive Phytoplankton Bloom Induced by an Ecosystem-Scale

Iron Fertilization Experiment in the Equatorial Pacific Ocean. *Nature*, vol.383, No.6600, pp.495-501, ISSN 0028-0836

Coale, K. H.; Johnson, K. S.; Chavez, F. P.; Buesseler, K. O.; Barber, R. T.; Brzezinski, M. A.; Cochlan, W. P.; Millero, F. J.; Falkowski, P. G.; Bauer, J. E.; Wanninkhof, R. H.; Kudela, R. M.; Altabet, M. A.; Hales, B. E.; Takahashi, T.; Landry, M. R.; Bidigare R. R.; Wang, X. J.; Chase, Z.; Strutton, P. G.; Friederich, G. E.; Gorbunov, M. Y.; Lance, V. P.; Hilting, A. K.; Hiscock, M. R.; Demarest, M.; Hiscock, W. T.; Sullivan, K. F.; Tanner, S. J.; Gordon, R. M.; Hunter, C. N.; Elrod, V. A.; Fitzwater, S. E.; Jones, J. L.; Tozzi, S.; Koblizek, M.; Roberts, A. E.; Herndon, J.; Brewster, J.; Ladizinsky, N.; Smith, G.; Cooper, D.; Timothy, D.; Brown, S. L.; Selph, K. E.; Sheridan, C. C.; Twining, B. S. & Johnson, Z. I. (2004) Southern Ocean Iron Enrichment Experiment: Carbon Cycling in High- and Low-Si Waters. *Science*, vol.304, No.5669, pp.408-414, ISSN 0036-8075

Colman, B. & Rotatore, C. (1995) Photosynthetic Inorganic Carbon Uptake and Accumulation in Two Marine Diatoms. *Plant, Cell and Environment*, vol.18, No.8, pp.919-924, ISSN 0140-7791

Cosgrove, J. & Borowitzka, M. (2011) Chlorophyll fluorescence terminology: An Introduction, In: *Chlorophyll a Fluorescence in Aquatic Sciences: Methods and Applications*, D. J. Suggett, O. Prášil & M. A. Borowitzka (Eds.), pp. 1-17, Springer, ISBN 978-90-481-9267-0, Dordrecht, Netherlands

DiTullio, G. R.; Grebmeier, J. M.; Arrigo, A. R.; Lizotte, M. P.; Robinson, D. H.; Leventer, A.; Barry, J. P.; VanWoert, M. L. & Dunbar, R. B. (2000) Rapid and Early Export of *Phaeocystis antarctica* Blooms in the Ross Sea, Antarctica. *Nature*, Vol. 404, No.6778, pp.595-598, ISSN 0028-0836

Duce, R. A. & Tindale, N. W. (1991) Atmospheric Transport of Iron and its Deposition in the Ocean. *Limnology and Oceanography*, vol.36, No.8, pp.1715-1726, ISSN 0024-3590

Enríquez, S. & Borowitzka, M. A. (2011) The use of the fluorescence signal in studies of seagrasses and macroalgae, In: *Chlorophyll a Fluorescence in Aquatic Sciences: Methods and Applications*, D. J. Suggett, O. Prášil & M. A. Borowitzka (Eds.), pp. 187-208, Springer, ISBN 978-90-481-9267-0, Dordrecht, Netherlands

Falciatore, A. & Bowler, C. (2002) Revealing the Molecular Secrets of Marine Diatoms. *Annual Review of Plant Biology* vol.53, pp.109-130, ISSN 1040-2519

Falkowski, P. G. (2000) Rationalizing Elemental Ratios in Unicellular Algae. *Journal of Phycology* vol.36, No.1, pp.3-6, ISSN: 0022-3646

Falkowski, P. G., Barber R. T. & Smetacek, V. (1998) Biogeochemical Controls and Feedbacks on Ocean Primary Production. *Science*, vol.281, No.5374, pp.200-206, ISSN 0036-8075

Falkowski, P.; Scholes, R. J.; Boyle, E.; Canadell, J.; Canfield, D.; Elser, J.; Gruber, N.; Hibbard, K.; Hogberg, P.; Linder, S.; Mackenzie, F. T. Moore, B.; Pedersen, T.; Rosenthal, Y.; Seitzinger, S.; Smetacek, V.; & Steffen, W. (2000) The Global Carbon Cycle: A Test of Our Knowledge of Earth as a System. *Science* vol.290, No.5490, pp.291-296, ISSN 0036-8075

Falkowksi P. G.; Katz M. E.; Knoll A. H.; Quigg A.; Raven J. A.; Schofield O. & Taylor F. J. R. (2004) The Evolution of Modern Eukaryotic Phytoplankton. *Science*, vol.305, No.5682, pp.354-360, ISSN 0036-8075

Falkowski P. G. & Raven J. A. (2007) *Aquatic Photosynthesis* (second edition), Princeton University Press, ISBN-10: 0-691-11551-6, Princeton, New Jersey, United States

Field, C. B.; Behrenfield, M. J.; Randerson, J. T. & Falkowski, P. (1998) Primary Production of the Biosphere: Integrating Terrestrial and Oceanic Components. *Science*, vol.281, No.5374, pp. 237-240, ISSN 0036-8075

Fogg G. E. & Thake B. (1987) *Algal cultures and phytoplankton Ecology* (third edition), The University of Wisconsin Press, ISBN 0-299-10560-1, Madison, Wisconsin, United States

Gallagher J. (1980) Population Genetics of *Skeletonema costatum* (Bacillariophyceae) in Narragansett Bay. *Journal of Phycology* vol.16, No.3, pp.464-474, ISSN 0022-3646

Geider, R. J. & La Roche, J. (2002) Redfield Revisited: Variability of C:N:P in Marine Microalgae and its Biochemical Basis. *European Journal of Phycology*, vol.37, No.1, pp.1-18, ISSN 0967-0262

Geider, R. J.; Delucia, E. H.; Falkowski, P. G.; Finzi, A. C.; Grime, J. P.; Grace, J.; Kana, T. M.; La Roche, J.; Long, S. P.; Osborne, B. A.; Platt, T.; Prentice, I. C.; Raven, J. A.; Schlesinger, W. H.; Smetacek, V.; Stuart, V.; Sathyendranath, S.; Thomas, R. B.; Vogelmann, T. C.; Williams, P.; & Woodward, I. F. (2001) Primary Productivity of Planet Earth: Biological Determinants and Physical Constraints in Terrestrial and Aquatic Habitats. *Global Change Biology*, vol.7, No.8, pp.849-882, ISSN: 1354-1013

Genty, B.; Briantais, J-M. & Baker, N. R. (1989) The Relationship between the Quantum Yield of Photosynthetic Electron Transport and Quenching of Chlorophyll Fluorescence. *Biochimica Biophysica Acta*, vol.990, No.1, pp.87-92, ISSN: 0006-3002

Glud, R. N.; Ramsing, N.B. & Revsbech, N. P. (1992) Photosynthesis and Photosynthesis-Coupled Respiration in Natural Biofilms Quantified with Oxygen Microsensors. *Journal of Phycology*, vol.28, No.1, pp.51-60, ISSN 0022-3646

Glud, R. N.; Kuhl, M.; Kohls, O. & Ramsing, N. B, (1999) Heterogeneity of oxygen production and consumption in a photosynthetic microbial mat as studied by planar optodes. *Journal of Phycology*, vol.35, No.2, pp.270-279, ISSN 0022-3646

Glud RN, Tengberg A, Kuhl M, Hall POJ, Klimant I, Host G (2001) An *in situ* instrument for planar O_2 optode measurements at benthic interfaces. *Limnology and Oceanography*, vol.46, No.8, pp.2073-2080, ISSN 0024-3590

Gower, J. & King, S. (2008) Satellite Images Show the Movement of Floating Sargassum in the Gulf of Mexico and Atlantic Ocean. In: *Nature Precedings*, hdl:10101/npre.2008.1894.1 : Posted 15 May 2008, 20.07.2011, Available from: http://precedings.nature.com/

Grinbalt, Y. P. & Dubinsky, Z. (2011) The study of phytoplankton photosynthesis by photoacoustics, In: *Chlorophyll a Fluorescence in Aquatic Sciences: methods and Applications*, D. J. Suggett, O. Prášil & M. A. Borowitzka (Eds.), pp.311-315, Springer, ISBN 978-90-481-9267-0, Dordrecht, Netherlands

Han, M. S.; Furuya, K.; & Nemoto, T. (1992) Species-Specific Productivity of *Skeletonema costatum* (Bacillariophyceae) in the Inner Part of Tokyo Bay. *Marine Ecology Progress Series*, vol.79, No.3, pp.267-273, ISSN 0171-8630

Hancke, K. & Glud, R. N. (2004) Temperature Effects on Respiration and Photosynthesis in Three Diatom-Dominated Benthic Communities. *Aquatic Microbial Ecology*, vol.37, No.3, pp.265-281, ISSN 0948-3055

Hansell, D. A.; Carlson, C. A.; Repeta, D. J. & Schlitzer, R. (2009) Dissolved Organic Matter in the Ocean: A Controversy Stimulates New Insights. *Oceanography*, vol.22, No.4, pp.190-201, ISSN 1042-8275

Ho T-Y.; Quigg, A.; Finkel, Z. V.; Milligan, A. J.; Wyman, K.; Falkowski, P. G.; Morel, F. M. M. (2003) The Elemental Composition of Some Marine Phytoplankton. *Journal of Phycology*, vol.39, No.6, pp.1145-1159, ISSN 0022-3646

Howarth R. B. (1988) Nutrient limitation of net primary production in marine ecosystems. *Annual Review of Ecology and Systematics*, Vol.19, pp.89-110, ISSN 0066-4162

Jeffey, S. W.; Mantoura, R. F. C. & Wright, S. W. (Eds.) (1997) *Phytoplankton Pigments in Oceanography: Guidelines to Modern Methods*, UNESCO, ISBN-10 9231032752, Paris, France

Johnston, A. M. & Raven. J. A. (1996) Inorganic Carbon Accumulation by the Marine Diatom *Phaeodactylum tricornutum*. *European Journal of Phycology*, vol.31, No.3, pp.285-290, ISSN 0967-0262

Karl, D. M. (2002) Hidden in a Sea of Microbes. *Nature*, vol.415, No.6872, pp.590-591, ISSN 0028-0836

Kirk, J. T. O. (1983) *Light and Photosynthesis in Aquatic Ecosystems*. Cambridge University Press, ISBN 0-521-33654-6, Cambridge, United Kingdom

Klausmeier, C. A.; Litchman, E. & Levin, S. A. (2004) Phytoplankton Growth and Stoichiometry under Multiple Nutrient Limitation. *Limnology and Oceanography*, vol.49, No.4, pp.1463-1470, ISSN 0024-3590

Knoll, A. H. (2003) *Life on a Young Planet*. Princeton University Press, ISBN 0-691-00978-3, Princeton, New Jersey, United States

Kolber, Z.S.; Barber, R. T.; Coale, K. H.; Fitzwater, S. E.; Greene, R. M.; Johnson, K. S.; Lindley, S. & Falkowski, P. G. (1994) Iron Limitation of Phytoplankton Photosynthesis in the Equatorial Pacific Ocean. *Nature*, vol.371, No.6493, pp.145-149, ISSN 0028-0836

Kolber ZS, Prášil O, Falkowski PG (1998) Measurements of Variable Chlorophyll Fluorescence Using Fast Repetition Rate Techniques: Defining Methodology and Experimental Protocols. *Biochimica Biophysica Acta*, vol.1367, No.1-3, pp.88-106, ISSN 0005-2728

Kolber, Z. S.; Van Dover, C. L.; Niederman, R. A. & Falkowski, P. G. (2000). Bacterial Photosynthesis in Surface Waters of the Open Ocean. *Nature*, vol.407, No.6801, pp.177-179, ISSN 0028-0836

Kromkamp, J. C. & Forster, R. M. (2003) The Use of Variable Fluorescence Measurements in Aquatic Ecosystems: Differences Between Multiple and Single Measurement Protocols and Suggested Terminology. *European Journal of Phycology*, vol.38, No.2, pp.103-112, ISSN 0967-0262

Kühl M, Glud RN, Ploug H, Ramsing NB (1996) Microenvironmental Control of Photosynthesis and Photosynthesis-Coupled Respiration in an Epilithic Cyanobacterial Biofilm. *Journal of Phycology*, vol.32, No.5, pp.799-812, ISSN 0967-0262

Lapointe, B. E. (1995) A Comparison of Nutrient-Limited Productivity In Sargassum natans from Neritic vs. Oceanic Waters of the Western North-Atlantic Ocean. *Limnology and Oceanography* vol.40, No. 3, pp.625-633, ISSN 0024-3590

Lassen, C.; Glud, R. N.; Ramsing, N. B. & Revsbech, N. P. (1998) A method to improve the spatial resolution of photosynthetic rates obtained by oxygen microsensors. *Journal of Phycology*, Vol.34, No.1, pp.89-93, ISSN 0022-3646

Lee R. E. (2008) *Phycology* (fourth edition) Cambridge University Press, ISBN 978-0-521-68277-0, Cambridge, United Kingdom

Leonardos, N. & Geider, R. J. (2004) Responses of Elemental and Biochemical Composition of *Chaetoceros muelleri* to Growth Under Varying Light and Nitrate: Phosphate Supply Ratios and their Influence on Critical N : P *Limnology and Oceanography*: vol.49, No.6, pp. 2105-2114, ISSN 0024-3590

Liu, H. B.; Nolla, H. A. & Campbell, L. (1997) *Prochlorococcus* Growth Rate and Contribution to Primary Production in the Equatorial and Subtropical North Pacific Ocean. *Aquatic Microbial Ecology*, vol.12, No.1, pp.39-47, ISSN 0948-3055

Longhurst, A.; Sathyendranath, S.; Platt, T.; & Caverhill, C. (1995) An Estimate of Global Primary Production in the Ocean from Satellite Radiometer Data. *Journal of Plankton Research*, vol.17, No.6, pp.1245-1271, ISSN 0142-7873

Lorenzen, C. J. (1966) A Method for the Continuous Measurement of *in vivo* Chlorophyll Concentration. *Deep-Sea Research*, Vol.13, pp.223-227

Luz B, Barkan E (2000) Assessment of Oceanic Productivity with the Triple-Isotope Composition of Dissolved Oxygen. *Science*, vol.288, No.5473, pp.2028-2031, ISSN 0036-8075

Luz B, Barkan E (2009) Net and Gross Oxygen Production from O_2/Ar, $^{17}O/^{16}O$ and $^{18}O/^{16}O$ Ratios. *Aquatic Microbial Ecology*, vol.56, No.2-3, pp.133-145, ISSN 0948-3055

Magaña, H. A.; Contreras, C; & Villareal, T. A. (2003) A historical assessment of *Karenia brevis* in the western Gulf of Mexico. *Harmful Algae*, vol.2, No.3, pp.163-171, ISSN 1568-9883

Mackey, M. D.; Mackey, D. J.; Higgins, H. W. & Wright, S. W. (1996) CHEMTAX – A Program for Estimating Class Abundances from Chemical Markers: Applications to HPLC Measurements of Phytoplankton. *Marine Ecology Progress Series*, vol.144, No.1-3, pp.265-283, ISSN 0171-8630

Mann, D. G. (1999) The species concept in diatoms. *Phycologia*, vol.38, No.6, pp.437-495, ISSN 0031-8884

Martin, J. H. (1992) Iron as a Limiting Factor in Oceanic Productivity, In: *Primary Productivity and Biogeochemical Cycles in the Sea*, P. G. Falkowski & A. D. Woodhead (eds.), pp.123-137, Plenum Press, ISBN 0306441926, New York, United States

Martin, J. H.; Coale, K. H.; Johnson, K. S.; Fitzwater, S. E.; Gordon, R. M.; Tanner, S. J.; Hunter, C. N.; Elrod, V. A.; Nowicki, J. L.; Coley, C. S.; Barber, R. T.; Lindley, S.; Watson, A. J.; Van Scoy K.; Law, C. S.; Liddicoat, M. I.; Ling, R.; Stanton, T.; Stockel, J.; Collins C.; Anderson, A.; Bidigare, R.; Ondrusek, M.; Latasa, M.; Millero, F. J.; Lee, K.; Yao, W.; Zhang, J. Z.; Friederich, G.; Sakamoto, C.; Chavez, F.; Buck, K.; Kolber, Z.; Greene, R.; Falkowski, P.; Chisholm, S. W.; Hoge, F.; Swift, R.; Yungel, J.; Turner, S.; Nightingdale, P.; Hatton, A.; Liss, P. & Tindale, N. W. (1994) Testing the Iron Hypothesis in Ecosystems of the Equatorial Pacific Ocean. *Nature*, Vol.371, No.6493, pp.123-129, ISSN 0028-0836

Maxwell, K. & Johnson, G. N. (2000) Chlorophyll Fluorescence – A Practical Guide. *Journal of Experimental Botany*, vol.51, No.345, pp. 659-668, ISSN 0022-0957

McClain, C. R. (2009) A Decade of Satellite Ocean Color Observations. *Annual Review of Marine Science*, vol.1, pp.19-42, ISSN 1941-1405

Middelburg, J. J.; Barranguet, C.; Boschker, H. T. S.; Herman, P. M. J.; Moens, T. & Heip, C. H. R. (2000) The Fate of Intertidal Microphytobenthos Carbon: An *in situ* ^{13}C Labeling Study. *Limnology and Oceanography*, vol.45, No.6, pp1224-1234, ISSN 0024-3590

Miller, C. B. (2004) *Biological Oceanography*, Blackwell Science Ltd, ISBN 0-632-05536-7, Malden, Massachusetts, United States

Miller, H. L. & Dunton, K. H. (2007) Stable Isotope ^{13}C and O_2 Micro-Optode Alternatives for Measuring Photosynthesis in Seaweeds. *Marine Ecology Progress Series*, vol.329, pp.85-97, ISSN 0171-8630

Millero, F. J. (2006) *Chemical Oceanography*, CRC Press, ISBN 0-8493-2280-4, Boca Raton, Florida, United States

Morán, X. A. G.; Fernández, E. & Peréz, V. (2004) Size-Fractionated Primary Production, Bacterial Production and Net Community Production in Subtropical and Tropical Domains of the Oligotrophic NE Atlantic in Autumn. *Marine Ecology Progress Series*, vol.274, pp.17-29, ISSN 0171-8630

Nagasaki, K.; Tomaru, Y.; Katanozaka, N.; Shirai, Y.; Nishida, K.; Itakura, S.; & Yamaguchi, M. (2004) Isolation and Characterization of a Novel Single-Stranded RNA Virus Infecting the Bloom-Forming Diatom *Rhizosolenia setigera*. *Applied and Environmental Microbiology*, vol.70, No.2, pp.704-711, ISSN 0099-2240

Parsons, T. R.; Maita, Y. & Lalli, C. M. (1984) *A Manual of Chemical and Biological Methods for Seawater Analysis*, Pergamon Press Ltd., ISBN 0-08-030287-4, Oxford, United kingdom

Perkins, R. G.; Oxborough, K.; Hanlon, A. R. M.; Underwood, G. J. C. & Baker, N. R. (2002). Can Chlorophyll Fluorescence be used to Estimate the Rate of Photosynthetic Electron Transport within Microphytobenthic Biofilms? *Marine Ecology Progress Series* vol.228, pp.47-56, ISSN 0171-8630

Perkins, R. G.; Kromkamp, J. C.; Serôdio, J.; Lavaud, J.; Jesus, B.; Mouget, J. L.; Lefebvre, S. & Forster, R. M. (2011) The Application of Variable Chlorophyll Fluorescence to Microphytobenthic Biofilms, In: *Chlorophyll a Fluorescence in Aquatic Sciences: Methods and Applications*, D. J. Suggett, O. Prášil & M. A. Borowitzka (Eds.), pp. 237-275, Springer, ISBN 978-90-481-9267-0, Dordrecht, Netherlands

Platt, T. (1972) Local Phytoplankton Abundance and Turbulence. *Deep-Sea Research*, vol.19, No.3, pp.183-187

Quigg, A.; Finkel, Z. V.; Irwin, A. J.; Rosenthal, Y.; Ho T.-Y.; Reinfelder, J. R.; Schofield, O.; Morel, F. M. M. & Falkowski, P. G. (2003) The Evolutionary Inheritance of Elemental Stoichiometry in Marine Phytoplankton. *Nature*, vol.425, No.6955, pp.291-294, ISSN 0028-0836

Redfield, A. C. (1958) The Biological Control of Chemical Factors in the Environment. *American Scientist*, Vol.46, pp.205-221

Revsbech, N. P. & Jørgensen, B. B. (1983) Photosynthesis of Benthic Microflora Measured with High Spatial Resolution by the Oxygen Microprofile Method: Capabilities and Limitations of the Method. *Limnology and Oceanography*, vol.28, No.4, pp.749-756, ISSN 0024-3590

Revsbech, N. P.; Nielsen, J. & Hansen, P. K. (1988) Benthic Primary Production and Oxygen Profiles, In: *Nitrogen cycling in Coastal marine Environments*. T. H. Blackburn & J. Sørensen (Eds.), pp.69 – 83, John Wiley & Sons, ISBN 0-471-91404-5, Chichester, United Kingdom

Ryther, J. H. & Dunstan, W. M. (1971) Nitrogen, phosphorus, and eutrophication in the coastal marine environment. *Science* vol.171, No.3975, pp.1008-1013, ISSN 0036-8075

Sarthou, G.; Timmermans, K. R.; Blain, S. & Treguer, P. (2005) Growth, Physiology and Fate of Diatoms in the Ocean: A Review. *Journal of Sea Research*, vol.53, no.1-2, pp.25-42, ISSN 1385-1101

Schreiber, U.; Schliwa, U. & Bilger, W. (1986) Continuous Recording of Photochemical and Nonphotochemical Chlorophyll Fluorescence Quenching with a New Type of Modulation Fluorometer. *Photosynthesis Research*, vol.10, No.1-2, pp.51-62, ISSN 0166-8595

Serôdio, J. (2004) Analysis of Variable Fluorescence in Microphytobenthos Assemblages: Implication of the Use of Depth Integrated Measurements. *Aquatic Microbial Ecology*, vol.36, No.2, pp.137 -152, ISSN 0948-3055

Sharp, J. H. (1977) Excretion of Organic Matter by Phytoplankton: Do Healthy Cells Do It? *Limnology and Oceanography*, vol.22, No.3, pp.381-399, ISSN 0024-3590

Steeman-Nielsen, E. (1952) The Use of Radioactive Carbon (^{14}C) for Measuring Organic Production in the Sea. *Journal of Du Conseil International Pour l'Exploration de la Mer*, vol.18, pp.117-140.

Sterner, R. W. & Elser, J. J. (2002) *Ecological Stoichiometry: The Biology of Elements from Molecules to the Biosphere*, Princeton University Press, ISBN 0-691-07491-7, Princeton, New Jersey, United States

Suggett, D.; Kraay, G.; Holligan, P.; Davey, M.; Aiken, J. & Geider, R. (2001) Assessment of Photosynthesis in a Spring Cyanobacterial Bloom by use of a Fast Repetition Rate Fluorometer. *Limnology and Oceanography*, vol.46, No.4, pp.802-810, ISSN 0024-3590

Suggett, D. J.; Prášil, O. & Borowitzka, M. (2011) Preface, In: *Chlorophyll a Fluorescence in Aquatic Sciences: Methods and Applications*, D. J. Suggett, O. Prášil & M. A. Borowitzka (Eds.), pp. v-xi, Springer, ISBN 978-90-481-9267-0, Dordrecht, Netherlands

Sugget, D. J.; Oxborough, K.; Baker, N. R.; Macintyre, H. L.; Kana, T. D. & Geider, R. J. (2003) Fast Repetition Rate and Pulse Amplitude Modulation Chlorophyll *a* Fluorescence Measurements for Assessment of Photosynthetic Electron Transport in Marine Phytoplankton. *European Journal of Phycology*, vol. 38, no.4, pp.371-384, ISSN 0967-0262

Suggett, D. J.; MacIntyre, H. L.; Kana, T. M. & Geider, R. J. (2009) Comparing Electron Transport with Gas Exchange: Parameterising Exchange Rates Between Alternative Photosynthetic Currencies for Eukaryotic Phytoplankton. *Aquatic Microbial Ecology*, vol. 56, no.2-3, pp.147-162, ISSN 0948-3055

Sundareshwar, P. V.; Morris, J. T.; Koepfler, E. K. & Fornwalt, B. (2003) Phosphorus Limitation of Coastal Processes. *Science*, vol.299, no.5606, pp.563-565, ISSN 0036-8075

Suttle, C. A.; Chan, A. M. & Cottrell, M. T. (1990) Infection of Phytoplankton by Viruses and Reduction of Primary Productivity. *Nature*, vol.347, no.6292, pp. 467-469, ISSN 0028-0836

Sverdrup, H. U. (1953) On Conditions for the Vernal Blooming of Phytoplankton. *Journal du Conseil Permanent International pour l'Exploration de la Mer*, vol.18, pp.287-295

Thamdrup, B. & Dalsgaard, T. (2008) Nitrogen Cycling in Sediments, In: *Microbial Ecology of the Oceans* (second edition). D. L. Kirchman (Ed.), pp. 527-568, Wiley-Blackwell, ISBN 978-0-470-04344-8, Hobokon, New Jersey, United States

Thornton, D. C. O. (2002). Diatom Aggregation in the Sea: Mechanisms and Ecological Implications. *European Journal of Phycology*, vol.37, no.2, pp.149-161, ISSN 0967-0262

Thornton, D. C. O. (2009) Effect of Low pH on Carbohydrate Production by a Marine Planktonic Diatom (*Chaetoceros muelleri*). *Research Letters in Ecology*, Article ID 105901, doi:10.1155/2009/105901

Thornton, D. C. O.; Santillo, D. & Thake, B. (1999) Prediction of Sporadic Mucilaginous Algal Blooms in the Northern Adriatic. *Marine Pollution Bulletin*, vol. 38, no.10, pp. 891-898, ISSN 0025-326X

Thornton, D. C. O.; Dong, L. F.; Underwood, G. J. C. & Nedwell, D. B. (2002) Factors Affecting Microphytobenthic Biomass, Species Composition and Production in the Colne Estuary (UK). *Aquatic Microbial Ecology*, vol.27, no.3, pp.285-300, ISSN 0948-3055

Thingstad T. F., Rassoulzadegan F. (1995) Nutrient Limitations, Microbial Food Webs, and Biological C-Pumps – Suggested Interactions in a P-limited Mediterranean. *Marine Ecology Progress Series, vol.*117, no.1-3, pp. 299-306, ISSN 0171-8630

Trujillo, A. P. & Thurman, H. V. (2005) *Essentials of Oceanography* (eighth edition), Pearson Prentice Hall, ISBN 0-13-144773, Upper Saddle River, New Jersey, United States

Tsuda, A.; Takeda, S.; Saito, H.; Nishioka, J.; Nojiri, Y.; Kudo, I.; Kiyosawa, H.; Shiomoto, A.; Imai, K.; Ono, T.; Shimamoto, A.; Tsumune, D.; Yoshimura, T.; Aono, T.; Hinuma, A.; Kinugasa, M.; Suzuki, K.; Sohrin, Y.; Noiri, Y.; Tani, H.; Deguchi, Y.; Tsurushima, N.; Ogawa, H.; Fukami, K.; Kuma, K. & Saino, T. (2003) A Mesoscale Iron Enrichment in the Western Subarctic Pacific Induces a Large Centric Diatom Bloom. *Science,* vol.300, no.5621, 958-961, ISSN 0036-8075

Tyrrell, T. (1999) The Relative Influences of Nitrogen and Phosphorus on Oceanic Primary Production. *Nature,* vol.400, no.6744, pp.525-531, ISSN 0028-0836

Underwood, G. J. C. (2002) Adaptations of Tropical Marine microphytobenthic Assemblages Along a gradient of Light and Nutrient Availability in Suva Lagoon, Fiji. *European Journal of Phycology*, vol.37, no.3, pp.449-462, ISSN 0967-0262

Underwood, G. J. C. & Kromkamp, J. (1999) Primary Production by Phytoplankton and Microphytobenthos in Estuaries. *Advances in Ecological Research,* vol.29, pp.94-153, ISSN 0065-2504

Westberry, T.; Behrenfeld, M. J.; Siegel, D. A. & Boss, E. (2008) Carbon-Based Primary Productivity Modeling with Vertically Resolved Photoacclimation. *Global Biogeochemical Cycles* 22, no.2, DOI: Gb2024 10.1029/2007gb003078, ISSN 0886-6236

White, A. J. & Critchley, C. (1999) Rapid Light Curves: A New Fluorescence Method to Assess the State of the Photosynthetic Apparatus. *Photosynthesis Research*, vol.59, no.1, pp.63-72, ISSN 0166-8595

Williams, P. J. leB. & Robertson, J. I. (1989) A Serious Inhibition Problem from a Niskin Sampler During Plankton Productivity Studies. *Limnology and Oceanography*, vol.34, no.7, pp.1300-1305, ISSN 0024-3590

Wiltshire, K. H. (2009) Pigment Applications in Aquatic Systems, In: *Practical Guidelines for the Analysis of Seawater*, O. Wurl (Ed.), pp. 191-221, CRC Press, ISBN 978-1-4200-7306-5, Boca Raton, Florida, United States

Yoshikawa, T. & Furuya, K. (2004) Long-Term Monitoring of Primary Production in Coastal Waters by an Improved Natural Fluorescence Method. *Marine Ecology Progress Series*, vol.273, pp.17-30, ISSN 0171-8630

Zehr, J. P. & Paerl, H. W. (2008) Molecular Ecological aspects of Nitrogen Fixation in the marine Environment, In: *Microbial Ecology of the Oceans* (second edition), D. L. Kirchman (Ed.), pp. 481-525, Wiley-Blackwell, ISBN 978-0-470-04344-8, Hobokon, New Jersey, United States

15

The Plant–Type Ferredoxin-NADP⁺ Reductases

Matías A. Musumeci, Eduardo A. Ceccarelli
and Daniela L. Catalano-Dupuy
*Instituto de Biología Molecular y Celular de Rosario (IBR), CONICET, Facultad de
Ciencias Bioquímicas y Farmacéuticas, Universidad Nacional de Rosario
Argentina*

1. Introduction

Ferredoxin-NADP⁺ reductases (FNRs, EC 1.18.1.2) constitute a family of hydrophilic, monomeric enzymes that contain non-covalently bound FAD as prosthetic group. These flavoenzymes deliver NADPH or low potential one-electron donors (ferredoxin, flavodoxin, adrenodoxin) to redox-based metabolisms in plastids, mitochondria and bacteria. The main physiological role of the chloroplast FNR is to catalyze the final step of photosynthetic electron transport, namely, the electron transfer from the ferredoxin (Fd), reduced by photosystem I, to NADP⁺ (Eqn. 1) (Shin & Arnon, 1965). This reaction provides the NADPH necessary for CO_2 assimilation in plants and cyanobacteria.

FNRs also participate in others electron transfer metabolic processes as nitrogen fixation, isoprenoid biosynthesis, steroid metabolism, xenobiotic detoxification, oxidative-stress response and iron-sulfur cluster biogenesis (Carrillo & Ceccarelli, 2003, Ceccarelli et al., 2004, Medina & Gomez-Moreno, 2004, Rohrich et al., 2005, Seeber et al., 2005). Eqn. 1 represents the electron flow through FNR as it occurs in the photosynthetic electron chain. However, the physiological direction of the reaction catalyzed by FNRs involved in the other pathways is opposite, i.e. toward the production of reduced Fd. On this basis, FNRs are sometimes classified as autotrophic (photosynthetic FNRs) and heterotrophic (all other FNRs) (Aliverti et al., 2008, Arakaki et al., 1997).

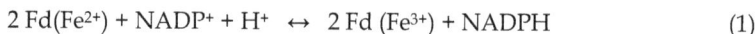

$$2 \, Fd(Fe^{2+}) + NADP^+ + H^+ \leftrightarrow 2 \, Fd \, (Fe^{3+}) + NADPH \qquad (1)$$

Some bacteria and algae posses the FMN-containing flavodoxin (Fld), that is able to efficiently replace Fd as the electron partner of FNR in different metabolic routes, including photosynthesis (Razquin et al., 1996). In cyanobacteria, Fld expression is induced under condition of iron deficit, when the [2Fe-2S] cluster of Fd cannot be assembled (Razquin et al., 1996). In other prokaryotes, flavodoxins are constitutively expressed, or induced by oxidants (Zheng et al., 1999). Both Fld and FNR participate in the detoxification of reactive oxygen species in aerobic and facultative bacteria (Krapp et al., 2002, Zheng et al., 1999).

The FNR displays strong preference for NADP(H) and is a very poor NAD(H) oxidoreductase. In contrast, various redox compounds, including complexed metals and aromatic molecules, can replace Fd or Fld as electron acceptors *in vitro*, in the so-called diaphorase activity (Avron & Jagendorf, 1956). All FNR-mediated reactions can thus be interpreted consisting of two-steps: hydride exchange with pyridine nucleotide (Eqn. 2) and electron transfer to and from the other partner (Eqn. 3).

$$NADPH + H^+ + FNR_{ox} \leftrightarrow NADP^+ + FNRH_2 \tag{2}$$

$$FNRH_2 + nA_{ox} \rightarrow FNR_{ox} + nA_{red} \tag{3}$$

A, electron acceptor $n = 1, 2$

The capability of FNRs to exchange electrons between mono and bi-electronic substrates is due to the prosthetic group FAD, which can exist in three redox states: oxidised, as radical reduced by one electron (semiquinone) and completely reduced (hydroquinone) by two electrons (Dudley et al., 1964). The active chemically moiety of the FAD is the isoalloxazine (Figure 1). Besides, the isoalloxazine ring can be protonated, providing a wide opportunity for tautomers (Heelis P.F., 1982).

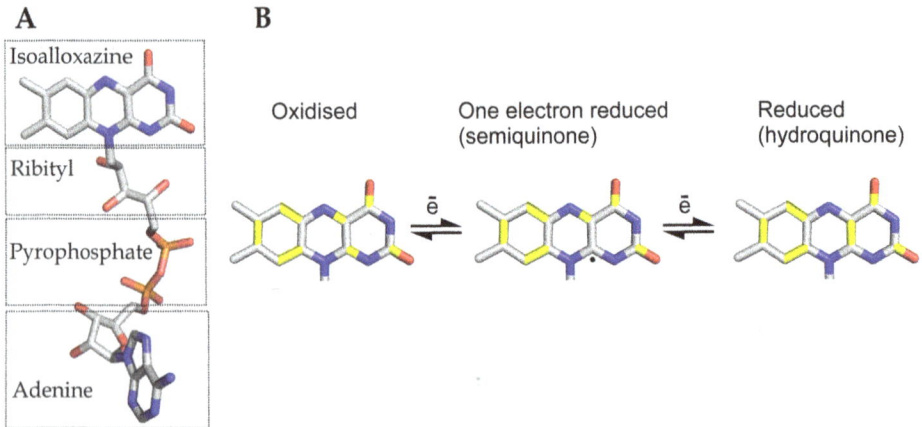

A   **B**

Fig. 1. Main structural traits of FAD. A) View of FAD as found in the pea FNR. B) Redox states adopted by the FAD. In blue, red and white are showed nitrogen, oxygen and carbon atoms respectively; the yellow colour depicts double bonds.

The protein environment modulates the redox potential of the isoalloxazine in such manner that in photosynthetic tissues it promotes the production of NADPH from reduced ferredoxin (Aliverti et al., 2008).

Two great families of FAD containing proteins displaying FNR activity have evolved from different and independent origins. The enzymes from mitochondria and some bacterial genera are members of the structural superfamily of disulfide oxidoreductases whose prototype is glutathione reductase. A second group, comprising the FNRs from plastids and most eubacteria, constitutes a unique family, the plant-type FNRs, totally unrelated in sequence with the former (Aliverti et al., 2008, Ceccarelli et al., 2004). In spite of their different origins, flavoproteins of the two FNR families display similar modes of NADP(H) docking and catalysis and can exchange electron partners (ferredoxin, flavodoxin, adrenodoxin) *in vitro* (Faro et al., 2003, Jenkins et al., 1997, Ziegler & Schulz, 2000). Members of the plant-type group can be readily identified by the presence of clusters of highly conserved residues, three of them belonging to the FAD (FMN) domain, and the remaining to the NADP(H) region (Arakaki et al., 1997, Bruns & Karplus, 1995, Ceccarelli et al., 2004, Karplus et al., 1991). This group can be further classified into a plastidic and a bacterial class, which differ not only in their sequences, but also in the environment of the active site, FAD conformations and catalytic efficiencies (Carrillo & Ceccarelli, 2003). Plastidic FNRs display high catalytic efficiencies (turnover numbers in the range 100-600 s^{-1}), whereas bacterial reductases are much less active (Ceccarelli et al., 2004). In plants and cyanobacteria, optimization for FNR catalytic efficiency might be related to the demands of the photosynthetic process that requires a very fast electron flow to sustain CO_2 fixation rates. In organisms growing on heterotrophic metabolisms or anoxygenic photosynthesis, FNR is involved in pathways that proceed at a much lower pace, acting as a shuttle between the abundant NAD(P)H pool and the low potential electron carriers.

This chapter will focus on structural and functional aspects of the ferredoxin-NADP(H) reductase in association with the metabolic process of photosynthesis. Special attention will be pay to techniques and approaches that could help to appreciate the importance of the enzyme and to understand, conceive and/or execute research on FNR.

2. FNR in the photosynthetic electron transport

The generation of reducing power is crucial for all biosynthetic processes within chloroplasts. The main source of reduction equivalents is the light-driven photosynthesis. The FNR is a key enzyme of photosynthetic electron transport. FNR transfers electrons between the one-electron carrier ferredoxin and the two-electron carrier NADP(H) at the end of the photosynthetic electron transport chain. FNR also participates in others relevant processes as the electron cyclic flow around the photosystem I and in the control of the NADPH/NADP⁺ homeostasis of stressed chloroplasts (Palatnik et al., 1997).

Photosynthetic electron flow is driven by two photochemical reactions catalyzed by photosystem II (PSII) and photosystem I (PSI), which are linked by the electron transport chain (Figure 2). Linear electron transport starts with the photo-induced water oxidation catalyzed by PSII. Electrons are transferred from PSII through the plastoquinone pool to cytochrome b_6f. The electron transport is coupled to proton translocation into the thylakoid lumen, and the resulting pH gradient drives the ATP synthase to produce ATP. Next, electrons move from cytochrome b_6f to the soluble electron carrier plastocyanin and then to PSI, which acts as light-driven plastocyanin ferredoxin oxidoreductase. Ultimately these electrons can be used by ferredoxin-NADP⁺ reductase to produce NADPH that together with the ATP generated by the ATP synthase will drive the Calvin cycle for CO_2 assimilation (Rochaix, 2011).

The primary function of PSI is to reduce NADP⁺ to NADPH, which is then used in the assimilation of CO_2 (Setif, 2006, Vishniac & Ochoa, 1951). In plants, it occurs via reduction of the soluble [2Fe-2S] ferredoxin (Fd) by PSI. Subsequent reduction of NADP⁺ by Fd_{rd} is catalyzed by FNR (Arakaki et al., 1997). In most cyanobacteria, and algae under low iron conditions, Fld, in particular Fld_{sq}/Fld_{hq}, substitutes for the Fd_{ox}/Fd_{rd} pair in this reaction (Eqn. 4) (Bottin & Lagoutte, 1992, Medina & Gomez-Moreno, 2004). Two Fld_{sq} molecules transfer two electrons from two PSI molecules to one FNR. FNR becomes fully reduced through formation of the intermediate, FNR_{sq}, and later transfers both electrons simultaneously to NADP⁺ (Eqn. 5) (Medina, 2009).

It was demonstrated that FNR is one of the rate-limiting steps in photosynthesis, and controls the balance between the demand for redox equivalents and photosynthetic activity under a wide range of environmental conditions (Hajirezaei et al., 2002). It gives to this enzyme relevance as a possible target for crop improvement and treatment of weeds.

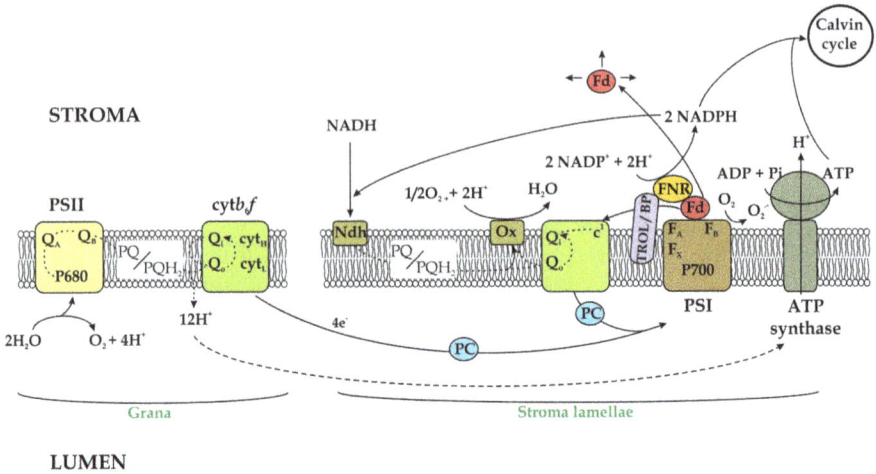

Fig. 2. Photosynthetic electron transport. The thylakoid membrane with the different component of the electron transport chain is shown. Electron transport pathways are shown by dotted lines with arrows to indicate the direction of electron flow. P680 is the PSII primary electron donor; Q_A and Q_B are the primary electron acceptors of PSII. Once Q_B has accepted two electrons, is released into the plastoquinone pool (PQ). PQH_2, reduced plastoquinone pool; PC, plastocyanin; Ndh, NADH/NADPH dehydrogenase complex; Ox, oxidase. P700 is the PSI primary electron donor. F_X, F_A, and F_B are internal [4Fe-4S] centers of the PSI. Adapted from (Rochaix, 2011).

In addition, ferredoxin donates electrons to other pathways such as sulfur and nitrogen assimilation, to thioredoxins, which regulate carbon assimilation and to the cyclic electron transfer process (Rochaix, 2011, Yamamoto et al., 2006). Many studies have suggested the involvement of FNR also in cyclic electron flow around PSI (Figure 2) besides its role in the linear electron transport of photosynthesis. In cyclic electron flow, electrons are transferred from PSI to cytochrome b_6f complex via Fd, with a concomitant formation of proton gradient. Consequently, cyclic electron transfer produces ATP without accumulation of NADPH (Mulo, 2011, Shikanai, 2007).

$$PSI_{rd} + Fld_{sq} \rightarrow PSI + Fld_{hq} \qquad (4)$$

$$NADP^+ + 2\, Fld_{hq} \xrightarrow{\quad FNR \quad} NADPH + 2\, Fld_{sq} \qquad (5)$$

While cyclic electron flow is well documented in cyanobacteria, unicellular algae, and C4 plants, it is only recently that its importance has been recognized and studied in C3 plants (Joliot & Joliot, 2006). In C3 plants, cyclic electron flow is especially important under specific stress conditions such as low CO_2, high light, drought, or during dark to light transitions. Under these conditions, cyclic electron flow allows for acidification of the thylakoid lumen, which is required for inducing ATP synthesis and for triggering non-photochemical quenching, which in turn down-regulates PSII. Cyclic electron flow is also important when photorespiration is active because more ATP is required for CO_2 fixation under these conditions (Osmond, 1981). Several routes for cyclic electron flow have been proposed. In the first, ferredoxin transfers its electrons through FNR to NADP⁺ and the NAD(P)H dehydrogenase (Ndh) complex and ultimately to the plastoquinone pool (Figure 2). However, in plants, the level of this complex is probably too low for mediating sufficient cyclic electron flow required for ATP production under steady state conditions (Burrows et al., 1998). FNR, which is found both in the stroma and associated with thylakoid membranes (through TROL or the FNR binding protein, see below), has been proposed to modulate partitioning between the cyclic and linear electron pathways (Rochaix, 2011). In the second pathway, ferredoxin is thought to transfer electrons to the plastoquinone pool through a ferredoxin–plastoquinone oxidoreductase (Cleland & Bendall, 1992), an enzyme that has however not yet been identified. This function has been recently assigned to the FNR (Szymanska et al., 2011). In the third, ferredoxin may interact directly with cytochrome b_6f and transfer its electrons to cytochrome c', a new component identified in the crystal structure of the cytochrome b_6f complex, using a Q-cycle derived mechanism (reviewed in Joliot & Joliot, 2006, Shikanai, 2007).

It has been proposed that the transhydrogenase activity of the FNR may function *in vivo* as an intra-chloroplastic source of NADH (Chopowick & Israelstam, 1971, Krawetz & Israelstam, 1978). However, the presence of NADH in chloroplasts is most likely to be related to the malate/oxalacetate shuttle (Carrillo & Vallejos, 1987, Krause & Heber, 1976).

3. Purification and characterization of FNR

3.1 Purification procedures
FNRs can be obtained using transgenic expression in *Escherichia coli* cells or from biological samples such as plant leaves, plant roots or cyanobacteria. Nevertheless, the purification procedures further applied to obtain FNR are similar in any case. In this section we will discuss how to obtain the soluble protein extracts from diverse sources, and how to apply different procedures to obtain pure FNR.

3.1.1 Preparation of soluble protein extracts from FNR transgenic expression in *Escherichia coli*
High throughput preparations of FNRs can be obtained from recombinant expression in *E. coli* cells by using vectors of the pET (Novagen, USA) or pQE (QIAGEN, USA) series. For the culture of FNR-expressing *E. coli* cells, Luria-Bertani or 2YT (16 g/l tryptone, 10 g/l yeast extract, 5.0 g/l NaCl) media are recommended. The expression conditions may vary depending on the features of the FNR being studied. Usually, for the wild-type FNR from plants cloned in vectors under the control of the *Lac* promoter 0.2–1 mM IPTG is used for the induction of protein expression during 2 to 5 h at 25-37°C (Aliverti et al., 1990, Catalano-Dupuy et al., 2006, Onda et al., 2000). For the *Anabaena variabilis* FNR no inductor was applied and the culture maintained overnight at 30°C (Tejero et al., 2003). For unstable FNR mutants, longer induction periods (10-14 h) at low temperature (18–20°C) are recommended. In these cases, lower inductor concentration (0.1 mM IPTG) gives better yields (Musumeci et al., 2008, Musumeci et al., 2011). Nevertheless, extremely long induction periods (20 h or

more) are detrimental for the quality of the obtained FNR due to the incorporation of a modified flavin or an improper protein folding (Figure 3A).

After the induction, *E. coli* cells that over-expressed FNR are harvested by centrifugation and resuspended in an appropriate buffer solution (25-50 mM Tris-HCl, pH 7.5–8.0, 100–150 mM NaCl). Addition of 5 mM benzamidine hydrochloride or 1 mM phenylmethylsulfonyl fluoride (PMSF) can be used to inhibit unwanted FNR proteolytic degradation. Cells are disrupted by sonication or French Press at 60 MPa. After 20 min incubation with DNase and RNase, centrifugation at 40,000 X *g* (60 min, 4 °C) is applied in order to remove cell debris and to obtain the soluble protein extract.

3.1.2 Preparation of soluble protein extracts from photosynthetic tissues

The biochemical purification of FNR from biological tissues was neglected with the advent of the recombinant DNA technology, which ensures the production of high amounts of enzyme in lesser times. However, the biochemical purification from photosynthetic tissues can be applied for some specific purposes. Several methodologies have been described in the literature for the purification of the FNR from paprika leaves (Dorowski et al., 2000), spinach leaves (Grzyb et al., 2004), soluble extracts of wheat chloroplasts (Grzyb et al., 2008) and from different plant types roots (Green et al., 1991, Morigasaki et al., 1990, Onda et al., 2000). FNR can be also purified from cyanobacteria cell cultures grown in BG 11 medium as described (Sancho et al., 1988).

3.1.3 Purification of FNR from soluble protein extract

All purification protocol steps should be carried out at 4°C to ensure good quality preparations of FNR. Soluble protein extract can be subjected to different purification procedures including precipitation with ammonium sulphate, affinity-, anion exchange- and hydrophobic-chromatography.

A typical purification protocol for FNR is as follows: The soluble extract is load onto a dye (reactive red or cibacron blue) affinity chromatography matrix. The FNR remains bound to the resin by interaction with the adenosine-like structure of the dye. After washing off the unabsorbed material, bound FNR is eluted with a linear gradient of NaCl or NADP+ (Aliverti et al., 1990, Carrillo & Vallejos, 1983, Dorowski et al., 2000). The FNR containing fractions are dialyzed and applied to an ionic exchange chromatography, such as DEAE Sepharose column. After extensive washing, a gradient of NaCl in the same buffer is applied. The FNR containing fractions are pooled and dialyzed. If a further purification is needed, the sample can be applied to a phenyl-Sepharose column. The FNR is eluted with a linear decreasing gradient of $(NH_4)_2SO_4$ (Dorowski et al., 2000). Affinity chromatography also can be applied by using immobilized ferredoxin. This methodology has been used for purification purposes (Sakihama et al., 1992) and also for FNR analysis (Onda & Hase, 2004). Purification procedures using specific tags as glutathion *S*-transferase (Serra et al., 1993) or polihystidines (Catalano-Dupuy et al., 2006) have been reported. In these cases, a recognition site for specific proteases (as factor Xa, thrombin, TEV protease) is inserted between the FNR sequence and the tag to allow removal of the carrier protein if necessary. Amino terminal extensions are preferred to the carboxy-terminal due to the involvement of this latter region in substrate binding and catalysis.

For Ni-NTA affinity chromatography the protein extract containing the His-tagged FNR is loaded onto a column containing the resin. After intensive washing with Tris-HCl buffer the fusion protein is eluted with 100-200 mM imidazole in the same buffer. Then, the histidine tag can be removed by digestion with the appropriate protease and an additional Ni-NTA chromatography (Catalano-Dupuy et al., 2006). Further purification steps have been applied to improve enzyme quality after the Ni-NTA affinity chromatography (Tejero et al., 2003).

3.1.4 Long term storage conditions
FNR frozen at -20°C retain the activity for several months (Zanetti et al., 1984). FNR can be stored for longer periods (years) at -70°C in 50 mM Tris-HCl, pH 8.0, 150 mM NaCl. Addition of 1 mM 2-mercaptoethanol or DTT avoid covalent aggregation of some FNRs (Nakajima et al., 2002).

3.2 Properties of the purified FNR
3.2.1 FNR spectroscopic properties
The transition energies of the isoalloxazine portion of FAD are susceptible to variations by the near environment. These perturbations are detectable by different spectroscopic techniques, which provide value information about the structure and the functional performance of the prosthetic group and the flavoenzyme.

3.2.2 Analysis of FNR by UV-visible spectroscopy
The isoalloxazine, the structural component of the flavin involved in electron transfer, absorbs light in the UV and visible spectral range giving rise to the yellow appearance of flavin and flavoproteins (Latin: *flavus* = yellow) (Macheroux, 1999). In solution, the FAD shows a spectrum with maxima at 380 and 450 nm. The typical plant FNR spectrum displays peaks at 385 and 456 nm (Figure 3A, green line). The interaction of the protein scaffold with the isoalloxazine induces a decrease of the maximum energy levels of the isoalloxazine orbitals that results in spectral shifts to higher wavelengths. The presence of typical peaks or shoulders in the spectrum may indicate chemical variations in the flavin or in its protein environment (Figure 3A, red and blue lines), or the persistence of a bound nucleotide to the enzyme (Piubelli et al., 2000).

The UV-visible absorption spectroscopy can be used with different purposes. Thus, a simple UV-visible spectrum provides information about the quality of the FNR enzyme. Aggregates or contaminants that cause light dispersion may result in an absorbance increase at wavelengths below 600 nm (Figure 3A, grey line). Knowledge of the enzyme extinction coefficient can be helpful to assess the stoichiometry of the flavin and protein moiety. In general, one flavin per polypeptide is found. Substoichiometric amounts of flavin may indicate depletion of the flavin during purification and handling.

The UV-visible spectroscopy can be used to study the binding of substrates, substrate analogues and inhibitors (Aliverti et al., 1995, Medina et al., 2001, Paladini et al., 2009, Piubelli et al., 2000). Titration with the compound of interest yields the dissociation constant and information about changes that occur in the active site environment (Figure 3B) (Paladini et al., 2009). Besides, the spectrum shape can provide useful information about the nature of the interaction with substrates. Experimental data indicate that when NADP⁺ is present at saturating concentrations, the degree of nicotinamide ring occupancy of the binding site is 14-15% for pea FNR, as revealed by the extinction coefficient value of the peak near 510 nm in differential spectra elicited by pyridine nucleotide binding to the various pea FNR forms (Piubelli et al., 2000). This value has been used to analyse nicotinamide occupancy when NADP+ is bound to different mutant FNR enzymes (Figure 3C).

On the other hand, the FNR spectral UV-visible properties also depend on the redox state of the flavin. These spectral data are very useful to analyse the flavoprotein function (Macheroux, 1999, Nogues et al., 2004).

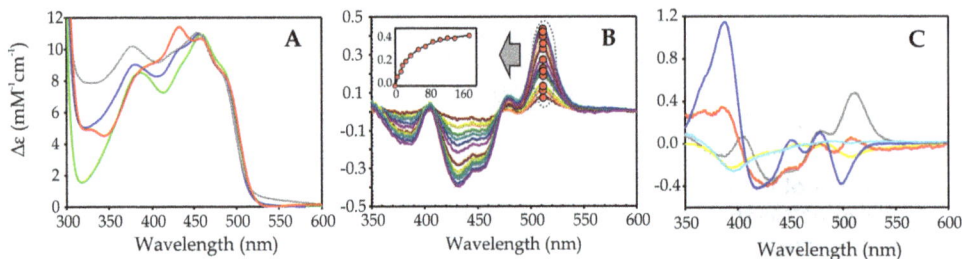

Fig. 3. Analysis of FNR by UV-visible spectroscopy. A) Absorption spectra of pea FNR obtained after 6 (green, good quality), 19 (blue) and 25 (red) hours of induction in *E. coli*. FNR preparation containing insoluble aggregates (grey line). B) Differential spectra obtained upon titration of wild-type FNR with NADP+. Inset: absorbance change at 515 nm plotted as function of NADP+ concentration. C) Differential spectra elicited by NADP+ binding to different FNR mutants (grey, yellow, cyan, red and blue corresponds to the spectra of wild-type, C266A, C266AL266A and C266M FNR forms respectively (Musumeci et al., 2008)).

3.2.3 Application of fluorescence spectroscopy

Plant FNRs display typical fluorescence spectra with emissions at 525-535 nm, which are obtained by exciting the sample at 450-460 nm (Figure 4A). The fluorescence of bound FAD is known to be largely quenched in the native oxidised form of wild-type FNR being 0.5-0.75% of that of free FAD in solution (Forti, 1966; Shin, 1973). For fluorescence measurements, the FNR sample should be of the highest purity and diluted into a buffer solution of desired pH (e.g., Tris-HCl or sodium phosphate). Particulate matter should be removed by brief centrifugation at high speed. Passing the sample/buffer solution through a gel filtration column may also remove particles and improve the quality of collected data (Munro A.W. & Noble, 1999). Besides, this procedure is needed to remove the excess of free FAD that can interfere in the measurements.

It is possible to examine the interaction between FNR and substrates such as simple molecules as NADP+ or proteins as ferredoxin or flavodoxin by analysis of the FAD fluorescence. NADP+ binding induces an increase of FAD fluorescence which can be employed to study the interaction of FNR with its substrate (Figure 4A) (Musumeci et al., 2008). Similarly, quenching of FNR tryptophan fluorescence by ferredoxin or flavodoxin is used to investigate the Fd:FNR complex formation (Davis, 1990). The main advantage of this procedure is that samples with low concentration can be used.

The degree of flavin accessibility to the solvent molecules can be estimated by titration of the FNR sample with a dynamic quencher, as KI (Figure 4B). The steeper graph of FAD fluorescence quenching versus quencher concentration indicates higher accessibility of the prosthetic group in the FNR enzyme (Paladini et al., 2009).

Fluorescence also can be applied to study the FNR affinity for FAD (Figure 4C). Although FNRs have high affinity for FAD, they can slowly exchange the bound prosthetic group with the FAD in solution. Thus, it is possible follow the process of FAD dissociation by measuring the increase in FAD fluorescence of a FNR sample at different times. The obtained data can be used to calculate the FNR:FAD complex dissociation constant or the FAD dissociation rate constant (k_{off}) (Musumeci et al., 2011, Paladini et al., 2009).

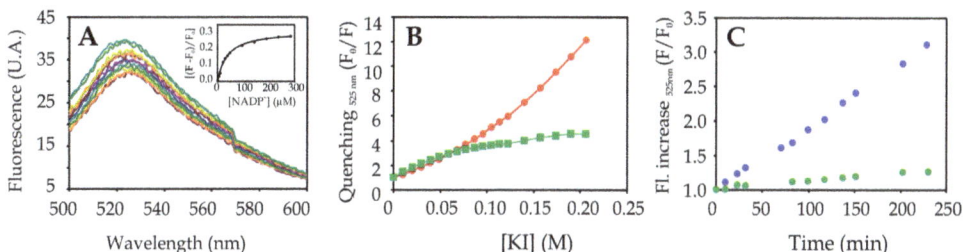

Fig. 4. Application of FAD fluorescence to study structural features of FNR. A) Fluorescence spectra obtained upon titration with NADP⁺. Inset: The increase of fluorescence at 525 nm was plotted against NADP⁺ concentration, and then K_d value can be obtained by adjusting the plot to the sigmoidal function. B) Study of FAD accessibility by titration with a dynamic quencher (KI). Pea Y308S mutant in which the tyrosine facing the isoalloxazine was replaced by a serine shows an increase in the flavin exposure (red) with respect to wild-type FNR (green) (Paladini et al., 2009). C) Fluorescence in solution as a function of time of wild-type FNR (green) and C266M mutant (blue). The k_{off} for FAD of the mutant is higher than the one obtained for wild-type enzyme.

3.2.4 Use of CD spectroscopy

CD spectroscopy can be used for the structural analysis of plant FNRs. The far UV spectral region (240–190 nm or lower) correspond to the peptide bond chromospheres. The absorption consists of a weak but broad $n{\to}\pi^*$ transition centred around 210 nm and an intense $\pi{\to}\pi^*$ transition around 190 nm (Munro A.W. et al., 1999). Figure 5A shows a typical CD spectrum in the far-UV of a plant FNR properly folded. The spectrum can be used to assess the conservation of secondary structure of FNR and how it is modified by mutations or external factors such as temperature, salts or solvents. Besides, the far-UV region allows to calculate the content of different secondary structure elements (α-helix, β-sheet and random coil), which can be done using available software and servers (Deleage & Geourjon, 1993, Whitmore & Wallace, 2004).

On the other hand, the aromatic residues are the main source of signals in the near UV region. These spectra can be used to obtain relevant information on perturbations in the tertiary structure (Figure 5B). Peaks in the near-UV and visible region (350–600 nm) are assigned to the bound FAD and can be applied to analyze perturbations of the prosthetic group environment (Musumeci et al., 2011).

The denaturation process of FNR can be followed by monitoring the CD signal at 220 nm whilst the temperature of the sample or the denaturant concentration is gradually increased (Figure 5C). A sigmoidal curve is obtained from which parameters associated to the transition and the stability of FNR can be calculated (Musumeci et al., 2008, Musumeci et al., 2011).

By measuring CD spectra it is possible to assess whether the cofactor binding site in a FNR is more sensitive to the denaturant than the overall enzyme structure. The decrease of the ellipticity in the far UV region measures the disruption of secondary structure, while a similar change at the visible region can be used to assess the integrity of the binding site of the flavin cofactor (Munro A.W. et al., 1999).

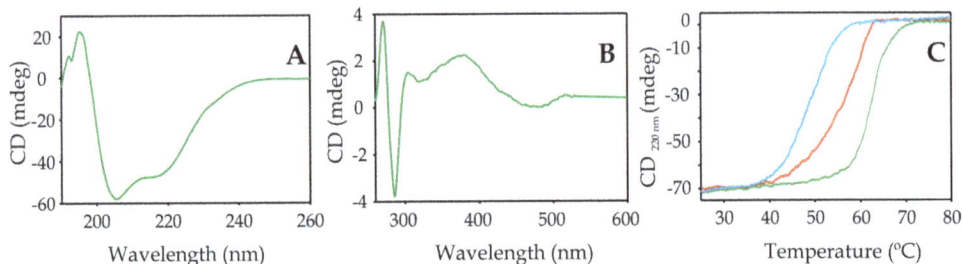

Fig. 5. Structural study of FNRs by CD spectroscopy. A and B) Typical CD spectrum in the far-UV (A) and near-UV (B) regions of plant FNR properly folded (pea FNR). C) Thermal unfolding of different FNRs. From these curves it is possible to estimate the parameters associated to FNR stability. Green: pea FNR; red: *E. coli* FNR; blue: *Xanthomonas axonopodis* FNR.

For CD spectroscopy the FNR sample concentration should be in the range of 0.5–3 μM for the far UV and 20-30 μM for the near UV. It is essential to minimize the absorption due to other components in the mixture (buffers, electrolytes, detergents, etc.), which can be achieved using 0.1 cm path length cuvettes. For the analysis of the far UV region of the FNR CD spectrum (190 nm or lower) suitable buffer systems as phosphate or borate, should be chosen and high chloride concentrations (> 50 mM) should be avoided (Munro A.W. et al., 1999).

3.2.5 Application of RMN to study FNRs

RMN studies can be employed to assess the mobility of different FNR regions and to identify the amino acids residues that interact with nucleotide and protein substrates.

Using this technique the formation of the catalytic competent complex of FNR and substrates was analysed, providing evidences of the movement of the carboxy-terminal region of the enzyme (Maeda et al., 2005). On the other hand, analysis of H/D exchange has provided relevant information about how FNR is modulated by pH, and that the NADP$^+$ binding domain is made of two subdomains with different dynamic behaviours. The flexible subdomain may be important for controlling the binding affinity of FNR for NADP$^+$ (Lee et al., 2007)

3.3 Natural and artificial substrates and Kinetic properties

The optimum pH for FNR activity is in the range of 7.5-8.0 (Masaki et al., 1982, Melamed-Harel et al., 1985) which is coincident value with the pH of the chloroplast stroma during the light step of photosynthesis (Heldt et al., 1973, Werdan et al., 1975). The FNR isolectric point varies in the range of 5.5–6.8 (Grzyb et al., 2008), thus FNRs would be negatively charged at the stromal pH which would enable it to interact with redox partners. With respect to the photosynthetic reaction, the mechanism of electron transfer from reduced ferredoxin to NADP$^+$ was characterized *in vitro* and it was postulated to occur as a two-substrate process (Carrillo & Ceccarelli, 2003). The first step involves the access of the

NADP⁺ to the catalytic site, followed by the binding of reduced ferredoxin, which reduces FNR to the semiquinone state. Then, the oxidised ferredoxin is released and the binding of a second ferredoxin molecule carrying one electron proceeds in order to completely reduce the FNR. Finally, the FNR transfers two electrons to NADP⁺ and produces NADPH. The reverse reaction, which involves the electron transfer from NADPH to ferredoxin can be followed *in vitro* by using cytochrome *c* as final electron acceptor in a coupled assay known as cytochrome *c* reductase activity.

Different activities have been associated with both soluble and membrane-bound FNR. Besides Fd, several other electron acceptors can participate in the reductase-mediated oxidation of NADPH, according to the Eqn. 3 (see the Introduction section), in which A_{ox} and A_{red} are the oxidised and reduced forms of the electron acceptor, respectively. The term n equals one or two depending whether the oxidant behaves as a one- (ferricyanide, cytochromes) or two-electron carriers (NAD⁺, 2,6-dichlorophenol indophenol). A summary of the different reactions catalyzed by FNR is given in Table 1. To measure these reactions *in vitro* the commercially available reduced nucleotide can be employed. An alternative source of NADPH can be obtained using a regenerating system containing NADP⁺, glucose- 6-phosphate (0.3 mM) and glucose-6-phosphate dehydrogenase (1 unit/ml).

Steady-state measurements have demonstrated that the diaphorase reactions proceeds through a two-step transfer "ping-pong" mechanism for either ferricyanide, indophenols dyes and tetrazolium salts (Forti & Sturani, 1968, Masaki et al., 1979, Nakamura & Kimura, 1971, Zanetti & Curti, 1981, Zanetti & Forti, 1966). The same kinetic mechanism holds for the transhydrogenase activity (Böger, 1971a, Shin & Pietro, 1968). Diaphorase activity is probably devoid of physiological meaning in most cases, but it has paid an enormous service to the understanding of FNR function and catalytic mechanism. Recently, emerging functions of these activities have been proposed for bacterial FNRs (Takeda et al., 2010, Yeom et al., 2009). Moreover, some of these artificial reactions might have technological relevance for bioremediation and pharmaceutical industry (Carrillo & Ceccarelli, 2003, Cenas et al., 2001, Tognetti et al., 2007).

Electron donor	Electron acceptor	Activity	k_{cat} (s⁻¹)
Ferredoxin	NADP⁺	Fd-NADP(H) reductase	500
NADPH	Potassium ferricyanide indophenols viologens tetrazolium salts	Diaphorase	100–500 (depending on the electron acceptor)
NADPH	Fd-cytochrome *c*	cytochrome *c* reductase	50-100
NADPH	NAD⁺	Transhydrogenase	5-10
NADPH	O₂, Fd-O₂, Fld-O₂	Oxidase	0.5

Table 1. Activities catalyzed by Ferredoxin-NADP⁺ reductase

Reduction of cytochrome *c* shows a strict requirement for ferredoxin. The reaction is most often described as consisting of two hemi-reactions: FNR-catalyzed reduction of ferredoxin by NADPH, and the subsequent reoxidation of the iron-sulfur protein by cytochrome *c*. It has been also proposed that the electrostatic complex Fd-cytochrome *c* is the true substrate for the reductase. Significant rates of oxygen uptake have been found to occur along with citochrome c reductase activity under aerobic conditions (Carrillo & Vallejos, 1987). The oxidase activity of FNR is very low (Carrillo & Vallejos, 1987, Gomez-Moreno et al., 1994). This reaction is enhanced several-fold by different electronic acceptors, including one-electron reduced ferredoxin or flavodoxin, viologens, nitroderivates and quinones, that can readily engage in oxygen-dependent redox cycling leading to superoxide formation (Gomez-Moreno et al., 1994, Shah & Spain, 1996). In addition, FNR catalyses the transhydrogenation between NADPH and NAD$^+$ (Böger, 1971b).

The stopped-flow technique allows to study fast enzyme kinetic events in the range of milliseconds. This approach has been employed to characterize pre-steady state processes which involve the interaction and electron transfer between oxidised FNR and NADPH, reduced FNR and NADP$^+$ and reduced ferredoxin or flavodoxin and oxidised FNR (Anusevicius et al., 2005, Hurley et al., 1995, Martinez-Julvez et al., 1998, Medina et al., 2001, Paladini et al., 2009, Tejero et al., 2007).

The first reduction of pea FNR by Fd produces a semiquinone form too fast to be measured by rapid mixing techniques (Carrillo & Ceccarelli, 2003) and references therein. Fortunately, this step was characterized for *Anabaena* FNR (Hurley et al., 1995, Martinez-Julvez et al., 1998). The electron transfer processes that involve the dissociation of oxidised ferredoxin, the binding of reduced ferredoxin and flavin reduction occurs more slowly and have been characterized in wild-type and mutant enzymes (Medina, 2009, Nogues et al., 2004).

4. Crystal structure of FNR and its complex with natural substrates

The understanding of the FNR structure–function relationships available nowadays derives from a combination of extensive biochemical studies with the structural analysis of wild-type and mutant FNR enzymes from a variety of sources.

Three-dimensional models of oxidised and fully reduced forms of different FNRs have been reported (Bruns & Karplus, 1995, Correll et al., 1993, Deng et al., 1999, Dorowski et al., 2000, Karplus et al., 1991, Kurisu et al., 2001). A resolution of up to 1.05 Å has been recently obtained for the enzyme from maize (Tronrud et al., 2010). A close related enzyme from the photosynthetic organism *Anabaena variabilis* has been crystallized and their structure resolved (Serre et al., 1996).

All FNRs are similar at the topology level. They are made up of two structural domains, each containing approximately 150 amino acids (Fig. 6A). The amino terminal of ca. 150 residues form a β-barrel FAD-binding domain (Fig. 6A, pink) whereas the carboxy-terminal region includes most of the residues involved in NADP(H) binding and displays a characteristic α-helix/β-strand fold (Fig. 6A, light blue). The FAD in plastidic FNRs is bound in an extended conformation through hydrogen bonds and van der Waals contacts (Bruns & Karplus, 1995, Serre et al., 1996), which is primarily obtained by the interaction of the AMP moiety of FAD with a strand-loop-strand β-hairpin motif of the protein that is partially absent in bacterial FNRs. The isoalloxazine stacks between the aromatic side chains of two tyrosine residues, Tyr89 on the *si*-face and Tyr308 on the *re*-face (Fig. 6B, numbers as in pea FNR). The phenol ring of Tyr308 is the carboxy terminus in pea FNR as in all plastidic type FNRs and is coplanar to the flavin in such a way as to maximize π-orbital overlap. Consequently, the NADP(H) nicotinamide should displace the terminal tyrosine for productive binding and catalysis (Fig. 6B) (Deng et al., 1999, Karplus et al., 1991, Musumeci et al., 2008, Tejero et al., 2005). Most of the isoalloxazine moiety is shielded from the solution but the edge of the dimethyl-benzyl ring is exposed and participates in electron transfer to other protein substrates (Bruns & Karplus, 1995).

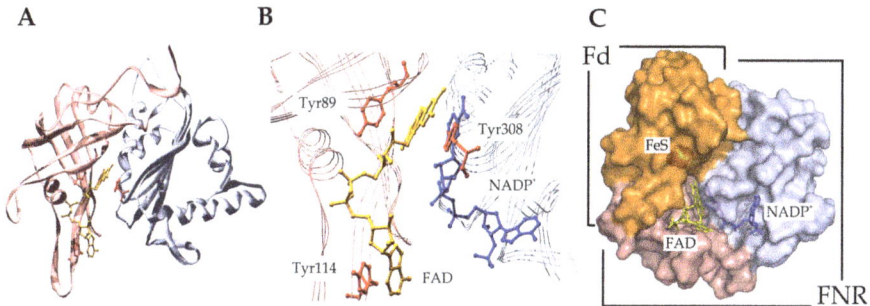

Fig. 6. Computer graphic based on X-ray diffraction data for FNR. A) Pea FNR. B) Detailed view of the isoalloxazine ring system and the NADP$^+$ binding in the FNR-Y308S mutant C) Representation of the superposition of the maize leaf bipartite FNR:Fd complex and the pea FNR:NADP(H) complex. The FAD binding domain of FNR is shown in pink, the NADP(H) binding domain in light blue, the FAD prosthetic group in yellow, the NADP$^+$ in blue and Fd in orange. See text for details.

Other crystal structures of wild-type FNRs with NADP$^+$ bound have been obtained. In these structures the P-AMP portion of NADP$^+$ is properly located and the nicotinamide is far from the catalytic site (Bruns & Karplus, 1995, Nascimento et al., 2007, Serre et al., 1996). It has been suggested that this NADP$^+$ conformations may represent different intermediate steps that occur during substrate binding and catalysis (Serre et al., 1996, Tejero et al., 2005).

Structural studies of binary complexes of oxidised FNR and Fd were resolved by X-ray crystallography for both the *Anabaena* and maize pairs (Kurisu et al., 2001, Morales et al., 2000). The [2S–2Fe] cluster of Fd and the FAD of FNR are in the appropriate proximity (6.0 Å) for electron transfer. Fd binds to a concave region of the FAD domain of maize FNR burying about 5% and 15% of the total surface areas of FNR and Fd, respectively, using in maize and *Anabaena* complexes the same FNR surface region for the interaction with Fd (Kurisu et al., 2001).

5. Interaction of FNR with biological membranes

During the early research on FNR, it was observed that the photosynthetic enzyme was mainly located on the outer surface of the thylakoid membrane (Berzborn, 1968, Berzborn, 1969), in agreement with its important roles in the photosynthetic electron transport. For years, the nature of the interaction of the enzyme with the membrane has been matters of research and debate. Three pools of FNR, soluble, loosely- and tightly- bound to the membrane are dynamically interchanged and may have a metabolic relevance (Carrillo & Vallejos, 1982, Matthijs et al., 1986, Mulo, 2011). In *Arabidopsis* and maize instead of a unique FNR which equilibrates between the membrane and the stroma, distinct FNRs have been identified (Hanke et al., 2005). These FNR isoforms display differential partitioning between the stroma and the membrane in response to their tissue localization and metabolic state of the chloroplast (Grzyb et al., 2008, Hanke et al., 2005, Lintala et al., 2007, Okutani et al., 2005, Zanetti & Arosio, 1980).

FNR solubilisation from the membrane plays a role in maintaining the NADPH/NADP+ homeostasis of the stressed plastid. The enzyme changes its function from a membrane-bound NADPH producer to a soluble NADPH consumer to overcome the accumulation of NADPH. The excess of reducing power might otherwise increase the risk of oxidative damage through the production of hydroxyl radicals (Palatnik et al., 1997).

The persistence of a fraction that remains tightly bound to the membrane has drive to several research groups to search for the existence of an internal membrane protein that acts as binding site for the reductase. The soluble and membrane-bound forms of FNR show different allotropic properties which may reflect conformational changes of the enzyme upon binding to the membrane (Carrillo et al., 1981, Schneeman & Krogmann, 1975). In this way, the membrane component may serve to transmit internal thylakoid conditions as pH changes and membrane fluidity to the bound reductase, which then changes its kinetics behaviour. Several years ago a polypeptide of 17.5 kDa was identified as the binding protein for the reductase (Ceccarelli et al., 1985, Chan et al., 1987, Vallejos et al., 1984). The 17.5-FNR complex was detected in different higher plants, including C3, C4, and Crassulacean acid metabolism species (Soncini & Vallejos, 1989). During solubilisation of FNR under stress condition the reductase-binding protein was released together with FNR, suggesting that it might be the target of some regulation of the membrane bound state (Palatnik et al., 1997). The reductase-binding protein was identified to be the same than the PsbQ like protein (Soncini & Vallejos, 1989). This localization challenges its participation as a FNR binding protein. It has been proposed that the PsbQ may also contribute to the integrity of grana thylakoids stacks (Anderson et al., 2008). Moreover, PsbQ-like homologs have been identified as essential members of the chloroplast NAD(P)H dehydrogenase complex in *Arabidopsis* (Yabuta et al., 2010). The reductase-binding protein (PsbQ like protein) may probably has a ubiquitous function on the thylakoid membrane structure. That makes this topic an interesting issue to pursue.

More recently, a thylakoidial transmembrane protein with a rhodanase like structure was identified as binding site for FNR (Juric et al., 2009). This integral membrane protein contains a conserved carboxy-terminal domain that interacts with high affinity with the FNR enzyme. Three imperfect of such domain were found in one of the proteins of the translocon at the inner envelope membrane of chloroplasts Tic62. Tic62 was proposed as a redox sensor of the translocon and may possibly act as a regulator during the translocation process and serves as a docking site for FNR. Moreover, Tic62 was found bound to thylakoids, soluble in the chloroplast stroma and attached to the inner membrane of the organelle envelope (Benz et al., 2009, Peltier et al., 2004).

An unquestionable evidence of the role of TROL protein in FNR binding is that its absence in knockout plants disables FNR from being tethered to the membrane. In these plants, a decrease of the electron transfer rates under high light intensity was detected (Juric et al., 2009). Nevertheless, the authors suggest that TROL is an important, but not exclusive site for FNR since the enzyme was found in a number of other protein complexes.

It was recently observed by crystallographic studies that the FNR-binding motif is a polyproline type II helix, which is located in the carboxy terminus of Tic62 (Alte et al., 2010). The polyproline type II helix mediates self-assembly of the FNR dimer without influence on the active sites of the enzyme. Interestingly, this type of secondary structure was detected on the crystal structure of the PsbQ (the 17,5 kDa identified as the reductase binding protein) on residues Pro9-Pro10-Pro11-Pro12, which forms a typical left-handed helix (or a polyproline type II structure) (Balsera et al., 2005). These structural elements have been implicated in protein–protein interactions in various cytosolic signal transduction pathways and eucariotic proteins. In addition, the carboxy terminus of TROL may have a similar structure as observed by sequence comparison (Juric et al., 2009). Thus, all proteins that have been identified up to date to firmly interact with FNR contain this structure.

FNR interacts with different integral components of the thylakoid membrane. Several year ago Wagner et al by using flash kinetic spectroscopy demonstrated that the enzyme undergoes a very rapid rotational diffusion on the thylakoid membrane surface (Wagner et al., 1982). This movement was significantly reduced upon addition of Fd due to the formation of a ternary complex between the reductase and the photosystem I protein complex. Direct interaction with subunit of PSI was also detected (Andersen et al., 1992), but the relevance or significance of this interaction is unknown. As in the case of PSI, FNR has also been detected in citochrome b_6f preparations (Clark et al., 1984, Zhang et al., 2001). FNR associated with the cytochrome b_6f complex can participate in the cyclic electron transport as PSI-plastoquinone or NADPH-plastoquinone oxidoreductases, which can explain the presence of FNR in citochrome b_6f preparations (Szymanska et al., 2011).

Some authors have identified FNR in the Ndh complex. The NAD(P)H dehydrogenase (Ndh) complex in chloroplast thylakoid membranes functions in cyclic electron transfer and in chlororespiration. Ndh is composed of at least 15 subunits, including both chloroplast- and nuclear-encoded proteins (Suorsa et al., 2009). As already mentioned, among the detected subunits of the Ndh complex is the PsbQ which was found associated with FNR. However, the amount of Ndh complex is a minor component of the chloroplast and it is present at ~1.5% of the level of PSII on a molar basis (Burrows et al., 1998) and probably makes a small contribution to the total cyclic electron transport measured *in vivo* (Joliot et al., 2004).

FNR is an abundant protein in chloroplast. Moreover, protein concentration in the chloroplast stroma was determined to be within the range of 520 to 730 g/dm³ for spinach and pea (Lilley, 1983). FNR structure is suitable for interaction with membrane and proteins. It is therefore likely to detect many interactions of the FNR with other structures inside the chloroplast, many of which may have metabolic relevance.

6. Concluding remarks

For over six decades much effort has been devoted to the elucidation of the structure, catalytic mechanism and regulation of the FNR. Similarly, enormous progress has been made in understanding the insertion and importance of this enzyme on the various metabolic processes in plants and other organisms. The ability to produce FNR by transgenic expression and the modern techniques available nowadays have provided the means to analyze more deeply this flavoprotein. However, multiple aspects of this enzyme are still unknown as the structural bases for the increase of its catalytic competence and the ways to change the nucleotide substrate specificity. Furthermore, the enzyme may be an attractive target for the development of new herbicides and for the design of bactericides. Likewise, the ability of FNR to metabolize xenobiotics and environmental pollutants remain to be exploited. The widespread distribution of FNR will allow in the near future to increase the knowledge of the enzyme by comparative analysis of FNRs from different backgrounds and to advance in the study and manipulation of this key enzyme in the photosynthetic process.

7. Acknowledgment

This work was supported by grants from CONICET, Agencia de Promoción Científica y Tecnológica (ANPCyT) and Fundación Bunge y Born, Argentina.

8. References

Aliverti, A., Bruns, C. M., Pandini, V. E., Karplus, P. A., Vanoni, M. A., Curti, B., & Zanetti, G. (1995). Involvement of serine 96 in the catalytic mechanism of ferredoxin-NADP+ reductase: structure--function relationship as studied by site-directed mutagenesis and X-ray crystallography. *Biochemistry 34*, 8371-8379.

Aliverti, A., Jansen, T., Zanetti, G., Ronchi, S., Herrmann, R. G., & Curti, B. (1990). Expression in *Escherichia coli* of ferredoxin:NADP+ reductase from spinach. Bacterial synthesis of the holoflavoprotein and of an active enzyme form lacking the first 28 amino acid residues of the sequence. *Eur J Biochem 191*, 551-555.

Aliverti, A., Pandini, V., Pennati, A., de Rosa, M., & Zanetti, G. (2008). Structural and functional diversity of ferredoxin-NADP(+) reductases. *Arch. Biochem. Biophys. 474*, 283-291.

Alte, F., Stengel, A., Benz, J. P., Petersen, E., Soll, J., Groll, M., & Bolter, B. (2010). Ferredoxin:NADPH oxidoreductase is recruited to thylakoids by binding to a polyproline type II helix in a pH-dependent manner. *Proc. Natl. Acad. Sci. U. S. A 107*, 19260-19265.

Andersen, B., Scheller, H. V., & Moller, B. L. (1992). The PSI-E subunit of photosystem I binds ferredoxin:NADP+ oxidoreductase. *FEBS Lett 311*, 169-173.

Anderson, J. M., Chow, W. S., & De Las, R. J. (2008). Dynamic flexibility in the structure and function of photosystem II in higher plant thylakoid membranes: the grana enigma. *Photosynth. Res. 98*, 575-587.

Anusevicius, Z., Miseviciene, L., Medina, M., Martinez-Julvez, M., Gomez-Moreno, C., & Cenas, N. (2005). FAD semiquinone stability regulates single- and two-electron reduction of quinones by *Anabaena* PCC7119 ferredoxin:NADP+ reductase and its Glu301Ala mutant. *Arch. Biochem. Biophys. 437*, 144-150.

Arakaki, A. K., Ceccarelli, E. A., & Carrillo, N. (1997). Plant-type ferredoxin-NADP+ reductases: a basal structural framework and a multiplicity of functions. *FASEB J. 11*, 133-140.

Avron, M. & Jagendorf, A. T. (1956). A TPNH diaphorase from chloroplast. *Arch. Biochem. Biophys. 65*, 475-490.

Balsera, M., Arellano, J. B., Revuelta, J. L., De Las, R. J., & Hermoso, J. A. (2005). The 1.49 A resolution crystal structure of PsbQ from photosystem II of *Spinacia oleracea* reveals a PPII structure in the N-terminal region. *J. Mol. Biol. 350*, 1051-1060.

Benz, J. P., Stengel, A., Lintala, M., Lee, Y. H., Weber, A., Philippar, K., Gugel, I. L., Kaieda, S., Ikegami, T., Mulo, P., Soll, J., & Bolter, B. (2009). *Arabidopsis* Tic62 and ferredoxin-NADP(H) oxidoreductase form light-regulated complexes that are integrated into the chloroplast redox poise. *Plant Cell 21*, 3965-3983.

Berzborn, R. (1968). On soluble and insoluble chloroplast antigens. Demonstration of ferredoxin-NADP-reductase in the surface of the chloroplast lamellar system with the aid of specific antibodies. *Z Naturforsch B 23*, 1096-1104.

Berzborn, R. J. (1969). Studies on the surface structure of the thylakoid system of chloroplasts using antibodies against ferredoxin-NADP-reductase. *Z Naturforsch B 24*, 436-446.

Böger, P. (1971a). Relationship of transhydrogenase and diaphorase activity of ferredoxin-NADP⁺ reductase with photosynthetic NADP⁺ reduction. *Z Naturforsch B 26*, 807-815.

Böger, P. (1971b). Einfluß von Ferredoxin auf Ferredoxin-NADP-Reduktase. *Planta 99*, 319-338.

Bottin, H. & Lagoutte, B. (1992). Ferredoxin and flavodoxin from the cyanobacterium *Synechocystis* sp PCC 6803. *Biochim. Biophys. Acta 1101*, 48-56.

Bruns, C. M. & Karplus, P. A. (1995). Refined crystal structure of spinach ferredoxin reductase at 1.7 A resolution: oxidised, reduced and 2'-phospho-5'-AMP bound states. *J. Mol. Biol. 247*, 125-145.

Burrows, P. A., Sazanov, L. A., Svab, Z., Maliga, P., & Nixon, P. J. (1998). Identification of a functional respiratory complex in chloroplasts through analysis of tobacco mutants containing disrupted plastid ndh genes. *EMBO J. 17*, 868-876.

Carrillo, N. & Vallejos, R. H. (1987). Ferredoxin-NADP+ oxidoreductase. In: *The Light Reactions (Topics in photosynthesis)*, edited by J. Barber, pp. 527-560. Amsterdam-New York-Oxford: Elsevier.

Carrillo, N. & Ceccarelli, E. A. (2003). Open questions in ferredoxin-NADP⁺ reductase catalytic mechanism. *Eur. J. Biochem. 270*, 1900-1915.

Carrillo, N., Lucero, H. A., & Vallejos, R. H. (1981). Light modulation of chloroplast membrane-bound ferredoxin-NADP⁺ oxidoreductase. *J. Biol. Chem. 256*, 1058-1059.

Carrillo, N. & Vallejos, R. H. (1982). Interaction of Ferredoxin-NADP⁺ Oxidoreductase with the Thylakoid Membrane. *Plant Physiol 69*, 210-213.

Carrillo, N. & Vallejos, R. H. (1983). Interaction of ferredoxin-NADP⁺ oxidoreductase with triazine dyes. A rapid purification method by affinity chromatography. *Biochim. Biophys. Acta 742*, 285-294.

Catalano-Dupuy, D. L., Orecchia, M., Rial, D. V., & Ceccarelli, E. A. (2006). Reduction of the pea ferredoxin-NADP(H) reductase catalytic efficiency by the structuring of a carboxyl-terminal artificial metal binding site. *Biochemistry 45*, 13899-13909.

Ceccarelli, E. A., Arakaki, A. K., Cortez, N., & Carrillo, N. (2004). Functional plasticity and catalytic efficiency in plant and bacterial ferredoxin-NADP(H) reductases. *Biochim. Biophys. Acta 1698*, 155-165.

Ceccarelli, E. A., Chan, R. L., & Vallejos, R. H. (1985). Trimeric structure and other properties of the chloroplast reductase binding protein. *FEBS Letters 190*, 165-168.

Cenas, N., Nemeikaite-Ceniene, A., Sergediene, E., Nivinskas, H., Anusevicius, Z., & Sarlauskas, J. (2001). Quantitative structure-activity relationships in enzymatic single-electron reduction of nitroaromatic explosives: implications for their cytotoxicity. *Biochim. Biophys. Acta 1528*, 31-38.

Chan, R. L., Ceccarelli, E. A., & Vallejos, R. H. (1987). Immunological studies of the binding protein for chloroplast ferredoxin-NADP⁺ reductase. *Arch. Biochem. Biophys. 253*, 56-61.

Chopowick, R. & Israelstam, G. F. (1971). Pyridine nucleotide transhydrogenase from *Chlorella. Planta 101*, 171-173.

Clark, R. D., Hawkesford, M. J., Coughlan, S. J., Bennett, J., & Hind, G. (1984). Association of ferredoxin-NADP+ oxidoreductase with the chloroplast cytochrome *b-f* complex. *FEBS Letters 174*, 137-142.

Cleland, R. E. & Bendall, D. S. (1992). Photosystem I cyclic electron transport: Measurement of ferredoxin-plastoquinone reductase activity. *Photosynthesis Research 34*, 409-418.

Correll, C. C., Ludwig, M., Bruns, C. M., & Karplus, P. A. (1993). Structural prototypes for an extended family of flavoprotein reductases: comparison of phthalate dioxygenase reductase with ferredoxin reductase and ferredoxin. *Protein Sci 2*, 2112-2133.

Davis, D. J. (1990). Tryptophan fluorescence studies of ferredoxin:NADP+ reductase indicate the presence of tryptophan in or near the ferredoxin binding site. *Arch. Biochem. Biophys. 276*, 1-5.

Deleage, G. & Geourjon, C. (1993). An interactive graphic program for calculating the secondary structure content of proteins from circular dichroism spectrum. *Comput. Appl. Biosci. 9*, 197-199.

Deng, Z., Aliverti, A., Zanetti, G., Arakaki, A. K., Ottado, J., Orellano, E. G., Calcaterra, N. B., Ceccarelli, E. A., Carrillo, N., & Karplus, P. A. (1999). A productive NADP+ binding mode of ferredoxin-NADP+ reductase revealed by protein engineering and crystallographic studies. *Nat. Struct. Biol. 6*, 847-853.

Dorowski, A., Hofmann, A., Steegborn, C., Boicu, M., & Huber, R. (2000). Crystal structure of paprika ferredoxin-NADP+ reductase - implications for the electron transfer pathway. *J. Biol. Chem. 276*, 9253-9263.

Dudley, K. H., Ehrenberg, A., Hemmerich, P., & Müller, F. (1964). Spektren und Strukturen der am Flavin-Redoxsystem beteiligten Partikeln. Studien in der Flavinreihe IX. *Helv. Chim. Acta 47*, 1354-1383.

Faro, M., Schiffler, B., Heinz, A., Nogues, I., Medina, M., Bernhardt, R., & Gomez-Moreno, C. (2003). Insights into the design of a hybrid system between *Anabaena* ferredoxin-NADP+ reductase and bovine adrenodoxin. *Eur. J. Biochem. 270*, 726-735.

Forti, G. & Sturani, E. (1968). On the structure and function of reduced nicotinamide adenine dinucleotide phosphate-cytochrome *f* reductase of spinach chloroplasts. *Eur. J. Biochem. 3*, 461-472.

Gomez-Moreno, C., Medina, M., Hurley, J. K., Cusanovich, M., Markley, J., Cheng, H., Xia, B., Chae, Y. K., & Tollin, G. (1994). Protein engineering for the elucidation of the mechanism of electron transfer in redox proteins. *Biochem Soc. Trans. 22*, 796-800.

Green, L. S., Yee, B. C., Buchanan, B. B., Kamide, K., Sanada, Y., & Wada, K. (1991). Ferredoxin and Ferredoxin-NADP+ Reductase from Photosynthetic and Nonphotosynthetic Tissues of Tomato. *Plant Physiology 96*, 1207-1213.

Grzyb, J., Malec, P., Rumak, I., Garstka, M., & Strzalka, K. (2008). Two isoforms of ferredoxin:NADP(+) oxidoreductase from wheat leaves: purification and initial biochemical characterization. *Photosynth. Res. 96*, 99-112.

Grzyb, J., Waloszek, A., Latowski, D., & Wieckowski, S. (2004). Effect of cadmium on ferredoxin:NADP+ oxidoreductase activity. *J. Inorg. Biochem. 98*, 1338-1346.

Hajirezaei, M. R., Peisker, M., Tschiersch, H., Palatnik, J. F., Valle, E. M., Carrillo, N., & Sonnewald, U. (2002). Small changes in the activity of chloroplastic NADP(+)-

dependent ferredoxin oxidoreductase lead to impaired plant growth and restrict photosynthetic activity of transgenic tobacco plants. *Plant J. 29*, 281-293.

Hanke, G. T., Ookutani, S. , Satomi, Y., Takao, T., Suzuki, A., & Hase, T. (2005). Multiple iso-proteins of FNR in *Arabidopsis*: evidence for different contributions to chloroplast function and nitrogen assimilation. *Plant, Cell & Environment 28*, 1146-1157.

Heelis P.F. (1982). The photophysical and photochemical properties of flavins (isoalloxazines). *Chem. Soc. Rev. 11*, 15-39.

Heldt, W. H., Werdan, K., Milovancev, M., & Geller, G. (1973). Alkalization of the chloroplast stroma caused by light-dependent proton flux into the thylakoid space. *Biochim. Biophys. Acta 314*, 224-241.

Hurley, J. K., Fillat, M., Gomez-Moreno, C., & Tollin, G. (1995). Structure-function relationships in the ferredoxin/ferredoxin: NADP⁺ reductase system from *Anabaena. Biochimie 77*, 539-548.

Jenkins, C. M., Genzor, C. G., Fillat, M. F., Waterman, M. R., & Gomez-Moreno, C. (1997). Negatively charged *Anabaena* flavodoxin residues (Asp144 and Glu145) are important for reconstitution of cytochrome P450 17alpha-hydroxylase activity. *J. Biol. Chem. 272*, 22509-22513.

Joliot, P., Beal, D., & Joliot, A. (2004). Cyclic electron flow under saturating excitation of dark-adapted *Arabidopsis* leaves. *Biochim. Biophys. Acta 1656*, 166-176.

Joliot, P. & Joliot, A. (2006). Cyclic electron flow in C3 plants. *Biochim. Biophys. Acta 1757*, 362-368.

Juric, S., Hazler-Pilepic, K., Tomasic, A., Lepedus, H., Jelicic, B., Puthiyaveetil, S., Bionda, T., Vojta, L., Allen, J. F., Schleiff, E., & Fulgosi, H. (2009). Tethering of ferredoxin:NADP⁺ oxidoreductase to thylakoid membranes is mediated by novel chloroplast protein TROL. *Plant J. 60*, 783-794.

Karplus, P. A., Daniels, M. J., & Herriott, J. R. (1991). Atomic structure of ferredoxin-NADP⁺ reductase: prototype for a structurally novel flavoenzyme family. *Science 251*, 60-66.

Krapp, A. R., Rodriguez, R. E., Poli, H. O., Paladini, D. H., Palatnik, J. F., & Carrillo, N. (2002). The flavoenzyme ferredoxin (flavodoxin)-NADP(H) reductase modulates NADP(H) homeostasis during the soxRS response of *Escherichia coli. J. Bacteriol. 184*, 1474-1480.

Krause, G. H. & Heber, U. (1976).*The intact chloroplast* , edited by H. Barber, pp. 171-214. Elsevier, Amsterdam.

Krawetz, S. A. & Israelstam, G. F. (1978). Kinetics of pyridine nucleotide transhydrogenase from *Chlorella. Plant Science Letters 12*, 323-326.

Kurisu, G., Kusunoki, M., Katoh, E., Yamazaki, T., Teshima, K., Onda, Y., Kimata-Ariga, Y., & Hase, T. (2001). Structure of the electron transfer complex between ferredoxin and ferredoxin-NADP⁺ reductase. *Nat. Struct. Biol. 8*, 117-121.

Lee, Y. H., Tamura, K., Maeda, M., Hoshino, M., Sakurai, K., Takahashi, S., Ikegami, T., Hase, T., & Goto, Y. (2007). Cores and pH-dependent dynamics of ferredoxin-NADP⁺ reductase revealed by hydrogen/deuterium exchange. *J. Biol. Chem. 282*, 5959-5967.

Lilley, R. (1983). Chloroplast metabolism: the pathways of primary carbon metabolism in C3 plants. *Plant, Cell & Environment 6*, 329-343.

Lintala, M., Allahverdiyeva, Y., Kidron, H., Piippo, M., Battchikova, N., Suorsa, M., Rintamaki, E., Salminen, T. A., Aro, E. M., & Mulo, P. (2007). Structural and functional characterization of ferredoxin-NADP⁺-oxidoreductase using knock-out mutants of *Arabidopsis*. *Plant J. 49*, 1041-1052.

Macheroux, P. (1999). *Flavoprotein Protocols*, edited by Stephen K.Chapman & Graeme A.Reid, Springer Science.

Maeda, M., Lee, Y. H., Ikegami, T., Tamura, K., Hoshino, M., Yamazaki, T., Nakayama, M., Hase, T., & Goto, Y. (2005). Identification of the N- and C-terminal substrate binding segments of ferredoxin-NADP⁺ reductase by NMR. *Biochemistry 44*, 10644-10653.

Martinez-Julvez, M., Hermoso, J., Hurley, J. K., Mayoral, T., Sanz-Aparicio, J., Tollin, G., Gomez-Moreno, C., & Medina, M. (1998). Role of Arg100 and Arg264 from *Anabaena* PCC 7119 ferredoxin-NADP⁺ reductase for optimal NADP⁺ binding and electron transfer. *Biochemistry 37*, 17680-17691.

Masaki, R., Wada, K., & Matsubara, H. (1979). Isolation and characterization of two ferredoxin-NADP⁺ reductases from *Spirulina platensis*. *J Biochem (Tokyo) 86*, 951-962.

Masaki, R., Yoshikawa, S., & Matsubara, H. (1982). Steady-state kinetics of oxidation of reduced ferredoxin with ferredoxin-NADP⁺ reductase. *Biochim. Biophys. Acta 700*, 101-109.

Matthijs, H. C., Coughlan, S. J., & Hind, G. (1986). Removal of ferredoxin:NADP⁺ oxidoreductase from thylakoid membranes, rebinding to depleted membranes, and identification of the binding site. *J. Biol. Chem. 261*, 12154-12158.

Medina, M. (2009). Structural and mechanistic aspects of flavoproteins: photosynthetic electron transfer from photosystem I to NADP⁺. *FEBS J. 276*, 3942-3958.

Medina, M. & Gomez-Moreno, C. (2004). Interaction of Ferredoxin-NADP⁺ Reductase with Its Substrates: Optimal Interaction for Efficient Electron Transfer. *Photosynth. Res. 79*, 113-131.

Medina, M., Luquita, A., Tejero, J., Hermoso, J., Mayoral, T., Sanz-Aparicio, J., Grever, K., & Gomez-Moreno, C. (2001a). Probing the determinants of coenzyme specificity in ferredoxin-NADP⁺ reductase by site-directed mutagenesis. *J. Biol. Chem. 276*, 11902-11912.

Melamed-Harel, H., Tel-Or, E., & Pietro, A. S. (1985). Effect of Ferredoxin on the Diaphorase Activity of Cyanobacterial Ferredoxin-NADP⁺ Reductase. *Plant Physiology 77*, 229-231.

Morales, R., Kachalova, G., Vellieux, F., Charon, M. H., & Frey, M. (2000). Crystallographic studies of the interaction between the ferredoxin-NADP⁺ reductase and ferredoxin from the cyanobacterium *Anabaena*: looking for the elusive ferredoxin molecule. *Acta Crystallogr D Biol Crystallogr 56*, 1408-1412.

Morigasaki, S., Takata, K., Suzuki, T., & Wada, K. (1990). Purification and Characterization of a Ferredoxin-NADP⁺ Oxidoreductase-Like Enzyme from Radish Root Tissues. *Plant Physiol 93*, 896-901.

Mulo, P. (2011). Chloroplast-targeted ferredoxin-NADP(+) oxidoreductase (FNR): Structure, function and location. *Biochim. Biophys. Acta 1807*, 927-934.

Munro A.W., Kelly S., & Price N. (1999). Fluorescence analysis of Flavoproteins. In: *Flavoprotein protocols* , edited by G. A. R. Stephen K.Chapman, pp. 111-129. Humana press. NJ, USA.

Munro A.W. & Noble, M. A. (1999) Circular Dichroism studies of flavoproteins. In: *Flavoprotein protocols* edited by G. A. R. Stephen K.Chapman, pp. 25-48. Humana Press. NJ, USA.

Musumeci, M. A., Arakaki, A. K., Rial, D. V., Catalano-Dupuy, D. L., & Ceccarelli, E. A. (2008). Modulation of the enzymatic efficiency of ferredoxin-NADP(H) reductase by the amino acid volume around the catalytic site. *FEBS J. 275*, 1350-1366.

Musumeci, M. A., Botti, H., Buschiazzo, A., & Ceccarelli, E. A. (2011). Swapping FAD Binding Motifs between Plastidic and Bacterial Ferredoxin-NADP(H) Reductases. *Biochemistry 50*, 2111-2122.

Nakajima, M., Sakamoto, T., & Wada, K. (2002). The complete purification and characterization of three forms of ferredoxin-NADP(+) oxidoreductase from a thermophilic cyanobacterium *Synechococcus elongatus*. *Plant Cell Physiol 43*, 484-493.

Nakamura, S. & Kimura, T. (1971). Studies on spinach ferredoxin-nicotinamide adenine dinucleotide phosphate reductase. Kinetic studies on the interactions of the reductase and ferredoxin and a possible regulation of enzyme activities by ionic strength. *J. Biol. Chem. 246*, 6235-6241.

Nascimento, A. S., Catalano-Dupuy, D. L., Bernardes, A., de Oliveira, N. M., Santos, M. A., Ceccarelli, E. A., & Polikarpov, I. (2007). Crystal structures of *Leptospira interrogans* FAD-containing ferredoxin-NADP⁺ reductase and its complex with NADP⁺. *BMC. Struct. Biol. 7*, 69.

Nogues, I., Tejero, J., Hurley, J. K., Paladini, D., Frago, S., Tollin, G., Mayhew, S. G., Gomez-Moreno, C., Ceccarelli, E. A., Carrillo, N., & Medina, M. (2004). Role of the C-terminal tyrosine of ferredoxin-nicotinamide adenine dinucleotide phosphate reductase in the electron transfer processes with its protein partners ferredoxin and flavodoxin. *Biochemistry 43*, 6127-6137.

Okutani, S., Hanke, G. T., Satomi, Y., Takao, T., Kurisu, G., Suzuki, A., & Hase, T. (2005). Three maize leaf ferredoxin:NADPH oxidoreductases vary in subchloroplast location, expression, and interaction with ferredoxin. *Plant Physiol 139*, 1451-1459.

Onda, Y. & Hase, T. (2004). FAD assembly and thylakoid membrane binding of ferredoxin:NADP⁺ oxidoreductase in chloroplasts. *FEBS Lett. 564*, 116-120.

Onda, Y., Matsumura, T., Kimata-Ariga, Y., Sakakibara, H., Sugiyama, T., & Hase, T. (2000). Differential interaction of maize root Ferredoxin:NADP(+) oxidoreductase with photosynthetic and non-photosynthetic ferredoxin isoproteins. *Plant Physiol 123*, 1037-1046.

Osmond, C. B. (1981). Photorespiration and photoinhibition : Some implications for the energetics of photosynthesis. *Biochim. Biophys. Acta 639*, 77-98.

Paladini, D. H., Musumeci, M. A., Carrillo, N., & Ceccarelli, E. A. (2009). Induced fit and equilibrium dynamics for high catalytic efficiency in ferredoxin-NADP(H) reductases. *Biochemistry 48*, 5760-5768.

Palatnik, J. F., Valle, E. M., & Carrillo, N. (1997). Oxidative stress causes ferredoxin-NADP⁺ reductase solubilization from the thylakoid membranes in methyl viologen-treated plants. *Plant Physiol 115*, 1721-1727.

Peltier, J. B., Ytterberg, A. J., Sun, Q., & van Wijk, K. J. (2004). New functions of the thylakoid membrane proteome of *Arabidopsis thaliana* revealed by a simple, fast, and versatile fractionation strategy. *J. Biol. Chem. 279*, 49367-49383.

Piubelli, L., Aliverti, A., Arakaki, A. K., Carrillo, N., Ceccarelli, E. A., Karplus, P. A., & Zanetti, G. (2000). Competition between C-terminal tyrosine and nicotinamide modulates pyridine nucleotide affinity and specificity in plant ferredoxin-NADP(+) reductase. *J. Biol. Chem. 275*, 10472-10476.

Razquin, P., Fillat, M. F., Schmitz, S., Stricker, O., Bohme, H., Gomez-Moreno, C., & Peleato, M. L. (1996). Expression of ferredoxin-NADP$^+$ reductase in heterocysts from *Anabaena* sp. *Biochem. J. 316 (Pt 1)*, 157-160.

Rochaix, J. D. (2011). Regulation of photosynthetic electron transport. *Biochim. Biophys. Acta 1807*, 375-383.

Rohrich, R. C., Englert, N., Troschke, K., Reichenberg, A., Hintz, M., Seeber, F., Balconi, E., Aliverti, A., Zanetti, G., Kohler, U., Pfeiffer, M., Beck, E., Jomaa, H., & Wiesner, J. (2005). Reconstitution of an apicoplast-localised electron transfer pathway involved in the isoprenoid biosynthesis of *Plasmodium falciparum*. *FEBS Lett. 579*, 6433-6438.

Sakihama, N., Nagai, K., Ohmori, H., Tomizawa, H., Tsujita, M., & Shin, M. (1992). Immobilized ferredoxins for affinity chromatography of ferredoxin-dependent enzymes. *J. Chromatogr. 597*, 147-153.

Sancho, J., Peleato, M. L., Gomez-Moreno, C., & Edmondson, D. E. (1988). Purification and properties of ferredoxin-NADP$^+$ oxidoreductase from the nitrogen-fixing cyanobacteria *Anabaena variabilis*. *Arch. Biochem. Biophys. 260*, 200-207.

Schneeman, R. & Krogmann, D. W. (1975). Polycation interactions with spinach ferredoxin-nicotinamide adenine dinucleotide phosphate reductase. *J. Biol. Chem. 250*, 4965-4971.

Seeber, F., Aliverti, A., & Zanetti, G. (2005). The plant-type ferredoxin-NADP$^+$ reductase/ferredoxin redox system as a possible drug target against apicomplexan human parasites. *Curr. Pharm. Des 11*, 3159-3172.

Serra, E. C., Carrillo, N., Krapp, A. R., & Ceccarelli, E. A. (1993). One-step purification of plant ferredoxin-NADP$^+$ oxidoreductase expressed in *Escherichia coli* as fusion with glutathione *S*-transferase. *Protein. Expr. Purif. 4*, 539-546.

Serre, L., Vellieux, F. M., Medina, M., Gomez-Moreno, C., Fontecilla-Camps, J. C., & Frey, M. (1996). X-ray structure of the ferredoxin:NADP$^+$ reductase from the cyanobacterium *Anabaena* PCC 7119 at 1.8 A resolution, and crystallographic studies of NADP$^+$ binding at 2.25 A resolution. *J. Mol. Biol. 263*, 20-39.

Setif, P. (2006). Electron Transfer from the Bound Iron–Sulfur Clusters to Ferredoxin/Flavodoxin: Kinetic and Structural Properties of Ferredoxin/Flavodoxin Reduction by Photosystem I. In. *Photosystem I: The LightDriven Plastocyanin:Ferredoxin Oxidoreductase*, edited by J. H. Golbeck, pp. 439-454. Springer, Dordrecht. Berlin

Shah, M. M. & Spain, J. C. (1996). Elimination of nitrite from the explosive 2,4,6-trinitrophenylmethylnitramine (tetryl) catalyzed by ferredoxin NADP$^+$ oxidoreductase from spinach. *Biochem. Biophys. Res. Commun. 220*, 563-568.

Shikanai, T. (2007). Cyclic electron transport around photosystem I: genetic approaches. *Annu. Rev. Plant Biol. 58*, 199-217.

Shin, M. & Arnon, D. I. (1965). Enzymatic mechanisms of pyridine nucleotide reduction in chloroplast. *J. Biol. Chem. 240*, 1405-1411.

Shin, M. & Pietro, A. S. (1968). Complex formation of ferredoxin-NADP$^+$ reductase with ferredoxin and with NADP$^+$. *Biochem. Biophys. Res. Commun. 33*, 38-42.

Soncini, F. C. & Vallejos, R. H. (1989). The chloroplast reductase-binding protein is identical to the 16.5-kDa polypeptide described as a component of the oxygen-evolving complex. *J. Biol. Chem. 264*, 21112-21115.

Suorsa, M., Sirpio, S., & Aro, E. M. (2009). Towards characterization of the chloroplast NAD(P)H dehydrogenase complex. *Mol. Plant 2*, 1127-1140.

Szymanska, R., Dluzewska, J., S'lesak, I., & Kruk, J. (2011). Ferredoxin:NADP$^+$ oxidoreductase bound to cytochrome b_6f complex is active in plastoquinone reduction: Implications for cyclic electron transport. *Physiologia Plantarum 141*, 289-298.

Takeda, K., Sato, J., Goto, K., Fujita, T., Watanabe, T., Abo, M., Yoshimura, E., Nakagawa, J., Abe, A., Kawasaki, S., & Niimura, Y. (2010). *Escherichia coli* ferredoxin-NADP$^+$ reductase and oxygen-insensitive nitroreductase are capable of functioning as ferric reductase and of driving the Fenton reaction. *Biometals 23*, 727-737.

Tejero, J., Martinez-Julvez, M., Mayoral, T., Luquita, A., Sanz-Aparicio, J., Hermoso, J. A., Hurley, J. K., Tollin, G., Gomez-Moreno, C., & Medina, M. (2003). Involvement of the pyrophosphate and the 2'-phosphate binding regions of ferredoxin-NADP$^+$ reductase in coenzyme specificity. *J. Biol. Chem. 278*, 49203-49214.

Tejero, J., Peregrina, J. R., Martinez-Julvez, M., Gutierrez, A., Gomez-Moreno, C., Scrutton, N. S., & Medina, M. (2007). Catalytic mechanism of hydride transfer between NADP+/H and ferredoxin-NADP+ reductase from Anabaena PCC 7119. *Arch. Biochem. Biophys. 459*, 79-90.

Tejero, J., Perez-Dorado, I., Maya, C., Martinez-Julvez, M., Sanz-Aparicio, J., Gomez-Moreno, C., Hermoso, J. A., & Medina, M. (2005). C-terminal tyrosine of ferredoxin-NADP$^+$ reductase in hydride transfer processes with NAD(P)+/H. *Biochemistry 44*, 13477-13490.

Tognetti, V. B., Monti, M. R., Valle, E. M., Carrillo, N., & Smania, A. M. (2007). Detoxification of 2,4-dinitrotoluene by Transgenic Tobacco Plants Expressing a Bacterial Flavodoxin. *Environmental Science & Technology 41*, 4071-4076.

Tronrud, D. E., Berkholz, D. S., & Karplus, P. A. (2010). Using a conformation-dependent stereochemical library improves crystallographic refinement of proteins. *Acta Crystallogr. D. Biol. Crystallogr. 66*, 834-842.

Vallejos, R. H., Ceccarelli, E., & Chan, R. (1984). Evidence for the existence of a thylakoid intrinsic protein that binds ferredoxin-NADP$^+$ oxidoreductase. *J. Biol. Chem. 259*, 8048-8051.

Vishniac, W. & Ochoa, S. (1951). Photochemical reduction of pyridine nucleotides by spinach grana and coupled carbon dioxide fixation. *Nature 167*, 768-769.

Wagner, R., Carrillo, N., Junge, W., & Vallejos, R. H. (1982). On the conformation of reconstituted ferredoxin:NADP$^+$ oxidoreductase in the thylakoid membrane. Studies via triplet lifetime and rotational diffusion with eosin isothiocyanate as label. *Biochimica et Biophysica Acta (BBA) - Bioenergetics 680*, 317-330.

Werdan, K., Heldt, H. W., & Milovancev, M. (1975). The role of pH in the regulation of carbon fixation in the chloroplast stroma. Studies on CO_2 fixation in the light and dark. *Biochim. Biophys. Acta 396*, 276-292.

Whitmore, L. & Wallace, B. A. (2004). DICHROWEB, an online server for protein secondary structure analyses from circular dichroism spectroscopic data. *Nucleic Acids Res. 32*, W668-W673.

Yabuta, S., Ifuku, K., Takabayashi, A., Ishihara, S., Ido, K., Ishikawa, N., Endo, T., & Sato, F. (2010). Three PsbQ-like proteins are required for the function of the chloroplast NAD(P)H dehydrogenase complex in *Arabidopsis*. *Plant Cell Physiol 51*, 866-876.

Yamamoto, H., Kato, H., Shinzaki, Y., Horiguchi, S., Shikanai, T., Hase, T., Endo, T., Nishioka, M., Makino, A., Tomizawa, K., & Miyake, C. (2006). Ferredoxin limits cyclic electron flow around PSI (CEF-PSI) in higher plants-stimulation of CEF-PSI enhances non-photochemical quenching of Chl fluorescence in transplastomic tobacco. *Plant Cell Physiol 47*, 1355-1371.

Yeom, J., Jeon, C. O., Madsen, E. L., & Park, W. (2009). Ferredoxin-NADP⁺ reductase from *Pseudomonas putida* functions as a ferric reductase. *J. Bacteriol. 191*, 1472-1479.

Zanetti, G., Aliverti, A., & Curti, B. (1984). A cross-linked complex between ferredoxin and ferredoxin-NADP⁺ reductase. *J. Biol. Chem. 259*, 6153-6157.

Zanetti, G. & Arosio, P. (1980). Solubilization from spinach thylakoids of a higher molecular weight form of ferredoxin-NADP⁺ reductase. *FEBS Lett 111*, 373-376.

Zanetti, G. & Curti, B. (1981). Interactions between ferredoxin-NADP⁺ reductase and ferredoxin at different reduction levels of the two proteins. *FEBS Lett 129*, 201-204.

Zanetti, G. & Forti, G. (1966). Studies on the triphosphopyridine nucleotide-cytochrome *f* reductase of chloroplasts. *J. Biol. Chem. 241*, 279-285.

Zhang, H., Whitelegge, J. P., & Cramer, W. A. (2001). Ferredoxin:NADP⁺ oxidoreductase is a subunit of the chloroplast cytochrome b_6f complex. *J. Biol. Chem. 276*, 38159-38165.

Zheng, M., Doan, B., Schneider, T. D., & Storz, G. (1999). OxyR and SoxRS regulation of fur. *J. Bacteriol. 181*, 4639-4643.

Ziegler, G. A. & Schulz, G. E. (2000). Crystal structures of adrenodoxin reductase in complex with NADP⁺ and NADPH suggesting a mechanism for the electron transfer of an enzyme family. *Biochemistry 39*, 10986-10995.

Permissions

The contributors of this book come from diverse backgrounds, making this book a truly international effort. This book will bring forth new frontiers with its revolutionizing research information and detailed analysis of the nascent developments around the world.

We would like to thank Mohammad Mahdi Najafpour, for lending his expertise to make the book truly unique. He has played a crucial role in the development of this book. Without his invaluable contribution this book wouldn't have been possible. He has made vital efforts to compile up to date information on the varied aspects of this subject to make this book a valuable addition to the collection of many professionals and students.

This book was conceptualized with the vision of imparting up-to-date information and advanced data in this field. To ensure the same, a matchless editorial board was set up. Every individual on the board went through rigorous rounds of assessment to prove their worth. After which they invested a large part of their time researching and compiling the most relevant data for our readers. Conferences and sessions were held from time to time between the editorial board and the contributing authors to present the data in the most comprehensible form. The editorial team has worked tirelessly to provide valuable and valid information to help people across the globe.

Every chapter published in this book has been scrutinized by our experts. Their significance has been extensively debated. The topics covered herein carry significant findings which will fuel the growth of the discipline. They may even be implemented as practical applications or may be referred to as a beginning point for another development. Chapters in this book were first published by InTech; hereby published with permission under the Creative Commons Attribution License or equivalent.

The editorial board has been involved in producing this book since its inception. They have spent rigorous hours researching and exploring the diverse topics which have resulted in the successful publishing of this book. They have passed on their knowledge of decades through this book. To expedite this challenging task, the publisher supported the team at every step. A small team of assistant editors was also appointed to further simplify the editing procedure and attain best results for the readers.

Our editorial team has been hand-picked from every corner of the world. Their multi-ethnicity adds dynamic inputs to the discussions which result in innovative outcomes. These outcomes are then further discussed with the researchers and contributors who give their valuable feedback and opinion regarding the same. The feedback is then collaborated with the researches and they are edited in a comprehensive manner to aid the understanding of the subject.

Apart from the editorial board, the designing team has also invested a significant amount of their time in understanding the subject and creating the most relevant covers. They scrutinized every image to scout for the most suitable representation of the subject and create an appropriate cover for the book.

The publishing team has been involved in this book since its early stages. They were actively engaged in every process, be it collecting the data, connecting with the contributors or procuring relevant information. The team has been an ardent support to the editorial, designing and production team. Their endless efforts to recruit the best for this project, has resulted in the accomplishment of this book. They are a veteran in the field of academics and their pool of knowledge is as vast as their experience in printing. Their expertise and guidance has proved useful at every step. Their uncompromising quality standards have made this book an exceptional effort. Their encouragement from time to time has been an inspiration for everyone.

The publisher and the editorial board hope that this book will prove to be a valuable piece of knowledge for researchers, students, practitioners and scholars across the globe.

List of Contributors

Arthur M. Nonomura, Barry A. Cullen and Andrew A. Benson
Scripps Institution of Oceanography, University of California San Diego, USA

Masato Baba
Graduate School of Life and Environmental Sciences, University of Tsukuba, Tsukuba, Ibaraki, Japan
CREST, JST, Japan

Yoshihiro Shiraiwa
CREST, JST, Japan

Alexander A. Ivlev
Russian Agrarian State University – "MSKHA of K.A.Timirjazev", Russian Federation

Roghieh Hajiboland
Plant Science Department, University of Tabriz, Iran

H.Z.E. Jaafar and Mohd Hafiz Ibrahim
Universiti Putra Malaysia (UPM), Malaysia

Vladimir I. Chikov and Svetlana N. Batasheva
Kazan Institute of Biochemistry and Biophysics of the Russian Academy of Sciences, Russia

Sheng Shu, Shi-Rong Guo and Ling-Yun Yuan
College of Horticulture, Nanjing Agricultural University, Key Laboratory of Southern Vegetable Crop Genetic Improvement, Ministry of Agriculture, Nanjing, China

Kumarakurubaran Selvaraj
Crops and Livestock Research Centre, Agriculture and Agri-Food Canada, Charlottetown, PE, Canada
Department of Biology, University of Prince Edward Island, Charlottetown, PE, Canada

Bourlaye Fofana
Department of Biology, University of Prince Edward Island, Charlottetown, PE, Canada

Carmen Arena
Department of Structural and Functional Biology, University of Naples Federico II, Italy

Luca Vitale
Istituto per I Sistemi Agricoli e Forestali del Mediterraneo, (ISAFoM – CNR), Italy

Jeffrey W. Touchman and Yih-Kuang Lu
Arizona State University, USA

Shinji Masuda
Center for Biological Resources & Informatics, Tokyo Institute of Technology, Japan

Nikolaos E. Ioannidis and Kiriakos Kotzabasis
Department of Biology, University of Crete, Heraklion, Crete, Greece

Josep Maria Torné and Mireya Santos
Departament de Genètica Molecular, Centre for Research in Agricultural Genomics, CRAG-CSIC-IRTA UAB, Barcelona, Spain

Geraldo Rogério Faustini Cuzzuol and Camilla Rozindo Dias Milanez
Universidade Federal do Espírito Santo, Brasil

Daniel Conrad Ogilvie Thornton
Department of Oceanography, Texas A & M University, College Station, Texas, USA

Matías A. Musumeci, Eduardo A. Ceccarelli and Daniela L. Catalano-Dupuy
Instituto de Biología Molecular y Celular de Rosario (IBR), CONICET, Facultad de Ciencias Bioquímicas y Farmacéuticas, Universidad Nacional de Rosario, Argentina